# CHALLINOR'S DICTIONARY OF GEOLOGY

# CHALLINOR'S DICTIONARY OF GEOLOGY

EDITED BY

**ANTONY WYATT**

SIXTH EDITION

CARDIFF: UNIVERSITY OF WALES PRESS
NEW YORK: OXFORD UNIVERSITY PRESS
1986

Published in Great Britain by The University of Wales Press,
6 Gwennyth Street, Cathays, Cardiff CF2 4YD

Published in the United States by Oxford University Press, Inc.,
200 Madison Avenue, New York, New York 10016



**Library of Congress Cataloging-in-Publication Data**

Challinor, John
Challinor's Dictionary of geology.

Rev.ed. of: A dictionary of geology. 5th ed. 1978
Includes index.
1. Geology—Dictionaries I. Wyatt, Antony
II. Challinor, John. Dictionary of geology
III. Title. IV. Title: Dictionary of geology.
QE5.C45 1986 550'.3'21 85–21828

ISBN 0–19–520505–7
ISBN 0–19–520506–5 (pbk.)

FIRST PUBLISHED IN 1961
REPRINTED, WITH MINOR CORRECTIONS, 1962
SECOND EDITION 1964
THIRD EDITION 1967
FOURTH EDITION 1973
FIFTH EDITION 1978
SIXTH EDITION 1986

PRINTED IN GREAT BRITAIN BY
THE EASTERN PRESS LIMITED

## FROM THE PREFACE TO THE FIRST EDITION

THERE appears to be room among works of geology for one that will probe the subject by examining the meaning and usage of names and terms that stand for the more significant things, facts, and concepts of the science. This small book is an essay towards a critical and historical review of a selected ABC of the subject.

## PREFACE TO THE FIFTH EDITION

THIS edition is again thoroughly revised and considerably enlarged, and the author is most grateful to the Director of the University of Wales Press and his Staff, and to the Printers, for their invaluable help in its production. The many new quotations add to his indebtedness to authors and publishers for their kind acquiescence in his drawing so freely on their works.

J. C.

*Aberystwyth*
*March 1978*

## PREFACE TO THE SIXTH EDITION

'JACK' CHALLINOR (as he is known to his friends) produced a large volume of amendments and new text for this edition, before advancing years prompted him to pass the task on. Most of this work has been included in this new edition, as well as further material gathered by the present editor. As far as possible the original style and concept have been followed, adding to the indebtedness to authors and publishers for their kind acquiescence in our drawing so freely on their works. We are most grateful to the Director of the University of Wales Press, his Staff, and to the Printers, for their invaluable help in the production of this edition.

A. R. W.

*Aberystwyth*
*July 1985*

# NOTES

QUOTATIONS are given with several purposes: to define the term (in the original definition or otherwise), to give a representative example of the use of the term (often either in a 'classic' or a very recent work), or to show the term in a variety of usages. In addition to these citations, references are given to significant definitions or discussions; an original use is commonly implied. These quotations and references are to works in the English language and most of them are to British works; surname of author, short title of book or abbreviated title of periodical, and year are given. For the names of minerals, rocks, and biological (fossil) groups, and where a work in a foreign language is indicated, authors and dates only are given. In the case of biological groups, no references are given for the names of phyla or groups not dominantly fossil, or for the names of any vertebrate or plant groups.

The names of the biological groups of fossils are plural nouns considered as being in the Latin language, though most of them are derived from Greek words and some are, in fact, the same as Greek plurals. In the etymologies of most of these names it is to be understood that the entry begins with the words: 'modern Latin plural from . . .'.

Abbreviations: in etymologies, f. from, G. Greek, L. Latin, pl. plural; in tables, s.l. sensu lato, s.s. sensu stricto.

A classified index will be found at the end of the dictionary.

# ABBREVIATIONS OF TITLE OF PERIODICALS

| | |
|---|---|
| *A* | *Archaeologia* |
| *AAAG* | *Annals of the Association of American Geographers* |
| *AAPG* | *American Association of Petroleum Geologists (Publication)* |
| *AC* | *Archaeologia Cambrensis* |
| *AdG* | *Advances in Geophysics* |
| *AdS* | *Advancement of Science* |
| *AG* | *American Geologist* |
| *AGI* | *American Geological Institute* |
| *AJS* | *American Journal of Science* |
| *AM* | *American Mineralogist* |
| *AMG* | *Amateur Geologist* |
| *Amg* | *Amgueddfa (National Museum of Wales)* |
| *AMNH* | *Annals and Magazine of Natural History* |
| *ANSP* | *Academy of Natural Sciences of Philadelphia (Publication)* |
| *Ant* | *Antiquity* |
| *AP* | *Annals of Philosophy* |
| *AREPS* | *Annual Review of Earth and Planetary Sciences* |
| *AS* | *Annals of Science* |
| *BA* | *British Association (Report, etc.)* |
| *BAAPG* | *Bulletin of the American Association of Petroleum Geologists* |
| *BBM* | *Bulletin of the British Museum (Natural History)* |
| *BGS* | *Bulletin of the Geological Survey of Great Britain* |
| *BGSA* | *Bulletin of the Geological Society of America* |
| *BGSI* | *Bulletin of the Geological Survey of Ireland* |
| *BGSUC* | *Bulletin of the Geological Sciences, University of California* |
| *BJHS* | *British Journal for the History of Science* |
| *BKSGS* | *Bulletin of the Kansas State Geological Survey* |
| *BM* | *British Museum (Publication)* |
| *BNRC* | *Bulletin of the National Research Council (America)* |
| *BO* | *Boreas* |
| *BPDMHS* | *Bulletin of the Peak District Mines Historical Society* |
| *BPSW* | *Bulletin of the Philosophical Society of Washington* |
| *BR* | *Biological Reviews* |

| | |
|---|---|
| *BRG* | *British Regional Geology (Institute of Geological Sciences)* |
| *BSEUSS* | *Bulletin of the South-eastern Union of Scientific Societies* |
| *BSS* | *Bulletin of the Soil Survey of England and Wales* |
| *BSSA* | *Bulletin of the Seismological Society of America* |
| *BUSGS* | *Bulletin of the United States Geological Survey* |
| *BV* | *Bulletin Volcanologique (Naples)* |
| *C* | *Ceredigion* |
| *CG* | *Colliery Guardian* |
| *CGA* | *Circular of the Geologists' Association* |
| *CGe* | *Chemical Geology* |
| *CGS* | *Circular of the Geological Society of London* |
| *CLG* | *Cement, Lime and Gravel* |
| *CM* | *Canadian Mineralogist* |
| *CMP* | *Contributions to Mineralogy and Petrology* |
| *Cmp* | *Compass* |
| *CN* | *Cape Naturalist* |
| *CRS* | *Cymmrodorion Record Series* |
| *D* | *Doklady—Earth Science Section* |
| *Dsc* | *Discovery* |
| *E* | *Endeavour* |
| *EG* | *Economic Geology* |
| *EMG* | *East Midland Geographer* |
| *EMJ* | *Engineering and Mining Journal* |
| *ENPJ* | *Edinburgh New Philosophical Journal* |
| *EPSL* | *Earth and Planetary Science Letters* |
| *ER* | *Edinburgh Review* |
| *ES* | *Earth Science* |
| *ESR* | *Earth-Science Reviews* |
| *FS* | *Field Studies* |
| *G* | *Geography* |
| *GS* | *Geoscience Canada* |
| *GCA* | *Geochemica et Cosmochimica Acta (Geochemical Society)* |
| *Geol* | *Geology (Association of Teachers of Geology)* |
| *GeolJ* | *Geological Journal* |
| *Geoly* | *Geology (Geological Society of America)* |
| *Geot* | *Geotimes* |
| *GJ* | *Geographical Journal* |
| *GJRAS* | *Geophysical Journal of the Royal Astronomical Society* |
| *GM* | *Geological Magazine* |
| *Gp* | *Geophysics* |
| *GR* | *Geographical Review* |

| | |
|---|---|
| *GS* | *Geological Survey of Great Britain (Publication)* |
| *GSASP* | *Geological Society of America, Special Paper* |
| *GSSR* | *Geological Society (London) Special Report (or Publication)* |
| *GSNJ* | *Geological Survey of New Jersey* |
| *Gst* | *The Geologist* |
| *GT* | *Geology Teaching* |
| *Gt* | *Géotechnique* |
| *HS* | *History of Science* |
| *ICQ* | *International Congress of the Quaternary (Report)* |
| *IGC* | *International Geological Congress (Report)* |
| *IGR* | *International Geology Review* |
| *IGS* | *Indiana Geological Survey (Publication)* |
| *IGSc* | *Institute of Geological Sciences (Publication)* |
| *IMNHAJ* | *Isle of Man Natural History and Antiquarian Journal* |
| *Is* | *Isis* |
| *JACS* | *Journal of the American Ceramic Society* |
| *JAP* | *Journal of Applied Physics* |
| *JG* | *Journal of Geology* |
| *JGe* | *Journal of Geological Education* |
| *JGI* | *Journal of Glaciology* |
| *JGS* | *Journal of the Geological Society (London)* |
| *JGSA* | *Journal of the Geological Society of Australia* |
| *JHI* | *Journal of the History of Ideas* |
| *JIP* | *Journal of the Institute of Petroleum* |
| *JIWE* | *Journal of the Institution of Water Engineers* |
| *JMBA* | *Journal of the Marine Biological Association* |
| *JP* | *Journal of Paleontology* |
| *JPRSNSW* | *Journal and Proceedings, Royal Society of New South Wales* |
| *JPt* | *Journal of Petrology* |
| *JRAS* | *Journal of the Royal Agricultural Society* |
| *JRSA* | *Journal of the Royal Society of Arts* |
| *JSBNH* | *Journal of the Society for the Bibliography of Natural History* |
| *JSP* | *Journal of Sedimentary Petrology* |
| *JSS* | *Journal of Soil Science* |
| *JWAS* | *Journal of the Washington Academy of Science* |
| *KGS* | *Kentucky Geological Survey (Publication)* |
| *L* | *Lithos* |
| *Let* | *Lethaia* |

| | |
|---|---|
| *LM* | *Lonsdale Magazine* |
| *LMGJ* | *Liverpool and Manchester Geological Journal* |
| *M* | *Microscopy* |
| *MC* | *Montgomeryshire Collections* |
| *MG* | *Mercian Geologist* |
| *MGJ* | *Mining and Geological Journal* |
| *MgM* | *Mining Magazine* |
| *MGS* | *Memoirs of the Geological Survey of Great Britain* |
| *MGSA* | *memoirs of the Geological Society of America* |
| *MGSI* | *Memoirs of the Geological Survey of India* |
| *MGSL* | *Memoirs of the Geological Society of London* |
| *MJ* | *Museums Journal* |
| *MM* | *Mineralogical Magazine* |
| *MMLPS* | *Memoirs of the Manchester Literary and Philosophical Society* |
| *MN* | *Midland Naturalist* |
| *MnS* | *Mineralogical Society (Publication)* |
| *Mp* | *Micropalaeontology* |
| *MPS* | *Monographs of the Palaeontographical Society* |
| *MRC* | *Mineral Resources, Canada* |
| *MS* | *Manuscript* |
| *MsM* | *Mines Magazine* |
| *N* | *Nature* |
| *NB* | *New Biology* |
| *NGM* | *National Geographic Magazine* |
| *NIW* | *Nature in Wales* |
| *NP* | *New Phytologist* |
| *NPJ* | *Nicholson's Philosophical Journal* |
| *NSJFS* | *North Staffordshire Journal of Field Studies* |
| *Nst* | *The Naturalist* |
| *NZJST* | *New Zealand Journal of Science and Technology* |
| *OE* | *Open Earth* |
| *OED* | *Oxford English Dictionary* |
| *OGJ* | *Oil and Gas Journal* |
| *OJS* | *Ohio Journal of Science* |
| *ONM* | *Ocean and National Magazine* |
| *P* | *Palaeontology* |
| *PA* | *Palaeontological Association (Publication)* |
| *PAAAS* | *Proceedings of the American Academy of Arts and Sciences* |
| *PAC* | *Pollen Analysis Circular* |
| *P-AG* | *Pan-American Geologist* |

| | |
|---|---|
| *PAPS* | *Proceedings of the American Philosophical Society* |
| *PBNS* | *Proceedings of the Bristol Naturalists' Society* |
| *PBSA* | *Proceedings of the British Speleological Association* |
| *PBSNH* | *Proceedings of the Boston Society of Natural History* |
| *PCE* | *Physics and Chemistry of the Earth* |
| *PCNFC* | *Proceedings of the Cotteswold Naturalists' Field Club* |
| *PE* | *Photogrammetric Engineering* |
| *Pet* | *Petroleum* |
| *PGA* | *Proceedings of the Geologists' Association (London)* |
| *PGS* | *Proceedings of the Geological Society of London* |
| *PICE* | *Proceedings of the Institute of Civil Engineers* |
| *PLGS* | *Proceedings of the Liverpool Geological Society* |
| *Pln* | *Palynology* |
| *PM* | *Philosophical Magazine* |
| *PNAS* | *Proceedings of the National Academy of Science* |
| *PPP* | *Palaeogeography, Palaeoclimatology, Palaeoecology* |
| *PPSG* | *Proceedings of the Philosophical Society of Glasgow* |
| *PR* | *Physics Review* |
| *PRS* | *Proceedings of the Royal Society of London* |
| *PRPSE* | *Proceedings of the Royal Physical Society of Edinburgh* |
| *PRSE* | *Proceedings of the Royal Society of Edinburgh* |
| *PRST* | *Proceedings of the Royal Society of Tasmania* |
| *PSL* | *Physical Society of London (Report)* |
| *PSR* | *Popular Science Review* |
| *PSWIE* | *Proceedings of the South Wales Institute of Engineers* |
| *PT* | *Philosophical Transactions (Royal Society of London)* |
| *PUS* | *Proceedings of the Ussher Society* |
| *PYGS* | *Proceedings of the Yorkshire Geological Society* |
| *QJ* | *Quarterly Journal of the Geological Society of London* |
| *QJEG* | *Quarterly Journal of Engineering Geology* |
| *QJS* | *Quarterly Journal of Science* |
| *QMJ* | *Quarry Managers' Journal* |
| *QR* | *Quarterly Review* |
| *QtnR* | *Quarternary Research* |
| *RASMNGS* | *Royal Astronomical Society Monthly Notices—Geophysics Supplement* |
| *RGMSN* | *Royal Geological and Mining Society of the Netherlands* |
| *RIC* | *Royal Institution of Cornwall (Report)* |
| *RPCGS* | *Report of Progress, Canada Geological Survey* |
| *RSC* | *Royal Society of Canada (Publication)* |
| *RSI* | *Report of the Smithsonian Institution* |

| | |
|---|---|
| *S* | *Science* |
| *SA* | *Systematics Association (Publication)* |
| *ScA* | *Scientific American* |
| *Sd* | *Sedimentology* |
| *S-ENA* | *South-eastern Naturalist and Antiquary* |
| *SEPM* | *Society of Economic Paleontologists and Mineralogists (Publication)* |
| *SG* | *Sedimentary Geology* |
| *SGM* | *Scottish Geographical Magazine* |
| *SGS* | *Saskatchewan Geological Society (Publication)* |
| *S-HBN* | *Stechert-Hafner Book News* |
| *SJG* | *Scottish Journal of Geology* |
| *SP* | *Science Progress* |
| *SPGS* | *Summary of Progress of the Geological Survey of Great Britain* |
| *SPGSA* | *Special Papers of the Geological Society of America* |
| *SU* | *Stanford University (Publication)* |
| *SvP* | *Soviet Physics* |
| *T* | *Tuatara (New Zealand)* |
| *TAIME* | *Transactions of the American Institute of Mining Engineers* |
| *TCNS* | *Transactions of the Cardiff Naturalists' Society* |
| *TCPS* | *Transactions of the Cambridge Philosophical Society* |
| *TCRG* | *Transactions of the Cave Research Group* |
| *TEGS* | *Transactions of the Edinburgh Geological Society* |
| *TGS* | *Transactions of the Geological Society (London)* |
| *TGSG* | *Transactions of the Geological Society of Glasgow* |
| *TIBG* | *Transactions of the Institute of British Geographers* |
| *TIME* | *Transactions of the Institution of Mining Engineers* |
| *TIMM* | *Transactions of the Institution of Mining and Metallurgy* |
| *TLBS* | *Transactions of the Liverpool Biological Society* |
| *TLGA* | *Transactions of the Leeds Geological Association* |
| *TNDFC* | *Transactions of the Newbury District Field Club* |
| *TNS* | *Transactions of the Newcomen Society* |
| *TNSFC* | *Transactions of the North Staffordshire Field Club* |
| *Tp* | *Tectonophysics* |
| *TRDS* | *Transactions of Royal Dublin Society* |
| *TRGSC* | *Transactions of the Royal Geological Society of Cornwall* |
| *TRSE* | *Transactions of the Royal Society of Edinburgh* |
| *TSA* | *Transactions of the Society of Arts* |
| *TSANHS* | *Transactions of the Shropshire Archaeological and Natural (History Society)* |

| | |
|---|---|
| *TSME* | *Transactions of the Society of Mining Engineers* |
| *TWNC* | *Transactions of the Worcestershire Naturalist's Club* |
| *TWNFC* | *Transactions of the Woolhope Naturalists' Field Club* |
| *USGS* | *United States Geological Survey (Publication)* |
| *USLG* | *United States Library of Congress (Publication)* |
| *UTP* | *University of Texas Publications* |
| *WGQ* | *Welsh Geological Quarterly* |
| *WRR* | *Water Resources Research (America)* |

# LIST OF PREFIXES AND SUFFIXES

## and the Greek or Latin words from which they are derived

| | |
|---|---|
| a-, an- | G. a-, *an-*, expressing absence |
| -acea, -aceae | L. pl. of *-aceus* (*-a, -um*), of the nature of |
| -aceous | L. *-aceus,* of the nature of |
| ad- | L. *ad,* to, towards, sometimes expressing addition or intensification |
| -ales | G. *hales,* assembled |
| ana- | G. *ana,* upwards, towards, anew; with a sense of increase |
| anti- | G. *anti,* against, opposite |
| archae(o)- | G. *archaios,* ancient |
| -arium | L. *-arium,* a place for |
| auto- | G. *autos,* self, self-same |
| bi- | L. *bi,* two |
| bio- | G. *bios,* life |
| blast(o)-, blast, -blastic | G. *blastos,* growth |
| caino-, caeno-, kaino-, ceno-, -cene | G. *kainos,* recent |
| cata-, kata- | G. *kata,* downwards, away, against, intensive |
| chrono-, -chron, -chronous, -chronic, -chronal | G. *chronos,* time |
| circum- | L. *circum,* around |
| clast(o)-, -clast, -clastic, -clasis, -clas | G. *klastos,* broken in pieces |
| clino-, -cline, -clinal | G. *klinō,* make to slope, be sloping |
| co-, com-, con- | L. *cum,* with |
| de- | L. *de,* off, from, away, down |
| di- | G. *di-,* two |
| dia- | G. *dia,* across, right through |
| dis- | L. *dis-,* apart, in pieces, in two |
| eo- | G. *éós,* the dawn |
| ep(i)- | G. *epi,* upon, after |
| eu- | G. *eu,* well, readily, rightly |
| ex-, e- | G. and L. *ex,* out of |
| -fer, -ferous | L. *fero,* bear, carry |
| ge(o)- | G. *ge,* the earth |

| | |
|---|---|
| geno-, -gen, -gene | G. *genos*, kind, race |
| -genous, -genic, -geny, -genetic, -genesis | G. *genesis*, origin |
| -gram | G. *gramma*, that which is delineated or written |
| graph(o), -graph, -graphy, -graphic, -graphical | G. *graphō*, delineate, write |
| hemi- | G. *hemi*, half |
| hetero- | G. *heteros*, different |
| holo- | G. *holos*, whole, complete |
| homo- | G. *homos*, the same |
| homoeo- | G. *homoios*, resembling |
| hyper- | G. *huper*, above, beyond, very |
| hyp(o)- | G. *hupo*, under, somewhat |
| -id | G. *-ides*, son of |
| -ida, idae, idea | modern L. pl. f. G. *-idēs* |
| in- | L. *in-*, expressing negative or privation |
| inter- | L. *inter*, between |
| intra- | L. *intra*, within |
| iso- | G. *isos*, equal |
| -ite, -ites, -itic | G. *ites*, belonging to; often a conventional ending for a particular kind of mineral, rock, or fossil |
| kaino- | see caino- |
| kata- | see cata- |
| lith(o)-, -lit, -lite, -lith, -litic, -lithic | G. *lithos*, a stone |
| -logy | G. *logos*, a word, discourse, science |
| macro- | G. *macros*, long, large |
| mega- | G. *megas*, great |
| meso- | G. *mesos*, middle |
| meta- | G. *meta-*, back again, by way of change |
| -meter | G. *metron*, a measure |
| micro- | G. *micros*, a small |
| mono- | G. *monos*, single, alone |
| morpho-, morph, -morphic, -morphism, -morphy | G. *morphe*, form |
| ne(o)- | G. *neos*, new, young |
| ob- | L. *ob*, against |
| -odont | G. *odous*, *odont-*, a tooth |
| -oid, -oidal | G. *eidos*, form, appearance |
| -oidea | modern L. pl. f. G. *eidos* |
| onto- | G. *on*, *ont-*, being |

| | |
|---|---|
| -orium | L. *orium,* a place for |
| orth(o)- | G. *orthos,* straight, right |
| -ose | L. *osus,* abounding in |
| palaeo-, paleo- | G. *palaios,* old |
| pan- | G. *pan,* all |
| para- | G. *para,* beside, along with |
| -ped | L. *pes, ped-,* a foot |
| peri- | G. *peri,* round about |
| petr(i)-, -(o)- | G and L. *petra,* a rock |
| -phyta | modern L. pl. f. G. *phuton* |
| phyto-, -phyte | G. *phuton,* a plant |
| -pod | G. *pous, pod-,* a foot |
| -poda | modern L. pl. f. G. *pous, pod-* |
| poly- | G. *polus,* many |
| post- | L. *post,* after |
| pre- | L. *prae,* before |
| pseudo- | G. *pseudo,* lie, be false |
| semi- | L. *semi-,* half |
| -sphere | G. *sphaira,* a ball |
| sub- | L. *sub,* under, nearly |
| super- | L. *super,* above, beyond |
| syn- | G. *sun,* together, alike |
| tecto-, tecton-, -tectonic | G. *tektōn,* a builder |
| tetra- | G. *tetra,* four |
| tri- | G. and L. *tri-,* three |
| ultra- | L. *ultra,* beyond |
| uni- | L. *unus,* one |
| xeno-, -xene | G. *xenos,* a stranger |
| zo(o)-, zoic, -zoan | G. *zaō* (*zōō*), to live; *zōon,* an animal |
| -zoa | modern L. pl. = G. pl. of *zoon* |

# A

**Aa.** A Hawaiian name (pronounced 'ahah') for one of the two chief forms of lava emitted from volcanoes of the Hawaiian type, the other being pahoehoe. 'The second form of the lavas is called by the natives a-a, and its contrast with pahoehoe is about the greatest imaginable. It consists mainly of clinkers sometimes detached, sometimes partially agglutinated together with a bristling array of sharp, jagged, angular fragments of a compact character projecting up through them. The aspect of one of these aa streams is repellent to the last degree, and may without exaggeration be termed horrible. For one who has never seen it, it is difficult to conceive such superlative roughness' (Dutton, *USGS*, 1883). One eruption may emit aa in one direction and pahoehoe in another, or both types may occur in the same lava-stream. These two names are applicable to lavas of any geological age, in so far as the physical structure characterizing them can be recognized. The two forms are often referred to as block lava and ropy lava.

**a-axis.** See TECTONIC AXIS.

**Aberystwyth Grits.** A stratigraphical formation (or 'series' in the looser sense) of alternating hard, light grey, sandstone ('grit') bands, from a fraction of an inch to a foot or two thick, and shaly, somewhat cleaved, mudstones. It outcrops as a crescentic belt along about 40 km of the Cardiganshire coast, north and south of Aberystwyth. The thickness is unknown as the tectonic structure of the formation itself and its structural relations to other formations have not been made out with any certainty. Nor have the inland geographical limits been precisely defined. Graptolites (chiefly *Monograptus priodon*, *M. spiralis*, and *M. nodifer*) and an occasional specimen of the brachiopod *Coelospira hemispherica* occur, but rather sparsely. The formation provides an extended succession within the *Monograptus turriculatus* zone, with, perhaps, the next higher, the *M. crispus* zone, of Upper Llandovery (Silurian) age. Sedgwick (*QJ*, 1847) named an Aberystwyth Group of rocks, placing it above his Bala but far below Murchison's Llandeilo. Keeping (*QJ*, 1881) defined and named the Aberystwyth Grits but had the structure of this part of the country upside down. Jones (*QJ*, 1909) placed the formation in its proper structural and stratigraphical position. Wood and Smith (*QJ*, 1959) have described the details of lithology and bedding and discussed the conditions of sedimentation.

**Ablation.** Surface wastage and removal; used particularly for the evaporation and melting of the surface of ice. 'The actual waste or melting of the ice at its surface. . . . the absolute "ablation" of the ice (as it has been termed by M. Agassiz)' (Forbes, *ENPJ*, 1846, reprinted in *Theory of Glaciers*, 1859). The term is also used in other connexions, e.g. : 'It is believed that ablation of the larger meteorites can proceed by (a) direct vaporization, or by (2) the sweeping away of a molten surface layer' (Maringer, *GCA*, 1960). For removal of material by the wind 'deflation' is the usual term. Grabau (*Stratigraphy*, 1913) uses the term for a large part of the general process of erosion, defining it as 'the separation of [rock] material from the main mass'; but his classification of the erosive processes is involved and unsatisfactory. [L. *ablatus*, borne away.]

**Abrade.** To wear away by rubbing or scraping. Hence 'abrasion'. It would usually imply the small-scale action which is part of the more general action of corrasion or attrition. 'So long as the particles [in a stream] are borne along in suspension, they will not abrade each other, but remain angular. . . . Issuing from Lake Erie, the Niagara approaches its fall by a series of rapids. The water leaves the lake with hardly any appreciable sediment. . . . The vast body of clear water leaping and shooting over the sheets of limestone in the rapids. . . . The geologist can observe that these rocks have been comparatively little abraded . . . the smoothed and striated surface left by the ice-sheet of the Glacial period can still be traced upon them' (Geikie, *Textbook*, 1903). 'The work of the natural sandblast, or "abrasion". . .' and '. . . a rock surface extraordinarily smooth and even polished in parts [by glacial action], though not planed, and thus affords evidence of abrasion' (Cotton, *Climatic Accidents*, 1942). Its use in connexion with larger-scale action is exemplified in 'abrasion platform' (see WAVE-CUT PLATFORM). The term has been used for the whole process of mechanical erosion: 'The siliceous part of the mountains has not been chemically dissolved; it has been only abraded and worn away' (Playfair, *Illustration*, 1802). An 'abrasive', as a noun, is 'any rock, mineral, or substance which, owing to its superior hardness, toughness, consistency, or other properties, is suitable for grinding, cutting, polishing, scouring, or similar use' (Nelson and Nelson, *Dictionary of Applied Geology*, 1967). [L. *abrado*, scrape off.]

**Absolute age.** See CHRONOLOGY. Holmes has argued (*N*, 1962) that 'absolute age' is philosophically 'a meaningless term'; but nevertheless it seems that we can use it without practical impropriety if we take as our unit of measurement the present length of the year.

**Absorption.** Particularly in the following connexion: 'Absorption is the measure of the quantity of water which a rock will absorb when immersed, and is measured by the difference between the dry weight of a sample and the weight after soaking' (Blyth and de Freitas, *A Geology for Engineers*, 1974). The greater the porosity, the greater the absorption.

**Abyssal. 1.** Of the unfathomable depths of the sea; for the bathymetric zone beyond and below the bathyal zone. ('Abysmal' is sometimes used, e.g. by A. Geikie in his textbooks.) **2.** Of the illimitable depths of the earth; for very deep-seated igneous rocks. [G. *abussos*, bottomless.]

**Abyssal deposits.** 'The sediments of the deep oceanic regions include some types, e.g. the organically derived oozes, the ultimate origins of which were recognized when they were first discovered. The origin of others, such as the deep-sea sands, has been substantially clarified in the last 15 years. But the genesis of the deep-sea lutites (red clay etc.) is still quite uncertain' (Nicholls, *PGS*, 1967).

**Accessory.** For minerals as constituents of a rock. **1.** In an igneous rock: those minerals occurring in such small quantities that their presence or absence is not significant when considering the mineral composition of the rock for classification purposes. Contrasted with the 'essential' minerals for such purposes. 'The original minerals of the crystalline rocks may be conveniently divided into two groups: those which are always present in every sample of the particular rock [rock-type], which may be called the "essential" constituents; and those which occur with greater or less frequency in addition to the essential ones, which may be called "accessory" or accidental constituents' (Judd, *QJ*, 1886). 'The accessory minerals of the granitic rocks of the English Lake District' (Rastall and Wilcockson, *QJ*, 1915). **2.** In a sedimentary rock: for those discrete minerals that occur in very small amounts compared with the few, perhaps one only, that make up almost the entire bulk of the rock. 'Accessory minerals of sandstones' (Hatch and Rastall, *Sedimentary Rocks*, 1965). The accessory minerals (in whatever kind of rock) happen to be mostly heavier than the main constituents and thus provide the 'heavy minerals'.

**Accidental.** For one special usage see XENOLITH.

**Accommodation structures.** Relatively small structures, particularly folds, that are considered as having accommodated themselves to space made available (and requiring to be filled) during the production of larger structures.

**Accordant drainage.** 'A systematic relationship apparent between rock-type and structure on the one hand, and surface drainage pattern on the other' (Monkhouse, *Dictionary of Geography*, 1965). A purely descriptive term, applying on a large or a small scale. It might be the result of a regional uplift (see CONSEQUENT DRAINAGE) or become initiated upon, or adapted to, a surface of already-formed underlying structure and varied outcrop.

**Accordant junctions.** 'Every river appears to consist of a main trunk, fed from a variety of branches, each running in a valley proportioned to its size, and all of them together forming a system of vallies, communicating with one another, and having such a nice adjustment of their declivities that none of them join the principal valley on too high or too low a level; a circumstance which would be infinitely improbable if each of these vallies were not the work of the stream that flows in it' (Playfair, *Illustrations*, 1802). 'Though exceptions to Playfair's law of accordant junctions may be found, they are all capable of explanation in such a way as not to contradict the principle' (Cotton, *Landscape*, 1948).

**Accordant summits.** See HILL-TOP SURFACE.

**Accretion.** Used loosely for, e.g., the enlargement of land which is being gained from the sea and the accumulation of sediment on a flood-plain. But in geology it seems specially appropriate for a distinctive kind of growth of rock-material exemplified by the coral reef: the coral polyp builds its skeleton by 'secretion', the reef grows by the 'accretion' of these skeletons (and those of associated organisms).

**Accretion hypothesis.** Any hypothesis that assumes that the earth has grown from a small nucleus by the gradual addition of solid bodies or particles.

**Accretionary prism.** The wedge of sediments found on the arc side of an oceanic trench, consisting of slices of sediment scraped off the descending oceanic plate, together with sediments derived from the arc. Interpretation of seismic records suggests a complex structure, with each new slice thrust under previously accreted slices, and with the dip of the slices increasing with distance from the trench.

**Acicular.** Needle-like; particularly of crystals. [L. *acus*, a needle.]

**Acid. 1.** Applied to an igneous rock rich in silica to the extent of its forming more than about two-thirds of its mass. Such a rock is usually oversaturated with silica so that free silica, quartz, is present. The chief types are granite, adamellite, granodiorite, microgranite, rhyolite, dacite. 'Nearly all the important rock-forming minerals are silicates, or compounds of silica with one or more metallic oxides. In these compounds silica may be regarded as playing the part of an acid, while the metallic oxides act as bases. Hence rocks rich in silica are spoken of as "acid", and those poor in silica but rich in metallic oxides as "basic"' (Rastall, *Textbook*, 1941). 'Rocks rich in silica were first spoken of as acid and called "acidites" by Von Cotta (1864)' (Johannsen *Petrography*, 1939). 'Acidic' is sometimes preferred. Either term may refer to a plagioclase felspar towards the albite end of the series. (Attempts that have been made to distinguish uses of 'acid' from uses of 'acidic' do not seem satisfactory.) **2.** As applied to soils, see pH VALUE.

**Acme.** In palaeontology, the culmination in abundance and variety (usually both together) which occurs at some time during the history of a species, genus, family, &c. 'Trilobites in acme and decline' (Swinnerton, *Palaeontology*, 1947). (G. *acme*, a culmination.]

**Acritarch.** 'The term acritarch is a non-committal term for organic-walled marine microplankton commonly found in Palaeozoic marine strata and less commonly in later sediments. They are of unknown affinity but are most probably the cysts of unicellular algae. Their test is made of sporopollenin, a highly resistant substance also found in the spores and pollen coats of higher plants' (Richardson and Rasul, *PGA*, 1979). See DINOFLAGELLATES. [G.*akrita*, uncertain, *arche*, origin.]

**Actinolite.** See AMPHIBOLE. A common mineral in metamorphic rocks, such as the actinolite-schists, derived from basic and ultrabasic rocks. (Kirwan, 1794.) [G. *aktis*, a ray.]

**Active volcano.** See VOLCANO.

**Actualism.** 'On the Continent the term "uniformitarianism" never won favour, but was gradually replaced by "actualism", which conveys much more appropriately the real meaning of Hutton's inspired appeal to "actual causes": the principle that the same processes and natural laws prevailed in the past as those we can now observe or infer from observations' (Holmes, *Physical Geology*, 1965). 'Contemplating the present operations of the globe, we may perceive the actual existence of those productive causes, which are now laying the foundation of land in the unfathomable regions of the sea, and which will, in time, give birth to future continents . . . We have now got to the end of our reasoning; we have no data further to conclude immediately from that which actually is: but we have got enough; we have the satisfaction to find, that in nature there is wisdom, system, and consistency' (Hutton, *Theory of the Earth*, 1795).

**Adamellite.** Granite, in the wide sense, in which alkali felspar and plagioclase (excluding albite), usually oligoclase, occur in about equal amounts. Intermediate between typical granite and granodiorite. (Cathrein, 1890; Brögger, 1895.) [Mt. Adamello, N. Italy.]

**Adaptive radiation.** 'The speciation that may ensue when an ancestral central stock sends evolutionary branches into different environmental niches is what has been called adaptive radiation' (George, *Evolution in Outline*, 1951). 'The writer has termed this principle of *embranchment* of Lamarck, or of divergence of Darwin, the "law of adaptive radiation" (1902)' (Osborn, *Age of Mammals*, 1910).

**Adinole.** A rock, rich in albite, produced metasomatically in contact metamorphism. 'The albitization of the rocks bordering certain basic intrusions, whereby an argillaceous sediment is converted to an "adinole". The igneous rocks responsible for this transformation are themselves rich in soda, and are generally interpreted as normal dolerites which, after their first consolidation, have been albitized by the action of "juvenile" liquid carrying sodic compounds. The same liquid solutions have invaded the adjacent rocks for a few feet from the contact, not merely along fissures but by intimate permeation, and have there brought about metasomatic changes of a radical kind' (Harker, *Metamorphism*, 1950). Though now usually regarded as a rock, the name was introduced as a mineral-name (Beudant, 1832). The rock (or mineral) is compact. [G. *adinos*, compact.]

**Adobe.** A Spanish word for a sun-dried brick, used primarily by the Mexicans for bricks made from the yellowish calcareous (Pleistocene) clay which occurs not only in that country but also over wide areas of the more arid regions of N. America. Russell (*GM*, 1889) extended the term, from the brick (and the clay making it), to the deposit as a whole, which he described.

**Adsorption.** Covers a variety of processes in which absorption and adherence are to some extent combined. **1.** In chemistry; the condensation of gas films on the surfaces of solids, particularly metals. **2.** In connexion with soils; the concentration of salts (particularly those of calcium and potassium) in colloids surrounding more or less decomposed mineral particles. **3.** In the compaction of sediments it refers to the thin films of water surrounding the grains and somewhat rigidly held by them (Sugden, *GM*, 1950).

**Aegerene.** (Aegerite, Acmite.) See PYROXENE. A green mineral, showing strong pleochroism. Occurs in soda-rich igneous rocks. 'J. Esmark, 1835, after Aegir, a Scandinavian god of the sea' (Chester, *Names of Minerals*, 1896).

**Aeolian.** (Eolian.) Produced by, or borne by, the wind; particularly as 'aeolian erosion', 'aeolian deposits' (dunes, etc.). [L. *Aeolus*, god of the winds.]

**Aeolianite.** An aeolian deposit of a former age, indurated to some degree (e.g. by calcareous material).

**Aeon.** (Eon.) *OED* has 'An age, or the whole duration, of the world, or of the universe'. A very long period in the history of the earth, longer than an era ('the Precambrian aeon', Schopf, *E*, 1975, 'Phanerozoic eon', Brazivnas, *Geoly*, 1975). There is some tendency to use the term for a definite period of time, one thousand million years ('billion' having a double meaning), the age of the earth then being '4.6 aeons'. See TIME UNIT ('Ga').

**Aerolite.** See METEORITE. Originally (early 19th century) any meteorite: 'By its connexion with such subjects as the origin of aerolites . . . geology becomes associated with astronomical speculations' (Buckland, *Vindiciae Geologicae*, 1820). 'The remarkable fact of the fall of foreign bodies, called aerolites, or meteoric stones, from the atmosphere' (Mantell, *Wonders of Geology*, 1857). [G. *aer, aero-*, air.]

**Aeromagnetic survey.** An instrumental magnetic survey of the land-structure made from a moving aeroplane. 'Interpretation of aeromagnetic maps' (Vaequier and others, *MGSA*, 1951). *Aeromagnetic Map of Great Britain* (*IGSc*, 1972/75). 'Aeromagnetic survey' (Collar and Patrick in Moseley, ed., *Geology of the Lake District*, 1978). See GEOPHYSICS (2).

**Agate.** See CHALCEDONY. [G. *achatēs*.]

**Age.** In a specialized stratigraphical sense, the time equivalent of a stage. See CHRONOSTRATIGRAPHY.

**Age and origin.** The two chief questions that arise in connexion with most geological facts are (a) the time or period, on the geological scale, when the thing (mineral, rock, fossil, stratum, structure, &c.) was formed, and (b) the source from which it was derived or the cause of its present state; in a phrase, its 'age and origin'. 'Age and origin [of sandstone injections near Cromarty]' (Waterston) and 'The age and origin of the Red Sea graben' (Tromp; both *GM*, 1950).

**Age of a rock. 1.** The age of an igneous rock dates from the time when the magma was intruded, or extruded, and when it became solidified. These two events are in themselves quite different, but solidification is presumed to have occurred so soon after intrusion (or extrusion) that they are taken to be, in a geological sense, of the same age. **2.** In the case of a sedimentary rock, it is 'most important to distinguish between the age of the grains and the age of the deposit' (Sorby, *QJ*. 1880). **3.** For a metamorphic rock we have to distinguish, in particular, between the age of formation of the original rock and the age of its metamorphism. For instance, what is the 'age' of the rocks forming the N. Caernarvonshire slate belt? The constituent grains were derived from Precambrian or early Cambrian rocks, the sediments were deposited in Cambrian times, the resulting hardened mudstones were formed later, and the slates are of the age of the end-of-Silurian orogeny. And what is the 'age' of a large conspicuous glacial erratic of Shap Granite to be seen in the Vale of York? For the radiometric ages of rocks see RADIOACTIVITY AGE-METHOD.

**Age of mountains.** 'A mountain range has many ages. The surface is of present origin, the product of continuing erosion. The rocks were raised at a past time. There are sediments laid at one time, deformed at another, and invaded by plutonic intrusions of lava at another. And there may have been more than one such event in history' (Kay and Colbert, *Stratigraphy and Life History*, 1965).

**Age of reptiles.** See MESOZOIC. 'In 1831 Gideon Algernon Mantell, an English physician living in Lewes, published a three-column communication in a local paper, the *Sussex Advertiser*, entitled "The Age of Reptiles"' (Colbert, *The Age of Reptiles*, 1965). The Permian might appropriately be included in this 'age'.

**Age of the earth.** 'A short account of the cause of the saltness of the ocean, and of the several lakes that emit not rivers, with a proposal, by help thereof, to discover the age of the World' (Halley, *PT*, 1715).

*The Age of the Earth and other Geological Studies* (Sollas, 1905). *Age of the Earth* (Holmes, 1913, 1937). 'The age of the earth' (Holmes, *E*, 1947, *RSI*, 1948). Recent estimates, by radiometric methods, give the earth-moon system, and probably the solar system as a whole, as having been formed 4,600 million years ago. The oldest known rocks on the surface of the earth (which are of a metamorphic character, in Greenland) are about 3,725 million years old. 'It is an interesting commentary on modern science that the age of the Earth has been determined, although its origin remains a baffling problem' (Whipple, *Earth, Moon, and Planets*, 1968).

**Agglomerate.** A mass of large pieces of rock (of any shape) mixed with finer material. Practically exclusively as (in full) 'volcanic agglomerate', formed of volcanic or country rock in the neck, or as part of the cone, of a volcano as a result of explosive action. 'A white tufaceous agglomerate [Santorin]' (Lyell, *Principles*, 1868). [L. *agglomero*, gather into a heap.]

**Agglutinate.** To cement. Seldom used; but it seems particularly appropriate for a very hard breccia or for a certain condition of pebbles in a conglomerate: 'All these different stones are thrown together and agglutinated with a cement so hard, that it is exceedingly difficult to separate them with a hammer, which in general rather breaks than disjoins them' (Faujas Saint Fond, *Voyage en Angleterre* [&c.], 1797, trans., 1799). See quotation under BONE BRECCIA. The term is also used, rather in the sense of 'weld', for the state of cohesion of the particles in such a rock as an ignimbrite; an 'agglutinite' being a pyroclastic rock formed from pieces or particles which were soft or fluid (through heat) when they fell and thus cohered. An 'agglutinate' is any agglutinated rock. There is a usage in palaeontology: e.g. the tests in one group of foraminifera are composed of 'agglutinated' foreign particles.

**Aggrade.** 'So long as deposition on the flood plain kept pace with the deposition in the channel, both would rise, but their relation to each other would not be altered. The valley bottom would be built up steadily; or, if we may coin a word to designate a process for which a name is needed, the valley bottom would be "aggraded"' (Salisbury, *GSNJ*, 1893, quoted in *AGI Glossary*, 1957). But this special usage has not been strictly followed by later writers, many of whom have used 'aggrade', in connexion with river action, whenever they have felt the need for a word contrasting with 'degrade'.

**Aggregate.** An ordinary word used in geology as required, but especially in describing a rock as an aggregate of particular minerals. See ASSEMBLAGE (4). The condition of 'aggregation' has been used for texture: 'The aggregation of the rock varies; it is compact, with smooth texture; granular; granular and crystalline; oolitic [&c.]' (Phillips, *Yorkshire*, 1836). In economic geology an 'aggregate' is broken stone, sand, gravel, &c. bound together by cement to form concrete. [L. *aggrego*, gather together into one body.]

**Agnatha.** Jawless vertebrates, a class of the superclass Pisces. Today they are represented by the lamprey (a cyclostome) but among fossils they are important, comprising the armoured ostracoderms of the Old Red Sandstone. [G. *gnathos*, a jaw.]

**Agricultural geology.** Geology applied to agriculture. *The Application of Geology to Agriculture* (Whitley, 1843). 'On the agricultural geology of England and Wales' (Trimmer, *JRAS*, 1851). *Agricultural Geology*; title of books by Marr (1903), Rastall (1916), and later authors. Sometimes called 'Agrogeology'.

**Agrogeology.** See AGRICULTURAL GEOLOGY.

**Air-heave structure.** A primary structure in a sandstone due to the upward motion of a pocket of air entrapped when the sand was exposed on a beach (of the open sea, lagoon, or lake). The structure, averaging a few inches across, is characterized by strong up-doming of the lamination with a core of unlaminated sandstone. (Reade, *GM*, 1884; Emery, *JSP*, 1945; Stewart, *BAAPG*, 1956, with use of the term; Rolfe, *GM*, 1960.)

**Alabaster.** See GYPSUM. 'from G. αλαβαστϱον [*alabastron*] said to be the name of a town in Egypt, Pliny, 77' (Chester, *Names of Minerals*, 1896).

**Albian.** See CRETACEOUS SYSTEM. [Aube (or Alba) river, France.]

**Albite.** Sodium felspar, sodium aluminium silicate; an end member of the plagioclase series, also a member of the alkali felspars. 'A name originally used by Berzelius [early 19th century] for a felspar from Fahlun in Sweden' (Miers, *Mineralogy*, 1929). See FELSPAR and PLAGIOCLASE. [L. *albus*, white.]

**Albitization.** The replacement, by albite, of a mineral in a rock, usually another felspar. This may be brought about (**1**) during the final stages of the production of a reaction series on the cooling of a magma, or (**2**) in metamorphism, chiefly metasomatism (see ADINOLE). 'The albitization of basic plagioclase felspars' (Bailey and Grabham, *GM*, 1909).

**Algal limestone.** See CALCAREOUS ALGAE.

**Algonkian.** The later Precambrian. See ARCHAEAN. 'The name has reference to the land of the Algonkian Indians—the Lake Superior-Ontario-Hudson Bay country—where these rocks have long been studied' (Pirsson and Schuchert, *Textbook*, 1915). (Walcott, 1889.)

**Alkali.** The following table shows the chemical meaning of the terms alkali, alkaline earth, alkali-metal; together with some others.

| *Metal* | | *Oxide* | | *Hydroxide* | |
|---|---|---|---|---|---|
| lithium | alkali metals | | | | |
| sodium | alkali metals | soda | alkalies | caustic soda | bases |
| potassium | alkali metals | potash | alkalies | caustic potash | bases |
| rubidium | alkali metals | | | | |
| caesium | alkali metals | | | | |
| (ammonium | alkali metals | | | | |
| calcium | | lime (quicklime) | alkaline earths | slaked lime | |
| strontium | | | alkaline earths | | |
| barium | | | alkaline earths | | |

**Alkali felspar.** The potassium and sodium felspars, orthoclase &c. and albite, and the intimate mixtures and solid solutions of the two, perthite &c. See FELSPAR.

**Alkali rock.** An igneous rock rich in any of the alkali-metals (sodium, potassium, and, more rarely, lithium) as shown by the presence of the alkali-felspars, felspathoids, and sodium-bearing mafic minerals. 'Problems of alkali rock genesis' (Tilley *QJ*, 1957). *The Alkaline Rocks* (Sorensen, ed., 1974).

**Allochemical. 1.** Pertaining to the material of a chemically-characterized sedimentary rock, particularly a limestone, this material having accumulated as clastic bits and particles (fossils, shell debris, and ooids included) called 'allochems'. Most allochems have undergone some degree of transport at some stage in their history. Cf. ORTHOCHEMICAL. 'Allochems' (Folk in Harn, ed., *Classification of Carbonate Rocks*, 1962). **2.** Pertaining to a change in the chemical composition of a rock, i.e. metasomatism.

**Allochthonous.** See AUTOCHTHONOUS. (1, 3, 4, 5). (Naumann, Germany, 1858; Gümbel, Germany, 1884.)

**Allogenic.** (Allothigenic.) Originating elsewhere; chiefly for the ordinary mineral particles and clastic constituents of a sedimentary rock which have been derived from older formations and transported to form the new rock. The opposite of 'authigenic'. (Kaskovsky, 1880.)

**Allotriomorphic.** Having a shape not natural to itself; for a mineral, in an igneous rock, that on crystallization was not able to form its own proper crystal faces but had to adapt itself to the faces of other minerals already crystallized. Also called 'anhedral'. Idiomorphic (euhedral) is the opposite condition. See ANHEDRAL. [G. *allotrios*, belonging to another.]

**Alluvium.** Sand, silt, and (especially) mud brought down by rivers in flood and deposited on the temporarily submerged land, the flood-plain or alluvial plain. A stream issuing into a main valley or onto an open plain may form an 'alluvial fan', which may contain coarse material (gravel). However, the term has been, and is, used in wider senses. The *OED* includes deltaic deposits and quotes : 'Our earth, where alluviums are made in some places and the sea gains upon the land in others' (*PT*, 1665). 'In his survey Mr. Smith fortunately succeeded in detecting, and defining more accurately than has yet been done by most writers on the subject, the Alluvial or Gravelly Matters, composed of the ruins of strata, which he found so plentifully distributed over the surface of England, but forming quite a distinct class from the regular Strata' (Farey, *Derbyshire*, 1811). 'A part of the series which I deduced from observations made in the south-eastern part of England, is as follows, beginning with the uppermost. 1. Alluvium, consisting of gravel, loam, sand, &c., and forming the surface or soil' (Webster, *TGS*, 1814). 'The formation of new lands on the banks of rivers and lakes by the alluvial depositions they carry down' (Conybeare and Phillips, *England and Wales*, 1822). Lyell (*Principles*, 1833) defined alluvium as including the coarse material (boulders, stones, and gravel) that may be laid down in the bed of the stream, and also deltaic deposits; while in the *Elements* (1838) he even included material drifted by marine currents. Holmes (*Nomenclature of Petrology*, 1928) defines it as 'a general term for all detrital deposits resulting from the operations of modern rivers, thus including the sediments laid down in river-beds, flood-plains, lakes, fans at the foot of mountain slopes, and estuaries'. 'Alluvial' is a perfectly good adjective but 'to alluviate', in the sense of to lay down alluvium, is hardly a good verb. The plural, now seldom used, can be '-ia' or-iums'. [L. *alluvio*, an overflowing.]

**Alpides.** The fold-structure mountains produced by the Alpine orogeny; in a more comprehensive sense, the whole Alpine orogenic belt.

**Alpine.** Of the many uses of 'of or appertaining to the Alps' the most important in geology is to denote the orogeny that culminated in the Miocene period of the Tertiary, forming the Alpine system of mountain ranges which still exist as such. See MOUNTAIN SYSTEM.

**Alpine chain.** See ANDEAN CHAIN.

**Alternation.** Much in use for those very common stratigraphical sequences where, on a larger or a smaller scale, two or more lithological types alternate over and over again. Small-scale alternation of two types of bed are well exemplified in the Aberystwyth Grits and the Blue Lias; large-scale alternation in the Millstone Grit. Early illustrations are those of Whitehurst (*Strata in Derbyshire*, 1778) and Forster (*Section of Strata*, 1809), both from the Carboniferous. The term is sometimes specially used for a kind of vertical transition between one lithological formation and another, where the new lithology appears as separate beds, becoming more and more frequent: 'The three groups [Old Red Sandstone, Carboniferous Limestone, Coal Measures] pass into one another by [either] gradual approximation of character, or repeated alternation of deposits' (Phillips, *Yorkshire*, 1836; principle more fully stated in his *Geology*, 1855).

**Amber.** A yellow translucent substance, being originally resin which exuded from wounded pines and other trees. As it hardened it often enclosed insects &c. which had been caught on the sticky surface. Found chiefly in the Oligocene of the S. Baltic, and thence washed up on the east coast of England. 'Amber of a deep yellow, found at the foot of a cliff near Whitby. The person who gave me this had been long enquiring into the nature of this body, and he assured me he had several times observed young flies and gnats in amber that he took up on these and the Scarborough shores' (Woodward, *Catalogue*, 1729). Apart from its ornamental value, its included fossils are of great interest to the palaeontologist. '[Early] Cretaceous amber from Alaska' (Lagenheim and others, *BGSA*, 1960). [name transferred from its original use for what is now called 'ambergris', f. French *ambre gris*, f. Arabic *anbar*, a wax-like ash-coloured substance secreted by the sperm-whale.]

**Ammonite.** In the widest sense, any member of the order Ammonoidea in which the septal suture is 'ammonitic' (lobes and saddles more or less finely divided) or 'ceratitic' (lobes denticulate, saddles smooth). All these are post-Carboniferous. Stricter senses would exclude (1) Permo-Triassic forms with ceratitic sutures and/or (2) forms abnormally coiled.

The large Lias ammonite, *Arietites bucklandi*, from Keynsham, Somerset, was noticed by Leland (c. 1538) and Camden (1586) as 'stones faschioned like serpents', and was described by Hooke as 'some kind of Nautili' (*Micrographia*, 1665). Hooke gave a careful description of this fossil, with its 'very pretty kind of sutures, most curiously shap'd in the manner of leaves; all these sutures, by breaking some of these stones, I found to be the termini or boundings of diaphragms or partitions'. Beautiful drawings of this and other ammonites are given in his posthumously published *Discourse of Earthquakes*, 1705. Childrey (*Natural Rarities*, 1661) refers to Lias ammonites in Gloucestershire as 'serpentine stones, or snake stones, as they call them there-about.' 'We find plenty in the County of Oxford, of different colours, figures, sizes, but all curled up within themselves, and therefore by the ancients called *Cornua Ammonis*, for that they resembled the curled horns of the ram, worshipp'd by the name of Jupiter Ammon in the desarts of Africa' (Plot, *Oxfordshire*, 1677). Morton (*Northamptonshire*, 1712) says: 'To these I shall assign the name of *Conchae Ammoniae*, and to the stones formed in them of *Ammonitae* . . . the *Ammonites* originally formed in the *Ammonia*'. This distinction was not generally adopted, *Ammonites* (sing.) and *Ammonitae* (pl.) being used for all the fossils, whether the original shell, an internal mould, or an external impression (e.g. Woodward, *Catalogue*, 1729). *Ammonites* (proposed by Brugière in 1789) came to be widely used as a generic name, e.g. by the Sowerbys (*Mineral Conchology*, 1812 onwards), but as such it has now been officially suppressed. Conybeare and Phillips (*England and Wales*, 1822) also speak of 'the ammonites' as an anglicized word, in the plural; so that here we have an early use of the English name 'ammonite'. And: 'The main bed of alum shale abounds with petrifactions, among which the ammonites (the well known Whitby "snake-stones") hold a conspicuous place' (Young, *Yorkshire Coast*, 1822). See Nelson, 'Ammonites: Ammon's horns into cephalopods' (*JSBNH*, 1968).

In France, in the 17th and 18th centuries, the rock we now know as oolite was called 'ammonite' or 'ammite' because its grains looked like sand (see etymology, below). Neither 'ammonite' nor 'ammite' ever seem to have been used for a sandstone; the nearest being 'psammite'. [G. *Ammon*. Ammon was the N. African representation of the god Jupiter or Zeus, shown in statues as having elaborately coiled horns. The form of these horns may itself have been suggested by the ammonite shell. The name of the god is derived from G. *ammos*, sand.]

**Ammonoidea.** An order of the class Cephalopoda, phylum Mollusca, having a chambered elongate-conical or tubular shell usually coiled in a symmetrical (plane) spiral. Some ammonoids are straight, or bent rather than coiled, and a few genera (e.g. *Turrilites*, Cretaceous) are coiled like a gastropod. The partitioning septa, where they meet the shell to form the sutures, are folded into lobes and saddles which may be further subdivided. There is a calcareous tube (siphuncle) which pierces the septa at their outer (in the Clymeniina, the inner) margins in the plane of symmetry. The form of the suture and the position of the siphuncle distinguish the ammonoid shell from the nautiloid. They range from the Devonian to the Cretaceous. The order comprises the following suborders: Anarcestina, Clymeniina, Goniatitina, Prolecanitina, Ceratitina, Phylloceratina, Lytoceratina, Ammonitina. (Zittel, 1884.) [See AMMONITE.]

**Amorphous.** Formerly much used in a general way for 'bodies devoid of regular form' (Lyell, *Principles*, 1833), rock bodies 'occurring in a continuous mass, without stratification, cleavage, or other division into similar parts' (*OED*). It is now chiefly used in mineralogy, 'to describe the complete absence of crystalline structure, a condition found in the natural glasses but rare in minerals' (Read, *Rutley's Mineralogy*, 1953). [G. *amorphos*, formless.]

**Amphibia.** A class of the subphylum Vertebrata. The modern groups, frogs, toads, newts, and salamanders, are seldom fossilized but have been found from the Jurassic onwards. Very different forms occur in the Carboniferous and Permian, the Stegocephalia (=Labyrinthodontia). [*G. amphibios*, living both on land and in water.]

**Amphibole.** A family of rock-forming minerals, mainly silicates of magnesium, calcium, and iron. Hydroxyl (OH) is always present. Most crystallize in the monoclinic system. Hornblende (magnesium, iron, aluminium, and calcium) is the commonest member. Among others are actinolite (magnesium, iron, and calcium), tremolite (magnesium and calcium), glaucophane (magnesium, iron, aluminium, and sodium), and riebeckite (iron and sodium). 'R. J. Haüy, 1797, f. G. αμφίβολος [*amphibolos*] ambiguous, because so easily mistaken for other minerals' (Chester, *Names of Minerals*, 1896). (The correct pronunciation is in three, not four syllables.)

**Amphibolite.** A metamorphic rock largely composed of an amphibole mineral, hornblende and a plagioclase felspar being the usual associates. Such rocks constitute the 'amphibolite facies' in metamorphism.

**Amygdaloidal.** Almond-like; for a lava in which gas bubbles (vesicles) have become filled with secondary minerals, often oval and light coloured, suggesting almonds in a cake. 'Toadstone contains bladder-holes, like the scoria of metals, or Iceland lava, some filled with spar, others only in part, and others again are quite empty' (Whitehurst, *Strata in Derbyshire*, 1778). 'The toad-stone of Derbyshire is of the amygdaloides species' (Hutton, *TRSE*, 1788). 'Amygdaloidal earths' (Maton, *Western Counties*, 1797). 'Amygdaloidal: when composed of a compact ground with cavities which have been filled up with another mineral substance' (Bakewell, *Geology*, 1813). 'The vesicles may be of microscopic minuteness, but are generally quite visible to the naked eye, and are often large and conspicuous. Sometimes these cavities have been subsequently filled up with calcite, quartz, agate, zeolites or other mineral deposition. As the kernels thus produced are frequently flattened or almond-shaped ("amygdales"), owing to elongation of the steam-holes by movement of the lava before its consolidation, the rocks containing them are said to be "amygdaloidal"' (Geikie, *Ancient Volcanoes*, 1897). 'Amygdaloid' as an adjective is the same as 'amygdaloidal', but is usually a substantive meaning amygdaloidal rock (e.g. Lyell, *Elements*, 1838). [G. *amugdalē*, an almond.]

**Anagenesis.** In evolution, particularly organic evolution, the progressive acquirement of new characters. (The opposite of catagenesis.)

**Anamorphism.** Metamorphism in which chemically complex minerals are produced at the expense of simpler ones. Most metamorphism, in the usual sense, is anamorphism. (Van Hise, 1904.)

**Anatexis.** 'Assimilation and re-melting of rocks with the aid of granitic magma' (Tyrrell, *Petrology*, 1929, where the meanings of 'anatexis' and 'palingenesis' are discussed). Hence 'anatectic' and 'anatectite'. (Cf. 'syntexis'.) (Sederholm, 1907). [G. *tēktos*, melted.]

**Anatomy.** The 'anatomy' of a deposit or rock-body is its intimate composition and structure. 'Planning for water supply, excavations, foundations and waste disposal must take into account the "anatomy" of the whole [glacial] deposit' White, *IGC*, 1972). 'The anatomy of a large submarine slump' (Dingle, *JGS*, 1977). 'The anatomy of a batholith' (Pitcher, *JGS*, 1978).

**Ancient volcano.** See VOLCANO. 'On the remains of ancient volcanoes on the north coast of Cornwall' (Whiteley, *RIC*, 1849). 'The ancient volcanoes of the Highlands' (Judd, *QJ*, 1874). *Ancient Volcanoes of Great Britain* (Geikie, 1897). 'North Wales displays in all the clarity of deep-seated erosion some of the finest examples in the world of ancient volcanoes' (Smith and George, *BRG*, *N.Wales*, 1935).

**Andalusite.** Aluminium silicate, $Al_2SiO_5$. A product of the low-grade thermal metamorphism of argillaceous rocks. It is an 'anti-stress mineral', the 'stress mineral' of the same composition being kyanite. The high-grade metamorphic equivalent of andalusite is sillimanite. Chiastolite is a variety of andalusite containing carbonaceous inclusions regularly arranged crosswise, well-known from the 'chiastolite-slate' in the metamorphic aureole of the Skiddaw granite. Cordierite, with iron and magnesium, is produced where the clay minerals of the original rock contain these elements, the stress mineral equivalent being staurolite. (Delamétherie, 1798.) [Andalusia, Spain.]

**Andean chain.** 'Aubouin's distinction of Andean from Alpine chains (*IGC*, 1972) is confirmed. Andean chains are characterised by andesite volcanics and tonalite batholiths, whereas Alpine chains contain ophiolite belts and sedimentary flysch. The differences may reflect their different tectonic setting' (Cobbing, *JGS*, 1978).

**Andesine.** See PLAGIOCLASE. Occurs particularly in the igneous rocks of intermediate composition, such as Andesite. (Abich, 1841). [Andes mts.]

**Andesite.** A fine-grained igneous rock, of intermediate composition, usually porphyritic, composed essentially of plagioclase (excluding albite) in excess of alkali felspar, and mafic minerals, either hornblende or augite dominant. Approximately the fine-grained equivalent of diorite. Extrusive, sometimes as minor intrusions. (Von Buch, 1835.) [Andes mts.]

**Andesite line.** 'The boundary between the basalts of the oceanic crust and islands, and the andesite-dacite-rhyolite volcanic rocks of the Circum-Pacific belt, is called the "Andesite Line", and it turns out to be essentially the boundary between the continental crust with sial and the oceanic crust without sial, and to follow the ocean trenches. . . . The Andesite Line around the Pacific, and its continuation around the Indonesian arc is, as Lester King puts it, "one of the fundamental geological boundaries of the earth". It has turned out to be the line where oceanic lithosphere is subducted along Benioff Zones. . . . Map of the Pacific Ocean to show the Andesite Line' (Holmes, *Physical Geology*, 1978).

**Angiospermae.** The flowering plants, one of the main classes of the plant kingdom, and the one that has been dominant since about the end of the Cretaceous. As a conspicuous part of the world's flora they range from the beginning of the Cretaceous, though a few remains have been found in the Jurassic. 'Fossil evidence and angiosperm ancestry' (Hughes, *SP*, 1961). [G. *aggeion*, a vessel, *sperma*, a seed.]

**Anhedral.** See ALLOTRIOMORHIC. [G. *hedra*, a base.]

**Anhydrite.** See GYPSUM. (Werner, 1804, *anhydrit.*) [G. *hudōr*, water.]

**Animalia.** The Animal Kingdom. Includes the following phyla: Protozoa, Porifera, Coelenterata, Echinodermata, Annelida, Brachiopoda, Bryozoa, Mollusca, Arthropoda, Chordata. The first three are in order of increasing complexity or organization and the Arthropoda are the most highly organized of the non-chordate (practically=invertebrate) animals. The phyla listed between are hardly to be arranged in any such order. The Chordata, almost entirely composed of the subphylum Vertebrata, are on a different scale of organization. Zoologists recognize a few other phyla, all of worms, of no interest to the palaeontologist. [L.pl., animals.]

**Anisotropic.** See ISOTROPIC. 'Shear failure of anisotropic rocks' (Jaeger, *GM*, 1960). (The negative word 'anisotropic' refers to the positive character.)

**Ankylosauria.** A suborder of the order Ornithischia, one of the two dinosaurian orders of Mesozoic reptiles. Herbivorous, quadrupedal, very strongly armoured with thick bony plates that completely encased the back and sides of the body and tail (they have been called 'reptilian tanks'). Cretaceous. See DINOSAURS. [genus *Ankylosaurus* (G. *agkulos*, crooked, *saura*, a lizard).]

**Annelida.** The phylum of segmented worms, mostly marine. Some live in tubes, which they secrete, and when these tubes are calcareous they are readily fossilizable,

*Serpula* being a common genus from the Silurian to the Present. [L. *an(n)ellus*, a little ring.]

**Anorogenic.** Not orogenic; without the presence of orogeny, not caused by orogeny.

**Anorthite.** Calcium felspar, calcium aluminium silicate; an end member of the plagioclase series. 'A name used by G.Rosé (in 1823) for the glassy crystals found in the volcanic ejections of Vesuvius' (Miers, *Mineralogy*, 1929). See FELSPAR and PLAGIOCLASE. [name in allusion to the non-rectangular crystal symmetry of the plagioclase felspars.]

**Anorthosite.** A monomineralic igneous rock composed of a calcic plagioclase felspar; a gabbro practically free from mafic constituents. (Sterry Hunt, 1863.) [French *anorthose*, plagioclase felspar.]

**Antagonism.** 'The principle of antagonism in the Earth's evolution. It is a rather well-known fact that two types of effect are implicated in the evolution of the Earth's surface features: endogenic effects and exogenic effects. [It seems likely] that the two types of effects are active at the same time and the momentary state of the Earth's surface is nothing but the instantaneous result of the antagonistic action of endogenic and exogenic effects' (Scheidegger, *Tp*, 1979). See HEIGHT OF THE LAND.

**Antecedent.** Used particularly for a river or whole drainage system. An antecedent river is one which, by its power of erosion, has been able to keep its course while uplift and a new geological structure have developed across it. Thus, as seems to be exemplified in the Himalayas, such a river may rise behind a mountain range and flow through it in a gorge which it has cut like a gigantic saw as the land pressed upwards. The river is antecedent to the relief and structure. Thus 'antecedent drainage' as a whole. 'I have endeavoured above to explain the relation of the valleys of the Uinta Mountains to the stratigraphy, or structural geology, of the region, and, further, to state the conclusion reached, that the drainage was established antecedent to the corrugation or displacement of the beds by faulting and folding. I propose to call such valleys "antecedent valleys"' (Powell, *Colorado River*, 1875, quoted by Mather and Mason, *Source Book*, 1939). 'The idea of antecedent rivers had occurred to several observers who gave it no name, and unnamed it gained no currency; but it became popular when Powell named it' (Davis, *Essays*, 1909).

**Anthozoa.** A class, of the phylum Coelenterata, that includes the corals and sea-anemones, the former occurring abundantly as fossils. The chief orders of fossil corals are the Scleractinia (=Hexacoralla) and the extinct (Palaeozoic) groups Rugosa and Tabulata. These three are placed in the subclass Zoantharia. The modern corals belonging to the subclass Octocoralla occur only very sparingly as fossils. [G. *anthos*, a flower.]

**Anthracite.** Coal of high rank, being very rich in carbon (over 90%); burns slowly, has a high lustre, and does not soil the fingers. (Davy, 1812.) 'The Greek philosopher Theophrastus, a pupil of Aristotle, knew coal and called it anthrax, a name from which our word anthracite has been derived' (Van Krevelen and Schuyer, *Coal Science*, 1957).

**Anticlinal ridge.** A ridge whose structure is anticlinal. The term is sometimes used with the implication that the ridge was formed as a direct result of the earth movement which at the same time formed the anticline.

**Anticlinal valley.** See VALLEY OF ELEVATION.

**Anticline.** Primarily, an arch-fold in stratified rocks. Hence 'anticlinal'. (The opposite terms are 'syncline', 'synclinal'.) The term anticline is sometimes applied, with doubtful propriety, to any fold that has older (originally underlying) rocks embraced by younger, and the term syncline to the reverse disposition, irrespective of the present attitude of the fold as a whole. See ANTIFORM. 'Anticline' is sometimes used for an anticlinal arrangement of other geological features, such as metamorphic zones (e.g. Read and Farquhar, *QJ*, 1956). 'The thermal structure of the Highlands is essentially that of a simple anticline of thermal "surfaces"' (Kennedy, *GM*, 1948). 'The fault-plane itself has been anticlinally folded' (Lees, *QJ*, 1952).'Otley notices [*LM*, 1820] that in Borrowdale the cleavage-planes dip to the north, and in Langdale to the south, which would indicate that the syncline of the Scawfell group of fells may coincide with an anticline of the cleavage-planes' (Marr, *Lake District*, 1916).

Today the nouns 'anticline', 'syncline' seem more fundamental than the adjectives 'anticlinal', 'synclinal', but the adjectives were the first to be used. 'These [coal] basins are divided from one another by lines which may be termed "anticlinal lines", formed by the saddles of the strata' (Buckland and Conybeare, *TGS*, 1824). And Lyell (*Principles*, 1833), of the Weald: 'In the centre are seen the Hastings sands, forming an anticlinal axis, on each side of which the other formations are arranged with an opposite dip'. Eventually the nouns creep in; thus Murchison has: 'The southern portion of this anticlinal [May Hill]' (*Silurian System*, 1839), and Page (*Textbook*, 1856): 'The strata are spoken of as forming an anticline or saddleback'. 'The general anticlinal structure of the surface rocks of the area was known to John Farey, who in 1806 produced in the form of a diagram a geological section across the Weald. Several manuscript copies of this were made, one of which is in the possession of the Geological Survey' (*BRG, Wealden District*, 1965). An early published section of an anticline is that by Watson (*Strata of Derbyshire*, 1811). The conception of an eroded anticline is well illustrated in the following words by John Michell (*PT*, 1760): 'Let a number of leaves of paper, of several different sorts or colours, be pasted upon one another; then bending them up together into a ridge in the middle, conceive them to be reduced again to a level surface, by a plane so passing through them, as to cut off all the part that had been raised'. Farey (*PM*, 1814) uses the terms 'strata-ridge line' and 'strata-trough line'.

'Anticline in Coal Measures, Saundersfoot, Pembrokeshire' (pl. in Holmes, *Physical Geology*, 1965/78), one of the most spectacular, and most instructively dissected, anticlines to be seen in Britain.

**Anticlinorium** (pl. **-ia**). 'An upward bend of the crust, or geanticlinal, is of itself an elevation; and such an elevation is an "anticlinorium"' (Dana, *Manual*, 1875). But later writers have used 'anticlinorium' (and 'synclinorium') in senses very different from Dana's; and their usages, which are convenient, have been followed ever since. Accordingly an anticlinorium is now defined as a compound anticline, that is, one in which the limbs are themselves thrown into folds; usually applied to structures on a large scale. The derivation now appears to be from the L. suffix *-orium*, 'a place for', rather than from the G. *oros*, 'a mountain'.

**Anticlise.** An upwarp of a platform that has become otherwise rigid. Typically the platform is beneath the sea and a cover of sediments is being deposited on it as the warping proceeds. A similar downwarp is a syneclise. (Shatsky, Russia, 1945, 1957; Tomkeieff, *PGS*, 1958.) [G. *klisis*, a bending.]

**Antiform.** An anticlinal arrangement of strata, but excluding any implication that the structure is simple and the right way up. 'Antiform' can be used for other anticlinal arrangements, e.g. 'antiform of cleavage' (Shackleton, *QJ*, 1957).

The following are some further remarks on the use of 'anticline' and 'antiform'. Both terms are, in their widest senses, applicable to any kind of arch-shaped arrangement of strata. 'Anticline' is etymologically felicitous and much the older established term, while 'antiform' may probably be said to be short for 'in the form of an anticline', thereby implying that it would be preferred to 'anticline' when the structure was not, or might not be, an 'anticline' in some particularly true or strict sense. The chief cases and corresponding usages may be analysed as follows. A. Condition known. (a) A simple arch-fold with the beds the right way up: 'anticline' invariably used. (b) An arch-fold, but with the beds inverted: 'antiform' would be used, with the statement that it was an original syncline turned upside down. (c) A composite arch-fold, the axial surfaces and limbs of several super incumbent folds being arched: 'antiform' would be used, with the structure described. (d) In a multiple fold (composed of a pile of recumbent folds with limbs and axes bent arch-wise), a component fold that was originally an anticline (older beds embraced by newer): 'anticline' used. ('Antiform' would be appropriate for each component fold, whether originally an anticline or a syncline, 'anticline' and 'syncline' being used for the component folds according to the original attitudes.) B. Condition not known. 'Antiform' would be used, either without qualification or with the uncertainty expressed. The same considerations, in all the above cases, apply in the case of 'syncline' and 'synform'.

'In common tectonic practice, an anticline has come to be understood as a fold with a core of previously underlying rocks, and a syncline as a fold with a core of previously overlying rocks . . . the authors have found it convenient to be able to use words that have an entirely geometrical significance. Thus in the following pages "antiform" means a fold that closes upwards, "synform" means a fold that closes downwards' (Bailey and McCallien, *TRSE*, 1937).

**Anti-stress mineral.** See STRESS MINERAL.

**Antithetic.** Various definitions of a structural 'antithesis' are given, all to do with faulting; of these the conception that seems the most to deserve a term is that of a system of faults that throw in a direction opposite to that of the dip of the strata (e.g. Mackin, *AJS*, 1960).

**Apatite.** An ubiquitous accessory mineral in igneous rocks of all kinds. Essentially, calcium phosphate. Hexagonal system, well-developed crystals in the prism-pyramid form being common. Usually bluish in specimens but colourless in thin section. (Werner, 1786). [G. *apataō*, deceive, because it had been mistaken for other minerals.]

**Apex.** 'The tip, top, peak, or pointed end of anything' (*OED*); thus applicable, in structural geology, to the top point of a dome structure (in any one stratal surface), but hardly to a line.

**Aphanitic.** For an igneous rock, or for the groundmass of one, in which the constituent minerals cannot be distinguished by the naked eye. Includes the microcrystalline and cryptocrystalline states. A rock wholly aphanitic is an 'aphanite', a name given by Haüy in 1822 to compact rocks of dioritic composition. [G. *aphanēs*, invisible.]

**Aplite.** A very leucocratic, fine-grained, igneous rock of simple composition, the commonest type being granite-aplite, made up almost entirely of quartz and felspar, often with saccharoidal or micrographic texture. Usually occurs as veins in the parent granite itself or penetrating the surrounding rocks. A term long in use. 'The origin of aplites' (Wells and Bishop, *PGA*, 1954). [G. *haploos*, simple; thus the rock might more properly be called 'haplite'.]

**Apophysis** (pl. **-es**). In geology, reserved for an offshoot from an igneous intrusion or a mineral vein. [G. *apo*, away from, *phusis*, origin.]

**Apparent dip.** See DIP. The term is also sometimes used for the dip in cross-bedding, as distinct from the dip of the formation as a whole.

**Apparent throw.** See THROW.

**Applanation.** See PLANATION. (Cairnes, *AJS*, 1912.)

**Applied geology.** See GEOLOGY and ECONOMIC GEOLOGY.

**Aptian.** See CRETACEOUS SYSTEM. [Apt, S. France.]

**Aptychus** (pl. **-i**). A fossil in the form of a pair of plates (true aptychus) or a single plate ('anaptychus' being the stricter, more correct term here), horny or calcareous, which closed the apertures of some ammonite shells. As a proper name, it is in the parataxonomic category of genus. [G. *a-ptuchē*, a fold.]

**Aquagene.** Produced under water. 'Volcanic tuff [formed] entirely beneath water—a rock type for which the term "aquagene tuff" is herein proposed' (Carlisle, *JG*, 1963).

**Aqueo-igneous.** 'A term applied to the mode of origin of certain minerals and rocks, the formation of which from a molten state [has] been influenced by the presence of water dissolved or intermixed in the magma' (Joly, *Surface History of the Earth*, 1930).

**Aqueous.** This term, strictly meaning watery, is used in geology chiefly for deposits (and the resulting consolidated rocks) laid down in water ('sub aqueous'). These include sediment, shelly or other organic accumulations, and chemical precipitates. 'The Neptunists hold the rocks, here enumerated, and also granite, to be produced by aqueous deposition' (Playfair, *Illustrations*, 1802). 'Aqueous sedimentary rocks' (Geikie, *Textbook*, 1903). Also used more generally: 'The aqueous agents are incessantly labouring to reduce the inequalities of the earth's surface' (Lyell, *Principles*, 1830).

**Aquiclude.** A rock-body virtually impermeable to water. In most hydrogeological contexts it would be implied that the rock-body was, in addition, impervious.

**Aquifer.** A water-bearing reservoir rock, such as a particular formation or stratum.

**Aquitanian.** See PALAEOGENE. [Aquitania, a district of Gaul.]

**Arachnida.** In the wide sense (Chelicerata), a division of the phylum Arthropoda, including the classes Arachnida in the narrow sense (spiders, scorpions, mites) and Merostomata, the aquatic members, mostly extinct. Spiders and scorpions do occur as fossils, exceptionally; e.g. in the Coal Measures (spiders) and the Triassic (scorpions). [G. *arachnēs*, a spider.]

**Aragonite.** A rather soft mineral, calcium carbonate, $CaCO_3$; usually white; orthorhombic system. (Calcite is another, and more stable, mineral of the same composition.) The name was in use by the latter part of the 18th century. [Aragon (region), Spain.]

**Archaean.** The crystalline rocks, igneous, metamorphic, or migmatitic, of Precambrian age. 'The term "Archaean" was introduced by J. D. Dana in 1872 [*AJS*] to designate the formations older than the Cambrian' (Krishnan, *India and Burma*, 1956). At that time all the Precambrian rocks were thought to be of that type. When such overlying sedimentary formations as the Torridonian and Longmyndian were found to be of Precambrian age, the term became extended to include them (e.g. Geikie, *Textbook*, 1903; Jukes-Browne, *Stratigraphical Geology*, 1912). The name has now reverted to its restricted meaning, the overlying sedimentary formations of Precambrian age being grouped together, usually as Algonkian, sometimes as Proterozoic, sometimes as Precambrian in a restricted sense. 'There is now a marked tendency in international [Precambrian] geology to draw a clear distinction between the ancient highly metamorphic continental blocks and the overlying little-altered and obviously sedimentary but unfossiliferous groups. These may be conveniently called the Archaean and Algonkian respectively. Of these the Lewisian and Torridonian of Scotland give a simple example' (Rastall, *GM*, 1944). We cannot say that the Archaean and Algonkian in one region are respectively the time-equivalents of the Archaean and Algonkian in another. 'It was proposed by the United States Geological Survey to reserve the term "Archaean" for all the essentially igneous rocks that underlie the Pre-Cambrian sedimentary formations, and to embrace these sedimentary formations under the general designation of "Algonkian". But we now know that the "Archaean" series includes various sedimentary intercalations, and that the "Algonkian" is actually pierced by portions of the "Archaean" masses' (Geikie, *Textbook*, 1903). 'The Archaean rocks of the Rodil district, South Harris' (Davidson, *TRSE*, 1943).

**Archaeocyatha.** Now recognized as a separate phylum; exclusively Cambrian. The calcareous skeleton is typically conical, composed of an inner and an outer cup with varied structural elements between. *Archaeocyatha from Antarctica and a Review of the Phylum* (Hill, 1965). *Archaeocyatha* (Teichert, ed., in *Treatise on Invertebrate Paleontology, part E*, 1972). (Billings, 1861; Vologdin, 1937.) [G. *kuathos*, a wine-cup.]

**Archaeornithes.** One of the two subclasses of the class Aves, comprising the two genera *Archaeopteryx* and *Archaeornis*, each known from (practically) one individual only, from the Lithographic Limestone (Kimmeridgian) of Solenhofen, Bavaria. Apart from having feathers, several important features are reptilian; teeth in the jaws, claws on the wings and a jointed tail. They are a perfect, but all too rare, example of an evolutionary link between two major groups. 'They might well have been taken for reptiles, were it not for the impressions of feathers which surround them on the stone slabs on which they are preserved' (Romer, *Vertebrate Palaeontology*, 1945). [genus *Archaeornis* (G. *ornis*, a bird).]

**Archaeozoic.** A term proposed by Dana (*Corals*, 1875) to be used for that period of time when it might be supposed that some of the simplest kinds of life were in existence. Beyond referring to some part of the Precambrian, it has no agreed meaning, some authors applying it to an early part, some to the middle, and yet others to the latest part.

**Ardmillan series.** The series of beds, nearly 900m. thick, that represents, in the Girvan region of Scotland, the upper part of the Caradocian together with the Ashgillian, in the shelly facies. Some of the beds are highly fossiliferous, with distinctive species. Equivalent to the Hartfell Shales in the graptolitic facies. (Lapworth, *QJ*, 1882.) [Ardmillan, Ayrshire.]

**Arena.** 'Another feature of Uganda scenery is the existence of large comparatively low-lying but undulating areas more or less completely enclosed by the hill ranges. I have called those areas "arenas". . . . The arenas are domes which, by downward corrosion from their summits, have suffered more or less reversal of their apical contours' (Wayland, *Some Facts and Theories relating to the Geology of Uganda*, 1920). These morphological features in Uganda are strikingly arena-like and appear to be peculiar to that part of the world, but 'arena' might perhaps be extended in a general way to any eroded structural dome that gives a relatively shallow central area surrounded by a higher rim of more resistant rocks. In this extended sense 'arena' would be the dome-equivalent of the anticlinal 'valley of elevation'.

**Arenaceous.** Sandy. For sediments and sedimentary rocks. 'Among the most indurated rocks, many are arenaceous' (Playfair, *Illustrations*, 1802). 'Arenaceous—Sandy' (Lyell, *Principles*, 1833). 'The great arenaceous formation of red and variegated sandstones, constituting the Bunter Sandstein of continental geologists' (Murchison, *Silurian System*, 1839). [L. *arena*, sand.]

**Arenig series.** (Arenigian.) The lowest series of the Ordovician system. The characteristic trilobite fauna includes, for example, *Ogygiocaris selwyni*. The graptolites constitute the Dichograptid fauna (Bulman, *P*, 1958). The name is sometimes used to include the succeeding series, the Llanvirn; either the whole or (more often) the lower half of it. (Sedgwick, *PGS*, 1843, *QJ*, 1852, *British Palaeozoic Rocks*, 1855; Fearnsides, *QJ*, 1905.) [Arennig (Arenig) mountain, Meirionnydd district of Gwynedd.]

**Arenilitic.** 'Of or pertaining to sandstone' (*OED*). 'Of arenilitic or sandstone soils' (Kirwan, *Geological Essays*, 1799).

**Arenite.** An arenaceous rock, a sandstone, a psammite. 'Classification of arenites' (Crook, *AJS*, 1960).

**Argillaceous.** Clayey. For sediments and sedimentary rocks. '*Argilla*, clay [for a specimen of fireclay]' (Woodward, *Catalogue*, 1729). 'There is ironstone, which is commonly found among the argillaceous strata attendant upon fossil coal' (Hutton, *TRSE*, 1788). 'The sand becoming finer above, then argillaceous, and at last changing into real clay' (Playfair, *Illustrations*,

1802). 'The Upper Ludlow subdivision consists essentially of thin-bedded, lightly coloured, and very slightly micaceous sandstones, in some parts highly argillaceous, and in others calcareous' (Murchison, *Silurian System*, 1839). [L. *argilla*, clay.]

**Argillite.** An argillaceous rock, a mudstone, a pelite; now, particularly a highly indurated unlaminated mudstone. In use in the 18th century (e.g. Kirwan, *Geological Essays*, 1799). 'Slate and some other kinds of argillite' (Playfair, *Illustrations*, 1802).

**Arkose.** A coarse-grained sandstone rich in undecomposed felspar, presumably derived from a granite. 'The "arkose" of central France' (Murchison, *Silurian System*, 1839). 'Tertiary formation of the Limagne d'Auvergne . . . the sandstones and conglomerates . . . M. Brongniart, who gave them the name of "Arkose". . . some beds consist of a conglomerate of worn pebbles and fragments of granite [&c.] . . . other beds consist of a quartzoze grit formed of separate crystals of quartz, mica, and felspar, evidently formed *in situ* from the disintegrated materials of the granite on which they rest, and from which they are scarcely to be distinguished' (Scrope, *Extinct Volcanos of Central France*, 1858). 'The great bulk of the Torridonian formation consists of more or less coarse-grained arenaceous sediments in the form of felspathic grits and sandstone (arkose)' (*MGS, NW. Highlands*, 1907). 'The term was introduced by Brongniart (1823) in an attempt to limit the use of the word sandstone (*grès*) which he felt included too many diverse rocks. His definition of the term, literally translated, is, "arkoses are composed of large grains of glassy quartz and of feldspar, mixed together unequally and including as fortuitous constituents mica, clay, often kaolin, etc." Why Brongniart chose the word "arkose" is not known. He does not indicate the derivation. The present writer tentatively suggests that the term "arkose" was taken from the Greek *archaios* meaning ancient or primitive' (Oriel, *AJS*, 1949). 'Arkosic' is used as the adjective.

**Armorican.** Applied by Suess (as *Das armoricanische Gebirge* in *Das Antlitz der Erde*, 1888) to the mountain building in western Europe in late Palaeozoic times, corresponding to his Variscan of central Europe. 'This is the great pre-Permian range of western Europe. The traces of its interior and presumably most elevated zone lie in Brittany [the Armorica of the Romans] and the Vendée; for this reason we give these fragments the general name of the "Armorican chain"' (Suess, 1888,trans., 1906). In Britain 'the intensive mountain building movements of the Hercynian orogeny were mainly confined to the area south of the Armorican Front (a line extending approximately from Dungeness to South Pembrokeshire)' (Hains and Horton, *BRG, Central England*, 1969).

**Armoricanoid.** See CALEDONIAN.

**Artesian well.** A well which reaches a water-bearing stratum (aquifer) through a cover of impervious rock and in which the water rises, perhaps to overflow, by hydrostatic pressure. The water in the aquifer must somewhere be considerably above the level of the point where the well reaches it, if the water is to rise in the well above that point. The geological situation of London with regard to the London Clay and the Chalk is admirably adapted for the sinking of artesian wells. 'Artesian wells are well known as borings, by which water rises to, and even above the surface' (De la Beche, *Manual*, 1831). Observations were added by Lyell to his *Principles* in the 3rd edition, 1834, on 'the boring of what are called by the French "Artesian wells", because the method has long been known and practised in Artois', and the matter was thoroughly expounded by Buckland in his Bridgewater Treatise on *Geology*, 1836.

**Arthrodira.** See PLACODERMI.

**Arthropoda.** The most highly organized phylum of invertebrate animals. The body is segmented, the segments not being all alike, but forming groups, and bearing jointed appendages. The body has a chitinous covering (exoskeleton) which may be hardened over certain parts by carbonate or phosphate of lime, and thus become fossilizable. The phylum includes the Crustacea, Arachnida in the wide sense

(=Chelicerata), Trilobita, Myriapoda (centipedes and millepedes, not important fossils), and Insecta. Arthropod is now considered to be a grade of organization. 'There is overwhelming support for the view that the major arthropodan taxa are not monophyletic in origin. The arthropods thus become a grade of organization independently reached a number of times. Apparently distinct phyla are the Chelicerata, the Crustacea, and the Uniramia (Onychophora, Myriapoda, Hexapoda). The status of the Trilobita and of many other early arthropods cannot be assessed at present' (Manton and Anderson, *SA* (12), 1979). [G. *arthron*, a joint.]

**Articulata.** One of the two classes of the phylum Brachiopoda, the valves opening and closing by means of two projections, teeth, on the pedicle valve, fitting into sockets on the brachial valve. The shell is calcareous. The class was formerly divided into two orders, Protremata and Telotremata, but is now usually grouped directly into 'suborders' of which some are the Orthoidea, Dalmanelloidea, Pentameroidea, Strophomenoidea, Productoidea, Rhynchonelloidea, Atrypoidea, Spiriferoidea, and Terebratuloidea. (Huxley, 1869.) [L. *articulatus*, distinctly jointed.]

**Artifact.** A product of human workmanship; particularly a flint implement made by 'early man'. [L. *ars, art-*, skill, *facio*, make.]

**Arundian.** See DINANTIAN. [L. *arundo*, a hobbyhorse; Hobbyhorse Bay, South Pembrokeshire district of Dyfed.]

**Arvonian.** The volcanic series of rocks of the country (Arvon) between the Menai Strait and Snowdonia, of Precambrian or Lower Cambrian age; but originally in a wider sense, to include certain similar rocks in other parts of Wales, and the Borderland. (Hicks, *GM*, 1878/9, *QJ*, 1879; Greenly, *GM*, 1930, *QJ*, 1944.)

**Asbian.** See DINANTIAN. [Little Asby Scar, near Newbiggin-on-Lune, Cumbria.]

**Aseismic.** Without seismic activity; as in 'The Mesozoic development of aseismic continental margins' (Kent. *JGS*, 1977). Ductile deformation is 'aseismic'.

**Ash.** A general term, more specifically 'volcanic ash', long in use, for all the finer pyroclastic material shot out of a volcano. Volcanic ashes result from explosion, not burning. 'A shower of ashes in the archipelago' (*PT*, 1667). The term is used both for the products of recent volcanicity and for those of the volcanicity of a past geological age. 'Tuffs' are ashes consolidated either recently (by the action of water or as ignimbrites) and/or, in the case of volcanic material of a past age, by the action of other geological processes. 'Ash-flow tuffs: their origin, geologic relations, and identification' (Ross and Smith, *USGS*, 1961). 'Ash-flow tuffs' (Chapin and Elston, eds., *GSASP*, 1979). The term 'ash' is sometimes restricted to the recent loose material. 'Jukes and Forbes rode to the foot of the Arenig; Aveline, Williams, Gibbs, and I walked. Foggy on the top. Ash, ash, ash everywhere' (Ramsay, diary, 27 October 1846). 'There are many beds here [N. Wales] that I would once have considered altered rocks, which are in reality nothing but hard consolidated ashes. The word "ashes" does not imply "cinders", but often rather volcanic dust, which may be as fine as you like' (Ramsay, letter to Aveline, 1846: both extracts from Geikie's *Life*, 1895). See TEPHRA.

**Ashgill series.** (Ashgillian.) The uppermost series of the Ordovician system and the upper part of the Bala series (which is the Caradoc and Ashgill combined). It is characterized especially by its trilobites which are species of genera rare in the beds below; for instance, *'Encrinurus' sexcostatus*, *Staurocephalus globiceps*, and *Phillipsinella parabola*. (Sedgwick in Salter, *Cambrian and Silurian Fossils*, 1873; Marr, *GM*, 1892, *QJ*, 1905.) [Ashgill Beck, Cumbria.]

**Assemblage.** The following are some special usages. **1.** The fossils of a bed constituting its whole fauna and flora; all the different kinds and their relative abundance. 'Fossil association' refers to the same thing. It is an objective feature of the bed, but species and varieties that are considered to be good time-markers or good indicators of conditions of deposition are particularly noted. Time, conditions, and place are the selective factors determining the assemblage. See THANATOCOENOSE. Any derived fossils would constitute a quite distinct component of the total assemblage, or would be considered as constituting a separate assemblage. 'Formation of fossil assemblages' (Johnson, *BGSA*, 1960). 'On separate occasions (Craig, *AJS*, 1953; Hallam, *PT*, 1960) we have defined terms for different fossil associations and we hope that the following amalgam of our views, presented schematically, may prove acceptable' (Craig and Hallam, *P*, 1963). **2.** The fossils of one small group of organisms, particularly of a species-group from some particular horizon or locality. 'The variation in an assemblage of the *Caninia cornucopiae* plexus from the Middle Visean' (Hudson, *QJ*, 1944). **3.** A group of discrete fossils apparently belonging to one individual animal. See CONODONT. **4.** The minerals in a sample of rock, particularly an igneous or metamorphic rock, constituting its make-up; all the different kinds and their relative abundance. The 'assemblage' hardly includes the texture and fabric of the rock. An 'assemblage' is practically the same thing as an 'aggregate', but the former term implies a synthesis while the second seems to take the rock first and then to analyse it. **5.** The metamorphic rocks constituting a whole group. 'The suggestion is made that the Scottish metamorphic rocks should be divided, in the first instance, into the Lewisian Metamorphic Assemblage, the Moinian Metamorphic Assemblage, and the Dalradian Metamorphic Assemblage. Where there is no likelihood of confusion, or where there are frequent repetitions, the word metamorphic could be dropped and the terms Lewisian Assemblage, etc. used' (Anderson, *GM*, 1948). However, 'metamorphic mineral assemblage' is sometimes shortened to 'metamorphic assemblage', which might cause confusion.

**Assemblage biozone.** See BIOZONE.

**Assimilation.** In petrology; the full term being 'magmatic assimilation'. '"Magmatic assimilation" is the mutual solution of magma with solid rock, with other magma, or with foreign gas, so that a new magma with at least some approach to chemical homogeneity results' (Daly, *Igneous Rocks*, 1933). (Michel-Levy, 1893.)

**Asteroidea.** See ASTEROZOA. The five (or more) arms merge gradually into the central disk, or the whole is pentagonal. [G. *astēr*, a star.]

**Asterozoa.** A class of the phylum Echinodermata (subphylum Eleutherozoa) comprising the starfish proper (subclass Asteroidea) and the brittle-stars (subclass Ophiuroidea). The five-rayed symmetry characteristic of the phylum is particularly well marked, there being a central disk and (typically) five arms. Asterozoa are rather rare as fossils and occur usually as detached plates; but there are 'starfish beds', in which complete individuals occur. Also called Stelleroidea. They range from the Upper Cambrian.

**Asthenosphere.** 'The theory of isostasy shows that below the lithosphere there exists a thick earth-shell marked by a capacity to yield readily to long-enduring strains of limited magnitude. It is a zone between the lithosphere above and the centrosphere below, both of which possess the strength to bear, without yielding, large and long-enduring strains. Its reality is not lessened because it blends on the limits into these neighbouring spheres, nor because its limits will vary to some degree with the nature of the stresses brought upon it and to a large degree by the awakening and ascent of regional igneous activity. Its comparative weakness is its distinctive feature. It may then be called the sphere of weakness—the "asthenosphere"' (Barrell, *JG*, 1914). A more particularized concept of the 'asthenosphere' is now envisaged. See LITHOSPHERE. [G. *asthenēs*, weak.]

**Astogeny.** The growth of a colonial animal organism, such as a graptolite or bryozoan. [G. *astos*, a citizen.]

**Astrobleme.** A cryptoexplosion structure caused from outside the earth. See CRYPTOEXPLOSION STRUCTURE 'Meteorite impact scars or "astroblemes" (a word from Greek roots meaning "star" and "wound from a thrown object such as a javelin or stone")' (Dietz, *S*, 1960). Also called 'impact structure' and, in particular, 'impact crater'. *Impact and Explosion Cratering* (Roddy and others, eds., 1978).

**Astrogeology.** 'A discipline covering the overlap between geology and astronomy, including (**1**) the application of geologic principles and techniques to the study of celestial bodies; (**2**) the study of objects occurring on earth but of known or possible extra-terrestrial origin (e.g. meteorites, tektites); (**3**) the study of collisions of asteroids, meterorites, and comets with the earth, moon, etc.; and (**4**) the measurement of possible changes in the motions of the sun, earth, and moon and their effects on geologic time—day, lunar month, year, paleoclimatic cycles; and geologic processes such as tides' (Fairbridge, ed., *Encyclopedia of Atmospheric Sciences and Astrogeology*, 1967). '"Astrogeology" was thought to be a new word in August 1960 when the Astrogeologic Studies Group was organized within the [U.S.] Geological Survey. However, in 1960, it was already 84 years old' (Milton, 'Astrogeology in the 19th century', *Get*, 1969.) 'The recent works of Meunier . . . can give an idea of the significance of the new branch of astronomy, which may be called astrogeology' (Lesevich, in translation, *Positive Philosophy*, 1876, quoted by Milton). [G. *astron*, a star.]

**Atlantic suite.** 'A general term for the whole assemblage of alkali-rocks, directing attention to their distribution in and around the atlantic, to their association with the Atlantic type of coast-line, and more generally to their association with tectonic structures due to tension, fracture, and differential radial movements' (Holmes, *Nomenclature of Petrology*, 1928). '[There is] a very general correspondence of the areas of the alkali and sub-alkali groups respectively with the areas of the Atlantic and Pacific types of coast-line as defined by Suess . . . an Atlantic and a Pacific facies of eruptive rocks, corresponding with distinct phases in crust-movements of a large order' (Harker, *SP*, 1896). 'Among plutonic rocks, for example, the Atlantic branch includes the alkali-granites, syenites and nepheline syenites, essexites, theralites, picrites, etc.' (Harker, *Igneous Rocks*, 1909).

**Atmoclastic rock.** (Atmoclast.) A rock consisting of materials broken *in situ* by atmospheric weathering. (Grabau, *AG*,1904.)

**Atmosphere.** The continuous gaseous envelope surrounding the rocky-metallic earth-body and the hydrosphere.

**Atoll.** A coral-reef island of circular, elliptical, or horse-shoe shape, enclosing a lagoon. 'The chain of coral reefs and islets, called the Maldivas, situated in the Indian Ocean to the south-west of Malabar . . . each circle or atoll, as it is termed . . . (Lyell, *Principles*, 1832). 'The origin of reefs and atolls' (Holmes, *Physical Geology*, 1965/78). [probably Malayalam (language of Malabar) *adal*, uniting.]

**Atremata.** One of the two orders of the class Inarticulata of the phylum Brachiopoda. The pedicle opening is not a definite aperture but merely a gap shared by both valves. Cambrian to present day. (Beecher, 1891.) [G. *trēma*, *trēmat-*, an aperture.]

**Attitude.** The disposition or posture of a rock-body, structural unit, structural element, or lineation. Dip and strike define an attitude in relation to the horizontal, and a direction. A succession of strata may be the right way up (either horizontal or dipping), vertical, or inverted; a fold may be in the normal upright position, be recumbent, or be overturned; a fault has attitudes relative to the strike, dip, and vertical displacement of strata. 'Trend and plunge are used to define the attitudes of linear features. The trend of a linear feature is the compass direction of the vertical plane that includes the feature. . . . The plunge is the vertical angle between the feature and a horizontal line' (Compton, *Field Geology*, 1962).

**Attrition.** The wearing away of rocks by the impact or friction of rock fragments or particles; the mutual wear and tear of loose rock fragments or particles. 'Gravel is formed by the mutual attrition of stones agitated in water' and 'We have to consider the mountains as formed by the hollowing out of the valleys, and the valleys as hollowed out by the attrition of hard materials coming from the mountains' (Hutton, *Theory of the Earth*, 1795). 'The fragments on the shore polished by the attrition of the waves' (Maton, *Western Counties*, 1797, of serpentine at The Lizard). 'The smooth surface of the rocks in all waterfalls, their rounded surface, and curious excavations, are the most satisfactory proofs of the constant attrition which they endure' (Playfair, *Illustrations*, 1802). 'Pebbles and sand decrease in size by attrition' (Lyell, *Principles*, 1830). [L. *attero*, rub against, destroy.]

**Augen.** Lenticular crystals or crystal aggregates in metamorphic rocks. Hence 'augen-structure', 'augen-gneiss', 'augen-schist'. [German pl. *augen*, eyes.]

**Augite.** See PYROXENE. It occurs as black, prismatic, stumpy crystals (monoclinic system) and is the ubiquitous component of the more basic igneous rocks such as gabbro, dolerite, and basalt. 'Augite, a word borrowed from Pliny, was a name used before pyroxene and applied to the dark opaque crystals found in many basaltic lavas; these were distinguished from tourmaline and hornblende and recognized as a separate species by Werner about 1792' (Miers, *Mineralogy*, 1929). [G. *augē*, lustre.]

**Aureole.** 'The belt of metamorphosed rocks surrounding a plutonic intrusion, conveniently styled a "metamorphic aureole" . . . . It is to be borne in mind that the aureole seen is merely the section of the actual ground-surface of a three-dimensional aureole' (Harker, *Metamorphism*, 1950). Otley (*English Lakes*, 1823) briefly and correctly described the lithological and mineralogical characters of the rocks surrounding the Skiddaw granite at successive distances but did not envisage them as representing successive belts of decreasing metamorphism. The aureole (not then so called) surrounding the Dartmoor granite was referred to by De la Beche (*Manual*, 1831). 'The Skiddaw Granite and its metamorphism. . . . The metamorphic aureole' (Rastall, *QJ*, 1910). 'Contact metamorphism in south-eastern Dartmoor . . . the granite intrusion and its metamorphic aureole' (Fitch, *QJ*, 1932). 'The [Shap] granite and its associated thermal aureole' (Firman, *QJ*, 1957). 'The aureole of the Main Donegal Granite' (Pitcher and Read, *QJ*, 1960).

**Authiclastic rock.** See AUTOCLASTIC ROCK.

**Authigenic.** For the substance, parts, or ingredients of a rock that have originated 'in place', or for a whole rock-mass that has been so produced. Thus normal cementing material and contemporary accumulations of particular minerals are 'authigenic', but later infiltrating mineral veins are 'allogenic'. Rocks that are chemical precipitates (e.g. evaporates), or that result from organic growth (e.g. coral reefs), or coals formed from plant debris on the spot, are 'authigenic'. (Kalkovsky, 1880.)

**Autobreccia.** See BRECCIA (2, 3b).

**Autochthonous.** 'Autochthonous' and 'allochthonous' are contrasted terms meaning, primarily, a rock-mass formed where now found and one transported bodily, in orogeny, from elsewhere, respectively. The terms are also used for the constituents of rocks. The following are some applications. **1.** As regards rock constituents the terms are synonymous with 'authigenic' and 'allogenic'. They are used particularly for the vegetable matter making a coal seam—for that derived *in situ* and for that transported. Chemically precipitated material is autochthonous. **2.** For a residual deposit ('autochthon') produced in place by decomposition. **3.** For fossils representing organisms which actually lived where found, not having been transported after death (allochthonous fossils). **4.** The two contrasting terms have been used for magma, or the magmatic constituent of a migma; autochthonous, produced by liquefaction *in situ*, allochthonous, derived from some extraneous source. **5.** 'Autochthon' and 'autochthonous' have been used in connexion with Alpine stratigraphy and tectonics and are applicable to similar conditions elsewhere. The 'autochthon' is essentially a rock-succession that, as a whole, has not been translated by tectonic movement (it forms part of the foreland); but it may show folding, autochthonous folding, within itself. 'In tectonics, an autochthonous fold is one that is made of untravelled indigenous rocks, the rocks, that is of the country itself. In parautochthonous folds and thrust masses (Arnold Heim) the travel of the rocks has been considerable or great, according to ordinary standards, but small in the Alpine scale of magnitudes' (Bailey, *Tectonic Essays*, 1935). 'The beds which form the Foundation Unit of the [Tintagel] area show no sign of having suffered any extensive translation: they are autochthonous' (Wilson, *QJ*, 1950). The 'allochthon' (allochthonous folds and thrust masses) is the far-travelled ground. (Naumann, Germany, 1858; Gümbel, Germany, 1884.) [G. *chthōn*, the earth, the land, country, ground.]

**Autoclastic rock.** A rock formed by the breaking up of a part of a rock-mass within itself. The term 'autoclastic' is applied most usually to a crush-breccia or crush-conglomerate, but it may also be applied to an autobrecciated lava. See BRECCIA. Some authorities define it as a rock produced by friction of one rock-mass over another, which would include a fault-breccia; but in doing so they sometimes go far beyond this, perhaps allowable, extension, and consider ice as a rock and accordingly include subglacial moraine, thus calling by the same name totally different kinds of rock (as regards both character and origin). 'Authiclastic' is synonymous, and both 'autoclast' and authiclast' are used sometimes for the rock itself and sometimes for the individual fragments or 'pebbles'.

**Autogenetic.** Originated, established, or developed from within itself. Has been used for land-forms evolved without interference by earth-movement and for organic evolution originated or directed by internal potencies (one meaning of orthogenesis).

**Auto-intrusion.** **1.** At a late stage in the solidification of a (differentiating) magma, the drawing of residual liquid into rifts formed in the crystal mesh by some kind of stress. **2.** 'In the upper part of the Fingal's Cave Lava there are complex relations which may fitly be ascribed to auto-intrusion. Still fluid lava has evidently involved, entangled, and carried forward previously semi-consolidated portions (*MGS, Mull*, 1924) **3.** An intrusion of sedimentary rock-material from one part of a bed, or set of beds, in process of deposition, into another part; perhaps the usual way in which a sedimentary injection is formed. This has been specifically called an 'auto-injection' (Wood and Smith, *JSP*, 1958).

**Autolith.** A segregated body within an igneous rock, as distinct from a xenolith.

**Autometamorphism.** See AUTOMETASOMATISM.

**Autometasomatism.** Metasomatism of an already solidified igneous rock by residual fluids derived from the parent magma. 'Autometasomatism in the Lower Spilites of the Builth Volcanic Series' (Nicholls, *QJ*, 1958). Also called 'autometamorphism'.

**Avalanche.** Originally and normally: a fall of a mass of snow down a mountainside. However, it finds a useful application to a similar fall of loose rock-material where this is more sudden than the usual kind of landslide and more massive than what is usually implied in a 'rock-fall'. 'Debris avalanche deposits' (Rutland, *QJ*, 1967).

**Aves.** The class of the birds (subphylum Vertebrata) of which the distinguishing feature is, without exception, the possession of feathers. They range from the Jurassic. [L. pl., birds.]

**Avonian.** 'The term is conveniently retained for the Lower Carboniferous rocks, most of them typical "Carboniferous Limestone", of the British South-Western Province' (George, *JGS*, 1972). It is definable exactly in the Avon Gorge, near Bristol. (Vaughan, *QJ*, 1905.)

**Axial plane.** Of a fold; see AXIAL SURFACE.

**Axial surface.** The imaginary surface within a fold passing through and comprising all the hinges of the folded bedding surfaces. If only one bedding surface is being considered it may be defined as the surface which is everywhere equidistant from either limb; but in that case the axial surfaces of the several bedding surfaces will not necessarily coincide exactly. Though 'axial surface' must surely be the only really correct term, there is a strong and reasonable tendency to use 'axial plane' in spite of the 'plane' being nearly always curved, if only slightly. See SURFACE and PLANE.

**Axis in folding.** In the description of fold-structures an 'axis' implies some particular line; that the line is straight is also implied, but perhaps this need not be insisted on. Playfair (1802) and Lyell (1833) use the term. As such it is used in the following senses. (*a*) The line, in one bedding surface, running along the top (crest) of an anticline or along the bottom (trough) of a syncline. (*b*) The hinge of a folded bedding surface. (*c*) The normal vertical section of the axial surface of a fold. (*d*) Loosely or shortly, for the axial surface itself (thus not a line in this case). (*e*) 'For the tectonician, the most useful definition of fold-axis is that given by Wegmann (1929), based as it is on the usage of the Alpine structural geologists (especially Lugeon and Argand) since the 1890's. The axis of a fold is defined as the nearest approximation to the line which, moved parallel to itself in space, generates the fold' (Clark and McIntyre, *AJS*, 1951). It is here a direction rather than a line; it is our 'fold direction'; but if 'axis' is used, this (as 'axis of folding' or 'fold-axis') perhaps seems the best sense in which to use it. It may be said to be a generalization of the tectonic axes; considered in the plane of the earth's horizon, it is the 'axial trend'. It may also be noted that this 'axis' while being parallel to the axial surfaces within the folding, may have, within that restriction, any direction relative to the strike and dip of those surfaces. (*f*) The plan-projection of this line or direction (*e*); that is, the bearing of it. (*g*) An elongated 'area of differential movement'. 'Throughout Jurassic and Cretaceous times there occurred in various parts of Britain areas of differential movement which greatly influenced the character and amount of the sedimentation. One of these critical zones persisted during the Jurassic and Lower Cretaceous times in the vicinity of Market Weighton. . . . The Market Weighton axis' (Wilson, *BRG, E. Yorkshire*, 1948). Thus in the analysis of fold structures the term 'axis' is used in so many senses, none of them quite satisfactory, that it is perhaps best avoided (except as a 'tectonic axis' of space reference).

**Aymestry Limestone.** 'The central member of the Ludlow Rocks is a sub-crystalline, argillaceous limestone, which might have been termed the Ludlow Limestone; but as there are few good examples near that town, I have preferred naming it after the beautiful village of Aymestry, where the rock is fully and clearly laid open, and where its fossil contents have been elaborately worked out by my friend the Rev. T. T. Lewis' (Murchison, *Silurian System*, 1839, mentioned, *PGS*, 1833).

**Azoic.** Without life; of the ages which have provided no recognizable organic remains; of the ages without life at all. Whatever the exact shade of meaning this must be an unsatisfactory term, as what are, and what are not, recognizable fossils in the Precambrian rocks is largely debatable, and, as a further question, it is not known at what part of Precambrian time life appeared on the earth. (The term is not used merely for an unfossiliferous formation.)

# B

**Back slope.** The slope at the back of a scarp. A dip-slope is one case.

**Back-folding.** 'Back-folding is shown by folds with their fronts in the opposite direction to the main nappes' (Jeffreys, *Earthquakes and Mountains*, 1935).

**Backland.** See HINTERLAND (2).

**Badland topography.** 'The effects of raindrop impact and of scouring by concentrated wash in places where the run-off is gathered quickly into definite streams are easily demonstrated. Both are conspicuous on outcrops of sandy clay where these are bare or only sparsely protected by vegetation, as is the case in semi-arid regions, in the vast areas of "badlands", for example, in the western interior of North America. . . . A miniature landscape of innumerable closely spaced steep-sided valleys and ridges is thus developed' (Cotton, *Landscape*, 1948). 'Badlands. A name originally applied to an area of semi-arid climate in S. Dakota because it was so difficult to cross' (Monkhouse, *Dictionary of Geography*, 1965).

**Bagshot Sands.** See EOCENE. (Smith in Farey, *Derbyshire*, 1811, 'Sand of Bagshot-Heath'; Conybeare and Phillips, *England and Wales*, 1822, 'Bagshot Sand'.) [Bagshot, Surrey.]

**Bajocian.** See JURASSIC SYSTEM and INFERIOR OOLITE SERIES. [Bayeux (Bajocia), Normandy.]

**Baked contact.** See CHILLED CONTACT.

**Bala series.** Comprises the Caradoc and Ashgill series of the Ordovician system. The name is used in those regions, particularly N. Wales, where the history of investigation and practical convenience make it appropriate. It is characterized by its abundant fauna of (particularly) species of brachiopods and trilobites and, in the graptolitic facies, by certain sub-faunas of the Diplograptid fauna (Bulman *P*, 1958). (Sedgwick, *PGS*, 1838.) [Bala, Meirionydd district of Gwynedd.]

**Bala Volcanic series.** Comprises the Ordovician volcanic rocks of (roughly) Caernarvonshire which are there of Bala age (though the earliest may be Llandeilo); the associated intrusive rocks are also usually included. 'During Lower Bala times volcanoes erupted from a number of centres mainly situated in Caernarvonshire, where their products form the Bala Volcanic Series' (Smith and George, *BRG*, *N. Wales*, 1961). *The Bala Volcanic Series of Caernarvonshire* (Harker, 1889).

**Ball-and-socket jointing.** See JOINT.

**Ball clay.** 'The term "ball clay" derives from an early method of working the clays, in which they were cut into "balls" about 0.25m square. The modern usage of the term refers to a fine-grained highly plastic kaolinitic sedimentary clay, the higher grades of which fire to a white or near-white colour in an oxidising atmosphere. This contrasts with china clay, which was formed by the *in situ* alteration of granite' (Edwards, *PGA*, 1976). Also called 'pipe clay'. The main source of British supplies is in the Tertiary of the Bovey basin, Devon. *Ball Clays* (Scott, *MGS*, 1929).

**Balled-up structure.** A structure resulting from slumping. 'Knots of highly contorted silty muds lie isolated in other muds, and masses from a few inches up to several feet across have been "balled up" so that the outer layers are wrapped round the core somewhat in the manner of a snowball' (Jones, *QJ*, 1937).

**Ballstone.** A nodule or larger rounded lump of rock in a stratified formation; particularly (*a*) an ironstone nodule in the Coal Measures, and (*b*) a more or less crystalline concretion, often containing corals in the position of growth, in the Silurian limestones, especially in Shropshire. 'These large concretions are called "ball-stones" by the workmen, to distinguish them from the common beds, which they term "measures"' (Murchison, *Silurian System*, 1839). 'A study of ballstone in the Wenlock Limestone of Shropshire' (Crosfield and Johnston, *PGA*, 1914). 'Ballstone' may refer to the phenomenon in general; otherwise it is usually in the plural.

**Band.** The *OED* gives: 'A strip of any material flat and thin . . . . *Geol.* A stratum with a band-like section'. In that definition 'stratum or other similar rock-unit' would be better. The term is used particularly for a thin bed with a distinctive lithology or fauna. 'Brief experience of fossil-hunting suffices to show that organic remains are rarely scattered evenly through a deposit, but tend to occur in more or less defined bands' (Hawkins, *Invertebrate Palaeontology*, 1920). 'A band of iron ore' (Murchison, *Silurian System*, 1839). 'The bands' (Garwood, *QJ*, 1912). 'A shelly band in graptolitic shales' (Challinor, *GM*, 1928). 'Ludlovian biotite bearing bands' (Tucker, *GM*, 1960). See MARINE BAND and MARKER HORIZON.

**Banding.** The occurrence, usually by alternation, of layers of distinctive composition or texture, conspicuous in section. **1.** In sedimentary rocks, bedding of that nature. **2.** In igneous rocks, flow-banding, crystallization-banding, or any other kind of banding, in lavas or intrusions. **3.** In metamorphic rocks, foliation (including gneissic structure) or mylonite-banding.

**Banket.** See PLACER.

**Barbados earth.** See SILICEOUS ORGANIC DEPOSIT.

**Barchan.** A crescentic desert-dune; a Turkestan name.

**Barr series.** The series of beds, in the Girvan district of S. Scotland, which represents the lower part of the general Caradocian series. These beds are the equivalent, in the shelly facies, of the Glenkiln Shales of S. Scotland in the graptolitic facies; certain of them, particularly the Stinchar Limestone, are highly fossiliferous. (Lapworth, *QJ*, 1882.) [Barr, Ayrshire.]

**Barrel fold.** 'Between Brunt Heugh and the promontory of Slains Castle, Aberdeenshire, close barrel folds—the barrels formed by longitudinal exposure of the fold-hinges plunging gently north—are broken by prominent joints ("Edinburgh Rock" structure)' (Read and Farquhar, *QJ*, 1956).

**Barton beds.** See EOCENE. (Prestwich, *QJ*, 1847, 'Barton Clay'.) (Barton, Hampshire.]

**Barysphere.** 'The inaccessible heavy interior of the earth is known as the barysphere. This is followed outwardly by the lithosphere' (Tyrrell, *Petrology*, 1929). In this, the usual, sense it is synonymous with 'endosphere'. [G. *barus*, heavy.]

**Baryte.** (Barytes.) A white, rather heavy, mineral, sulphate of barium, $BaSO_4$; often occuring in veins in a massive crystalline form. (Karsten, 1800).

**Basal conglomerate.** The conglomerate often occurring in the basal part of a series where it overlies a surface of unconformity, representing, typically, a beach deposit spreading over the old land surface. If this spreading takes some considerable time, being associated with overlap, the conglomerate will be diachronous. 'In describing the vertical and horizontal strata of the Jed, no mention has been made of a certain pudding-stone, which is interposed between the two, lying immediately upon the one and under the other. When we examine the stones and gravel of which it is composed, these appear to have belonged to the vertical strata or schistus mountains' (Hutton, *Theory of the Earth*, 1795). '. . . the thin basal conglomerate of the Arenig resting with obvious angular discordance upon the Middle Cambrian mudstones' (Nicholas, *PGA*, 1939, describing an excursion to St. Tudwals' Peninsula).

**Basalt.** A fine-grained basic igneous rock composed essentially of the more calcic plagioclase felspars and pyroxene, often with olivine also; sometimes porphyritic, also sometimes vesicular or amygdaloidal. Approximately the fine-grained equivalent of a gabbro. It is usually extrusive, being by far the most abundant of all types of lava in the earth's crust; but it may also occur as minor intrusions. The true nature of basalt was established by Desmarest in France in his memoir *Sur l'origine et la nature du Basalte* (1774). 'Faujas Saint-Fond is well known as being one of the early demonstrators of the volcanic origin of basalt rocks, particularly in his *Recherches sur les volcans eteints du Vivarais et Velay* (1778). He had no difficulty in recognizing the basalts and other extrusive and minor intrusive rocks of the Glasgow and Edinburgh districts, the Oban district, and the Inner Hebrides (Staffa and Mull), and assumes their volcanic nature without argument, but he had no idea that they were of widely different geological ages' (Challinor, *As*, 1954). These Scottish examples were described in his *Voyage en Angleterre, en Écosse et aux îles Hébrides*, published in 1797. 'Basalt, like granite, has its problems' (Tilley, *QJ*, 1950). *Basalts* (Hess and Poldervaat, eds., 1967–1968). [*basaltes*, used in Roman times by, e.g., Pliny, though not necessarily always or exactly our basalt; probably f. *bsalt*, an Ethiopic name.]

**Basaltiform.** In the form of columnar basalt. 'These prismatic, or, as some may choose to call them, these basaltiform lavas [in the crater of volcano]' (Spallanzani, *Travels in the Two Sicilies*, 1792/7, trans., 1798, quoted by Mather and Mason, *Source Book*, 1939). 'The trap, near its walls, puts on a prismatic form, like the basalt of the Giant's Causeway. The composition of the rock is unlike any I had previously seen in "basaltiform" dykes' (Murchison, *Silurian System*, 1839). See COLUMNAR JOINTING.

**Basanite.** Alkali-olivine-basalt, containing a felspathoid. (Brongniart, 1813.) [G. *basanos*, the touchstone for gold (but the name became confused with basalt).]

**Base-level.** The term 'base-level of erosion' was introduced, in America, by Powell in 1875 (*Colorado River*), for a vague idea ambiguously expressed. It was for a slightly curved irregular surface below which a river and its tributaries were supposed to be able to cut no further. 'I take some liberty in using the term "level" in this connection, as it would, in fact, be an imaginary surface, inclining slightly in all its parts toward the lower end of the principal stream or having the inclination of parts varied in direction as determined by tributary streams'. But it was immediately realized that what wanted a term, and what the term 'base-level' itself would be admirably suited to denote, was something different, much more fundamental and much more easily imagined. W. M. Davis (*Essays*, 1909) has reviewed the whole matter. The following quotations will make the now accepted meaning of 'base-level' clear (Powell's original conception being excluded). 'An imaginary level surface, the level base with respect to which normal sub-aerial erosion proceeds'. 'The limit of sub-aerial erosion is the "level base" or "base-level" drawn through a land mass in prolongation of the normal sea-level surface . . . the base level must be more and more closely approached [and ever more and more slowly] as time is extended'. 'Various local or temporary controls of erosion: a rock ledge or a lake on a river course, the central basin of a dry interior basin, the surface of a lake in such a basin. Nothing can be simpler than to imagine a level surface passing through any of these controls, and rising or sinking as the control rises or sinks; and such a surface is naturally called a local or temporary base-level'. This simple conception seems to have been first fully expressed in words by Greenwood (*Rain and Rivers*, 1857).

**Basement.** See COMPOUND STRUCTURE.

**Basement complex.** See COMPOUND STRUCTURE.

**Basement control.** In a region of compound structure, the control of the structure of the basement over the structure of the cover. The structure of the basement may affect the sedimentation of the deposits (of the cover) being laid down on it; or reactivation, at a later time, of structures in the basement may affect the structure of the covering formations. 'Basement control of structures in the Mesozoic rocks in the Strait of Dover region' (Shephard-Thorn and others, *PT*, 1972).

**Basement gneiss dome.** A dome-shaped mass of basement gneiss, often with granite, having a cover of sedimentary or metasedimentary rocks (except where these have been removed by erosion); also known as a 'mantled gneiss dome'. The covering rocks are usually heavier than the basement rocks, so that the doming may be due to gravitational instability. 'The problem of mantled gneiss domes' (Eskola, *QJ*, 1948) 'The . . . Ntongamo gneiss dome [Uganda]' (Nicholson, *QJ*, 1965). 'Basement gneiss domes, Salta region, Norway' (Cooper and Bradshaw, *JGS*, 1980).

**Basic.** Applied to an igneous rock relatively poor in silica, that is, having less than about half its mass of that oxide. In contrast to 'acid' and grading into 'ultrabasic'. See ACID.

**Basic front.** See FRONT.

**Basic patch.** 'There occur in many plutonic rocks darker and finer-textured ovoid or irregularly rounded patches. They are possibly in some cases clots formed by the aggregation of some of the earlier-crystallized minerals at a stage when the main mass was still liquid, and they have often been referred to as "basic secretions". There is, however, good evidence to show that in many cases such more basic patches represent inclusions of foreign rocks, either igneous or sedimentary, which have been totally transformed by reaction with the igneous magma' (Harker, *Petrology for Students*, 1954). 'One of the most striking characters of the Shap Fell granite is the occurrence of distinct patches of darker colour and somewhat finer texture than the surrounding rock. These patches are abundant in the quarries, and may be well studied in the polished slabs and pillars used in building' (Harker and Marr, *QJ*, 1891). Although these features are of course three-dimensional, they are always seen as conspicuous surface patches; hence the familiar way of referring to them.

**Basin.** Applies generally to form; either to a physiographical feature, whether of the present day or (as a basin of deposition, a cuvette, for instance) in the geological past, or to a structural arrangement of the rocks. As it is important to keep these meanings clear, the latter, the tectonic meaning, is best defined separately as 'basin structure'. 'In applying the term "basin" to the physical regions containing the tertiary formations of Vienna and Styria we use the customary language of Geology. They were, however, nothing more than two deep bays in the ancient tertiary sea' (Murchison, *TGS*, 1832). 'After the post-Silurian movements had ceased part of the area was converted into a basin of deposition. This basin was termed Lake Cheviot by A. Geikie' (Pringle, *BRG*, *S. Scotland*, 1948). 'The inception, development, infilling, obliteration and final structural deformation of a basin of sedimentation' (Hodson, *Fossils*, 1961). 'Basins and swells' (Sellwood and Jenkyns, *JGS*, 1975, and discussion, *JGS*, 1976). See CUVETTE.

**Basin facies.** A stratal facies corresponding to sedimentation beyond the outer limits of a land-bordering submarine shelf. 'In NW. Yorkshire the Carboniferous Limestone is either composed of black limestones with shales, grey unbedded limestone, or well-bedded dark or light grey limestone. As such it can be related to the sedimentation and referred to as of outer or basin facies, marginal or reef facies and inner or massif facies' (Hudson, *PGA*, 1933). 'Though the Ludlow rocks of Usk belong essentially to the shelf facies characteristic of Shropshire and the Welsh Borderland generally, they show certain features which link them with the basin facies of central Wales' (Straw, *QJ*, 1958).

**Basin structure.** See BASIN and COAL-BASIN. A syncline in which the dips are towards a centre rather than towards a line. An elliptical downfold may be equally well called an elongated basin or a short syncline (with opposed pitch at each end).

It bears the same relation to a typical syncline as does dome structure to a typical anticline. 'Description of the mineral bason of the counties of Monmouth [&c.]' (Martin, *PT*, 1806). 'It appears that in this country there are two chalk basins, in a greater or lesser degree resembling that of Paris. One of them is called the Isle of Wight Basin; the other, the London Basin' (Phillips, *Mineralogy*, 1815). 'When masses or strata decline upon every side towards a certain point, they are said to be basin-shaped' (Greenough, *Geology*, 1819).

**Basin-range structure.** See BLOCK FAULTING.

**Basset.** An old term (referred to, for instance, by Plot in his *Staffordshire*, 1686), both as substantive and verb, but now nearly obsolete, for 'an outcrop' (often as 'basset edge') or 'to outcrop'. [probably from the word 'base'.]

**Bastard.** Used by quarrymen and miners for any inferior or impure rock or mineral, e.g. 'bastard limestone', a siliceous limestone incapable of being converted into quicklime when burnt in the kiln.

**Batholite.** (Batholith, Bathylith.) 'A large intrusive mass (100 square kilometers or more in areal exposure) structurally concordant or crosscutting, or a combination of both, and without a visible or inferable floor of older rocks' (Knopf in Poldervaart, *Crust of the Earth*, 1955). The term *batholithen* was used by Suess (*Das Antlitz der Erde*, 1888), and given as 'batholite' in the English translation (1904), which form was adopted by Harker (*Igneous Rocks*, 1909). 'The nature of batholiths' (Hamilton and Myers, *USGS*, 1967). 'The anatomy of a batholith' (Pitcher, *JGS*, 1978). For further discussion of the question of the alternative forms of the term see LACCOLITE. [G. *bathus*, deep.]

**Bathonian.** See JURASSIC SYSTEM and GREAT OOLITE SERIES. [Bath.]

**Bathyal.** Deep; applied to the moderately deep seas beyond the continental shelf and down to an indefinite depth, perhaps about 4000 m., approximately to the limit of the accumulation of the terrigenous deposits, beyond which are the abyssal depths.

**Bauxite.** A mineral which is a mixture of hydrated aluminium oxides; but the name is applied particularly to varieties of the rock laterite rich in hydroxides of aluminium which are the chief ores of the metal. Full discussion and history in 'Aluminium and its ores' (Eyles, *Dsc*, 1955). *Bauxites* (Valeton, 1972). (Dufrenoy, 1847.) [Les Baux, S. France.]

**b-axis.** See TECTONIC AXIS.

**Beach.** There are slightly differing senses: (*a*) = shore, (*b*) the shore with its covering of sand, shingle, &c., excluding a bare rocky shore, (*c*) this sand, shingle, &c., itself, and (*d*) the shingle only.

**Beach cusp.** 'Beach cusps are triangular ridges extending across the beach generally at right angles to the shore front. When most typically developed the beach cusp has the form of an isosceles triangle with its base parallel to the beach, but at its upper edge, and its apex near the water. The cusp may be broad, approaching in form an equilateral triangle, but more generally it is long, narrow and extremely acute, the sides sometimes appearing almost parallel' (Grabau, *Stratigraphy*, 1913; he refers to Johnson, *BGSA*, 1910). 'The formation of beach cusps' (Kuenen, *JG*, 1948).

**Bearing.** The horizontal direction in relation to a meridian. Obviously the concept is much required in describing the physical and structural features of a region. 'I have in that country [Grampian Highlands] traced a particular class of strata for near two hundred miles upon the bearing [strike], which is nearly from north-east to south-west and found both the bearing and declivity surprisingly regular' (Williams, *Mineral Kingdom*, 1789).

**Becke test.** A microscopical method of determing in thin section which of two materials in contact has the higher refractive index. In brief, on raising the objective a band of illumination moves into the material (mineral, liquid, or mountant) having the higher index. [Friederich Johann Becke, 1855-1931.]

**Bed.** A layer of rock. A thickness from a few centimetres to several metres is usually implied. It is essentially a feature of original deposition (though its character may be subsequently altered) and is nearly always a sedimentary rock (in the wide sense), but the term may be used in connexion with a land-deposited pyroclastic rock and even a lava-flow. Distinctive or structurally defined layers in an igneous instrusion are not usually called 'beds'; but if such layers, in some particular case, are regarded as due to a kind of 'deposition', by successive solidifications, the term may be applicable: 'Each rhythmic band of the interior of the Skaergaard gabbro is a graded bed' (Bailey, *TGSG*, 1958). 'Stratum' is the same as 'bed', but (in the singular) is less often used. See STRATUM. Both terms are words of a general nature; they are synonyms one being the translation of the other. '"Bed" is always applied as the English synonym of "stratum"' (Sedgwick, *TGS*, 1835). Greenough (*Geology*, 1819) long ago pointed out that it was 'injudicious to employ synonymes for the purpose of expressing contrast' (referring to a proposed distinction between 'bed' and 'stratum'). 'That bed of sand and cockle shells found in sinking a well' (Ray, *Philosophical Letters*, 1718). 'The bed or stratum of freestone worked here' (Smeaton, *Eddystone Lighthouse*, 1793). 'Perhaps the commonest term in geology' (Arkell and Tomkeieff, *Rock Terms*, 1953). In the formal hierarchy of lithostratigraphical grades (see LITHOSTRATIGRAPHY), 'bed' is the lowest grade (smallest unit). Quarrymen speak of 'the bed' of a rock as meaning the bedding.

**Bedded.** Made up of beds; occurring in the form of beds. When the beds are numerous and well marked the rock-mass is 'well-bedded'. 'The stones [of the Trias in Lincolnshire] ly yn the ground lyke a smothe table: and be bedded one flake under another' (Leland, *Itinerary*, *c*. 1538). There may be special applications, e.g.: 'The term "bedded" as applied to iron ores means sedimentary rocks high in metallic iron content and forming part of an established stratigraphical sequence, thus differentiating them from ore-bodies (e.g. haematite) found in lodes, pipes, veins, etc., in varying types of country rock' (Milner, *Sedimentary Petrography*, 1962).

**Bedding.** The property of a rock-mass of being composed of beds; the general physical and structural character of the beds, and their contacts, in a bedded rock-mass; the character of the small-scale lamination, and other depositional features of the kind, within individual beds, or within the rock-mass as a whole, e.g. 'graded bedding', 'cross bedding'. But although certain layers of land-deposited pyroclastics and lava-flows, and even of igneous intrusions, might be called 'beds', internal original structures of these 'beds' due to some process (such as flow), which had nothing to do with successive depositions, would not be called 'bedding'. See STRATIFY.

**Bedding cleavage.** Rock cleavage (typically, slatey cleavage) normally appears as a splitting at an angle to the bedding, being approximately parallel to axial surfaces of folds, and is thus obviously connected with the manner and time of production of these folds, a time clearly later than the time when the beds were laid down. In isoclinal folding the cleavage will, fortuitously, largely coincide with the bedding. It has been suggested (Roberts, *Geol.J.*, 1977) that a kind of cleavage ('bedding cleavage') may be produced that is essentially coincident with the bedding but also essentially due to subsequent pressure. 'It is considered here that the bedding cleavage is due to a rough alignment of flaky minerals during sedimentation [of the Skiddaw Slates] and which was subsequently more strongly aligned when chlorite developed under pressure at the end of the Silurian period, when the Skiddaw Slates were buried beneath a 8.5 km. (28,000 ft.) thick pile of Borrowdale Volcanic and Silurian strata'.

**Bedding interface.** See INTERFACE.

**Bedding surface.** A surface, conspicuous or inconspicuous (but usually implying the former), within a sedimentary rock mass, representing an original surface of deposition. When plane, or nearly so, it is a 'bedding-plane'; this latter term being often used even when the 'planes' are conspicuously bent. The bedding surface is the most important entity in both stratigraphy and the structure of the stratified rocks. See SURFACE and PLANE.

**Bedding-fault.** The result of bedding-plane slip. As regards the beds themselves this would perhaps hardly justify the term 'fault', but the slip may displace such things as veins, or faults previously formed, which cross the beds; these, certainly, are thereby faulted. 'Bedding-faults and related minor structures in Upper Valentian rocks near Aberystwyth' (Lewis, *GM*, 1946).

**Bedding-plane.** See BEDDING SURFACE.

**Bedding-surface slip.** (Bedding-plane slip.) Slip along bedding surfaces, normally due to the beds accommodating themselves to the process of folding. See SLIP-MARK.

**Bedrock.** The solid rock underlying any loose superficial material such as soil, alluvium, glacial drift or marine sediments. See SOLID GEOLOGY.

**Beef.** Used in geology for fibrous carbonate of lime, especially when it occurs as thin layers along the bedding (the fibres lying at right angles to the bedding). Seen particularly in the Lias and Purbeck of the Dorset coast; apparently a Purbeck quarrymen's term. 'Shales-with-"beef", a sequence in the Lower Lias of the Dorset coast' (Lang and Richardson, *QJ*, 1923).

**Before the present.** For events in human prehistory, and for any geological events that could be dated approximately to within a few thousand years, it is often convenient to use the initials B.P. (Before the Present) instead of B.C. or writing '— years ago'. B.P.=B.C. plus 2,000. See TIME UNIT.

**Belemnite.** See BELEMNOIDEA. In the older works (e.g. Woodward, *Catalogue*, 1729) referred to as *Belemnites* (sing.) or *Belemnitae* (plur.) 'From their form, by all naturalists called Belemnites, from the Greek word βέλεμνον [*belemnon*]' (Plot, *Oxfordshire*, 1677). 'Belemnites' is now the English plural. An early description is that by Brander (*PT*, 1754); a much fuller one that by Miller (*TGS*, 1826).

**Belemnoidea.** An extinct order (Trias to Eocene) of the subclass Dibranchiata (class Cephalopoda, phylum Mollusca). The typical members are the belemnites (family Belemnitidae in the wide sense) in which the internal shell is largely a solid cigar-shaped or dart-shaped 'guard' of calcite with a conical chambered hollow at one end. The other family is the Belemnoteuthidae. (Naef, 1912.) [G. *belemnon*, a dart.]

**Belt of disturbance.** A strip of country, being the outcrop of a zone of faulting or folding or both combined. '. . . the Careg Cennen belt of disturbance on the north crop that supposedly sinks to depths southwards underneath the coalfield' (George, *BRG, South Wales*, 1970).

**Belt of variables.** The belt of marine deposition extending from the coast (high water-mark) to a depth of about 180m, that is, corresponding roughly with the continental shelf (in the wide sense, to include the shore); passing into the 'mud belt' at the 'inner mud line'. 'The terrigenous deposits on the shoreward side of the mud belt cannot be divided into minor parallel belts by reference to variations in the size of their constituent fragments. On the contrary, the sea-floor of this inner belt is often marked by stretches of pebbles, sand, silt and mud, or even organic deposits, lying side by side with no very definite arrangement. Accordingly this inner belt adjoining the coast will be spoken of henceforth as the "belt of variables"' (Marr, *Deposition of the Sedimentary Rocks*, 1929). Some authors (e.g. Rastall, *Textbook*, 1941) exclude the 'shore belt' and 'storm beach' from the 'belt of variables', appearing not to realize that Marr had used the term 'continental shelf' to include the shore. 'Marr's "belt of variables", a term applied to the deposits that form at the present day around the coasts, possesses many characteristics which have been noted in ancient sedimentary formations. The chief rock types exhibiting these characteristics are conglomerates, sandstones, muds of every grade of fineness, and limestones, both bedded and in reef masses. The seas in which they were laid down have been termed "platform-" or "shelf-seas"' (Jones, *QJ*, 1938).

**Benioff zone.** The zone of seismicity which everywhere seems to coincide with the subduction zone envisaged in the theory of plate tectonics. 'Seismic evidence for crustal structure' (Benioff, *GSASP*, 1955). [H. Benioff, American seismologist.]

**Bennettitales.** An extinct, Mesozoic, group of plants belonging to the division Gymnospermae, having cycad-like trunk and leaves but bearing flower-like cones. [genus *Bennettites* (named after J. J. Bennett, 1801-1875).]

**Benthos.** The life of the seafloor (at any depth), sessile, creeping, or burrowing. Hence 'benthic', 'benthoic', or 'benthonic'. [G. *benthos*, the depths of the sea.]

**Bentonite.** 'According to C. S. Ross and E. V. Shannon (*JACS*, 1926) "Bentonite is a rock composed essentially of a crystalline clay-like mineral formed by devitrification and accompanying chemical alteration of a glassy igneous material, usually a tuff or volcanic ash". The clay mineral is montmorillonite. The name was originally given to the type of clay occurring in the Fort Benton shale (Cretaceous) at Rock Creek, Wyoming (W. C. Knight, *EMJ*, 1898). D. F.Hewitt (*JWAS*, 1917) recognized that bentonites contain minerals characteristic of volcanic rocks' (Milner, *Sedimentary Petrography*, 1962). 'It has the unique characteristic of swelling to several times its original volume when placed in water' (Grim, *Applied Clay Mineralogy*, 1962). 'Individual beds are usually quite thin, varying from about an inch to a few feet in thickness, but in spite of this, they remain surprisingly uniform in character over enormous areas' (Hatch and Rastall, *Sedimentary Rocks*, 1965). 'Bentonite beds result from volcanic ash falls which blanket whole depositional provinces simultaneously and provide time-rock markers which are much more dependable than index fossils' (Krumbein and Sloss, *Stratigraphy and Sedimentation* 1951). 'Bentonites are derived from the most violent of volcanic eruptions' (Fitch and Miller in Harland and others, eds., *The Phanerozoic Time-scale*, 1964). *Bentonites* (Grim and Guvan, 1978).

**Bergschrund.** 'Some valleys are terminated by a kind of mountain-circus with steep sides, against which the snow rises to a considerable height. As the mass is urged downwards, the lower portion of the snow-slope is often torn away from its higher portion, and a chasm is formed, which usually extends round the head of the valley. To such a crevasse the specific name "Bergschrund" is applied in the Bernese Alps' (Tyndall, *Glaciers of the Alps*, 1860).

**Bevel.** In geomorphology, refers to any surface that has the appearance of one that has been bevelled. It often refers, in particular, to the steep, or moderately steep, straight or convex 'coastal slope' above a sea-cliff. The significance and terminology of this feature are discussed by, among others, Arber (*GJ*, 1949, *GM*, 1951) and Savigear (*TIBG*, 1952, 1962).

**Biochron.** See BIOZONE (2).

**Bioclastic rock.** (Bioclast.) There are two quite different senses. **1.** Rocks produced from fragments or particles broken or pulverized from pre-existing rocks by organisms, such as plant-roots, earthworms, &c. Some would say that the chief bioclastic rocks were those, such as concrete, produced artificially by man. (Grabau, *AG*, 1904). **2.** Rocks composed of the broken remains of organisms, such as limestones composed of shell-fragments; one kind of biolith. (For discussion of these usages, and associated terms, see Thomas, *BAAPG*, 1960.)

**Biocoenose.** Primarily a neontological term for an assemblage of organisms living together as an interrelated community ('biocoenosis'). 'Life assemblage' is synonymous and is usually preferred in palaeontology. (Contrasted with 'thanatocoenose'.) [*G. koinos*, common.]

**Biodepositional.** See ORGANIC DEPOSIT. 'The term biodepositional (biodeposited) is suggested to describe a formation which by sight can be recognized as having been deposited, to a very large percentage, by living organisms' (Forgotson, *BAAPG*, 1966).

**Biofacies.** See FACIES. In a wide sense, the general facies of a fossil fauna or flora; the total biological characteristics of a sedimentary bed, formation, &c. However, it usually implies that this fossil character is being considered as the expression of local biological conditions, as distinct from considering it as an expression of the stage reached in the evolution of life during the course of geological time. Used particularly when beds presumed to have been formed during the same period of time are being contrasted as regards their fauna (or flora).

**Biogenic.** Of organic origin; for rocks, rock-constituents, and material such as wood, peat and shells. Bioturbation structures are 'biogenic'. See BIOLITH.

**Biogeochemistry.** 'Biogeochemistry is the name that has been given to a new subject concerned with the relationship between botany and sub-surface geology. Its techniques are applied in the search for buried mineral ores by making trace metal analyses on trees and other vegetation and, more recently, in the study of lead pollution' (Warren, *E*, 1972). The term may also refer to the chemistry of organically-formed rocks or even to the chemical composition of fossils themselves.

**Bioherm.** A reef formed of organic material, such as a coral reef; one kind of biolith. But it has been suggested that 'bioherm' should be confined to organically formed accumulations that lack resistance to wave-action, thus excluding an organic reef in the restricted sense of that term (see REEF). (Cumings and Shrock, *BGSA*, 1928.) [G. *herma*, a mound, a reef.]

**Biolith.** 'Biogenic rocks, or "bioliths" as Chr. G. Ehrenberg has termed them, are deposits of organic material, or material formed through the physiological activities of the organisms' (Grabau, *Stratigraphy*, 1913). There are two kinds, 'zooliths' (such as coral reefs, sponge reefs, Bryozoa reefs, shell limestones, crinoidal limestones, and lithified calcareous and siliceous oozes) and 'phytoliths' (such as algal deposits, lithified diatom ooze; and peat, lignite, and coal). Where the biolith is a more or less extensive bed it is sometimes called a 'biostrome'. Hence 'biolithogenesis', the origin and early development of bioliths: 'Biolithogenesis of *Microcodium* [a supposed algal biolith]' (Klappa, *Sd*, 1978).

**Bioseries.** An evolutionary series of fossils, whether of whole individuals or of specimens illustrating trends in particular features.

**Biosparite**, A sparite containing fossils or fossil fragments as allochems.

**Biosphere.** Comprises all the regions occupied, or potentially occupiable, by living organisms; on, and in the uppermost parts of, the lithosphere, in the hydrosphere, and in the atmosphere. Two vertically adjacent regions are particularly considered: (*a*) the relatively thin layer of atmosphere immediately above the ground in which animals and plants obviously have their main sphere of terrestrial life, and (*b*) the upper layers of the soil which also have a manifold living flora and fauna. 'Taken altogether, the diverse forms of life constitute an intimate and ever-changing network, clothing the surface with a tapestry that is nearly continuous' (Holmes, *Physical Geology*, 1944/65/78).

**Biostratigraphical unit.** A stratigraphical unit defined by its fossil content. See STRATIGRAPHICAL UNIT.

**Biostratigraphy.** The description and study of the fossil-content of strata from the stratigraphical, rather than from the palaeontological, point of view. **1.** 'Biostratigraphy is the utilisation of fossils for correlation in stratigraphy. . . . [It] provides for many Phanerozoic rocks the most useful and accurate method of correlation, and hence the most common step from lithostratigraphy to chronostratigraphy' (*Stratigraphical Guide*, *GSSR* (11), 1978). William Smith was the founder of biostratigraphy, the two chief works of the Smith epoch being Townsend's misleadingly entitled *Character of Moses Established for Veracity* (1813) and Smith's own *Strata Identified by Organized Fossils* (1816). 'The use of fossils in stratigraphy' (McKerrow, *JGS*, 1971). *Concept and Methods of Biostratigraphy* (Kauffman and Hazel, eds., 1977). The term was originally proposed by Louis Dollo in 1904 in a wider sense including what would now usually be distinguished as stratigraphical palaeontology. **2.** The use of fossils for inferring the conditions under which the strata were deposited is also sometimes taken as falling within the scope of biostratigraphy.

**Biostrome.** See BIOLITH. 'Biostromes in the Namurian Great Limestone of Northern England' (Johnson, *P*, 1958). [G. *strōma*, something spread out.]

**Biotic weathering.** See WEATHERING.

**Biotite.** In the wider sense, includes all the ferromagnesian micas, silicates of aluminium, potassium, magnesium, and iron, with hydroxyl (OH). Common in many types of igneous rock from acid to ultrabasic and very common in metamorphic rocks. (Haunsmann, 1847). [J. B. Biot, French scientist, 1774-1862, who first called attention to the optical differences in mica.]

**Bioturbation.** Disturbance of sediments (loose or more or less consolidated) by organisms, particularly burrowing organisms. 'Biologic energy can give a completely different character to sediments initially deposited by physical processes' (Warne, *JSP*, 1967).

**Biozone. 1.** In stratigraphical palaeontology. The total time-rock span (stratigraphical range) of a species, genus, order, &c. 'The life zone in geology has reference to the vertical range. Why cannot we call this a "biozone", using the term to signify the range of organisms in time as indicated by their entombment in the strata? Thus we might have the biozone of a species, of a genus, of a family, or of a larger group. Thus the biozone of the Trilobita would be, say, from Cambrian to Carboniferous. . . . The biozone of an ammonite species . . .' (Buckman, *GM*, 1902). **2.** In biostratigraphy. Biozones are the units of biostratigraphy and are recognized and specified only by their fossil content. The time interval represented by a biozone would be informally, a 'biochron', but this interval is indefinite. A biozone might, however, correspond approximately with a formal chronostratigraphical division, particularly a chronozone (defined by marker points). 'Reasonable tolerance in the geometrical [lateral] extension of biozonal boundaries must of course, be allowed; it cannot be expected that a biozonal index will appear in every cubic metre of rocks examined' (*Stratigraphical Guide*, *GSSR* (11), 1978). A biozone may be an 'assemblage biozone', characterized by a certain assemblage of taxonomic kinds (particularly species), or be based on the occurrence of one taxonomic kind (e.g. a particular ammonite species), either as regards its acme or its total range. See INDEX FOSSIL.

**Birefringence.** The property, in a mineral, of being doubly refracting.'With a few exceptions minerals cause a ray of light passing through them to be resolved into two components, vibrating in two different directions at right angles to one another, and refracted to different extents. This is the phenomenon of double refraction. . . . The most spectacular result is the production of interference colours when a mineral section is viewed between "crossed polarizers". This property is of the greatest value in the identification of minerals through their optical reactions under the microscope' (Hatch and Wells, *Igneous Rocks*, 1952).

**Birkhill Shales.** The highly condensed, famously graptolitic succession, in beds only some 25 m. thick, typically exposed in Dobb's Linn in the lonely Birkhill district north-east of Moffat, Dumfriesshire. The series comprises all but the topmost two or three zones, of eighteen, of the general Llandovery series, that is, up to and including the zone of *Rastrites maximus*. (Lapworth, *GM*, 1876, *QJ*, 1878.)

**Bitumen.** In the widest sense, any natural hydrocarbon ranging in state from rigid or highly viscous (asphalt etc.), through the less viscous (tarry) varieties, to liquid (petroleum); more particularly applied to the rigid and highly viscous varieties. The gaseous members of the series are 'natural gas'. [L.]

**Bituminous coal.** Ordinary household and coking coal, occurring in the Carboniferous coalfields. It is composed of four main types. 'These four distinguishable ingredients, all of which, in varying quantities, are to be found in most ordinary bituminous coals, I name provisionally as follows: Fusain, Durain, Clarain, Vitrain' (Stopes, *PRS*, 1919).

**Black mud.** See BLACK SHALE.

**Black shale.** 'Black shale is a commonplace lithology, represented in every geological system from early Pre-Cambrian to Recent, though in the more ancient rocks it has generally been metamorphosed to graphitic schist and in the youngest deposits its condition may be that of clay rather than shale. There is strong reason to suppose that living matter is indispensable to its formation, for it can be shown that the black pigment is of organic origin, made up of the principal biophile elements carbon, hydrogen, oxygen, nitrogen. Only in the metamorphosed rocks have the volatile elements been eliminated, leaving free carbon. . . . Black shale is the compressed and indurated rock formed by diagenesis from black mud. . . . Black mud can be observed forming at present in swamps, estuaries, lagoons, silted marine basins, and in hollows where stagnation prevails on the sea floor. . . . Stagnation and/or the rapid deposition of sediment in organically-productive waters leads to the preservation of carbon and other biophile elements' (Dunham, *AS*, 1961). 'Black shales' (Hallam, introducing specialized papers, *JGS*, 1980).

**Blackband ironstone.** 'A variety of clay-ironstone containing sufficient carbonaceous matter to allow of calcining without the addition of fuel in a separate charge' (Holmes, *Nomenclature of Petrology*, 1928).

**Blanket.** A spread of a rock-formation or deposit covering, discordantly the underlying rocks. Thus the Pre-Mesozoic rocks of the Midlands are largely buried under a blanket of New Red Sandstone, and glacial drift forms a blanket concealing the 'solid' rocks over wide areas of the country. '[In foreland folding] the sedimentary cover should be regarded as an inert blanket which accommodates itself to a shrinkage of its basement' (Lees, *QJ*, 1952). 'The Precambrian shields of the Sahara and Arabia are overlain by a thick laterally extensive Cambro-Ordovician sandstone blanket' (Selley, *JGS*, 1972).

**Blastoidea.** An extinct class of the phylum Echinodermata (subphylum Pelmatozoa), having a bud-shaped body with a few main plates regularly arranged and rather elaborate structural features arranged in five radial areas. The body is attached by a stem. Blastoids are not common, but occur in some variety in the Carboniferous; they range from the Ordovician to the Permian. (Say, 1825.) [G. *blastos*, a bud.]

**Blende.** The mineral zinc sulphide, ZnS, commonly dark brown and having a resinous lustre. The most important ore of zinc; occurs in veins. [An old German name, variously given as f. *blenden*, deceive, because it resembles lead ore but yields no lead, or f. *blenden*, dazzle, because of its lustre.]

**Block diagram.** In geology, a plane-diagram depicting the geology of a 'block' of country in three dimensions by some method of projection, isometric (for making measurements) or true perspective (for giving a more natural view). *Block Diagrams* (Lobeck, 1924). See GEOLOGICAL MODEL.

**Block faulting.** A system of faulting which divides a region into blocks of relative structural elevation and depression. These structural blocks ('fault blocks') may be reflected in the relief ('block mountains' or relief 'blocks') either directly, if erosion has not had time to plane them down, or indirectly as a result of differential erosion. Block faulting reflected in the relief is also called 'basin-range' (or 'basin-and-range') structure, particularly as regards part of the SW U.S.A., this term usually implying that the present relief is taken to be the direct result of the faulting. 'For more than 1,500 km. along the western Cordillera of North America, late Cenozoic extensional faulting has produced block-faulted basin-range structure characterized by alternating elongate mountain ranges and alluviated basins. The faulting follows older geologic patterns' (Stewart in Smith and Eaton, eds., *Basin-range Structure in Western North America*, 1978). See HORST.

**Block mountain.** See BLOCK FAULTING.

**Blow-hole.** A natural chimney, on a coast, reaching from the inner end of a cave to the surface of the ground above formed (probably) by the action of air suddenly and forcibly compressed. When formed, it is an outlet for such air, which may be loaded with spray.

**Blue mud.** The characteristic terrigenous mud of the oceanic slopes beyond the continental shelf, covering some 36 million square kilometres. The blue-grey colour is due to finely divided carbon and iron sulphides. (Murray and Renard, *Voyage of H.M.S. Challenger, 1872-76*, 1891 and earlier.)

**Blueschist.** See GLAUCOPHANE. 'A re-evaluation of the blueschist facies' (Wood, *GM*, 1979).

**Bocanne.** The phenomenon of a shale burning naturally at its outcrop. 'A note on the term bocanne' (Crickmay, *AJS*, 1967). (Selwyn, *RPCGS*, 1875.) [A French-Canadian, ultimately Guianan, word for smoke.]

**Bog iron ore.** A deposit of iron compounds, chiefly iron hydroxide, in a bog or swamp, produced by the oxidizing action of algae, bacteria, or the atmosphere. See LIMONITE.

**Bog oak.** 'The wood of oak preserved in a black state in peat bogs, etc.' (*OED*). 'Trees which fell and were buried in the peat with the onset of fen conditions are now commonly encountered in ploughing when, owing to wastage of the peat, the ground surface is lowered to within 2 or 3 feet of underlying clays (in the present instance Gault) on which the forest trees grew. This 36-foot trunk of bog oak was found at Blinkers Hill, Swaffham Prior fen' (Worssam and Taylor, *MGS, Cambridge*, 1969). Though strictly applying to oak only, the term is sometimes used in a general way for such remains of any large tree.

**Bomb.** As 'volcanic bomb'. 'I noticed volcanic bombs, that is, masses of lava which have been shot through the air whilst fluid and have consequently assumed a spherical or pear-shape' (Darwin, *Voyage of the Beagle*, 1845). The term is similarly defined by Lyell (*Principles*, 1833). The Precambrian volcanic agglomerate of Charnwood Forest is called the 'bomb rock' (Watts, *Ancient Rocks of Charnwood Forest*, 1947).

**Bone.** The material, largely a mixture of calcium phosphate and calcium carbonate, forming the hard skeletal parts (the 'bones') of vertebrate animals. The main bony skeleton is internal; but parts originate in the skin, forming external bony scales and plates or the teeth, the most readily fossilizable parts of the vertebrate skeleton. 'A brief relation of some strange bones lately digged up' (Somner, *Chartham News*, 1669; the bones being those of *Rhinoceros* from the river gravels).

**Bone bed.** A bed of rock largely composed of the fossil bones, teeth, &c., of vertebrates. Examples are the Ludlow Bone Bed (described as the 'fish bed' by Murchison, *Silurian System*, 1839, but called 'bone-bed' by him in *QJ*, 1853, and *Siluria*, 1854), the Rhaetic Bone Bed, 'the well-known "bone bed"' (Murchison, 1839), and the Wealden bone beds (Allen, *PGA*, 1949). 'Bone-beds' (Antia, *MG*, 1979).

**Bone breccia.** See BRECCIA (7). 'Accumulations of bones are formed in caves and other retreats by predaceous animals, and such accumulations in limestone caves may become agglutinated together by the drip of water from the roof, and so form "bone-breccias" ' (J. Geikie, *Outlines of Geology*,

1903). 'Osseous breccia' is synonymous (see OSSEOUS).

**Bone cave.** A cave yielding fossil bones, an ossiferous cave. 'Recent researches in bone-caves in Wales' (Hicks, *PGA*, 1885).

**Borrowdale Volcanic series.** 'Volcanic Series of Borrowdale. The rocks of this series seem almost wholly to be made up of volcanic material. This consists of beds of volcanic ash and breccia, alternating with ancient sheets of lava, and the whole traversed by dykes and masses of intrusive igneous rock. Fossils are altogether absent. . . . In substituting the name "Volcanic Series of Borrowdale" for that of "Green Slates and Porphyries" [Sedgwick] I follow the proposition of Profs. Harkness and Nicholson, as given in a paper by the latter in 1872 [*TEGS*]' (Ward, *MGS, Lake District*, 1876). The thickness is about 3000 m. and the outcrop occupies a broad belt stretching NE.-SW. across all the central part of the Lake District; rugged mountains are carved out of the rocks. The occurrence of the series between the Skiddaw Slates (Arenig) and the Coniston Limestone (Caradoc) shows that the rocks are of a general Llanvirnian-Llandeilian age. 'Borrowdale Volcanic Group' (Millward and others in Moseley, eds., *Geology of the Lake District*, 1978).

**Boss.** 'When a stock has a roughly circular outline, like the Shap granite of Westmorland, it is sometimes referred to as a boss' (Holmes, *Physical Geology*, 1944). 'An intrusion working its way up through solid rocks by "overhead stoping" must, if this action be sufficiently continued, acquire something of the vertical cylindrical form which seems to be implied in the term boss' (Harker, *Igneous Rocks*, 1909). Some small bosses may, in fact, be the intrusive parts of volcanic necks. The term is also used in its ordinary sense of a rounded prominence on the surface of the ground, usually composed of hard rock. Hutton (*Theory of the Earth*, 1795) refers to the intrusive necks of North Berwick Law, &c., as now 'variously embossing the surface of the earth'. The following table shows the relation between the terms 'boss', (volcanic) 'neck', and (volcanic) 'plug':

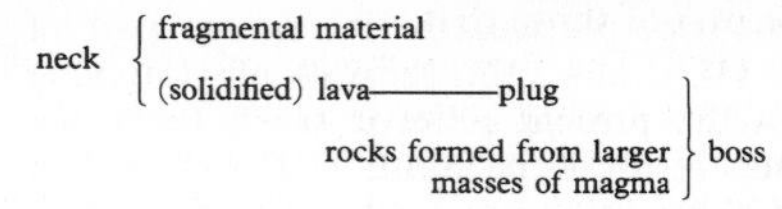


**Bostonite.** An igneous rock composed almost entirely of alkali-felspars with trachytic texture. Thus allied to trachyte. Occurs chiefly as dykes. (Rosenbusch, 1882.) [Boston, Mass., U.S.A.]

**Botryoidal.** Resembling a bunch of grapes; applied to a habit of growth in mineralogy.

**Bottomset beds.** Beds, usually of fine-grained material, laid out on a floor of deposition, over which the foreset beds of a delta are being built forwards.

**Boudinage.** When a relatively rigid bed, a sandstone or limestone, is enclosed between relatively plastic beds, and the whole subjected to folding, the rigid bed may be broken into portions, 'boudins', elongated along the strike, either quite detached or joined by pinched connexions. In the pinched parts, or where the boudins have been actually torn apart, quartz or calcite (according to the composition of the rock) is usually recrystallized. The term 'boudinage' for the structure was first used in Belgium by Lohest, 1908. (Walls, *GM*, 1937; Rast, *GM*, 1956.) [French *boudin*, a sausage.]

**Bouguer anomaly.** 'Suppose that we have a planet, most of whose surface is a sphere, but that extra mass is piled on a limited region to give a plateau. What is the effect on gravity? Clearly above the plateau the extra mass gives an extra attraction downwards. . . . A remarkable fact of observation, first noticed in the Andes by the Frenchman Bouguer [1749], and since extended to most of the major mountain systems, is that they produce much less disturbance of gravity than would be expected if they were simple added loads . . . the Bouguer anomaly (so called because Bouguer was surprised to find it)' (Jeffreys, *Earthquakes and Mountains*, 1935). Bouguer's observation was one of the first of those that led to the concept of isostasy. The measurement and study of Bouguer gravity anomaly variations is perhaps the most important of the geophysical methods used today to give evidence of deep structure (e.g. 'A gravity survey of the Lake District and the Vale of Eden', Bott. *JGS*, 1974).

**Boulder.** A large rounded stone; particularly a glacial erratic. [Short for boulder-stone, Middle English *bulderstone*.]

**Boulder-bed.** A boulder-bearing conglomerate or a tillite. In British geology the term is used chiefly for certain Dalradian beds, e.g.: 'The lithology and structure of a boulder-bed in the Dalradian of Mayo, Ireland' (Ewell, *QJ*, 1955). 'Kimmeridgian boulder-bed on the shore at Kintradwell, Sutherland' (pl. in Craig, ed., *Geology of Scotland*, 1965).

**Boulder-clay.** The most widespread and distinctive of the glacial deposits left behind on the melting of an ice sheet. It is a fine clay containing, more or less thickly distributed in it, pebbles and boulders of all sizes but most of them having a characteristic subangular shape, and many being conspicuously scratched by abrasion. 'Although the boulder-clay has been the subject of a vast literature since Agassiz first described it in 1837, it remains to some extent the "mysterious deposit" R. Chambers designated it. It is an accumulation *sui generis*' (Charlesworth, *Quaternary Era*, 1957). The term itself, well established in the eighteen-sixties, was used by J. Geikie (*Great Ice Age*, 1874) for a variety which he thought to be partly of marine origin, to distinguish it from the general 'till'. However, 'boulder-clay' is now a more usual term than the synonymous 'till' for the deposit in general. See quotation under DRIFT. Owen described this deposit ('clay marle') in SW. Wales in a MS. of 1599, ascribing it to 'Noes flood' (John, *JGl*, 1964).

**Boundary reference point.** See BOUNDARY STRATOTYPE.

**Boundary stratotype.** A section (stratotype) in which the boundary reference point that defines the base of a chronostratigraphical unit is selected. This reference point is termed a 'marker point' or, occasionally, a 'golden spike'. 'The definition of boundary stratotypes for systems is of prime importance. Boundaries of systems must by definition coincide with the appropriate boundaries of divisions of lower rank: the series, stages, and chronozones. Boundaries of series must by definition coincide with the appropriate boundaries of stages and chronozones; boundaries of stages must by definition coincide with those of chronozones. . . . The base of each chronozone must be defined at a marker point in a boundary stratotype section, and will be extended from there by correlation using all available means' (*Stratigraphical Guide*, *GSSR* (11), 1978).

**Bourne.** 'Springs which only flow at certain times of the year, or only after a prolonged spell of rainy weather, form temporary streams, which are known as bournes, nailbournes, winterbournes, woebournes, levants, and gypsies in different parts of the country. Some of them break out every year at the same spot; these may be called regular bournes, and are generally "winterbournes". Others only come into existence after a season of great and prolonged rainfall, and these may be termed occasional bournes. In some localities they are regarded with superstitious terror, and have been called "woe-bournes" under the idea that they only appeared when some disaster was about to happen' (Jukes-Browne, *MGS*, *Cretaceous Rocks of Britain*, 1904, referred to in Stamp, ed., *Geographical Terms*, 1961).

**Boxstone.** A spheroidal or rectangular box-like structure, typically containing fossils, of a hard substance consisting of sand cemented by iron hydroxide. It occurs chiefly as rolled nodules in the Pliocene of East Anglia, probably representing pre-Pliocene Tertiary Strata. 'I have spent a good deal of time in working at the nodules, which I propose to call "box-stones", since the name of "boxes" has been applied to those which exhibit the remains of a shell on being broken open by the phosphate-diggers of Suffolk' (Lankester, *QJ*,

1870). (Reid, *MGS*, *Pliocene*, 1890; Boswell, *GM*, 1915; Bell, *GM*, 1918.)

**Brachiopoda.** A phylum of marine animals having a shell in two pieces (valves) which can open (slightly) and close about a hinge-line beneath the beaks of the valves. One valve (the pedicle valve) is larger than the other (the brachial valve) but the whole shell is normally equilateral, having a plane of symmetry passing through the median line of each valve. The larger valve has a perforation (pedicle opening) just below the beak. Brachiopods, ranging from the Lower Cambrian to the present day, are far more abundant and varied as fossils than they are in the modern seas, being, in general, the commonest of all fossils throughout the Palaeozoic. There are two classes, Inarticulata and Articulata, and, within these, many distinctive groups. [G. *brachiōn*, an arm.]

**Brachy-anticline.** See DOME STRUCTURE. Hence, also, 'brachy-syncline'. [G. *brachus*, short.]

**Bracklesham beds.** See EOCENE. (Prestwhich, *QJ*, 1847; redefined by Fisher, *QJ*, 1862.) [Bracklesham, Hampshire.]

**Bradford Clay.** See GREAT OOLITE SERIES. (Smith in Warner, *Guide through Bath*, 1811, 'Clay'; Greenough, *Map*, 1819, 'Bradford Clay'.) [Bradford-on-Avon, Wiltshire.]

**Branchiopoda.** A subclass of the class Crustacea, phylum Arthropoda, comprising several groups, one of which includes the bivalved *Estheria* which lives in freshwater today and, occurring in rocks from the Old Red Sandstone onwards, is a valuable indicator of a freshwater facies. [G. *bragchia*, gills.]

**Brash.** An old term (current in the 17th century) for a mass or layer of fragments of rock, particularly a very stony subsoil. See CORNBRASH. [probably French *brèche* =Italian *breccia*.]

**Break** (substantive). Sometimes used for an interruption in (upward) stratigraphical continuity. 'Breaks in succession of the British strata' (Ramsay, *QJ*, 1863/4). 'Breaks in the succession' (Donovan, *Stratigraphy*, 1966). 'Kidston and Traquair, working on the fossil plants and fishes respectively, demonstrated that a marked palaeontological "break" occurred in the lower part of the Millstone Grit' (Macgregor and Macgregor, *BRG, Midland Valley of Scotland*, 1948). See DIASTEM.

**Break-away fault.** A fault on one side of which a rock-mass has become broken away and separated from a stationary block on the other side. The resulting gap would normally be filled with later rocks, sedimentary or volcanic. 'The Heart Mountain break-away fault, north-western Wyoming' (Pierce, *BGSA*, 1980), (Pierce, *USGS*, 1960).

**Breccia.** A rock composed of angular clastic fragments mixed with finer-grained-material. (Cf. conglomerate). There are various kinds. **1.** Deposits formed as scree, by surface creep, by glacial or pluvial action, or by some other subaerial process. **2.** A sedimentary rock (usually a limestone) brecciated by some process involved in, and penecontemporaneous with, its consolidation (a sedimentary 'autobreccia'). Includes the peculiar 'pseudobreccia' occurring particularly in the D zone of the Carboniferous Limestone; first noticed and named by Tiddeman (*MGS*, *Swansea*, 1907) and described by Dixon (*QJ*, 1911). **3.** 'Volcanic breccia'; (*a*) a pyroclastic rock, differing from a volcanic agglomerate in that there are not such large fragments and all are angular; (*b*) a lava broken up during cooling (a volcanic 'autobreccia'), either by the explosive escape of steam, &c., or by the hardening and breaking-up of the surface in the process of flow (the latter specified still further as 'flow-breccia'). Any volcanic breccia (or angular agglomerate) formed by explosive action, and whether composed of country rock, volcanic rock, or both, may be called an 'explosion breccia' (Richley, *QJ*, 1932). A volcanic ('explosion') breccia might not reach the surface; for instance it might be formed in 'an entirely subsurface cauldron subsidence fracture zone' (Dunham, *PGS*, 1968). 'Classification of volcanic breccias' (Fisher, *BGSA*, 1960; and Wright and Bowes, and Fisher, *BGSA*, 1963). **4.** Gash breccia. Typically seen in the Carboniferous Limestone of the S. Pembrokeshire coast. They 'appear to be the tumbled walls of caverns in Carboniferous Limestone eroded in Triassic times' (George, *BRG, S. Wales*, 1970). See also Thomas, *TIBG*, 1972. Similar gash-breccias are conspicuous in the Magnesian Limestone of the Durham coast. **5.** Intrusion breccia. A breccia consisting of fragments of country rock which have, apparently, been torn from the walls of a fissure (thus widening it) and carried upwards in a rush of gas-charged fluid connected with igneous activity. The fragments are not all sharply angular but are, rather, faceted and rounded as if by mutual attrition. 'An appinitic intrusion-breccia at Kilkenny, Maas, Co. Donegal' (Pitcher and Read, *GM*, 1952). **6.** Tectonic breccia, formed along a fault or a crush-belt: 'fault-breccia', 'crush-breccia'. See AUTOCLASTIC ROCK. The term 'brecciation' applies to the production of a tectonic breccia only. **7.** A breccia formed in some special way and in which the fragments are of some particular substance, e.g. 'bone breccia' found in some caves. (See OSSEOUS.) In early days, the angularity of the fragments, now definitely implied in the term, was not so much insisted on; thus Playfair (*Illustrations*, 1802): 'A breccia between the primary and secondary strata [at an unconformity] in which the fragments, whether round or angular, are always of the primary rock'. **8.** See EXPLOSION BRECCIA (2). Reynolds (*GM*, 1928) gives an elaborate classification of breccias. [adopted in the 18th century from Italian *breccia*, rubbish of broken walls.]

**Breccio-conglomerate.** See CONGLOMERATE. 'Polygenetic breccio-conglomerate. La Tete des Hougues, Jersey' (Bates and Kirkaldy, *Field Geology in Colour*, 1976).

**Brecciation.** See BRECCIA (6).

**Breconian.** see OLD RED SANDSTONE. (White and Toombs, *IGC*, 1948.) [Brecon Beacons, Powys.]

**Brick clay.** A clay or mudstone (including shale) of a particular composition suitable for making building-bricks. In Britain such rocks are especially abundant in the Upper Carboniferous shales, Keuper Marl, and Jurassic 'clays'; Pleistocene clays (including the brick-earth) are also used.

**Brick-earth.** A structureless silty clay, apparently resulting from various processes, occurring among the Pleistocene deposits of SE England used, particularly in former times, for making bricks. 'The Brick-earth . . . of the Sussex coast' (Austen, *QJ*, 1857).

**Brigantian.** See DINANTIAN. [Brigantes, the Celtic tribe that inhabited the part of northern England in which the boundary stratotype (at Janny Wood, near Kirkby Stephen, Cumbria) is situated.]

**Brittle deformation.** Deformation resulting from the breaking of a rock-mass by faulting or by a grinding process, as opposed to 'ductile deformation' due to the drawing out of rock-material under a process of plastic flow. 'Deformation in the brittle (seismic) mode is characteristic of the upper parts of fault systems, while at lower levels displacements are predominantly ductile (aseismic)' (Grocott, *JGS*, 1977). *Experimental Rock Deformation—The Brittle Field* (Paterson, 1978).

**Brockram.** Cumbrian term for breccia and adopted by geologists (beginning with Phillips, *Yorkshire*, 1836) for the breccias which are a notable feature of the Permo-Trias of Westmorland and Cumberland; at first restricted to those of the Penrith Sandstone of the Vale of Eden but later extended to include those of west Cumberland.

**Brown coal.** 'An intermediate stage in the conversion of peat into true coal; sometimes used as synonymous with lignite, the term is preferably restricted to materials whose vegetable origin is less evident than that of lignite' (Himus, *Dictionary of Geology*, 1954). 'Geology of the brown coals of Victoria' (Thomas and Baragwanath, *MGJ*, 1949/51).

**Bryozoa.** A phylum of minute marine animals, with a calcareous skeleton, forming compound colonies which may be branching, laminar, globular, or encrusting. Also called Polyzoa. [G. *bruon*, moss.]

**Building stone.** 'Any stone used in masonry construction, generally stone of superior quality that is quarried and trimmed or cut to form regular blocks' (*AGI Dictionary*, 1976). *British and Foreign Building . . . specimens in the Sedgwick Museum, Cambridge* (Watson, 1911). *Oxford Stone* (Arkell, 1947).

**Bullion.** A coal-ball; but the term includes, and is used specially for, the concretions, similar to the roof-nodule coal-balls, in Carboniferous rocks other than coal measures. 'The palaeontology and palaeoenvironment of some Namurian limestone "bullions"' (Holdsworth, *MG*, 1966).

**Bunter.** The lower half (in Britain) and lower third (in Germany, where first named) of the New Red Sandstone facies of the Triassic. The rocks are sandstones and conglomerates ('pebble beds'); unfossiliferous, except for the (derived) fossils in the pebbles. [German *bunt*, variegated.]

**Buried floor.** See COMPOUND STRUCTURE. 'The buried floor of eastern England' (Kent in Sylvester-Bradley and Ford, eds., *Geology of the East Midlands*, 1968).

**Buried forest.** Buried remains of trees. In the wide sense the term would include 'submerged forests', but it is more particularly restricted to occurrences in peat (formed either in lowland or upland areas) remaining above sea-level. 'On the buried forests and peat deposits of Scotland' (J. Geikie, *TRSE*, 1867).

**Buried landscape.** See PALAEOGEOGRAPHY (last sentence), EXHUME (Sylvester-Bradley quotation), and UNCONFORMITY (last paragraph). 'A buried Triassic landscape [Charnwood Forest]' (Watts, *GJ*, 1903).

**Burst.** 'An evolutionary burst may be defined as acceleration of stock diversification relative to the previous history of the stock' (Stubblefield, *QJ*, 1959). The question of the recognition of such bursts has been considered by Cooper and Williams (*JP*, 1952). Vaughan (*QJ*, 1915) used the term for the acme ('culmination') of a stock ('sheaf of parallel gentes which arrive at the same structural stages at the same time') which he visualized as being reached with a sudden and great increase in abundance, and consequent migration. Also called 'evolutionary explosion'.

**Butte.** See CIRCUMDENUDATION. [French.]

**Bysmalith.** (Bysmalite.) Essentially a laccolite (laccolith) by nature, but plug-like, being more localized and altogether steeper and more convex, with more acute doming of the overlying rocks which may be faulted, as well as bent, upwards. The Americans having substituted '-lith' for '-lite' in the names of rock-bodies, the term was proposed in this form (by Iddings, *JG*, 1898). Harker (*Igneous Rocks*, 1909) without comment writes 'bysmalite', uniform with 'laccolite'. (See LACCOLITE.) The original type bysmalith is Mt. Holmes, Yellowstone National Park, U.S.A. [G. *buo*, plug.]

**Bytownite.** See PLAGIOCLASE. 'The name was originally given to the felspar of a rock from Bytown in Canada which, however, ultimately proved to be a mixture of anorthite with quartz and other minerals' (Miers, *Mineralogy*, 1929). (Thomson, 1835.)

# C

**Caenozoic.** See CAINOZOIC.

**Caerfai beds.** Reddish and greenish conglomerates, sandstones, and shales, some 275 m. thick, unconformably overlying the Precambrian rocks in Pembrokeshire and succeeded by the Solva beds with Middle Cambrian fossils. They are thus no doubt of Lower Cambrian age, but are themselves almost entirely unfossiliferous. (Hicks, *PSR* and *PGA*, 1881, *GM*, 1892/4.) [Caerfai Bay, Preseli district of Dyfed.]

**Cainozoic.** (Kainozoic, Caenozoic, Cenozoic.) Of the ages of recent life; applied collectively to the stratigraphical divisions and all deposits above the Cretaceous and to the era of time they represent. This era extends from a time of the order of 70 million years ago up to the present day. The main fossil groups are, therefore, those that are potentially the main fossil groups of today, the lamellibranchs and gastropods together with the microfossils, the foraminifers. Among the subordinate groups are corals (hexacorals), echinoids, fish, and crustacea. However, in a reconstruction of Cainozoic life the various groups of mammals, and the angiosperms among plants, are so important—though occurring only exceptionally as fossils because of their dominantly land habitat—that the era is appropriately called the 'age of mammals', or the 'age of angiosperms'. See TERTIARY. (Phillips, *Palaeozoic Fossils of Cornwall* [&c.], 1841.)

**Calc-alkali.** For igneous rocks with both lime and alkali (usually soda), shown chiefly in the presence of a lime-bearing plagioclase felspar.

**Calcareous.** (Calcarious.) Containing, or partly composed of, calcium carbonate. The spelling -eous is established, but -ious is the more correct. 'Stones of the warm calcarious kind' (Plot, *Oxfordshire*, 1677, alluding to the Jurassic oolites). See quotation (Murchison) under ARGILLACEOUS. [L. *calcarius*, of or pertaining to lime.]

**Calcareous algae.** See THALLOPHYTA. 'Calcareous algae are those members of the various algal classes of the main botanical classification which form calcified structures, as opposed to the majority which do not, and they do not themselves form a separate intrinsic class' (Elliott, *QJ*, 1959). 'Calcareous algae have contributed even more substantially to the formation of Palaeozoic rock-masses than was suggested by Garwood in his 1931 Anniversary Address to the Society' (Stubblefield, *QJ*, 1960). 'The Lower Carboniferous rocks of the Bewcastle district are chiefly notable for the remarkable development of calcareous algae. In Carn Beck an algal limestone (about 1 ft. thick) is exposed in the form of a "nobbly" pavement' (Day, *MGS*, *Bewcastle*, 1970).

**Calc-flinta.** A very fine grained, flinty, lime-silicate rock produced by thermal metamorphism. First described from the St. Austell metamorphic aureole: 'Certain peculiar banded hornstones, which usually contain calc-silicates, and must therefore represent various forms of more or less calcareous sediment . . . first recognised as a distinct group of rocks by Mr. Ussher (1903) and subsequently termed calc-flintas by Mr. Barrow (1905)' (*MGS*, *Bodmin*, 1909).

**Calcic.** Of or containing the element calcium.

**Calcicole, Calcifuge.** Calcicole and calcifuge plant species are those that respectively prefer and abhor lime-bearing soils. They thus usually reflect the character of the underlying rock in this respect. Calcicole species indicate limestones and, to a less extent, calciferous basic igneous rocks and may be useful in geological mapping. The contrast in vegetation is often striking, e.g. between that over Carboniferous Limestone and adjacent Millstone Grit (Moss, *Vegetation of the Peak District*, 1913). Some notable calcicole species are Kidney Vetch, Salad Burnet, Rock Rose, Marjoram, Hoary Plantain, Biting Yellow Stonecrop; while among calcifuge species are Broom, Foxglove, Heath Bedstraw, Heather, Bilberry, Milkwort. [L. *colo*, inhabit, *fugio*, avoid.]

**Calciferous.** Yielding calcium carbonate. **1.** For a rock. **2.** For a series of strata containing beds of limestone; in particular, the Calciferous Sandstone series, forming the lower half or so of the Lower Carboniferous of Scotland.

**Calcilutite, Calcarenite, Calcirudite.** Clastic limestones corresponding in texture to lutite, arenite, and rudite respectively. These names are more general than calcite-mudstone &c. and imply that the rocks are entirely, or nearly entirely, composed of calcium carbonate.

**Calcite.** A rather soft mineral, calcium carbonate, $CaCO_3$; usually white; hexagonal system, occurring in the forms nailhead spar, dog-tooth spar, Iceland spar, &c. Common as veins, and the dominant constituent of limestones. 'The name was applied at first (Freiesleben, 1836, as *calcit*) to crystals of calcium carbonate pseudomorphous after celestite, from Sangerhausen. Afterwards applied to the common mineral that now bears its name, perhaps first by E. J. Chapman, *Practical Mineralogy*, 1843' (Chester, *Names of Minerals*, 1896). (Aragonite is another mineral of the same composition.) [L. *calx, calc-*, lime, limestone.]

**Calcite-mudstone.** The rock-type characteristic of the calcareous lagoon phases of the Carboniferous Limestone.'[There are] compact limestones, dolomites, and argillaceous rocks which, under the microscope, are seen to be composed, essentially, of exceedingly fine-grained, calcareous (that is, either calcitic or dolomitic), argillaceous, and siliceous material, and evidently represent consolidated muds of various compositions. For this reason, the calcareous members will be spoken of as calcite-mudstones or dolomite-mudstones, in order to distinguish them from limestones or dolomites of standard types. The calcite-mudstones include the rocks known as "chinastone-limestones", so called because their compactness and white weathered surface impart to them a porcellanous appearance' (Dixon, *QJ*, 1911). There is also 'calcite-siltstone' and 'calcite-sandstone' (Bathurst, *LMGJ*,1958).

**Calcrete.** A term suggested by Lamplugh (*GM*, 1902): 'In the drifts around Dublin, as in most places where in like manner limestone-debris enters largely into the composition of the superficial deposits, the sand and gravel beds are often cemented sporadically into hard masses by solution and redeposition of lime through the agency of infiltrating waters. . . . Moreover I have the hardihood to suggest that the term [calcrete] might be complemented by equivalents—"silcrete", for sporadic masses of loose material of the "grey-wether" type, indurated by a siliceous cement; and "ferricrete", when the binding substance is an iron-oxide'. The three terms are now used chiefly for calcareous, siliceous, and ferruginous duricrusts, particularly for those of Australia and Africa.

**Calc-sinter.** See SINTER.

**Caldera.** A large cauldron-shaped volcanic structure. This may be produced in various ways. 'Knowledge of how calderas are formed has been extended considerably during recent years, and the idea that many result simply from explosive decapitation of volcanic cones has been abandoned. Some calderas originate by engulfment when the feeding chambers of volcanoes are hurriedly drained either by copious discharge of lava and juvenile ejecta at the surface or by subterraneous withdrawal of magma; other calderas develop by ring-fracture or piecemeal stoping' (Williams, *QJ*, 1953). Examples are listed in Holmes. *Physical Geology*, 1978. There are several volcanic mountains named Caldere in the Canary Islands (Spanish) and others named Caldeira in the Azores (Portuguese). These include typical calderas. [Spanish f. late Latin. *caldaria*, a boiling-pot.]

**Caledonian.** In geology the chief use of this adjective is in its application to a region and epoch of mountain building. First used by Suess (*Das Antlitz der Erde*, 1888, English trans., 1906) as follows: 'These pre-Devonian mountains, which proceed from Norway and form the whole of Scotland we call the "Caledonian mountains" [*caledonische Gebirge*]. . . . The Caledonian mountains are continued through a great part of Ireland and Wales'. Later on in his work, Suess calls these mountains the Caledonides. The term 'Caledonian' included the earth-movements and structures as well as the mountains themselves. The general trend of the structures in the stumps of these old mountains—that is, the 'Caledonian trend'—is NE.-SW. Associated igneous intrusions are often called 'Caledonian' (e.g. Hatch and Wells, *Igneous Rocks*, 1965). 'The Caledonian orogenic belt of north-western Europe extends through Norway and west Sweden to the British Isles, where it occupies much of Scotland, north-western England, Wales and Ireland. The foundations of the Caledonian structure were laid in late Pre-Cambrian and early Palaeozoic times when considerable thicknesses of sediment began to accumulate in and near the area which was later raised to form the orogenic belt. From Ordovician to Devonian times this region was subjected to repeated phases of earth movement which led to folding and upheaval, while at the same time the crust became abnormally hot, metamorphism took place over considerable areas and granitic material was generated at depth. By the close of the Devonian Period, the building of the fold-mountain belt was virtually completed and the Caledonian orogeny had come to an end. In Britain, the Caledonides are made up of two rather different tracts lying side by side. In the north-western part of the structure lies the metamorphic belt, made up of Pre-Cambrian and Lower Palaeozoic rocks, which occupies the Scottish Highlands [south-east of the Moine Thrust] and north-western Ireland. To the south-east is the nonmetamorphic belt which makes the folded Lower Palaeozoic regions of the Southern Uplands, the Lake District, Wales and eastern Ireland' (Watson, *PGA*, 1963). 'Caledonoid' was suggested by Lapworth (BA *Handbook*, 1913) for a NE.-SW. tectonic direction (at least in Britain) of any age; at the same time Armoricanoid, Charnoid, and Malvernoid were suggested to denote E.-W., SE.-NW., and N.-S. directions, respectively, of any age. A recent discussion of the use of the term is in *PGS* 1965. See MOUNTAIN SYSTEM.

**Caledonides.** See CALEDONIAN. The name is also used in a comprehensive sense for the whole Caledonian orogenic belt. *The British Caledonides* (Johnson and Stewart, eds., 1963). *Caledonides of the British Isles Reviewed* (*GSSR*, 1979).

**Caledonoid.** See CALEDONIAN.

**Callovian.** See JURASSIC SYSTEM. [Kellaways (Callovium), Wiltshire.]

**Camber.** The condition of strata of being slightly arched, with the special implication that this is due to superficial disturbance under the influence of gravity. See SUPERFICIAL STRUCTURE. 'The term "cambering" is intended to describe the way in which the rock layer drapes the hill slopes and spurs, rather like a tarpaulin thrown over a railway truck' (Arkell, *Oxford*, 1947).

**Cambrian system.** The lowest of the fossil-bearing systems and thus the lowest of that part of the whole geological column which can be put in a regular order. There is no generally defined base; all that can be said is that where, in a local succession, there are strata with a fauna characteristic of the lowest series of the system, and either these or the (unfossiliferous) strata below them rest unconformably on still lower rocks, the base of the system is usually taken to be there represented by the strata immediately overlying the unconformity (but see Snyder, *JG*, 1947, Glaessner, *JGSA*, 1963, Razanov, *GM*, 1967, and Cowie in Harland and others, eds., *The Fossil Record*, 1967). The commonest fossils are brachiopods and trilobites, the latter being the more important. The Cambrian trilobites belong to groups which are nearly all confined to this system. The chief families are: Olenellidae, Redlichidae, Protolenidae, Pagetidae, Eodiscidae,

Dorypigidae, Conocoryphidae, Paradoxididae, Ptychopariidae, Ogygopsidae, Olenidae (subfamilies Oleninae and Leptoplastinae), Dikelocephalidae, Agnostidae (restricted). The brachiopods are nearly all Inarticulata (which have horny shells); particular species occur in particular beds but many of the genera persist into the Ordovician. Genera of other invertebrate groups also occur in certain beds and are restricted to them. The following are the series of the Cambrian system in Europe and eastern N. America:

| | | |
|---|---|---|
| Upper (*s.l.*) | Transition | Tremadocian |
| | Upper (*s.s.*) | Olenus series |
| | Middle | *Paradoxides* series |
| | Lower | *Olenellus* series |

The Transition series (Tremadocian) is sometimes in Britain, and usually outside Britain, placed in the Ordovician system. The term 'Upper Cambrian' is ambiguous unless it is stated whether or not it includes the Tremadocian. The classic ground of the Cambrian is in Wales where it is displayed in three areas; on the north side of the Snowdon syncline, over all the central part of the Harlech dome, and on both flanks of the St. David's anticline. The fossiliferous rocks of the Lower and of the Transition series are best exemplified in Shropshire.

'The Silurian rocks are underlaid by vast masses which rise up into the mountains of North, and western parts of South, Wales, and to these Professor Sedgwick, connecting his labours with those of Mr. Murchison, has assigned the name of "Cambrian System" ' (*PGS*, 1836). 'Sedgwick and Murchison have, in truth, "discovered" the Cambrian and Silurian systems, and given us a wide and magnificent field of ancient strata and palaeozoic life' (Phillips, *Geology*, 1855).

As it came to be defined by Sedgwick the system included all the rocks which contained fossils from the lowest up to the top of the Bala series, that is, the present Cambrian and Ordovician systems combined. It very soon began to be recognized that some of the commonest fossils in the upper part of the 'Cambrian', where it was being investigated by Sedgwick in N. Wales, were the same as those common also in parts of Murchison's Silurian system, but it was assumed that all this 'Cambrian' was below Murchison's Llandeilo series and the fossil evidence was interpreted as showing that the Silurian fauna extended downwards into much older rocks; but between 1842 and 1850 it was proved that the upper part of Sedgwick's Cambrian was stratigraphically equivalent to the lower part of Murchison's Silurian. The nomenclature of the Lower Palaeozoic systems was settled in 1879 when Lapworth instituted the Ordovician system, showing that, on geological grounds, three systems were appropriate. The top of the Cambrian was then taken at the top of the Tremadoc series. The Cambrian period is now estimated as beginning about 600 million years ago; but an agreed base to the system (i.e. a general Precambrian- Cambrian boundary) is elusive.

In the authoritative *Correlation of Cambrian Rocks in the British Isles* (*GSSR* (2), 1972) new names are given for the four divisions listed above: Lower-Comley series; Middle-St. David's series; Olenus series (Lingula flags)-Merioneth series; Tremadocian-Tremadoc series. These are particularly applicable in the regions after which they are named, but are also proposed as general series names throughout the British Isles. (Sedgwick, *PM*, 1835 and, later, *BA*, 1835.) [Cambria=Wales.]

**Cambrian-to-Recent.** See PHANEROZOIC.

**Cambro-Silurian.** The Cambrian, Ordovician, and Silurian collectively (i.e. Lower Palaeozoic) or some part of these collective systems for which the appropriate name is uncertain. Used particularly during the middle of the nineteenth century when there was the long controversy over the application of the names Cambrian and Silurian and before Ordovician had been proposed. 'On some new Cambro-Silurian fossils' (M'Coy, *AMNH*, 1851).

**Cannel coal.** A dull coal, with conchoidal fracture, somewhat akin to durain, largely composed of lycopod spores and bodies thought to be algal. It has a high ash, and a high volatile, content and contains fossils of water-dwelling animals (chiefly fish remains and minute crustaceans). Cannel coal appears to be sapropelic rather than humic in origin: 'It is found chiefly in Lancashire; its proper name is "candle coal", as it burns like a candle, but candles in that shire are called "cannels" ' (Kirwan, *Mineralogy*, 1796).

**Cannon-ball concretion.** In the Magnesian Limestone (Permian): 'A globular concretionary structure . . . seen in its most imposing form on some parts of the coast of Durham, where the whole cliff resembles a great irregular pile of cannon balls' (Sedgwick, *TGS*, 1829). In the Lower Calcareous Grit of the Corallian (Jurassic): 'Curious bands of large nodules of intensely hard, dark grey, argillaceous gritstone, some of which resemble cannon balls protruding from the larger doggers or grit bands' (Arkell, *Jurassic System*, 1933).

**Canyon.** 'Deep gorge with stream' (*OED*). 'Where the influence of air, rain, frost and general subaerial weathering has been slight, and the streams, supplied from distant sources, have had sufficient declivity, deep, narrow precipitous ravines or gorges have been excavated. The canyons of the Colorado are a magnificent example of this result' (Geikie, *Textbook*, 1882). The term is used for a similar submarine feature (see SUBMARINE CANYON). [Spanish *cañon*, a tube.]

**Cap rock.** 'Where salt domes have reached the surface the solvent effect of rainfall is to leave behind a mantle of insolubles. However, long before reaching the surface, a plug may encounter percolating groundwaters, which have a similar effect. The evaporite minerals lose more rock-salt than anything else by solution, while anhydrite (or gypsum, to which it is commonly altered) and insoluble impurities such as clay form a mantle, which is called the "cap-rock", over the core of salt' (Holmes, *Physical Geology*, 1978).

**Caradoc series.** (Caradocian.) The Ordovician series between the Llandeilo below and Ashgill above, forming the main part of the Bala series (which comprises the Caradoc and Ashgill). The shelly facies is characterized by many species of trilobites and brachiopods, such as *'Trinucleus' caractaci*, *'Orthis' actoniae*, *Rafinesquina expansa*, and *Sowerbyella sericea*. The Caradoc series as originally described by Murchison (*PM*, 1835 *Silurian System*, 1839) included beds which (in the early 1850's) became separated as the *'Pentamerus* beds' or May Hill Sandstone, and a little later still (in the late 1850's) as the Llandovery (Silurian). [Caradoc Hills, Salop.]

**Carbonaceous.** Coaly. [L. *carbo, carbon*, charcoal.]

**Carbonate.** A salt of carbonic acid. The most important mineral carbonates are calcite and aragonite ($CaCO_3$), dolomite ($CaCO_3.MgCO_3$), and siderite ($FeCO_3$). There is no corresponding adjective, 'carbonate' being used adjectivally as required; but see CALCAREOUS.

**Carbonate rock.** In the literal and widest sense, any rock composed mainly of a carbonate mineral (including, for instance, iron carbonate), but in practice the name is usually limited to rocks composed of the carbonate of calcium, or the double carbonate of calcium and magnesium (dolomite), or a mixture of both; that is, it is synonymous, in this limited sense, with limestone in the widest sense. The simple carbonate of magnesium, magnesite, is rare in the sedimentary rocks. *Classification of Carbonate Rocks* (Ham, ed., 1962). *Carbonate Rocks* (Chilingar and others, eds., 1967).

**Carbonate sediments.** Sediments (in a wide sense) composed preponderantly of calcium (including magnesium) carbonate. Modern marine carbonate sediments may be roughly grouped into three main classes: the deep-water calcareous oozes; reefs composed of coral and/or the calcareous hard parts of other organisms; continental-shelf limestones and lime-muds. *Carbonate sediments and their Diagenesis* (Bathurst, 1971). *Deep Water Carbonate Environments* (Cook and Enos, eds., 1977).

**Carbonatite.** 'The rocks now generally spoken of as carbonatites may be briefly described as rocks which, though in general mineral composition similar to limestones and marbles of known sedimentary origin, yet appear to behave as intrusive rocks and are closely associated with alkaline igneous rocks' (Smith, *QJ*,1956). *The Carbonatites* (Tuttle and Gittins, eds., 1966). *The Geology of Carbonatites* (Heinrick, 1966). (Brögger, Norway, 1921.)

**Carboniferous.** Coal-bearing; for coal bearing strata. 'By carboniferous soils I mean the various sorts of earth or stone among or under which coal is usually found' (Kirwan, *Geological Essays*, 1799). 'Clay-ironstone occurs interstratified with the carboniferous shales in the Tyrone Coal Districts' (Portlock, *Londonderry* [&c.], 1843). However, it is seldom used in this sense. Carboniferous, with capital C, means 'of or pertaining to the Carboniferous system' and it is this very common usage, no doubt, that has ousted the more general and appropriately applicable 'carboniferous'. [L. *carbo, carbon-*, charcoal, *fero*, bear.]

**Carboniferous Limestone series.** The lower half or so of the Carboniferous system is, very generally, predominantly limestone, so that 'Carboniferous Limestone series' ('Mountain Limestone' in the older works) is the familiar name for the more non-committal 'Lower Carboniferous'. However, ambiguities of nomenclature occur where, as in Scotland, most of this Lower Carboniferous is not limestone. Here 'Carboniferous Limestone Series' is usually restricted to the upper half of the Lower Carboniferous, where limestone bands are most common. Here, as anywhere else, 'Carboniferous limestone' (with capital C and small 1) means a limestone occurrying somewhere within the Carboniferous system. Fossils are in many places abundant; corals and brachiopods are the principal groups, but many others are common, especially molluscs and crinoids. Some well-known characteristic genera are, among corals, *'Zaphrentis', Caninia, Palaeosmilia, Lithostrotion, Dibunophyllum*; among brachiopods, *Productus (sensu lato), Schizophoria, Spirifer, Athyris*. All these brachiopods and the coral *Zaphrentis* occur outside the Carboniferous system, but particular species of each are confined to the system. The assemblages allow a classification into faunal zones, which are known by the letters, K, Z, C, S, D (from below upwards). This well-known formation is exemplified in its geological character and its scenic effects in, for instance, the Ingleborough district of Yorkshire, the Peak District of Derbyshire, the Gower peninsula of Glamorgan, and the Mendip Hills in Somerset. (Whitehurst, *Strata in Derbyshire*, 1778, 'Limestone and Toadstone'; Smith in Warner, *Guide through Bath*, 1811, 'Mountain Limestone', but placed above the Coal Measures; Smith, *Stratigraphical System*, 1817, 'Mountain Limestone', correctly placed; Conybeare and Phillips, *England and Wales*, 1822, 'Carboniferous Limestone'.)

**Carboniferous system.** The stratigraphical system which includes in the upper part nearly all the true coal seams of the world (that is, apart from beds of lignite). These strata were at first grouped together with the Old Red Sandstone and the whole called, by Buckland in 1818, the 'Great Coal and Mountain Limestone formation', the name 'Carboniferous' being given (again to the whole group) by Conybeare and Phillips in 1822 ('Medial or Carboniferous Order'). Lyell (1833) included, in his 'Carboniferous Group', strata ('grauwacke and transition limestone') even below the Old Red Sandstone. Murchison, in 1839, removed the Old Red Sandstone (nonmarine Devonian), giving it the status of a separate system, and later still, the marine Devonian was recognized. Thus the Carboniferous system overlies the Devonian system and is overlain by the Permian.

The fossils of the Carboniferous are abundant and belong to many groups of which the corals, crinoids, brachiopods, lamellibranchs, cephalopods, fish, and, of the plants, the pteridophytes and pteridosperms are the chief. The outcrop of the system in Britain occupies a larger area than that of any other system and also occupies large areas in other regions of the world.

The Carboniferous is divided into the Dinantian (Lower) and Silesian (Upper) subsystems. *A Correlation of Dinantian*

*Rocks in the British Isles* (*GSSR* (7), 1976). *A Correlation of Silesian Rocks in the British Isles* (*GSSR* (10), 1978). These are not exactly equivalent to the Mississippian and Pennsylvanian systems of American usage.

| Subsystem | Series |
|---|---|
| Silesian | Stephanian<br>Westphalian<br>Namurian |
| Dinantian | Viséan<br>Tournasian |

**Carbonification.** See COALIFICATION and quotation under RANK IN COAL.

**Carbonization.** The conversion of a carbonaceous substance into carbon by the removal of the other constituents. The term is chiefly used in connexion with the process of fossilization.

**Carious.** Decayed, irregularly weathered. 'Weathered Portland Stone. The head and shoulders of a figure [from] St. Paul's Cathedral . . . removed about 1920. Solution-weathering of the limestone has given rise to "carious" (or pitted) surfaces, on which abundant small fragments of shelly fossils may be seen' (label attached to an exhibit in the Geological Museum).

**Carstone.** Hard ferruginous sandstones of the Lower Cretaceous. [? Old English, *carr*, a rock. See Arkell and Tomkeieff, *Rock Terms*, 1953.]

**Cassiterite.** Tinstone, $SnO_2$; the ore of tin. Occurs chiefly in pneumatolytic veins, as in Cornwall. (Beudant, 1832.) [G. *kassiteros*, tin.]

**Cast. 1.** In palaeontology. In ordinary usage a cast is a model made in a mould, and this seems applicable in describing states of fossilization. Taking, for example, a lamellibranch shell, a cast is then a replica, a replacement, in secondary material (mineral or rock, usually the former) of the original shell, reproducing the shell (completely or incompletely) in size, shape, and surface features. A complete mould would be the hollow space (with its boundary surface) left by the dissolution of the shell, but dissolution and replacement would normally proceed more or less gradually. On this basis 'external mould' and 'internal mould' are definable as follows: external mould—the material (matrix) surrounding the exterior of the shell, the surface in contact with the shell (or its cast) showing in negative, as impressions, the features of the external surface of the shell; internal mould—the material (rock matrix or mineral, usually the former) filling the hollow interior of the shell, the surface showing in negative, as impressions, the features of the internal surface of the shell and the whole in size and shape reproducing the form of the hollow interior. But 'internal mould' is not very appropriate for the whole solid infilling of a hollow shell, in fact 'cast' ('internal cast') comes to mind and is often used; another term is 'core': 'External cast' is sometimes used for our 'external mould', though it would be hard to justify this; 'external impression' is always appropriate. Thus there is liable to be a confusion of the terms 'cast' and 'mould'. The terminology here recommended (in spite of some awkwardness) has been adopted by, for instance, the present writer (*GM*, 1928), and is apparently becoming standard, judging from recent textbooks and monographs. An early clear statement of the matter is given by Parkinson (*Fossil Organic Remains*, 1822), incorporating his own variant of the terminology: '"Casts" or "nuclei" of organic remains are formed by different mineral substances filling their cavities, and thus taking the impressions of their internal forms and markings. "Impressions" of the external surface are formed by investment by the surrounding matrix and by its subsequent induration. After this is accomplished, and the original substance removed, a "cavity" or "mould" is left in the matrix corresponding in its figure and markings with the removed substance. Any mineral matter being introduced into this mould, acquires, as it hardens, a renewal or "redintegration" of the external form of the original substance.' See also quotation under INCRUSTATION (2). 2. In sedimentology. See SOLE MARK.

**Cataclasis.** The crushing and granulation of a rock or mineral under stress; a form of pure dynamic metamorphism. Occurs particularly in hard, brittle rocks. Hence 'cataclastic', and 'cataclasite' for the resulting rock.

**Catagenesis.** In evolution, particularly organic evolution, the acquirement of characters which are taken to be retrogressive either because they repeat, perhaps in reverse order, those acquired and superseded during earlier progressive evolution (anagenesis) or because they show exaggerated or anomalous features.

**Catastrophe.** 'We hear of sudden and violent revolutions of the globe, of the instantaneous elevation of mountain chains, of paroxysms of volcanic energy, of general catastrophies and a succession of deluges, of the alternation of periods of repose and disorder, of the refrigeration of the globe, of the sudden annihilation of whole races of animals and plants' (Lyell, *Principles*, 1833).

**Catastrophism.** See UNIFORMITARIANISM. 'The doctrine known as Catastrophism—the myth of successive destructions of the face of tbe earth by violent and supernatural cataclysms' (Holmes, *Physical Geology*, 1965/78).

**Cauldron subsidence.** The subsidence of a cylindrical or conical mass of rock into underlying magma so that displaced magma flows upwards round and over the subsiding mass. The resulting intrusion, in a diagrammatic case, thus has the shape of an inverted very thick-bottomed drinking tumbler, being a ring-dyke in the part represented by the sides of the tumbler. The classic examples are those of Ben Nevis and Glen Coe (Clough and others, *QJ*, 1909, and *MGS*, 1916). 'Diagrammatic cross-sections of cauldron subsidences' (Richey, *BRG, Tertiary Volcanic Districts*, 1961).

**Cave.** Nearly all caves fall into two classes: (*a*) those formed by underground drainage and erosion in limestone formations, these channels having been later cut into by subaerial erosion and so exposed on, for instance, valley sides; (*b*) those formed in sea cliffs by the marine erosion of rocks that are well bedded and well jointed or that have within them belts of weakness such as crush or fault belts. The term 'cave' applies to either, or to any other kind of underground hollow with an opening to the air. The term 'cavern' is sometimes used; it seems to suggest a greater degree of size and enclosure. See SPELEOLOGY.

**Cave-earth.** 'In [certain] caves, it is commonly the case that the bones are imbedded in a red loamy matrix, to which the name of "cave-earth" has been given, and which appears to consist, in a great measure, of those portions of the limestone-rock that are insoluble in water charged with carbonic acid' (Evans, *Ancient Stone Implements*, 1897).

**Cavern.** See CAVE. *The Story of Treak Cliff Cavern, Derbyshire* (Ford, 1963).

**c-axis.** See TECTONIC AXIS.

**Cedar-tree structure.** An interdigitation of igneous sheets, intrusive or extrusive, tapering away from a centre, with sedimentary or other igneous rocks. 'The interdigitation [of the central-eruptive products] with the flood-basalts is so striking that the term "cedar-tree volcano" seems appropriate, following the "cedar-tree laccolith" of Holmes (1887)' (Walker, *QJ*, 1963).

**Celtic Sea.** 'The area between Ireland and Brittany, Cornwall, and the 100 fathom line is here called the "Celtic Sea", as suggested by Cooper and Vaux (*JMBA*, 1949)' (Stridee, *QJ*, 1963).

**Cement.** In the geological, as distinct from the industrial, sense: mineral matter deposited between rock-particles so that this and the particles together form a hard mass. Hence 'cemented', 'cementation'. '[Purbeck marble] is composed of shells closely cemented together by a calcareous spar' (Maton, *Western Counties*, 1797). 'The third cause of lapidification is "cementation"' (Kirwan, *Geological Essays*, 1799). See AGGLUTINATE.

**Cementstone.** Argillaceous limestone, and particularly septarian nodules, from which cement is or could be made. Cementstone bands are well-known from the London

Clay. (The Cementstone group is a stratigraphical division at the base of the Calciferous Sandstone series of the Scottish Lower Carboniferous; the beds are argillaceous dolomites.)

**Cenomanian.** See CRETACEOUS SYSTEM. [Cenomani, a Celtic people who inhabited the present department of Sarthe, France, of which the capital is Le Mans.]

**Cenozoic.** See CAINOZOIC.

**Central eruption.** Volcanic eruption at a centre as distinct from along a fissure; the usual type of eruption at the present day. Continued eruption at one centre builds volcanoes of the 'central type'. See FISSURE ERUPTION.

**Centric.** A texture, as in spherulites and ooliths, in which there is a regular arrangement of crystalline matter about a centre. (Becke, 1878.)

**Centroclinal.** Having dips towards a centre. See QUAQUAVERSAL.

**Cephalopoda.** One of the three chief classes of the phylum Mollusca, having a chambered shell usually coiled in a symmetrical (plane) spiral but in one group, and sometimes in the others, it is straight. The form of the shell distinguishes nearly all cephalopods from nearly all gastropods, but the constant distinguishing feature is the partitioning of the shell by septa into chambers. They range from the Cambrian (Tremadocian, possibly earlier) to the Present. There are three orders, the Nautiloidea, the Ammonoidea, and the Dibranchiata (the first two sometimes being grouped together as the Tetrabranchiata). [G. *kephalē*, the head.]

**Ceratopsia.** A suborder of the order Ornithischia, one of the two dinosaurian orders of Mesozoic reptiles. Herbivorous, quadrupedal; the skull greatly enlarged, forming a large frill extending back over the shoulders and with horns on the front part. Upper Cretaceous. [genus *Ceratops* (G. *keras, kerat-*, a horn).]

**Chadian.** See DINANTIAN. [Chad, the former saint and bishop of Lichfield, from whom the village of Chatburn, near Clitheroe, Lancashire, derives its name.]

**Chalcedony.** A very hard glassy—translucent or porcellanous—mineral, cryptocrystalline silica, $SiO_2$, variously coloured. Agate is specially variegated; flint and jasper are also varieties of chalcedonic silica. [Chalcedon, a Greek city, now Kadi-keui in Asia Minor at the entrance to the Bosphorus.]

**Chalk.** White, pure (or nearly pure) calcium carbonate, breaking into crumbly or powdery pieces, particularly as occurring as natural rock (see CHALK FORMATION) or as prepared for writing or drawing. 'Al this way goeth Chilternhilles, al the soile being a chalke clay' (Leland, *Itinerary, c.* 1538). 'Chalk, from Greenhyth in Kent' (Woodward, *Catalogue*, 1729). [Old English *cealc*, f. L. *calx*, *calc-*, lime, but=L. *creta*, and not f. G. *chalkos* (copper).]

**Chalk formation.** The most obviously distinctive of all the stratigraphical formations of NW. Europe, very conspicuous in Britain, both inland along its extensive downland outcrop and in the white precipitous cliffs where this outcrop is cut by the sea. It corresponds to the Upper Cretaceous. 'The Chalk is not, of course, by any means a homogeneous deposit; but microscopic analyses of samples from all horizons show that from 40 to 90 per cent. of the bulk consists of finely divided $CaCO_3$. In addition there are often large numbers of minute spheres of problematic origin, which some hold to be ooliths and others the remains of a planktonic organism of unknown affinities; also coccoliths, or small discoid plates which once built up the calcareous sheaths of unicellular planktonic algae; foraminifera in certain bands, but seldom commonly; sponge spicules; and lastly shells of macroscopic animals, mainly echinoderms and molluscs' (Arkell, *Oxford*, 1947). 'The petrology of the Chalk' (Hancock, *PGA*, 1975).

The Chalk in Britain is divided into Lower, Middle, and Upper Chalk, which correspond (exactly or approximately according to varying usage) with the Cenomanian, Turonian, and Senonian general series. There are many distinctive fossils. Some of those characteristic of the three divisons are: Lower Chalk, the echinoids *Discoidea cylindrica*, *Holaster subglobosus*, the lamellibranch *'Pecten' orbicularis*, the ammonite *Schloenbachia varians*, and the fish *Ptychodus decurrens*; Middle Chalk, the echinoid *Micraster corbovis*, the brachiopods *'Terebratula' semiglobosa* and *'Rhynchonella' cuvieri*, and the lamellibranch *Inoceramus labiatus*; Upper Chalk, the sponge *Ventriculites infundibuliformis*, the echinoids *Conulus albogalerus*, *Echinocorys scutata*, and *Micraster coranguinum*, the lamellibranch *Spondylus spinosus*, the belemnite *Belemnitella mucronata*, and the fish (teeth) *Lamna appendiculata*. (Strachey, *PT*, 1725.)

**Chalky boulder-clay.** 'We find the whole of East Anglia covered with an immense sheet of chalky boulder-clay—the Great Chalky Boulder-Clay—which spreads out in a sort of fan around the basin of the Wash, as if deposited by an immense glacier deploying from this centre' (Wright, *Quaternary Ice Age*, 1937). (Wood and Harmer, *Crag Mollusca, MPS*, 1872, 'Great chalky clay'; Harmer, *QJ*, 1877, *PGA*, 1902, 'Chalky boulder-clay'; Baden-Powell, *GM*, 1948; Bristow and Cox. *JGS*, 1973.)

**Chalybeate.** See MINERAL SPRING. [G. *chalubos*, steel.]

**Chalybite.** The mineral, iron carbonate ($FeCO_3$).

**Characteristic fossil.** A species or genus of fossil 'characteristic' of a stratigraphical formation, series, zone, &c. is one either confined to it or, if not quite that, specially abundant in it; a 'characteristic' fauna (or flora) is an assemblage of fossils that has been found to occur only within that formation, series, zone, &c. However, a general stratigraphical division that is taken to be a time-division, such as a series or zone, is defined by the fossils it contains, so that certain particular kinds (genera or species, one or more), or particular assemblages, come to be not merely found characteristic of what appears to be a time-division but to be deemed diagnostic of such a division. *Strata identified by Organized Fossils*, 'containing prints on coloured paper of the most characteristic specimens in each stratum' (William Smith, 1816). The first published account of the application of Smith's discovery, in detail and in a particular British region, was given by Parkinson for the London Basin (*TGS*, 1811). See STRATIGRAPHY. *Figures of Characteristic British Fossils* (Bailey, 1875). 'A list of characteristic fossils' (*BA*, 1924).

**Charnockite.** An Archaean rock-type, originally described from India. 'The Charnockite Series comprises a whole range of rocks, characterized by the presence of hypersthene, varying in composition from acid to the ultrabasic. The name was first given by T. H. Holland [*MGSI*, 1900] to the acid type (hypersthene granite) in memory of Job Charnock, the founder of Calcutta [whose tombstone is made of it]. The name has also been extended to the whole series' (Krishnan, *India and Burma*, 1956). The name has been further extended to rocks in parts of the world outside the Indian region. 'Charnockites of the type area near Madras' (Subramaniam, *AJS*, 1959). 'Charnockites of Ceylon' (Corray, *QJ*, 1962; in which paper, and the discussion on it, questions of nomenclature are aired). 'Charnockite and gneiss, N. Nigeria' (Oyawoye, *GM*, 1964). 'It was particularly appropriate that a section [of the International Geological Congress, India, 1964] should be devoted to charnockites, for this rock type was first described from India at the end of the nineteenth century. The Indian charnockites were originally considered to be igneous rocks, but in recent years many charnockites have been recognized as having a metamorphic origin' (report by Howie, *PGS*, 1965).

**Charnoid.** See CALEDONIAN.

**Chatter-marks.** A series of marks, pits, scratches, or grooves made on a rock surface by the movement over it of a mass having small hard projections. In most cases the moving mass would be either (**1**) a glacier, (**2**) a rock mass slipping along the bedding in tectonic adjustment, or (**3**) the side of a fault plane. In cases (2) and (3) chatter-marks might be produced on either or both of the two rock surfaces in contact according to the side on which the projections occurred.

**Chattian.** See PALAEOGENE. [Chatti, an ancient German people who inhabited the region of Cassel, Germany, which is the type area of the stage.]

**Chelicerata.** A class (or subphylum) of the phylum Arthropoda, = Arachnida in the wide sense. (Another name for the group is Arachnoidea.) [G. *chēlē*, a claw, *keras, kerat-*, a horn.)

**Chelogenic cycle.** 'Within the crust, we find fluctuations in the rate at which igneous and metamorphic minerals were created and in the incidence of continental drift, together with periodic displacements of the mobile or orogenic regions on an unusually large scale. . . . [It is suggested that] the three phenomena may be related to a cycle of events with a periodicity of 750-1,250 m.y. which has been repeated at least four times. . . . As the long-term cycles lead ultimately to the production of shields, the name "chelogenic cycle" is proposed (Gr., literally a tortoise, and thence the carapace of shields held over besieging soldiers while sapping fortifications)' (Sutton, *N*, 1963). See also Fitch and Miller, *N*, 1965, and Sutton, *PGA*, 1967.

**Chemical composition.** In each of the main kinds of sedimentary rock one mineral is dominant; thus quartz in a sandstone, a 'clay mineral' in a mudstone, calcite (or dolomite) in a limestone. With these are subordinate accessory minerals and a variable amount of interstitial material; and there are all degrees of mixtures of the main kinds of rock. The determination of the chemical composition of a sedimentary rock is thus a matter of simple chemical analysis. In igneous rocks the process is more difficult and the result not so directly linked with the mineralogy. It is usually expressed in terms of the proportions of oxides present. Tables of these abound in the works on petrology, but it needs to be explicitly emphasized that these give proportions by weight. The molecular proportions are obtained by dividing the weight-proportions by the molecular weight of the oxides. The actual minerals present are not given by the tables (see HETEROMORPHISM).

**Chemical deposit.** A deposit formed by chemical crystallization or precipitation; not one produced by mechanical transport and settling nor from the growth of organisms or the accumulation of organic remains. 'Deposits which are regularly crystallized, or which have a tendency to crystallization, and in which the action of mechanical causes cannot be traced' (Playfair, *Illustrations*, 1802). They include some limestones and ironstones, and all evaporites.

**Chemical fossil.** A sign of life made evident by the occurrence in the rocks of a chemical substance which (supposedly) could have been produced only through the agency of some kind of living organism.

**Chert.** Layers or irregular concretions of chalcedonic silica occurring, usually, in limestone formations. 'Chert and other productions of the limestone' (Whitehurst, *Strata in Derbyshire*, 1778). 'The nature and origin of chert in the Upper Greensand of Wessex' (Tresise, *PGA*, 1961). [Old English, *ceart*.]

**Chiastolite.** See ANDALUSITE. (Karsten, 1800). [G. *chiastos*, arranged crosswise.]

**Chilled contact.** The part of an igneous intrusion that cooled more rapidly than the rest owing to its having been chilled near its contact with the country rock. This more rapid cooling is made evident by the finer grain of the igneous rock in that part, and the nearer the contact the finer. Chilled contacts are usually most conspicuous in minor intrusions; such a rock may be glassy at its extreme limit. The country rock itself would usually show effects of thermal metamorphism, perhaps with a 'baked contact'. Chilled contacts are of course seen in the field at outcrop or in section, hence the usual terms 'chilled margin', 'chilled edge'.

**China-clay.** See KAOLIN.

**China-clay rock.** Granite in its most completely kaolinized form in which all the felspar has been converted to kaolin while the rest of the rock consists chiefly of quartz grains and a little white mica. The rock is so soft as to be easily crumbled in the fingers. (Howe, *MGS, Kaolin* [&c.], 1914.)

**China-stone. 1.** Partially kaolinized granite which is also free from the dark minerals, tourmaline and biotite, but which may have undergone some greisenization. It is a hard rock, but requires only grinding to fit it for manufacturing purposes. **2.** Several kinds of hard, whitish, very fine grained rock; (*a*) 'chinastone-limestone', see CALCITE-MUDSTONE; (*b*) a hard, compact, silicified rock occuring among Carboniferous strata (Farey, *Derbyshire*, 1811; Wedd, *MGS*, *Oswestry*, 1929); (*c*) rocks in the Ordovician of the Welsh Borderland, either (i) the result of contact alteration (Murchison, *Silurian System*, 1839), or (ii) an indurated volcanic ash, 'china-stone ash' (Watts, *PGA*, 1925).

**Chitin.** A horny organic substance, a nitrogenous carbohydrate, forming the skeletal material of many invertebrates, either alone (as in the graptolites and inarticulate brachiopods) or strengthened with calcareous material (as in many arthropods). It is relatively stable; even in the most ancient fossiliferous rocks it is sometimes found in an excellent state of preservation. [G. *chitōn*, a coat.]

**Chlorite.** A greenish mineral, silicate of iron, magnesium, and aluminium, with hydroxyl (OH). In igneous rocks it occurs as a secondary mineral, on the alteration of biotite, hornblende, &c. In sedimentary rocks it occurs in clays. It is especially common in metamorphic rocks. Hence 'chloritization', alteration to chlorite; with, in metamorphism, the production of 'chlorite-schist' ('green schist'). (Werner, 1789, *chlorit*). [G. *chlōros*, yellowish green.]

**Chondrichthyes.** Sharks and their allies, a class of the superclass Pisces. The main skeleton is cartilaginous and thus not fossilizable, but fin-spines and, particularly, teeth are quite common fossils. The class extends from the Devonian (not Downtonian) to the Present. [G. *chondros*, cartilage, *ichthus*, a fish.]

**Chondrite.** See CHONDRULE.

**Chondrule.** 'Chondrules are solid spheres or spheroids of silicate and metallic minerals (usually mixed) ranging in diameter from 0.5 to 3 mm. No terrestrial geological process produces structures which resemble chondrules more than superficially. Meteorites containing chondrules are called chondrites' (Wood, *N*, 1962). [G. *chondros*, a grain.]

**Chonolith.** (Chonolite.) 'May be thus defined: a discordant igneous body (*a*) injected into dislocated rock of any kind, stratified or not; (*b*) of shape and relations irregular; and (*c*) composed of magma passively squeezed into a subterranean orogenic chamber or actively forcing apart the country rocks' (Daly, *Igneous Rocks*, 1933). The filling of an actual or potential cavity is the essential conception, as Daly refers to chonoliths being formed by cauldron subsidence; and the etymology indicates this. Harker (1909), writing 'chonolite', notes: 'We adopt the English spelling in preference to the German'. See BYSMALITH, LACCOLITE. [G. *chōnē*, a mould.]

**Chordata.** The phylum of animals having throughout life or, at least, in an early stage of development, an internal supporting structure in the form either of a continuous flexible rod, the notochord, or of a series of jointed pieces, the vertebrae, running dorsally along the midline of the body. Those with vertebrae are the Vertebrata, by far the most important section. The classification and nomenclature may be set out thus:

| | | |
|---|---|---|
| Chordata | Protochordata | Stomochorda (= Hemichorda = Hemichordata) |
| | | Urochorda ( = Tunicata) |
| | | Cephalochorda (*Amphioxus*) |
| | Vertebrata | |

The Stomochorda, in which the graptolites are now tentatively placed, are themselves somewhat *incertae sedis* in the animal kingdom. [G. *chordē*, a gut.]

**Chromatography.** 'A convenient descriptive term for a number of physical methods used in chemistry and biology [and geology] for the separation and identification of mixtures of chemical compounds. Although the term unfortunately implies that colour (*chrōma*) is a necessary part of the detection procedure, no such limitation exists in fact, and the methods are perfectly general and applicable to coloured and uncoloured compounds alike' (*Encyclopaedia Britannica*, 1962). In the earlier decades of the present century the method was used particularly in the separation of plant pigments.

**Chron.** See CHRONOSTRATIGRAPHY.

**Chronology.** The science of time permeates the science of geology; geological chronology is 'Geochronology'. Superposition (locally) and fossil content (more widely) are the traditional criteria for establishing a time-order and time-correlation. 'The characters of fossils are not to be counterfeited by all the craft in the world, nor can they be doubted to be what they appear; and tho' it must be granted, that it is very difficult to read them, and to raise a Chronology out of them, and to state the intervalls of the times wherein such or such catastrophes and mutations have happened; yet 'tis not impossible but that much may be done' (Hooke, *Discourse of Earthquakes*, 1705). However, another method is now appearing on the horizon: 'We may anticipate that in the future determinations of the absolute ages of sediments will reduce the large uncertainties which to-day underlie their time correlation, and that eventually evidence derived from atomic disintegration will supersede that afforded by palaeontology' (Hawkes, *QJ*, 1957). See RADIOACTIVITY AGE-METHOD. 'Jurassic chronology: Lias' (Buckman, *QJ*, 1917) 'The denudation chronology of the dipslope of the South Downs' (Sparks, *PGA*, 1949). 'The glacial chronology of part of the middle Trent basin' (Clayton, *PGA*, 1953). 'Chronology of tectonic movements in the Appalachian region' (Rodgers, *AJS*, 1967). There is the distinction between a succession of events in their time-order (relative age) and the additional placing of these events on an absolute time-scale of years (absolute age).

**Chronomere.** See TIME-UNIT. [G. *meros*, a part.]

**Chronostratigraphy.** 'There are many branches of stratigraphy, depending on the particular features of rock strata under consideration. One of the most important is "chronostratigraphy" which deals with the age determination and age classification of strata. Its basic purpose is to interpret the history of the earth through the chronologic sequence of its rock strata' (Hedberg, *BGSA*, 1961). The following is from the *Guide to Stratigraphical Procedure* (*GSSR* (11), 1978): 'Chronostratigraphy (time-rock stratigraphy) is the definition of an internationally agreed standard chronostratigraphical hierarchy including the system, series, and stage, and the age-correlation of rocks with this standard scale, as accurately as possible and by means of all available methods. The standard chronostratigraphical scale must eventually be defined entirely by marker points in stratotype sections for the basal boundaries of all divisions'.

The following shows the hierarchy of the chronostratigraphical divisions and the corresponding divisions of geological time. Brackets denote divisional terms recognized for logical completeness but seldom employed in stratigraphical practice:

| chronostratigraphical divisions | corresponding divisions of geological time |
|---|---|
| (eonothem) | eon |
| (erathem) | era |
| system | period |
| series | epoch |
| stage | age |
| chronozone | (chron) |

**Chronotaxis.** The placing and correlation of rocks on the time-scale. 'Strata characterized by similar fossils are known to occupy corresponding positions in different stratigraphic sections. This condition is termed "homotaxis". The determination of actual and exact age equivalence, or "chronotaxis", is generally an unattainable ideal' (Weller, *Stratigraphic Principles and Practice*, 1960). [G. *taxis*, an arranging.]

**Chronozone.** The lowest (smallest) of the chronostratigraphical divisions. See CHRONOSTRATIGRAPHY and BOUNDARY STRATOTYPE.

**Circumdenudation.** Erosion (denudation) that, in dissecting a land mass, has left a part of the ground upstanding (a residual 'hill of circumdenudation') by having worked round it apparently fortuitously, such a hill not being obviously due to the outcrop of a resistant rock, though it may very likely be capped by a hard stratum. Term used by Jukes (*Manual*, 1862). Examples are seen in the outliers in front of the Jurassic escarpment and in Ingleborough and other similar hills in the Pennines. Prominent flat-topped hills of circumdenudation are seen in the 'buttes' of the western United States; the larger, flatter, remmants of the land-mass undergoing erosion being the 'mesas'.

**Cirque.** A feature of mountain sides and valley heads. 'Cirques—J. de. Charpentier was apparently the first to recognize the type and introduce the word (*Pyrenees*, 1823)—are hemispherical bowls. The encircling cliffs often sweep up to narrow knife-edges or sharply serrated crests. These cliff-bound stadia ("armchair" shaped hollows) are semi-circular or horseshoe shaped in plan. In the ideal form, a lake bounded on its outer side by an arcuate moraine, nestles in the bottom' (Charlesworth, *Quaternary Era*, 1957). The glacial origin of cirques was first propounded by Ramsay in 1860 (*Old Glaciers*); they are intimately related to regions of present or past glaciation. They are included, in various languages, under names which refer to high valleys in general; of these 'corrie' (or 'coire'), which includes the Scottish examples, is well established in Britain; 'cwm' includes the Welsh examples. The French '*cirque*' is the most definite and descriptive name for the feature and has been widely adopted. 'Similar but not homologous forms are to be seen in unglaciated countries. These "pseudo-cirques" have been described, for instance, from arid areas, such as the Arabian and African deserts, territories in which there is no downwash and where wind removes the debris resulting from insolation' (Charlesworth).

**Cirripedia.** Barnacles; a marine subclass of the class Crustacea, phylum Arthropoda. They range (apart from some doubtful fossils) from the Jurassic. [L. *cirrus*, a curl.]

**Clan.** 'If the 700 named varieties of eruptive rocks were in no sense manifestly related, the problem of origins would be, in very truth, a difficult one. However, field and laboratory observations without number prove the possibility of grouping into relatively few chemical series. For convenience these may be called "clans". Each clan is composed of families, distinguished less by chemical composition than by mode of field occurrence, by mineralogical composition, or by rock texture. The genetic problem, being most concerned with the explanation of chemical diversity, is simplified by recognition of the fact that the clans represent so many chemical groups, each of which contains syngenetic families. . . . Biotite granite is a species; all granites constitute a family; granites, granite porphyries, rhyolites etc. form a clan' (Daly, *Igneous Rocks*, 1933). 'The name applied to the family in the coarse-grained group becomes the name of the clan: thus amongst the chief ones are the Granite, Granodiorite, Syenite, Diorite, and Gabbro clans' (Hatch and Wells, *Igneous Rocks*, 1952).

**Clarain.** 'The term clarain was introduced by M. C. Stopes in 1919 [*PRS*] to designate the macroscopically recognisable bright lustrous constituent of coal, which, in contrast to vitrain, is intrinsically striated by dull intercalations. Nowadays the term is used to describe all finely striated bands of coal which have an appearance intermediate between vitrain and durain' (*Nomenclature of Coal Petrology*, 1957). [L. *claro*, make bright, -ain (to match 'fusain').]

**Class.** In the widest sense, from which 'classify' is derived, a 'class' is any rank or taxonomic category; but in the systematic classification of animals, and less rigorously of plants, recent or fossil, a 'class' is a division of a phylum, itself comprising one or more orders. In either sense, particular classes are named; e.g. the 'sedimentary class' of rocks, the 'class Cephalopoda'.

**Classification.** The general meaning of 'classification' is clear, but with natural objects the varying methods and purposes are to be borne in mind. The several kinds of geological material are, each of them, classifiable in various ways according to the basis selected. Thus minerals, according to chemical composition or crystal form. Igneous rock-types may be classified strictly according to their mineral composition and texture; that basis being chosen, the question as to how these types may be associated, or how they may be expressions of conditions of formation (intrusion or extrusion for instance), is irrelevant (in constructing the classification). In the classification of fossils it is important to distinguish between those groupings that are based entirely on degrees of likeness, the observable facts of composition, structure, and morphology (the morphological classification), and those that are meant to express the underlying, but not directly observable, biological relationship, the genealogical tree (the genetic classification). The first will be an approximation, perhaps a very close one, to the second, which may be deemed the more philosophical one, the one chiefly to be sought; but it would greatly hamper the freedom of truth-seeking with regard to the latter, if the former, with its nomenclature, were to be reshuffled every time a new, necessarily hypothetical, relationship were proposed as being the most likely one.

**Clastic.** Broken; for rock-fragments or rock-particles broken from pre-existing rocks and, normally, carried some distance away; also for the sediments and rocks composed of such material. Gravel and sand, and the corresponding rocks, are clastic (epiclastic), as are pyroclastic deposits. The term would, in practice, hardly include autoclastic rocks (in the sense recommended in this Dictionary). Mud is not, strictly, a clastic deposit, as it is chiefly formed of new minerals resulting from weathering (and some muds may even be autochthonous); but it may contain truly clastic constituents. Other rocks, such as limestones, may also contain clastic constituents; and a limestone made up of fragments of shells, &c., is a special type of clastic rock ('bioclastic', sense 2). The individual fragments and particles are sometimes called 'clasts'. [G. *klastos*, broken.] See FRAGMENTAL DEPOSITS.

**Clastic dyke.** See SEDIMENTARY INSERTION.

**Clay.** A weathered accumulation, a superficial deposit, a sediment, or an indurated rock characterized by an extremely fine grain and a certain mineral composition. Classifiers of sedimentary rock-types adopt various sizes of particles below which they take the particles to be of the 'clay grade', e.g. $\frac{1}{100}$, $\frac{1}{200}$, $\frac{1}{256}$ mm. The essential constituents are the 'clay minerals', hydrous silicates of alumina with, in some of them, iron or magnesium. These are minute flaky crystals produced (directly or at some remove) by chemical weathering of felspars and other destructible minerals of the igneous rocks. They are often so small that they can hardly be separated or optically examined, even under the highest powers of the microscope. With these a fine 'rock flour' is usually mixed; this is composed of finely pulverized rock-forming minerals, chiefly quartz. A third constituent is (solid) organic material; this naturally enters largely into the composition of clay-soils, but it is also usually deposited, in greater or lesser amount, with clay sediments. The fourth constituent is organic or mineral matter in the colloidal state. 'True' clay can have other substances mixed with it (e.g. calcium carbonate) and still be called 'clay', in a loose sense, if it has the typical fine grain; see quotation (Leland) under CHALK.

Clay is plastic when moist and forms mud when mixed with water. A sediment of clay thus begins as a mud and an indurated mud is called a mudstone rather than a 'clay' or a 'claystone', but the term 'clay' is incorporated in the names of such well-known stratigraphical formations as the Oxford Clay, Weald Clay, London Clay. [Old English *clæg*.]

**Clay mineralogy.** The study of the mineral content of clays. *Clay Mineralogy* (Grim, 1953). *The Clay Mineralogy of British Sediments* (Perrin, 1971). 'On the clay mineralogy of upper Eocene and Oligocene sediments in the Hampshire Basin' (Gilkes, *PGA*, 1978).

**Clay-ironstone.** Argillaceous ironstone, the iron compound being usually the carbonate, occurring as beds but often concretionary. Formed either by direct deposition or by replacement (it is often doubtful which). Associated particularly with the Coal Measures. See CARBONIFEROUS for an early use.

**Clay-slate.** An old term for slate (see SLATE). Now sometimes used for argillaceous slate to distinguish this from slate developed in volcanic ash.

**Clay-with-flints.** 'The clay-with-flint of W. Whitaker [*MGS*, 1861] is a comprehensive but ill-defined category of miscellaneous materials in which the proportions of clay, sand and flint vary greatly. "Foreigners" as a rule are lacking and the bedding is obscure. Bounded below by an uneven surface, it changes rapidly in thickness, frequently filling basin-like hollows and funnel-shaped "pipes" of considerable depth. It covers much of the interstream tracts in south England from Sussex to Hertford and Devon, and spreads over the Downs of north Kent and the Isle of Wight, its boundaries being indefinite on both Chalk and Tertiary outcrops. In age and origin, the deposit is probably composite, more than one process having acted on many kinds of strata. It represents the insoluble residue of the Chalk, reinforced with Tertiary waste which added quartz pebbles and rounded flints. Floods from melting snows and solifluxion distributed and redistributed it during the Glacial period' (Charlesworth, *Quaternary Era*, 1957). 'Clay-with-flints and associated pebbly clay and sand. These deposits are confined to the Chalk areas where they cap the high ground' (Sherlock, *BRG*, *London and Thames Valley*, 1962).

**Cleat.** 'Coal is noted for the regularity of its jointing. Usually there are two directions approximately at right angles to each other and to the bedding, one of which is better developed than the other, exhibits lustrous surfaces, and is specifically termed the "cleat" or "bord", the other being the "end" or "head". Because of the ease of hewing coal across the cleat, its direction influences the layout and working of coal-mines' (Hills, *Structural Geology*, 1963).

**Cleavage.** In geology this applies to certain cases of regular splitting, other than original bedding (but see BEDDING CLEAVAGE). **1.** Rock cleavage. (*a*) Slaty cleavage. A tectonic structure and, at the same time, a form of metamorphism, relatively slight and purely dynamic. It is developed particularly in fine-grained rocks, as a result of great lateral pressure, the planes of splitting usually lying nearly parallel to the axial planes of folds and independently of the bedding. The property appears to be due to (i) minute flaky minerals re-orientating themselves, with, probably, some shearing, (ii) the growth of new flaky minerals similarly orientated and (iii) the flattening of mineral particles by pressure. The rock is also hardened. Traces of the original bedding may often be seen on the cleavage surfaces and edges. This kind of cleavage grades into schistosity. For further history and discussion see Dennis, *Tectonic Dictionary*, 1967, and Siddans, *ESR*, 1972. See SLATE. (*b*) Spaced Cleavage. The splitting of a rock along closely spaced, but individually distinct, parallel planes. This kind of cleavage has been given various names, some indicating particular features, others with genetic implications. 'Fracture cleavage' and 'false cleavage' (a quarryman's term) are synonymous with the general term 'spaced cleavage'. 'Strain-slip cleavage' ('slip cleavage') implies some degree of shearing stress as the cause. 'Crenulation cleavage' exhibits wrinkles between the cleavage planes. 'Close-joints cleavage' visualizes the phenomenon as jointing on a small scale.

Macculloch (*Western Islands of Scotland*, 1819) gives a figure of stratification and cleavage, with a partial explanation. 'Beds of rock are occasionally subject to split into smaller laminae not parallel to the plane of stratification; thus: [giving a diagram]. This structure is called the cleavage of the bed' (Conybeare and Phillips, *England and Wales*, 1822). 'Most of these rocks [Borrowdale Volcanic Series] are of a pale bluish-grey colour. The fine pale-blue roofing slate occurs in beds (called by the workmen veins). The most natural position of the lamina or cleavage of the slate appears to be vertical; but it is to be found in various degrees of inclination, both with respect to the horizon, and the planes of stratification. The direction of slaty cleavage bears most commonly towards the north-east and south-west; while the dip or inclination varies' (Otley, *English Lakes*, 1823). 'I return to the description of the most general facts exhibited in slaty cleavage' (Sedgwick, *TGS*, 1835). 'On slaty cleavage' (Sharpe, *QJ*, 1847). 'Development of slaty cleavage' (Sorby, *QJ*, 1880).'On slaty cleavage and allied rock structures' (Harker, *BA*, 1885). 'Rock cleavage' (Leith, *BUSGS*, 1905). 'The relation of slaty cleavage to tectonics' (Wilson, *PGA*, 1946). 'Rocks that possess "cleavage" may be split into thin sheets along parallel or sub-parallel "cleavage planes" that are of secondary origin and are formed by the combined effects of metamorphism and deformation' (Hills, *Structural Geology*, 1963). It has lately come to be realized that some rock cleavage (oblique to the bedding) may be of sedimentary origin: 'Cleavage and many other deformation structures seen in the Lower Palaeozoic rocks of Wales are attributed to contemporaneous creep and decollement movement in soft, unlithified sediment on a palaeoslope' (Davies and Cave, *SG*, 1976). **2.** Mineral cleavage. The property of mineral crystals, due to their atomic structure, whereby they readily break parallel to certain planes.

**Cleavage-fan.** A fan-shaped disposition of the cleavage (the divergence may be upwards or downwards). The term is used chiefly when this is developed on a regional scale, and the disposition may then be related to an anticlinorium or synclinorium, but 'the "cleavage-syncline" or "cleavage-fan" [in the Snowdon district] appears to be wholly unrelated to the major folding' (Williams, *QJ*, 1927). 'J. Phillips (*BA*, 1857) notes that fan-like directions of cleavage planes were observed by Darwin, Studer, Rogers, and others' (Dennis, *Tectonic Dictionary*, 1967).

**Clinkstone.** See PHONOLITE.

**Clinometer.** An instrument for measuring inclinations from the horizontal, used in geology for measuring dips.

**Clint.** In a general sense, any hard, flinty rock; but applied particularly where such rock forms bare surfaces, especially to those of the Carboniferous 'limestone pavements', which are, in Britain, most extensive in the NW. Pennines. These surfaces are irregularly corrugated and furrowed by chemical weathering and are often fissured with the deeper and more regular 'grikes' which are chiefly determined by joints. '"Clints" and "grikes" produced by chemical weathering on a shelf of the Great Scar Limestone, above Malham Cove, Yorkshire' (pl. in Holmes, *Physical Geology*, 1965/78). [Danish and Swedish *klint*, rock.]

**Closed system.** See ISOCHEMICAL.

**Close.** Used in several senses in connexion with folding. (A.) As a verb. (1) The 'closing' of a fold is the convergence of the limbs towards the hinge (see quotation, Bailey and McCallien, under ANTIFORM); (2) a fold is sometimes said to be 'closed' when the bending of the strata has proceeded as far as it can without disruption; (3) 'A structure contour that passes all the way ("closes") around an anticline or dome [or a syncline or basin] . . . that is, the structure is "closed"' (White, *BUSGS*, 1922, quoted in Dennis, *Tectonic Dictionary*, 1967). (B.) As an adjective. The terms 'gentle', 'open', 'close', 'tight', and 'isoclinal' may be applied to folds according to the inter-limb angle. (Arbitrary limiting angles have been suggested by Fleuty, *PGA*, 1964.)

**Clunch.** An old local name for various stiff clays; particularly those occurring in the Coal Measures, Jurassic, and Cretaceous (especially the Chalk Marl). (See Arkell and Tomkeieff, *Rock Terms*, 1953.) [? Dutch *klont*, a lump, a clod.]

**Coal.** 'Coal is a combustible rock which had its origin in the accumulation and partial decomposition of vegetation' (Stutzer and Noé, *Coal*, 1940). Coals differ in the kinds, and manners of preservation, of plant materials (type). They vary in the degree of carbon concentration (rank), and in the amount of inorganic impurities (ash) they contain (grade). The chief ranks of ordinary coal are anthracite and bituminous coal; the various types of coal (clarain, durain, fusain, vitrain) are most clearly distinguishable in bituminous coal. (For terminology: Tomkeieff, *Coals and Bitumens* [&c.]: *Nomenclature and Classification*, 1954; Francis, *Coal*, 1954/61; *Nomenclature of Coal Petrology*, 1957; Stack and others, *Textbook of Coal Petrology*, 1975.) 'The word coal has been spelt in a great variety of ways: cole, coale, cool, cooles, coly—and long before it was applied [exclusively] to the black fuel dug from the earth it meant, in turn, wood and charcoal . . . until well into the 17th century' (North, *ONM*, 1938). In a poem, on Wales, written in Latin in the 14th century and translated in the 15th century (quoted by Bromehead, *PGA*, 1945) there is the following: '*Carbo sub terra cortice, Crescit viror in vertice*—Col groweth under lond, And gras above at the hond'. And George Owen in his *Description of Pembrokeshire, MS*, 1603 (*CRS*, 1892) refers to 'a vayne if not severalle vaynes of coales that followeth those of the lymestone . . . whether all the lande between these two vaynes [of limestone] should be stored with coales, I leave it to the judgement of the skillfull mynors'. 'A short history of coal' (section of *Hydrostaticks*, Sinclair, 1672). *Coal and Coal-bearing Strata* (Murchison and Westoll, eds., 1969). Whitehurst (*Strata in Derbyshire*, 1778) was among those who recognized the true nature of coal: 'Now since it appears that all strata accompanying coal universally abound with vegetable forms, it seems to indicate that all coals were originally derived from the vegetables thus enveloped in the stone or clay'. [Old English *col*.]

**Coal measures.** Strata with coal-seams, or the coal-seams themselves; sedimentary rocks of 'coal measures' facies. See COAL MEASURES and MEASURES. 'Cole veins or measures in these parts [S. Staffordshire]' (Dudley, *Metallum Martis*, 1665).

**Coal Measures.** The uppermost part of the Carboniferous system, being typically a succession of sandstones and shales, with coal-seams. Though 'coal measures' with productive coal-seams occur in the Lower Carboniferous of Scotland, the term 'Coal Measures' (with capital letters) does not include them, as it refers to a definite stratigraphical time division (chronostratigraphic unit), and whether or not there are coal-seams everywhere and at all parts of the succession. It is the equivalent of the combined Westphalian and Stephanian series of the Silesian subsystem (see CARBONIFEROUS SYSTEM). Fossils are chiefly plants and non-marine lamellibranchs but there are bands with marine fossils in the lower part. The base of the series, where it conformably overlies the Millstone Grit, is defined by the base of a marine band with the goniatite *Gastrioceras subcrenatum*. (Strachey, *PT*, 1725, 'Coal Clives', 'Coal'; Whitehurst, *Strata in Derbyshire*, 1778, 'Stone, Clay and Coal'; Michell, *MS*, 1788, 'Coal strata'; Jameson, *Geognosy*, 1808, 'Coal formation'; Farey, *Derbyshire*, 1811, 'Coal Measures'.)

**Coal-ball.** A concretion in a coal-seam or in the roof above. The former are the typical coal-balls: 'These occur as roughly spherical stones (either singly or in masses) embedded in the coal itself, and are therefore distinguished as "seam" nodules. They vary in size from that of a pea to two or three feet in diameter, but most frequently being of about the size of a potato. They consist of varying proportions of the carbonates of lime and magnesia, with certain impurities. Sections of these nodules show them to be composed of a medley of plant fragments [petrifactions] indicating that the debris accumulated *in situ*, in a swamp. "Roof" nodules occur in the shaly roof of the same seams and generally contain marine shells (goniatites) and an occasional plant fragment' (Crookall, *Coal Measure Plants*, 1929). 'On the present distribution and origin of the calcareous concretions in coal seams known as "coalballs"' (Stopes and Watson, *PT*, 1908). 'An example of the origin of coal-balls' (Evans and Amos, *PGA*, 1961). See quotation (Scott) under PETRIFACTION. Coalballs are also called 'bullions'.

**Coal-basin.** See COALFIELD. 'It is an observation as common as true, and which may justly challenge our admiration, that the mines in all parts of the world (I mean coal and kennel-mines) are always found in strata, shelving towards the center; or as the miners call it, dipping' (Leigh, *Lancashire* [&c.], 1700). 'The great coal basin of South Wales' (Buckland and Conybeare, *TGS*, 1824).

**Coalfield.** An area of outcrop of strata containing coal-seams, or an area beneath which coal seams occur concealed by overlying strata ('concealed coalfield') with the implication (usually) that the coal is workable. The chief coalfields (practically all those with true coal, not lignite) are of Carboniferous, mostly Upper Carboniferous, age (but the coalfields of Gondwanaland may be Permian). Within any conformable series, older rocks tend to outcrop in anticlinal, newer in synclinal, regions; as the Upper Carboniferous are the uppermost of the strata deposited before the Hercynian folding, they are almost everywhere preserved in synclinal regions; coalfields are thus almost everywhere structural coal-basins. Exposed coalfields are usually bordered either by the conformably underlying rocks, normally the Millstone Grit series, or by an edge of unconformably overlying Permo-Triassic (an edge which may also be faulted). 'There are fields, which are so large, that 'tis impossible to work the Coal so far to the dipp, it falling deep . . . so as to overtake its center, where it takes a contrary course, and yet the contrary cropp hath been wrought in several places, which is evident to be a part of the same body. The greatest field I know wherein this is conspicuous, is in Mid-lothian where their great-seam, or Main-coal, may be traced in the order following [showing that there was an elliptical coal-basin]' (Sinclair, *Short History of Coal*, 1672). 'The great Derbyshire and Yorkshire Coal-field' (Farey, *Derbyshire*, 1811). *The Coal-fields of Great Britain* (Hull, 1860/96). 'The South Wales coalfield is an exposed coalfield. Its Coal Measures, bent by earth-movement into basin form, present to observation the details of their upper surface cut across by erosion. Certain other British coalfields, such as those of Yorkshire and Durham, are semi-exposed. The Kent coalfield on the other hand is completely concealed. Here Coal Measures exist with an eroded upper surface deeply hidden under accumulations of later sediments' (Bailey, *JRSA*, 1944). *The Coalfields of Great Britain* (Trueman, ed., 1954).

**Coalification.** The gradual increase in carbon content of plant material as it passes through the 'coalification series' (or 'coal series'), peat, lignite, bituminous coal, anthracite. 'Carbonification' is synonymous.

**Coal-seam.** A stratum of coal. See COALFIELD. *The Nature and Origin of Coal Seams* (Raistrick and Marshall, 1948).

**Coast.** This term, and 'coastline', are variously defined, though their general meanings are clear; see Stamp, ed., *Geographical Terms*, 1961. This region, strip, edge, or line is obviously of the greatest importance to the geologist, for three separate reasons: it often provides the best natural exposed sections across the rocks; it shows, in action and recent effects, the geological processes of marine, and marine-induced, erosion; and it is a recorder of the more recent changes in sea-level. *The Sea Coast* (Wheeler, 1903). *Coast Scenery of N. Devon* (Arber, 1911). *Coastline of England and Wales* (Steers, 1946/64). *Coastal Development* (Davies, 1972). *Coastline of Scotland* (Steers, 1973).

**Coastal slope.** See BEVEL. (Challinor, *GM*, 1931.)

**Cobble.** A rounded stone (usually water-worn), especially one suitable for paving; thus between the size of a pebble and a boulder.

**Coble creep.** See CREEP. [R. L. Coble, *JAP*, 1963.]

**Coccolith.** 'Complexly constructed, button-shaped skeletal elements produced by chrysomonad flagellates called coccolithophores. In the living organism, the coccoliths are arranged in the form of a hollow sphere, called a coccosphere, just within the cell membrane. [They are] the most common calcareous nannofossils' (Hay in Kummel and Raup, eds., *Handbook of Paleontological Techniques*, 1965). See quotation under CHALK FORMATION. The range is from the Lias to the present day. 'Coccoliths from the English Chalk' (Black and Barnes, *GM*, 1959). 'The stratigraphical distribution of coccoliths' (Black, *PGS*, 1962). 'Coccoliths' (Black, *E*, 1965). 'Cretaceous coccoliths and associated nannofossils from France and the Netherlands' (Stover, *MP*, 1966). 'Taxonomic problems in the study of coccoliths' (Black, *P*, 1968). (Huxley, *Lay Sermons*, 1868). (Rhabdoliths are similar but spiny.) [G. *kokkos*, a grain.]

**Coelenterata.** (Coelentera.) A phylum of animals with tissues, not constituted into definite organs, arranged on a radially symmetrical plan. Includes as its main classes the Scyphozoa (jelly-fish), Hydrozoa, and Anthozoa (corals and sea-anemones). The corals are overwhelmingly the most important representatives of the phylum in the fossil record. [G. *koilos*, hollow, *enteron*, the intestine.]

**Cogenetic.** Of common origin. Applied particularly to the relationship of certain plutonic and volcanic rocks in space and time: 'the granitoid rocks of the central Andes have intruded the cogenetic volcanics' (Pitcher, *JGS*, 1978).

**Cognate.** Akin, having affinity, of common origin, allied in nature. Used in rather special petrological senses; thus for the relation between xenocrysts or xenoliths and the igneous rocks in which they occur when these are ultimately from the same parent magma (see XENOLITH), and between a mass of pyroclastic material and a lava flow. Hardly used at all in palaeontology. [L. *cognatus*, akin.]

**Collision zone.** Contacts between lithospheric plates in relative motion, involving such processes as subduction or transform faulting, represent 'collisions' between the two plates. Orogenic belts are produced by collision between two plates, whether the upper part of the plate margins consists of oceanic crust and continental crust, as along the western side of South America, or both margins of continental crust, as in the Alpine-Himalayan belt.

**Colluvial.** 'Colluvial. Under this head it is proposed to include those heterogeneous aggregates of rock detritus commonly designated as talus and cliff debris. The material of avalanches may also be classed here. Such result wholly from the transporting action of gravity. The deposits in themselves are comparatively limited in extent, ever varying in composition, and are composed of an indiscriminate admixture of particles of all sizes, from those as fine as dust to blocks it may be of hundreds of tons' weight. Such are necessarily limited to the immediate vicinity of the cliffs or mountains from which they are derived' (Merrill, *Rocks, Rock-weathering and Soils*, 1897). The noun is 'colluvium'. [L. *colluvies*, a mass of sweepings.]

**Colour index.** An index of the darkness of an igneous rock, the proportion of coloured minerals in it. This is taken as the volume per cent. of everything other than quartz, felspar, and felspathoids.

**Columnar jointing.** See JOINT. Lhwyd gave an early account of this structure among British igneous rocks in a letter to Ray in 1691: 'One naked precipice [on the Glyder Mountain, N. Wales] is adorned with numerous equidistant pillars, and these again slightly cross'd at certain joynts. We must allow a natural regularity in the frame of the rock, which the storms, by weathering render more conspicuous'. The famous columnar jointing in the Tertiary lavas of the Giant's Causeway, N. Ireland, was brought into notice a few years later (see Tomkeieff, *BV*, 1940). The equally famous occurrence in the Isle of Staffa ('Island of Staves'), W. Scotland, was fully described by Sir Joseph Banks (in Pennant's *Tour in Scotland*, 1774). 'In a columnar prismatic regular form the basalts appear in large masses or strata of close-lying columns in the castle-hill at Felsberg and Aldenburg, in the Maderstine near Gudensberg, in the Widelsberg near Wolfhagen, in the Holestine under the Dornberg, and, as I have been told, but never observed myself, in the Dornberg. . . . The prismatical columnar lavas or basaltes show varieties of forms. The Irish basaltes in the Giants causeway in the county of Antrim, appear in polygonal articulate columns or prisms' (Raspe, *An Account of some German Volcanos*, 1776, quoted by Mather and Mason, *Source Book*, 1939).

**Comagmatic.** From the same parent magma; applied (hypothetically) to a set of igneous rocks. The term is used both for a very general source of origin, such as one from which the rocks of a large petrographical province are supposed to have been derived, and also in referring to a particular restricted body of magma from which various rock types may have arisen through local differentiation.

**Comley Sandstone.** The Cambrian sandstones of S. Shropshire, poorly exposed but containing many fossils of Lower Cambrian (Lower Comley Sandstone) and Middle Cambrian (Middle Comley Sandstone) age. The fossils of the former are specially noteworthy, as elsewhere in Britain the Lower Cambrian series is very sparsely fossiliferous; they are here crowded into the top 1·5 m. of a total thickness of some 500 m., some of the chief species being the trilobites *Callavia callavei*, *Strenuella salopiensis*, *Eodiscus bellimarginatus*, the brachiopod *Obolella atlantica*, and *Hyolithellus micans* (*incertae sedis*). The Comley Sandstone passes downwards into a quartzite which rests unconformably on the Precambrian. (Lapworth, *GM*, 1888/91; Cobbold, *BA* and *QJ*, 1908/20.)

**Comley series.** See CAMBRIAN SYSTEM. [Comley, near Church Stretton, Salop.]

**Comminution.** The reduction in the general size of rock particles by mutual attrition.

**Compactability.** The degree to which a sediment can undergo compaction.

**Compaction.** Consolidation with reduction in volume, particularly the 'gravitational compaction' of sediments by the weight of an increasing load (over burden), the grains becoming more tightly and firmly packed. 'Gravitational compaction of clays and shales' (Hedberg, *AJS*, 1936). 'The consolidation of clays by gravitational compaction' (Skempton, *QJ*, 1969). *Compaction of Argillaceous Sediments* (Rieke and Chilingarian, eds., 1974).

**Competent bed.** A bed that has been strong enough, competent, to transmit pressure. A bed that, in a particular case of folding, has remained non-plastic; that is, one that has been bent perhaps purely elastically (in a very gentle fold), or, usually, that has undergone concentric folding. Tension cracks may tend to develop on the outer bends of the folds, particularly the anticlines. A bed may behave as absolutely competent, or as relatively competent with regard to the behaviour of stratigraphically adjacent beds. An incompetent bed is the contrasted case.

**Complementary folds.** In a series of folds any anticline is complementary to its adjacent synclines and any syncline to its adjacent anticlines. See quotation under ECHELON.

**Complex.** A more or less complicated association of rocks, as in 'basement complex', 'igneous complex', 'metamorphic complex', 'intrusive complex', 'tectonic-magmatic complex', 'orogenic complex'.

**Composite intrusion.** See MULTIPLE INTRUSION.

**Compound structure.** A regional structure comprising two or more parts separated by complete breaks such as unconformities. The typical form consists of two parts, a strongly folded or metamorphic 'basement' and a gently folded 'cover'; but the cover itself may be folded to any degree. Where the basement is highly complex in structure it is often called a 'basement complex' ('basal complex'), particularly when it is of Precambrian (Archaean) age. A basement (apart from the Precambrian) is usually designated by the name of the orogenic movements that produced its structure: Caledonian, Hercynian, &c. The upper surface of the basement in a compound structure is a 'platform' (with its 'platform cover') or 'floor' ('buried floor').

**Compressibility.** The ability of a substance to contract in volume under pressure. In geology we have to distinguish between the cases of (*a*) a solid of homogeneous chemical composition and physical state, such as a mineral, (*b*) a non-homogeneous compact solid, such as a piece of granite, and (*c*) a non-compact mass, such as a mud, which can be easily compressed by the elimination of pores. (Skempton, *QJ*, 1944.)

**Compression.** The act, the causes, the processes, and the multifarious results, of compression are obviously of prime concern to the geologist. 'A compression' usually refers to a form of fossilization, particularly a fossil plant.

**Compression-subsidence.** 'Far greater thicknesses of sediment have accumulated in geosynclines than can be accounted for by sedimentary loading alone; and we are accordingly forced to seek some other cause of depression. There is little doubt that we must find it in the fact that compression has already begun, bending down the floor of the growing trough and bringing the two margins or jaws nearer together. To this process we may apply the term "compression-subsidence". This process of narrowing and deepening proceeds, but the trough fills, on the whole, faster than it deepens, so that eventually it becomes full of sediment' (Wooldridge and Morgan, *Geomorphology*, 1937).

**Concealed.** This term finds its chief geological employment for formations or whole systems, and for structural surfaces such as unconformities, that are hidden beneath overlying rocks. It is particularly used for coalfields so hidden, either for concealed portions, such as those of the Yorkshire coalfield, or for a whole field, such as the Kent coalfield. 'In 1834 Conybeare wrote to the *London and Edinburgh Philosophical Magazine* "On the probable future extension of the coalfields at present worked". He showed that the structure of the coalfields east of the Pennines was such that the Coal Measures passed southwards or eastwards beneath a covering of newer strata. We are now so familiar with concealed coalfield exploration that it is useful to be reminded that someone had to enunciate the principle for the first time' (North, *PBNS*, 1956). 'On the possible extension of the Coal-Measures beneath the south-eastern part of England' (Godwin-Austin, *QJ*,1856). *Concealed Coalfield of Yorkshire and Nottinghamshire* (Gibson, *MGS*, 1926). 'Prospects of undersea coalfield extension in the North-East' (Hickling, *TIME*, 1950). *Concealed Mesozoic Rocks in Kent* (Lamplugh and others, *MGS*, 1923). *Concealed Coalfields* (Wills, 1956). Falcon (in Kent and others, eds., *Time and Place in Orogeny*, 1969) warns us: 'Only at relatively shallow depths where the rock succession is well known can concealed structure be predicted with confidence from surface geology alone'.

**Concentric folding.** Folding in which bedding surfaces form, ideally, concentric curves, concentric folds; it can be assumed only by competent beds. The compressive forces are accommodated by slipping along concentric shear-surfaces parallel to the bedding. Most of the slip is concentrated on the more marked bedding-surfaces but also occurs along bedding-surfaces within individual beds. A large slip along one surface may constitute a bedding-fault. The beds which fold concentrically retain a constant thickness. Above and below, concentric folding either dies out or is replaced by other structures. Concentric folding may be assumed exclusively by a series of competent beds of any thickness or may be assumed by the competent beds within a series made up of both competent and incompetent. Also called 'parallel folding'. The contrasting term is 'similar folding'.

**Concertina structure.** 'A sheet formed by the repeated folding of a bed on itself, after the manner of the bellows of a concertina when shut up' (Barrow, *QJ*, 1904). The sheet is formed entirely of the one bed and thus the structure differs from ordinary isoclinal folding, where distinct beds are involved. 'In some exposures the quartzites are reduplicated by superimposed recumbent-folds without infolds of other material. The structure is reminiscent of Barrow's "concertina-folding", [but here] the close-packed folds are not vertical but recumbent' (McIntyre, *GM*, 1951).

**Conch.** The shell of an invertebrate animal; practically, of a mollusc or a brachiopod. Hence 'conchite', an old name for a fossil shell: 'Conchites or cockle-stones' (Plot, *Oxfordshire*, 1677). 'Conchitic' means 'abounding in (fossil) shells' (*OED*). [L. *concha*, G. *kogchē*, a mussel, a shell.]

**Conchoidal.** For a fracture (in minerals and rocks) presenting smooth shell-like convexities and concavities. The term was in use in Playfair's day (1802).

**Conchology.** The science of shells; including that of fossil shells (Mollusca, Brachiopoda, and possibly some other groups). *Mineral Conchology of Great Britain* (Sowerby, 1812 onwards). *Fossil Conchology of Great Britain and Ireland* (Brown, 1849).

**Concordance.** Though in a general way synonymous with 'agreement' or 'being concordant' it seems applicable with special propriety to a case of correspondence between two quite different features, e.g.: 'There is a broad concordance of topography and rock-type' (Smith and George, *BRG*, *N.Wales*, 1961).

**Concordant.** As applied to an igneous intrusion: lying parallel with planes of stratification or foliation; contrasted with a discordant relationship. As applied to bedding: parallel bedding (Tyrrell, *Petrology*, 1929).

**Concretion.** 'Concretions are more or less spherical or irregular masses, varying from a few inches to several feet in diameter, which consist of material of different nature from the surrounding rock brought together [presumably] by percolating waters: in many cases the material was scattered through the rock and has been collected together. Thus flints may be regarded as concretions. In other rocks (especially in clays and shales) concretions of iron carbonate and of calcium carbonate may be found' (Trueman, *Introduction to Geology*, 1938). Most concretions have doubtless formed by growth outwards from a nucleus, and the term, in a narrow sense, may be said to imply this (cf. incretion, secretion). In any case, nothing is implied in the term as to whether the object was formed contemporaneously with the containing rock or subsequently. 'Concreted' is used in a rather more general sense. 'Stony concretions' (Playfair, *Illustrations*, 1802) and, in another part of the work, 'substances [in veins] must have concreted from a fluid state'. See CANNON-BALL CONCRETION, CONE-IN-CONE, NODULE, SEPTARIUM. In the Crackers Bed in the Lower Greensand of the Isle of Wight the concretions ('crackers') are highly fossiliferous. [L. *concretus*, grown together.]

**Condensed succession.** A stratigraphical succession (sequence) which is relatively thin because deposition was slow and perhaps intermittent during the time represented. The opposite of an extended succession. A remarkable example is the limestone Junction Bed in the Lias of the Dorset Coast. Though only 1·5m. thick it comprises the highest zone of the Middle Lias and the four lowest zones of the Upper Lias.

**Cone-in-cone.** A small-scale structure occurring in sedimentary rocks, particularly in the outer parts of some bands or (typically) concretions in argillaceous strata. The cones, which usually point inwards and have annular grooves, fit one into another. The structure was described by Henslow from Anglesey (*TCPS*, 1822); its origin is obscure. 'Cone in-cone structures' (Durrance, *PGA*, 1965).

**Cone-sheet.** 'A most important class of centralized intrusion of hypabyssal type was first detected by Dr. Harker in Skye, when he discovered and mapped in the Cuillin Hills a great group of centrally inclined sheets [*MGS*, 1904]. He demonstrated their circular distribution and regular inclination towards a focus. Such inclined sheets or cone-sheets, as they have been renamed in the "Tertiary Mull Memoir" [*MGS*, 1924] consist of relatively thin sheets, which, viewed as members of a suite, occupy conical fissures that have a common apex and common vertical axis. Generally the sheets are moderately open-spaced, but occasionally they may be so numerous and frequent as almost to obliterate the country-rock into which they have been intruded' (Thomas, *MGS*, *Ardnamurchan*, 1930). Cone-sheets occur in a ring complex associated with ring dykes. See RING COMPLEX.

**Confocal.** Having a common focus; applicable to a system of structural directions, particularly dips. 'The dips are confocal and steepen inwards towards a central area [of cauldron subsidence]' (Skelhorn and Elwell, *JGS*, 1971).

**Conformable.** Beds are conformable when they lie one on another in a regular manner, as a result of regular deposition; the deposition being either continuous or the pauses being represented merely by structurally indetectible non-sequences. They show 'conformity' throughout. Individual beds within a conformable series may vary in thickness or die out altogether when traced laterally, one kind of lithology in one place being replaced by another in another. A conformable series may, of course, be deformed as a whole to any degree by folding or faulting and still remain perfectly conformable within itself. 'Then proceed to consider the several strata . . . whether all the strata lye parallel and conformable to each other' (Woodward, *Method of Fossils*, 1728). 'The general conformity or parallelism between the several formations' (Conybeare and Phillips, *England and Wales*, 1822). Conformity implies an undisturbed relationship, not only structural parallelism, so that the term cannot be used, for instance, in the case of a thrust where the bedding, both above and below the thrust, happens to be parallel to the thrust. See UNCONFORMABLE.

**Conformity.** See CONFORMABLE.

**Conglomerate.** Literally, a mixture of rock-materials (differences in size being implied) gathered together into a coherent mass, particularly a more or less rounded mass. But in the technical meaning of the term this rather vague 'roundness' of the mass as a whole is lost sight of and instead we have an essential roundness of the larger constituents, as in Lyell's definition (*Principles*, 1833): 'Rounded water-worn fragments of rock, or pebbles, cemented together by another mineral substance, which may be of a siliceous, calcareous, or argillaceous nature'. An earlier name for a conglomerate in this sense was 'pudding-stone'. 'A course-grained gravel . . . contains rounded pebbles, which are two or three inches in diameter: it then approaches to a conglomerate pudding-stone [New Red Sandstone of Devon]. This conglomerate is in nearly horizontal strata, which probably extend eastward below the chalk' (Phillips, *England and Wales*, 1818). A conglomerate is then to be contrasted with a breccia in the modern sense. 'To what extent must the fragments be rounded before the term "breccia" is dropped and the term "conglomerate" applied instead? Usage is variable, but most workers apply the term breccia only to angular fragments; conglomerate is used for subangular and the better-rounded materials' (Pettijohn, *Sedimentary Rocks*, 1949). The identification of the pebbles in a conglomerate with parent rocks is obviously important as indicating sources of origin and establishing age relations. 'Intraformational conglomerates' (or breccias) may be formed contemporaneously, or penecontemporaneously, with the rest of the deposit; for instance, dried and hardened blocks or pellets of alluvial mud may be eroded and redeposited. See INTRACLAST. There is also a 'tectonic conglomerate', or 'crush conglomerate', formed of rounded or lenticular fragments; cf. tectonic breccia. 'Breccio-conglomerate' (or 'breccia-conglomerate') is a term sometimes used, but it is not always clear whether this means a rock in which the larger fragments may be said to be on average slightly angular, or a breccia 'conglomerated' into a rock-mass. [L. *glomeratus*, gathered into a mass, particularly a rounded mass.]

**Conglomeration.** The heaping together of diverse materials into one mass. (Would not ordinarily be used for the production of a conglomerate.)

**Congruous folds.** Minor fold plications, on the side of a major fold, obeying 'Pumpelly's rule', with the assumption that this implies that the axial planes of the minor folds are parallel to the axial plane of the major fold.

**Coniston Limestone.** A formation of predominantly calcareous rocks, a few hundred feet thick, outcropping along a belt of country running NE.-SW. across the southern part of the English Lake District and through the town of Coniston. The fossils are typical of the Caradocian. (Sedgwick, *QJ*, 1845.)

**Conjugate.** This adjective obviously has the primary meaning of 'joined together', but in various connexions it has special usages such as 'complementary', 'having the same origin'. Something of both these latter subsidiary meanings is contained in the application of the term to geological structures; thus to a joint system in which the joint sets are at a constant angle (particularly a right angle) one to another, to faults forming two or more parallel sets and produced at the same time, and (but here ideas are various and complicated and the term is perhaps best avoided) to certain relationships between the axial planes of contemporaneous folds.

**Connate.** **1.**='cognate'. In petrology, sometimes used specially for fluids derived from a magma, which are thus connate among themselves and connate with the magma. **2.** The term, as 'connate water', has been used for 'fossil sea-water', supposed to have been entrapped during the formation of a sedimentary rock.

**Conodont.** Conodonts are minute toothlike fossils, showing much variation in form, recorded from Cambrian to Trias (possibly later), of unknown body-position or function, belonging to some unknown kind of (extinct) animal. A number occasionally occur together as an assemblage, presumably belonging to one individual animal, but such an assemblage may include different forms. Their composition is largely calcium phosphate. Many forms are useful stratigraphical indices. The systematic position of the conodonts still remains one of the most perplexing problems in palaeozoology. (Pander, 1856; 'Annotated bibliography and index of conodonts', Ellison, *UTP*, 1962; *Conodonts*, Lindstrom, 1964; 'Conodonts: problematical index fossils', Jenkins, *Geol*, 1974.) 'The stratigraphic distribution of conodonts in the British Silurian' (Aldridge, *JGS*, 1975). [G. *kōnos*, a cone.]

**Conrad discontinuity.** 'Seismic evidence indicates that in continental regions, beneath a zone of superficial sediments, the crust can be divided into two layers, an upper crust often considered to be granitic or granodioritic in composition, and a lower crust probably of more basic composition. The boundary between the upper and lower crust, sometimes called the Conrad Discontinuity, is known only from the seismic evidence and is often difficult to define and variable in nature and in depth. In the floors of ocean basins, beneath the superficial sedimentary material, the 'granitic' layer is missing' (Harris in Gass and others, eds., *Understanding the Earth*, 1972). [V. Conrad who discovered it when studying an earthquake that occurred in the Austrian Alps in 1923.]

**Consanguinity.** 'Judd's conception of petrographical provinces has proved a singularly fertile one. It leads at once to the hypothesis that the igneous rocks of one province stand to one another in a real genetic relationship, community of characters being attributable to community of origin. For the presumed relationship implied in such resemblances among associated rock-types, Iddings (*BPSW*, 1892) first employed the convenient term consanguinity' (Harker, *Igneous Rocks*, 1909).

**Consequent drainage.** A river drainage system initiated as the consequence of uplift producing a region of subaerial erosion. The primary main streams, flowing in 'consequent valleys' (Powell, *Colorado River*, 1875, hence 'consequent streams'), will flow in the directions of tilting. In a normal case of uplift from beneath the sea these directions would be the directions of dip of recently deposited strata; the main streams would be dip streams of the first order. Tributaries of the first order are the 'subsequent streams' (Davis, 1889) and would normally be strike streams. A further order of tributaries would normally comprise secondary dip, and counter-dip, streams. Scarps and dip-slopes would tend to be formed and the drainage pattern while remaining strictly 'consequent' and 'accordant', might become complicated as it evolved.

**Consolidation.** The process of becoming solid as a mass, from a state of being in a more or less loose or fluid condition; particularly for a loose sediment becoming a firm and coherent, and more or less hard, rock by compaction under compression or by cementation. Hence 'consolidated'. 'Along the south side of Loch Ness, there are mountains formed of the debris of schistus and granite mountains, first manufactured into sand and gravel and then consolidated into a pudding-stone' (Hutton, *Theory of the Earth*, 1795). 'As the materials of the strata were originally loose and unconnected we must consider by what means they were consolidated into stone' (Playfair, *Illustrations*, 1802). 'If you take a quantity of mud, and place it under a weight which will squeeze the water out of it, you will find that it gets firmer. You can thus harden it by pressure. Again, if you place some sand under water which has been saturated with lime or with iron, or with some other mineral which can be dissolved in water, you will notice that as the water slowly evaporates it deposits its dissolved materials round the grains of sand and binds them together. Were you to continue this process long enough, adding more of the same kind of water as evaporation went on, you would convert the loose sand into a solid stone. In this case the hardening of the sediment into stone would be done by the process called infiltration. In one or other or both of these ways most of the sedimentary rocks have been hardened into the state in which we now find them' (Geikie, *Geology Primer*, 1873).

**Constitution of the earth.** Perhaps the most fundamental of all gological problems. Very little is known about the composition and physical constitution of the earth as a whole. The chief evidence we have is from the interpretation of seismograms, which reveal certain variations and, in particular, discontinuities, in the way successive concentric zones within the earth transmit earthquake waves. There are two main discontinuities; that separating the core from the mantle (the Gutenberg discontinuity), and the Mohorovičić discontinuity at a relatively small depth below the surface. These discontinuities are usually taken to imply a change in chemical composition. 'The constitution of the interior of the earth as revealed by earthquakes' (Oldham, *QJ*, 1906). *Internal Constitution of the Earth* (Gutenberg, ed., 1939/51). *Physical Constitution of the Earth* (Coulomb and Jobert, trans. Nairn, 1963).

**Contact.** The boundary surface between one rock body and another, such as that between an igneous intrusion and the adjacent country rock (if the passage is abrupt) and that between two beds or distinct parts in a mass of sedimentary rock. The nature of a contact, the contact relationship, is important in geological interpretation; the chief distinction being between one that is original and undisturbed (sometimes called 'normal' or 'natural', though neither of these terms is satisfactory) and one that is the result of dislocation (faulting). See JUNCTION. Most of the contacts in the first category are 'depositional' in kind. A fault is sometimes drawn at a contact where, prima facie, the mapping points to a simple unconformity and where no convincing evidence of faulting is brought forward. (Conspicuous examples of this are the 'Red Rock fault', along the western edge of the North Staffordshire coalfield, and the 'Malvern fault' along the eastern side of the Malvern hills.) 'Some Lower Palaeozoic contacts in Pembrokeshire' (Jones, *GM*, 1940).

**Contact metamorphism.** Metamorphism produced, by an igneous intrusion, in the adjacent and surrounding rocks. Though usually considered as purely thermal, part of the metamorphism is probably due to deformation directly related to the emplacement of the intrusion; 'deformation and heating have acted together' (Pitcher and Read, *QJ*, 1958). Any associated pneumatolytic action is usually included; the term certainly seems to require this inclusion for, as remarked by Harker (*Metamorphism*, 1950), 'the special effects [of pneumatolysis] are confined to the neighbourhood of an igneous contact, and are essentially dependent upon that situation'. Contact metamorphism was noticed by Sinclair in the Lothian coalfield (*Short History of Coal*, 1672): 'These gaes [dykes] that consist of whin-rock, render the coal next to it, as if it were already burnt'; by Whitehurst (*Strata in Derbyshire*, 1778): 'Another remarkable phenomenon accompanying the Derbyshire lava is that the stratum of clay lying under no. 6 toadstone is apparently burnt, as much as an earthen pot or brick'; and by Playfair (*Illustrations*, 1802): 'It is where whinstone takes the form of veins, intersecting the strata, that the induration of the latter is most conspicuous. The coast of Ayrshire, and the opposite coast of Arran, exhibit these veins in astonishing variety and abundance . . . the induration of the sides of these veins, in some cases, has been such, that the sides have become more durable than the vein itself; so that the whinstone has been worn away by the washing of the waves, and has left the sides standing up, with an empty space, like a ditch, between them'. 'Contact' metamorphism of a wide area surrounding a major intrusion is more appropriately, but not so conveniently, called 'thermal metamorphism', with or without accompanying pneumatolysis and dynamic effects. 'I must notice a beautiful group of crystalline slates, which are seen in Skiddaw forest, between the black slates and the granite of the Caldew. . . . I believe that this beautiful mineral group is nothing more than the Skiddaw slate, altered and mineralized by the long continued action of subterranean heat' (Sedgwick, letters to Wordsworth, 1842, in Wordsworth and Hudson's *Guide to the Lakes*). 'Rocks affected by contact-metamorphism, by which term we mean the alteration produced upon sedimentary rocks by the intrusion into them of igneous masses' (Bonney, *QJ*, 1886). 'Metamorphism by the Whin Sill, Falcon Clints, Teesdale. This section is one of the best in England, showing the metamorphism of a limestone by a basic igneous rock' (Garwood, description of a *BA* photograph, *c.* 1903). 'Contact metamorphism in south-eastern Dartmoor' (Fitch, *QJ*, 1932). The following are examples of contact metamorphism by dolerite sills, striking important in the controversy between the Neptunists and the Plutonists or Vulcanists (photographs in Bates and Kirkaldy, *Field Geology*, 1976): Old Red Sandstone at the foot of Salisbury Crags, Edinburgh, and Lias Clay at the top of Ramore Head, Portrush, Co. Antrim. See AUREOLE, THERMAL METAMORPHISM, ZONE (2), and quotation under REGIONAL METAMORPHISM.

**Contact mineral.** A mineral produced by contact metamorphism. 'Characteristic aluminous "contact-minerals"' (Harker and Marr, *QJ*, 1891).

**Contamination.** 'The chemical change of a magma, due to the solution of foreign material, solid, liquid or gas' (Daly, *Igneous Rocks*, 1933). Hence 'contaminated'. 'The contaminated tonalites of Loch Awe' (Nockolds, *QJ*, 1934).

**Contemporaneous.** Two special usages are the following. **1.** Contemporaneous deformation; deformation contemporaneous with, and involved in, the process of sedimentation. **2.** 'When a lava-flow occurs in the middle of a succession of sedimentary rocks it is said to be interbedded with them, and is spoken of as a "contemporaneous" volcanic rock' (Rastall, *Textbook*, 1941).

**Continental.** Applied specially to deposits; those laid down on land masses, including lacustrine and deltaic deposits, as distinct from marine deposits, laid down on the sea floor or along the shore.

**Continental crust.** See EARTH'S CRUST.

**Continental drift.** The hypothesis that the sial (continental) masses of the earth's crust not only float in the sima but also drift relatively to one another. In particular, it is the hypothesis that, since about Carboniferous times, what was one large continental mass (Pangaea) has split up and drifted apart, at different times in different parts. 'The suggestion that there might have been lateral displacements of the continental masses on a gigantic scale is generally ascribed to F. B. Taylor in America (1908) and to Alfred Wegener in Germany (1910). Actually, however, the same idea had occurred to Antonio Snider more than fifty years before in a book with the optimistic title *La Création et ses mystères dévoilés* (Paris, 1858)' (Holmes, *Physical Geology*, 1944; for further history and exposition see 'Reassembling the continents', chapter in 1978 edition). 'He who examines the opposite coast of the South Atlantic Ocean must be somewhat struck by the similarity of the shapes of the coastlines of Brazil and Africa. This phenomenon was the starting point of a new conception of the nature of the earth's crust and of the movements occurring therein; this new idea is called the theory of the displacement of continents, or, more shortly, the displacement theory, since its most prominent feature is the assumption of great horizontal drifting movements which the continental blocks underwent in the course of geological time and which presumably continue even to-day' (Wegener, *Origin of Continents and Oceans*, 3rd ed.1922, trans. 1924. Latest English edition, translated from 4th ed. 1929, with introduction by King, 1967; in this the above quotation does not occur). *Continental Drift* (Runcorn, ed., 1962). 'Continental drift' (Bullard, *QJ*, 1964). *A Symposium on Continental Drift* (Blackett and others, eds. for the Royal Society, 1965). *Continental Drift* (Garland, ed., *RSC*, 1966). *Continental Drift* (Tarling and Tarling, 1971). *Continental Drift: The Evolution of a Concept* (Marvin, 1973). *The Evolving Continents* (Windley, 1977/84). See SEA-FLOOR SPREADING, POLAR WANDERING, and PLATE TECTONICS.

**Continental margin.** The part of the sea-floor comprising the continental shelf, the continental slope, and the continental rise. *The Geology of Continental Margins* (Burk and Drake, eds., 1974). 'Processes and lithofacies in submarine slope, canyon, fan, and trench settings. . . . Study of continental margin sedimentation is in a state of explosive growth' (Stanley and Bertrand, *Geoly*, 1979).

**Continental rise.** 'Continental rises first were recognized [and named] as distinct physiographical units of continental margins little more than a decade ago [Heezen and others, *GSASP*, 1959]. Seismic refraction measurements have shown that these regions of the sea floor between the continental slope and the abyssal plains are underlain by very thick sediments, in places exceeding 6 km. In fact, continental rises are among the largest sedimentary structures of the earth' (Emery and others, 'Continental rise of eastern North America', *BAAPG*, 1970). 'The Lower Palaeozoic sediments of Wales compare with the accumulation in modern continental rises' (Davies and Cave, *SG*, 1976).

**Continental shelf.** The submarine shelf which extends along many coastlines, seaward from low water to about the 100-fathom line; or it may include the shore (see BELT OF VARIABLES). The edge of this shelf ('shelf edge') is a truer edge of a continent than the coast, and its outline is important particularly in questions of continental drift. 'Sediments of the continental shelves' (Shepard, *BGSA*, 1932). *The Continental Shelves* (Emery, 1969). *Geology of the North-west European Continental Shelf—The North Sea* (Pegrum and others, 1975).

**Continental slope.** The submarine slope bordering the continental shelf on its seaward side.

**Continental terrace.** Usually defined as the continental shelf and continental slope combined, but it has been taken to include 'the underlying crust' and thus to be a solid structure rather than merely a surface (Dunham, *QJ*, 1968). (The same double meaning is apparent in the use of 'continental rise'.)

**Continents and ocean basins.** 'The theory of the permanence of continents and ocean-basins was practically founded in 1846-7 by J. D. Dana (*AJS*), who held that their general forms "were to a great extent fixed in the earliest periods"; he regarded the continents as thick early consolidated blocks of the Earth's crust, and the ocean-basins as thinner parts of the crust that have sunk continuously' (Gregory, *QJ*, 1929). 'The impermanence of oceans' (Lees, *QJ*, 1953). 'The origin of ocean basins and continents' (Gilvarry, *N*, 1961). 'Permanency of the continents' (Carr, *N*,

1966). 'The origin of continents and oceans' (Harland, *GM*, 1969).

**Continuum.** This naturalized Latin word, used for a continuous, particularly a continuously variable, substance, quantity, or state, is applicable to, for instance, a series of minerals of continuously variable chemical composition (such as the plagioclase felspars) or, among fossils, to a series of continuously variable forms or to an evolving plexus.

**Contorted.** Folded in an intricate manner. 'Those contortions of the strata which, when on the great scale, are among the most striking and instructive phenomena of geology' (Playfair, *Illustrations*, 1802). 'Lulworth Cove, and the cliffs in its vicinity, present many objects highly deserving the attention of the geologist; among which the very extraordinary manner in which the Purbeck shell limestone is contorted, struck me particularly' (Webster in Englefield, *Isle of Wight* [&c.], 1816). 'Contorted gravels of terrace of the River Cam . . . produced in pre-existing gravel beds by the over-riding action of ice' (Chatwin, *BRG*, *East Anglia*, 1964). For the contorted appearance of bedding in a stratum, presumably due to conditions of deposition, the term 'convolute' is used.

**Contorted drift.** A glacial deposit seen particularly in E. Anglia. 'In many places movements in the ice-sheet resulted in folding and thrusting of the enclosed material. When the ice evaporated or melted at a slow rate these structures remained after the ice had disappeared; hence originated the term "contorted drift" for such deposits. Examples of contortions are to be seen at Ipswich and in the cliffs near Cromer' (Chatwin, *BRG*, *East Anglia*, 1964).

**Contourite.** 'The sediment deposited from contour following bottom currents' (Heezen and others, *S*, 1966). As regards this particular manner of formation it is thus contrasted with a downhill-deposited turbidite. (But a turbidite may be deposited by an 'along-slope turbidity current', Woodcock, *PGA*, 1976.) 'Contourites: their recognition in modern and ancient sediments' (Stow and Lovell, *ESR*, 1979).

**Contraction hypothesis.** See SHRINKING EARTH

**Control.** In its ordinary sense used in connexion with geological processes and conditions, as in, e.g., 'factors [temperature and pressure] controlling metamorphism' (Pitcher and Flinn, eds., *Controls of Metamorphism*, 1965), 'palaeogeographical controls have given rise to regional differences amongst the British trilobite faunas' (Allen in Sylvester-Bradley and Ford, eds., *Geology of the East Midlands*, 1968), *Evolution and Extinction Rate Controls* (Boucot, 1975), and 'Controls of Copper mineralization at Coniston' (Dagger, *GM*, 1977). In geological logic: preliminary speculation, suggestion, and inference are said to be 'controlled' by ascertained fact; e.g., tentative correlation of strata is controlled (more or less rigidly) by fossil content, tentative structural interpretation from surface exposure by deep boring (and see quotation from Lees under SECTION (3)), photogeological interpretation is checked by ground control (Hepworth, *QJ*, 1967).

**Convection theory.** There are several theories according to which major changes in the earth's constitution and in the earth's crust, such as the growth of the core, continental drift, orogeny (particularly the formation of a tectogene), are due to convection currents in the mantle. 'The operation of thermal convection currents within the mantle, as well as in the core, . . . is the most favoured hypothesis, at the present time, for accounting for the transport of oceanic lithosphere necessitated by sea floor spreading [with discussion and history]' (Holmes, *Physical Geology*, 1978).

**Convergence. 1.** In petrology, the production of two or more rocks, similar petrographically, along separate lines of petrogenesis; the rocks, in their final form, bearing only the stamp of a similar final act. This may occur particularly in metamorphism. **2.** In palaeontology, the production of similar forms, in stocks not specially nearly related, as a result of evolution working to the same end.

**Convolute bedding.** 'The present author proposes the term "convolute bedding" . . . convolute bedding in fine grit, south of Aberystwyth' (Kuenen, Netherlands, 1953). However, this purely descriptive term, always available, can hardly be specially 'proposed'. Its significance has recently aroused much interest: e.g. (in addition to Kuenen's papers), 'Significance of convolute lamination' (Haaf, Netherlands, 1956), 'Convolute bedding in Ludlovian rocks' (Holland, *GM*, 1959). 'Convolute lamination from the Lower Coal Measures of Yorkshire' (Davies, *Sd*, 1965). 'Convolute lamination in graded sand beds' (Allen, *JGS*, 1977).

**Coombe rock.** A Pleistocene deposit. 'The coombe-rock, a term G. A. Mantell first introduced geologically (*Fossils of the South Downs*, 1822), is a structureless mass of unrolled and unweathered flints, embedded in a matrix of chalky paste and disintegrated chalk. Its narrow tongues run into the valleys or "coombes" of Sussex and often pass into brickearth. It lies at various heights in positions unrelated to any present river-system. It mantles the major outcrop of the chalk north-west of London and spreads widely over the North Downs and through Kent and Sussex. It was formed by solifluxion (sheet-flowing or sludge-creep), not only in the coombes but over the whole of the dip slopes, aided by wind when the ground was frozen and unprotected by vegetation' (Charlesworth, *Quaternary Era*, 1957).

**Copper pyrites.** A brassy or copper-yellow mineral, sulphide of copper and iron, $CuFeS_2$; the principal ore of copper. Also called 'chalcopyrite' (f. G. *chalkos*, copper).

**Coprolite.** A phosphatic nodule supposed to be the excrement of a reptile. Coprolites occur most commonly in the Jurassic clays (particularly the Lias) and the Cretaceous greensands. 'Coprolites . . . the petrified faeces of saurian animals, whose bones are so numerous in the same strata with themselves' (Buckland, *TGS*, 1829). Phosphatic nodules of other origins, in the Cretaceous greensands, have also been called coprolites commercially or popularly. [G. *kopros*, dung.]

**Coquinite.** A 'conchitic' zoolith (see CONCH). 'Where shelly fossils dominate, the general term "coquinite" [for the rock] may be used' (Compton, *Field Geology*, 1962). Another name is 'coquina'.

**Coral.** Any member or group of the class Anthozoa that possesses a (calcareous) skeleton. More precisely, it is the calcareous skeleton itself, whether of a single organism, a compound organism, or a colony. 'Fossil corals and coralloid bodies' (Woodward, *Catalogue*, 1729). 'Between the tropics, islands are formed from the mere accumulation of corals; and it is the peculiarity of those regions to produce rocks that have not passed through the usual process of mineral consolidation' (Playfair, *Illustrations*, 1802).

**Coral mud.** A highly calcareous mud formed round coral islands. It may be carried out into deep water.

**Coral reef.** See REEF. *Structure and Distribution of Coral Reefs* (Darwin, 1842). 'The coral reefs of the Great Scar Limestones were formed in a shallow sea' (Sedgwick, letters to Wordsworth, 1842, in Wordsworth and Hudson's *Guide to the Lakes*). 'On the nature, origin and climatic significance of the coral reefs in the vicinity of Oxford' (Arkell, *QJ*, 1935).

**Corallian.** The Jurassic formation representing in Britain all but the lower third or so of the general Oxfordian stage. The most distinctive rocks are oolitic limestones and, particularly, coral limestones and calcareous sandstones (the Coral Rag and Calcareous Grit), but there are also clays, particularly the local clay facies in Bedfordshire, the Ampthill Clay. The formation, 60-90 m. where thickest, outcrops from the Yorkshire coast, south of Scarborough, to the Dorset coast at Weymouth, forming the Hambleton Hills in Yorkshire and passing near Oxford and through Wiltshire. Among the characteristic fossils are the corals *Isastraea explanata*, *Thamnasteria arachnoides*, and *Thecosmilia annularis*, the echinoids *Hemicidaris intermedia* and *Nucleolites scutatus*, the lamellibranch '*Pecten*' *lens*, and the gastropod *Pseudomelania heddingtonensis*. (Smith in Townsend,

*Character of Moses*, 1813, 'Calcareous Grit', 'Coral Rag'; Buckland in Phillips, *England and Wales*, 1818, 'Oxford Oolite'.)

**Cordaitales.** A group of extinct plants belonging to the division Gymnospermae, growing as trees; found in Carboniferous and Permian rocks. [genus *Cordaites* (named after A. C. J. Corda, 1809-49, a Czechoslovakian botanist).]

**Cordierite.** See ANDALUSITE. 'J. A. H. Lucas, 1813, after P. L. Cordier who had described it' (Chester, *Names of Minerals*, 1896).

**Core. 1.** Of a fold. The inner, central part of a folded mass, particularly of a compound fold that includes a folded unconformity or other structural break. The part within the surface of structural break, a part which itself may have been previously folded, is the 'core', the rest of the fold being the 'envelope'. **2.** Of granitization. The central, completely granitized, part of the whole regional body of rock undergoing the process towards granitization, as distinct from the magmatized-metamorphosed 'envelope'. **3.** Of the earth. The central sphere within the earth, of radius about half the earth's radius. Judging from the way seismic waves are transmitted through it, most of the core behaves as a fluid, but the inner part (of radius about one-third of the whole) seems to be solid. It is thought that the core is composed dominantly of iron or of a mixture of iron and nickel, so it is sometimes referred to as the 'nife' (cf. 'sial', 'sima'). 'Oldham [*QJ*, 1906] produced evidence for a central core differing greatly in physical properties from the outer earth shells. This conclusion was questioned for a time, but Oldham's discovery has been amply confirmed—the Oldham-Gutenberg discontinuity is a major feature of the earth's internal structure' (Hawkes, *QJ*, 1958). *The Earth's Core* (Jacobs, 1975).**4.** The solid cylinder of rock, or a part of it, brought to the surface by a hollow drilling bit. **5.** In palaeontology. See CAST. **6.** Of a volcano. (Walker, *QJ*, 1963.)

**Core-stone.** See quotations under TOR (Linton) and JOINT (Bakewell).

**Cornbrash.** A thin but extensive calcareous formation (only about 9 m., even where thickest) of which the lower part is classed, from its fossils, in the Bathonian series of the Middle Jurassic and the upper part in the Callovian series of the Upper Jurassic. Characteristic fossils of the Lower Cornbrash are *Holectypus depressus*, *Nucleolites clunicularis* (echinoids), *'Terebratula' intermedia*, *T'. obovata* (brachiopods), and the ammonite zone fossil *Clydoniceras discus;* while the Upper Cornbrash has *'Terebratula' lagenalis* (and other brachiopods), and the zone fossil *Macrocephalites macrocephalus*. 'Wiltshire dialect for brash suitable for growing corn, adopted as a formation name by William Smith [in Townsend, *Character of Moses*, 1813]' (Arkell and Tomkeieff, *Rock Terms*, 1953).

**Cornstone.** Cornstone *par excellence* is the 'concretionary cornstone' (M'Cullough, *TWNFC*, 1868), 'calcareous concretions embedded in marls and grading to solid concretionary limestones' (Allen). It occurs typically in the Old Red Sandstone and the New Red Sandstone, most abundantly in the former. Somewhat similar concretions in other formations have also been called 'cornstones'. The term 'cornstone' was used by Greenough (*Geology*, 1819, and *Map*, 1819), concretionary cornstone being clearly described by Buckland (*TGS*, 1821). Another kind of cornstone is the 'conglomeratic cornstone' (M'Cullough), 'marl and limestone fragments embedded in more or less sandy and calcareous matrices' (Allen). It was Murchison (*Silurian System*, 1839) who confused matters by applying the unqualified term 'cornstone' to these conglomeratic, as well as to the concretionary rocks. They, also, occur in the Old Red and New Red Sandstones. A full discussion is given by Allen (*GM*, 1960). [for 'corn' see *OED*.]

**Corona.** A zone of one mineral (or more) surrounding another. This may be a reaction rim.

**Corrasion.** The wearing away of rocks, particularly of those *in situ*, by the scraping action (friction and impact) of rock-particles being transported by the natural agencies of rivers, waves, ice, wind, &c. It would include any such work as might by done by the action of moving ice (being a solid) itself; and might be held to include also the loosening of the material of a rock mass (at the base of a sea cliff or forming a river bank) by the impact of rushing water itself. See EROSION. [L. *corrado*, scrape.]

**Correlation.** This term, with its general meaning, is applied as required. **1.** Its most usual application is in stratigraphy where, unless otherwise stated, it means equivalence in time. ' "Correlation" is the process by which stratigraphers attempt to determine the mutual time relations of local sections' (Dunbar and Rodgers, *Stratigraphy*, 1957). Discussion and history in 'The meaning of correlation' (Rodgers, *AJS*, 1959). Time-correlation of strata cannot be known otherwise than approximately, but fossil assemblages are found to be more or less reliable guides to geological age. 'A suggested correlation of the Coal Measures of England and Wales' (Trueman, *PSWIE*, 1933). 'Correlation by fossils' (Woodford in Albritton, ed., *Fabric of Geology*, 1963). See CHRONOLOGY and STRATA IDENTIFIED BY FOSSILS. **2.** Tectonic correlation is the recognition of one major tectonic element, particularly a fold, in two or more separated regions. 'A correlation of structures in the coalfields of the Midland Province' (Fearnsides, *BA*, 1933). See quotation under REFOLD. **3.** Metamorphic correlation may mean the equivalence between (*a*) the grade of metamorphism of rocks of different original composition, or (*b*) a metamorphic unit and its unmetamorphosed representative elsewhere.

**Corrie.** See CIRQUE. [Gaelic *coire*.]

**Corrosion.** The eating away of rocks by natural agencies. It usually refers to chemical rather than physical action and may mean (chemical) weathering only, or be used for any kind of erosive chemical or solvent action. In petrology, it is the eating away of the outer parts of crystals, particularly phenocrysts, by the solvent action of the residual magma. 'Calcareous earths are immediately dissolved by water; and though the quantity so dissolved be extremely small, the operation, by being continually renewed, produces a slow but perpetual corrosion by which the greatest rocks must in time be subdued' (Playfair, *Illustrations*, 1802).

**Corundum.** Crystallized alumina, $Al_2O_3$ notable for its extreme hardness. A product of the thermal metamorphism of argillaceous rocks. The gem-stones, ruby and sapphire, are coloured varieties. [*Kurund*, its Indian name.]

**Cosmogony.** The origin of the universe, particularly the exposition or discussion of a theory or theories as to this. It thus treats of a wider subject than geogony, but it is more often used than 'geogony' when in fact the earth alone is being considered. 'The Mosaick Cosmogony . . . supposes the waters to have encompass'd the Globe' (Whiston, *Theory of the Earth*, 1696). '*The Anatomy of the Earth*—a seventeenth century cosmogony' (North, *GM*, 1934). 'The progress of geological inquiry in Europe during the seventeenth century was marked by a characteristic feature—the development of a series of cosmogonical systems' (Geikie, *Founders of Geology*, 1905). [G. *kosmos*, the universe, *gignomai*, become.]

**Cosmology.** 'The theory of the universe as an ordered whole, and of the general laws which govern it' (*OED*). Geology in the widest sense, and philosophically considered, may be said to be a part of the still more general science of cosmology. 'Cosmology, where geography, geology and astronomy meet' (Bromehead, *PGA*, 1945).

**Country rock. 1.** The rock traversed by a mineral vein. 'I have used the word "country" in the sense in which it is used by the miner. If a miner be driving an adit in any other direction than that of the load, he says he is "driving through the country"' (Phillips, *TGS*, 1814). See quotation (Jones) under FAULT. **2.** The rock invaded by an igneous intrusion. 'The diverse postures assumed by the intruded rock-bodies and their attitude towards the "country"-rocks in which they are intruded' (Harker, *Igneous Rocks*, 1909).

**Courceyan.** See DINANTIAN. [Courceys, the barony including the Old Head of Kinsale, County Cork.]

**Course.** Apart from its ordinary use (as in 'course of a river') it was, particularly in early times, applied to a stratum; e.g. by Owen (1603) in tracing the 'course' (i.e. the outcrop) of limestone surrounding the S. Wales coalfield and by Sinclair (1672) in noting the regular 'course' (i.e. stratification) of the Coal Measures of Midlothian. Murchison describing the New Red Sandstone of Shropshire (1839) uses 'course' for the stratum itself, almost in the builder's sense.

**Cover.** See COMPOUND STRUCTURE.

**Crag.** In a special lithological and stratigraphical sense: the shelly sandstones of E. Anglia, the main stratigraphical divisions, from below upwards, being the Coralline Crag, Red Crag, and Norwich Crag. Until a few years ago these were all placed in the Pliocene, but now it is generally agreed that the Coralline Crag only should be so placed, the other two being Pleistocene. (Smith, *Stratigraphical System*, 1817, but wrongly placed; Conybeare and Phillips, *England and Wales*, 1822.) Crag fossils were first described by Samuel Dale (*History of Harwich and Dovercourt*, 1730). Probably from 'the Celtic word *cregga*, meaning a shell, because shells constitute a large part of some of the beds' (North, *Limestones*, 1930).

**Crag and tail.** A hard outstanding rock ('crag') which boldly fronted and obstructed a mass of moving ice, deflecting it along the sides of its more gently sloping 'tail'. 'The feature of crag and tail, first described by Hall ["craig and tail", attributed to some "diluvian" action] from the Midland Valley of Scotland (*TRSE*, 1815), usually centres about some hard igneous rock. The abundance of necks accounts for the very fine crags and tails to be seen in the Midland Valley of Scotland as well as for their early discovery there, e.g. Calton Hill, Castle Rock, and Arthur's Seat in Edinburgh, North Berwick Law and Traprain Law in the Lothians, Largo Hill in Fife, and Necropolis Hill, Glasgow. The impact side, which is scraped bare, is commonly steep or precipitous and, as at Castle Rock, has a horse-shoe shaped valley half encircling its base and extending leewards as lateral grooves which gradually diminish in cross-section, e.g. Princes Street Gardens and Grassmarket. The tail frequently descends from the very summit of the hill in a smooth, gentle slope; it may be solid or may consist of drift or preglacial soil preserved in the *morte-espace* of reduced ice-pressure or stagnation' (Charlesworth, *Quaternary Era*, 1957).

**Crater.** A dish-shaped, bowl-shaped, or funnel-shaped hollow, it being understood that this is produced by igneous or some other explosive agency. (Other funnel shaped hollows, such as a swallow hole, would be described as 'crater-like'.) The ordinary crater is the volcanic crater, the orifice of eruption, at the summit, or on the side of a volcano. Another kind of crater is the meteor crater, a scar formed by the explosive impact of a meteorite; Meteor Crater in Arizona is supposed to have been formed in this way. *Origin and Development of Craters* (Jaggar, *MGSA*, 1947). [G. *kratēr*, a bowl.]

**Craton.** Cratons are relatively immobile regions of the earth's crust between which geosynclines may develop. Schuchert and Dunbar (*Textbook*, 1933) gave the term in another form: 'the name "shield" was suggested by the fact that the surface of the ancient rocks arches up in gentle convexity like the surface of a medieval shield. A much more significant term is "kratogen"'. ('Kratogen' was being used in *Der Bau der Erde* by Kober in 1928.) [G. *kratos*, strength.]

**Creep. 1.** 'The phenomenon of a continuously increasing deformation under constant load is called "creep"' (Burgers and Burgers, Netherlands, 1935, quoted in Dennis, *Tectonic Dictionary*, 1967). **2.** 'Surface creep', 'soil creep' ; the creep downhill, under the force of gravity, of loose rock-material particularly on a steep slope, where it may cause, immediately underneath it, superficial deformation, terminal curvature. Includes solifluxion. 'Less obvious than landsliding, but at the same time far more general, indeed almost universal, is an imperceptible downhill movement of the waste-mantle of slopes that is continuously in progress. Working along with surface wash, which is effective during heavy rains, this movement, termed creep by W. M. Davis, is the cause of migration of much waste to lower levels before it is eventually carried off by permanent running streams' (Cotton, *Landscape*, 1948). **3.** A form of pressure solution mechanism occurring in metamorphism. This is a process of 'diffusive mass transfer of matter from grain boundaries which are subjected to high normal stress to less highly stressed grain boundaries. If the diffusion is predominantly through the grain the process is termed the Nabarro-Herring creep whereas if it is predominantly round the grain boundaries the flow is termed Coble creep' (McClay, conference paper, *JGS*, 1977).

**Crenulation cleavage.** See CLEAVAGE (1b). 'The formation of crenulation cleavage' (Cosgrove, *JGS*, 1976). (Rickard, *GM*, 1961.)

**Crest.** In connexion with folding: the line along the top of one bedding surface in an anticline. The crest of an anticline as a whole would probably mean the line on the surface of the ground about which the strata dip in opposite directions. In any case, the crest of a fold is to be distinguished from the hinge.

**Cretaceous.** Containing chalk; for strata, but seldom used, 'chalky' being preferred. Cretaceous, with capital C, is concerning the Cretaceous system. [L. *creta*, chalk.]

**Cretaceous system.** The uppermost of the three Mesozoic systems. 'Designated *terrains crétacés* by d'Omalius d'Halloy in 1822, the System takes its name from *creta*, the Latin for chalk. This is the most conspicuous rock-type in Europe, and Cretaceous chalks are also known from parts of North America and Western Australia. The system is here divided into two. . . . The lower Cretaceous was formerly separated by some American authors as a distinct system, the Comanchian (from the town of Comanche, Texas). The Upper Cretaceous corresponds roughly to the Chalk Formation of earlier authors, chalk (Saxon *cealc*, German *kalk*, meaning lime) having been used as a geological term in England since the Middle Ages and in print from the time of Martin Lister (*De Fontibus* . . ., 1684). Its white chalk cliffs gave England its first recorded name of "Albion"' (*A Correlation of Cretaceous Rocks in the British Isles*, *GSSR* (9), 1978; see this work for discussion of the base and top of the system and for all details as to nomenclature, chronostratigraphy, and correlation).

The system is classified as follows:

| *General stages* | *British formations (approximate correlations)* | |
|---|---|---|
| Maastrichtian | | |
| Senonian | Upper Chalk | |
| Turonian | Middle Chalk | |
| Cenomanian | Lower Chalk | |
| Albian | {Upper Greensand<br>Gault | |
| Aptian | Lower Greensand | } Speeton |
| Neocomian (s.l.) | Wealden | } Clay |

(D'Halloy, 1822.)

Porifera (sponges) are often common. Of the echinoids, the Spatangoida (heart-urchins) are specially characteristic, together with many Regularia. *Inoceramus* and *'Pecten'* are abundant lamellibranchs. Among the several characteristic ammonite superfamilies are the Hoplitaceae, the Acanthocerataceae, and the Turrilitaceae (uncoiled and helical genera). The Cretaceous system outcrops along the Yorkshire-Norfolk coastal area and then swings south-westwards into Wiltshire whence it continues to the Dorset coast, throwing off a broad arm eastwards into the Downs and Weald of Sussex and Kent. *Cretaceous Rocks of Britain* (Jukes-Browne, *MGS*, 1900/4). 'A Cretaceous time scale' (Hinte, *BAAPG*, 1976).

**Crinkle mark.** 'The slump structures of the *M. tumescens* flags [in N. Herefordshire] assume two main forms. The more obvious are folds of fairly large dimensions, which are quite comparable with slump structures described from other regions. The second form shows itself as a series of subparallel corrugations of the bedding-plane surface; and sectioning the rocks shows that these corrugations are a direct reflection of a crumpling of the internal lamination. The appearance of the bedding surface is so characteristic that the need is felt for a special name; it is here proposed to call the surface markings "crinkle marks"' (Williams and Prentice, *PGA*, 1957).

**Crinoidea.** A class of the phylum Echinodermata, subphylum Pelmatozoa, comprising the sea-lilies and featherstars of today. They were much more abundant and varied in past ages, ranging from the Tremadocian and being most prolific in the Silurian, Devonian, and Carboniferous. The test is globular, with arms, and in most cases it is fixed by a stoutly skeletonized stem. Broken fragments of stems are very abundant as fossils, sometimes forming masses of crinoidal limestone. Fossil crinoids were described by Beaumont (*PT*, 1676, 1683) under the title 'Rock-plants growing in the lead mines of Mendip Hills'. Miller's *Natural History of the Crinoidea or Lily-shaped Animals* (1821) is perhaps the first palaeontographical monograph, at least in Britain, restricted to one biological group. [G. *krinon*, a lily.]

**Crop.** See OUTCROP. 'The whole body of the metalls rises till they be at the very surface of the earth, which is here termed a "cropping". . . . At more than 100 paces distant the "crop" of a coal was found' (Sinclair, *Short History of Coal*, 1672). This specialized geological meaning of the term (as noun or verb) may be derived from its general meaning of 'cut off'; the strata (etc.) are 'cut off' by erosion at the ground surface, or, in a subcrop, by the unconformable formations.

**Cross-bedding.** 'Where the bedding planes within a bed are inclined more or less regularly to the separation planes between the beds, the arrangment is known as "cross-bedding". The alternative terms "false bedding" and "inclined bedding" are not recommended. False bedding might equally refer to pseudo-bedding; inclined bedding to any initial dip. "Current-bedding", being a genetic term, should include all bedding structures due to current action, but is commonly used for the small-scale ripple-like bedding of rapidly deposited sand. The term "ripple-bedding" is here preferred to this' (Hills, *Structural Geology*, 1963; with references). 'The classification of cross-stratified units' (Allen, *Sd*, 1963). 'Descriptive classification of cross-stratification' (Jacob, *Geoly*, 1973). 'A classification of climbing-ripple cross-lamination' (Allen, *JGS*, 1973). ('Cross-stratification' and 'cross-lamination' are terms appropriate when the condition is on the larger and the smaller scale respectively.)

**Cross-folding.** A system of folding in which there are two fold-trends, more or less at right angles. Usually one trend is dominant, the folds following the other trend being then termed 'cross-folds' (so that 'cross-folding' might apply in a restricted sense to these 'cross-folds' only). 'It deserves remark, that all the anticlinal axes are subject to cross rolls or undulations, of considerable amount, so as to break up each long axis into several short oval quaquaversal elevations' (Phillips, *Yorkshire*, 1836). 'Cross-folds' (Rast and Platt, *GM*, 1957). 'In the Highlands of Scotland the main folds of the Caledonian orogenic belt strike north-eastward, and are intersected by cross-folds striking roughly north-westward' (Holmes, *Physical Geology*, 1978).

**Crowstone.** A hard, brittle, pale-coloured quartzite (fine-grained with secondary silica), characteristic of the arenaceous rocks of the lower part of the Namurian (Carboniferous) series, particularly in north Staffordshire. The name (of obscure derivation) has been given to other rock-types in the Carboniferous, but the above is definitive. 'The "crowstones" of Staffordshire, Derbyshire and Cheshire' (Holdsworth, *NSJFS*, 1964).

**Crush-belt.** A belt of intensely crushed rock. 'The Great Glen is flanked on the south-east by a broad crush-belt characterized by intense cataclasis and localized mylonitization' (Eyles and MacGregor, *GM*, 1952).

**Crush-breccia.** See BRECCIA (6). 'Crush-breccia of slates of different colours [at Porthluney Cove, Cornwall] with small lenticles of igneous rocks and quartzite. This shows the general structure of the area, which is similar under the microscope or on a large scale with lenticles a hundred yards long' (Reid, *MGS*, *Mevagissey*, 1907).

**Crush-conglomerate.** See CONGLOMERATE. 'The crush-conglomerates of the Isle of Man' (Lamplugh, *QJ*, 1895, who seems to have established the term in this connexion; but recent work, by Gillott, *LMGJ*, 1956, shows that these particular conglomerates, or breccias, are probably not of tectonic origin).

**Crust.** See EARTH'S CRUST.

**Crustacea.** A class (mainly aquatic) of the phylum Arthropoda, in which the exoskeleton is frequently a hardened crust, particularly over the back of the animal. It includes the subclasses Branchiopoda, Ostracoda, Cirripedia and Malacostraca. The Trilobita, too, are sometimes included as a subclass. [L. *crusta*, the hard surface of a body.]

**Crust-Mantle boundary.** See EARTH'S CRUST. 'The crust-mantle boundary' (symposium papers, *JGS*, 1977).

**Cryopedology.** 'Cryopedology is suggested as a suitable name for the subscience concerned with the study, both theoretical and practical, of intensive frost action and permanently frozen ground [permafrost]' (Bryan, *AJS*, 1946). [G. *kruos*, frost, *pedon*, the ground, earth]

**Cryoturbation.** The geological action of frost; a term first suggested in Holland during the latter part of the 19th century. 'Large-scale superficial structures due to mass movements are distinguishable from surface structures due to cryoturbation by their much greater size and regional effect. Very shallow disturbances due to cryoturbation such as involutions, fossil frost wedges, patterned ground and the like, produced at or near the surface of perennially or seasonally frozen ground, are also found in the Midlands' (Hains and Horton, *BRG, Central England*, 1969). [G. *kruos*, frost, L. *turbo*, disturb.]

**Cryptocrystalline.** See CRYSTALLINE. [G. *kruptos*, hidden.]

**Cryptoexplosion structure.** 'Roughly circular structures with elevated craters consisting of materials forced up from below in disordered fashion were called "cryptovolcanic" by Branco and Fraas [Germany] in 1905. In the last decades it was realized that the impact of a giant meteorite may have caused the explosion. Dietz's term "cryptoexplosion structures" [*JG*, 1959] is useful; it leaves the cause unspecified. For the same reason, open craters surrounded by a low wall of ejected bedrock fragments should be called "explosion craters", not "meteor craters" ' (Bucher, 'Cryptoexplosion structures caused from without or from within the earth: "astroblemes" or "geoblemes"?', *AJS*, 1963, see also Dietz *ibid.*). It might be simpler to have the general term 'explosion structure' and the special term 'explosion crater'; or to have 'cryptoexplosion', instead of 'explosion', in both terms.

**Cryptographic.** See GRAPHIC.

**Cryptovolcanic structure.** 'We can define a "cryptovolcanic structure" as a natural explosion structure . . . related to volcanism but without the extrusion of volcanic rock or any marked hydrothermal effects' (Dietz, *AJS*, 1963). See CRYPTOEXPLOSION STRUCTURE.

**Cryptozoic.** See PHANEROZOIC. (Pre-Phanerozoic is synonymous.) The term is occasionally restricted to the Archaean.

**Crystal.** **1.** A homogenous chemical substance having, at least potentially, a definite geometrical form. This character is now known to be the expression of an internal atomic structure which is arranged on a definite plan, in definite space-lattices. In speaking of a 'crystal' it is often implied that part at least of the geometrical form is realized in the specimen, and in chemistry and mineralogy the term is sometimes defined in this sense; but in petrology and geology generally a 'crystal' implies 'having the essential crystal character' (having the crystalline structure), whether or not any of its proper faces are developed. **2.**=quartz; an obsolete but historically important meaning. 'Theophrastus, the pupil and successor of Aristotle, in his book on stones, mentions one called crystal (κρύσταλλος [*krustallos*]). This name, apparently, is derived from κρύος [*kruos*], icy cold, and στέλλειν [*stellein*], to contract. It would seem from this that the ancient Greeks believed crystal (modern quartz) to be a variety of super-cooled ice. Pliny the Elder (*c.* A.D. 60) says that crystal is "a substance which assumes a concrete form from excessive congelation", and goes on to explain that crystal is only to be found "in places where the winter snow freezes with the greatest intensity". Pliny's description of crystal, however, leaves no doubt that he applied this name to the crystalline variety of silica which is now called quartz. This meaning of the word crystal survived until almost the end of the 18th century and was then replaced by the name quartz' (Tomkeieff, *MM*, 1942). 'I stile them chrystalls, because many of them are composed of two hexagonal pyramids, and an intermediate column, likewise hexagonal, which according to Steno [1669] is the very definition of a chrystall' (Plot, *Staffordshire*, 1686). [G. *krustallos*, clear ice, quartz, a crystal.]

**Crystal system.** A primary division in the classification of crystals according to their form. Each of the six systems comprises several of the 32 possible symmetry classes. The crystal systems are: cubic, tetragonal, hexagonal, orthorhombic, monoclinic, triclinic.

**Crystal tuff.** See TUFF.

**Crystalline.** Having the essential crystal character (depending on a regular atomic structure); thus primarily for a crystal. Also for a rock composed of crystals, but here usage varies. An igneous rock (crystallized from a molten state) or a (recrystallized) metamorphic rock, composed entirely of various kinds of crystalline minerals, is so certainly and obviously 'crystalline' that the term tends to imply such a rock almost exclusively; but a quartzite, certain limestones, and particularly salt-deposits (evaporites), when entirely composed of contiguous crystals, must also be allowed to be 'crystalline'. An ordinary sandstone, though it may be predominantly composed of crystalline material (grains of quartz) is hardly spoken of as being itself a 'crystalline rock'. For igneous and metamorphic rocks, the following terms are used: 'macrocrystalline' or 'phanerocrystalline', the crystals large enough to be plainly seen with the naked eye; 'microcrystalline', the crystalline nature visible only under the microscope; 'cryptocrystalline', the crystalline nature only disclosed by the reaction of the aggregate to polarized light, the individual minerals being too small to be separately distinguishable. See also APHANITIC.

**Crystalline schists.** See REGIONAL METAMORPHISM. A useful general name, if rather a loose one. 'The crystalline schists and allied products of regional metamorphism have acquired their distinctive characters as the result of reconstitution of the substance, which took place in response to continued rise of temperature, but was further determined by definite mechanical conditions' (Harker, *Metamorphism*, 1950). 'In 1904 appeared the first edition of Grubenmann's great work *The Crystalline Schists* [*Die kristallinen Schiefer*]' (Turner and Verhoogen, *Igneous and Metamorphic Petrology*, 1960).

**Crystallinity. 1.** The property of being crystalline. **2.** The degree to which an igneous rock is crystalline or otherwise, that is, whether holocrystalline, glassy, or partly crystalline. **3.** The degree to which the crystalline nature of an igneous rock is developed (macrocrystalline, microcrystalline, or cryptocrystalline) or is apparent (phanerocrystalline or aphanitic).

**Crystallite.** Crystallites are embryo crystals (tiny globules, rods, or hair-like bodies) not referable to any definite mineral, and not having the polarizing property, occuring in glassy rocks. 'The word crystallite was first used by Sir James Hall to denote the lithoid substance obtained by him after fusing and then slowly cooling various "whinstones". Since its revival in lithology it has been applied to the minuter bodies above described' (Geikie, *Textbook*, 1903).

**Crystallize.** In geology, the intransitive sense 'to become crystalline in structure' is naturally much used. Hence 'crystallization'. See quotation (Playfair) under VEIN.

**Crystalloblastic.** For those textures and crystallographic forms in a metamorphic rock that are due to the new minerals growing, in the process of recrystallization in a more or less solid medium, by forcibly thrusting their way outward against the resistance of the rock substance they are replacing. A relatively strong mineral that succeeds in establishing its own crystal form against the resistance is 'idioblastic', a relatively weak mineral that has a foreign crystal shape forced upon it is 'xenoblastic'. If one new mineral makes crystals of much larger size than the other constituents of the rock, the structure is 'porphyroblastic' and the new crystals are 'porphyroblasts'. A texture, produced in metamorphism, resembling the igneous poikilitic is 'poikiloblastic' (Angus, *GM*, 1962). 'Granoblastic' is a mosaic arrangement of the crystals in a metamorphic rock.

**Crystallogenesis.** The origin and natural production of crystals.

**Crystallography.** 'The science of crystals' (Phillips, *Crystallography*, 1946); the study of their form, constitution, physical properties, and classification. It is especially concerned with crystal-form; that is, with the symmetry of crystals and the disposition of crystal-faces. *Origins of the Science of Crystals* (Burke, 1966).

**Crystallology.** (Crystallogy.) The whole study of crystals, morphological, physical, chemical, genetic. Includes crystallography, the other aspects of crystallology being more conveniently taken under the wider sciences of mineralogy and petrology. The term is thus seldom used.

**Cuesta.** See ESCARPMENT.

**Culm.** Coal dust or impure sooty coal, particularly in S. Wales and Devon. [? Old English *col*, coal.]

**Culm Measures.** The extensive Carboniferous formation occupying all the central part of the regional synclinal area of Devon and Cornwall, ranging stratigraphically from the upper part of the Carboniferous Limestone series to the lower part of the Coal Measures series. Lithologically the beds are chiefly shales and sandstones with, in some places, limestones and soft sooty coal (culm). (Sedgwick and Murchison, *BA*, 1836, *PGS*, 1837, *TGS*, 1840.)

**Culmination.** In full: 'tectonic culmination'. 'In the fold-mountains and foothills of the Zagros ranges of Iraq and South-West Persia . . . individual anticlines have lengths of up to 250 miles, following long straight courses and rising and falling into culminations and saddles' (Lees, *QJ*, 1952). 'The Teifi anticline . . . Ordovician (Bala) rocks emerge in the Plynlimon culmination' (George, *BRG*, *S. Wales*, 1970). The term is particularly used in describing tectonics of the Alpine type where a nappe is itself, as a whole, bent into a dome-shaped antiform (due perhaps to a regular main-folding and cross-folding of the nappes). The erosion of a nappe-culmination may result in the domed back of a lower nappe, or basement rocks, being revealed through the 'window' thus formed.

**Culture.** See PALAEOLITHIC.

**Cumulate.** 'In layered intrusions the lower levels contain minerals of presumed high-temperature crystallization and upwards the minerals are of progressively lower temperature type. This relationship, together with textural features, has led to the view that discrete crystals successively separated from the magma and accumulated, as a result of their greater density, at the bottom of the liquid, building up gradually to form a layered series. Crystal accumulation without remelting was advocated by Bowen for the origin of certain ultrabasic rocks (*Evolution of the Igneous Rocks*, 1928), and the term "accumulative rocks" has gradually appeared in petrological literature. Now that an origin by crystal accumulation has been widely accepted for certain eucrites, gabbros, and ferrogabbros as well as for various extreme rock types such as dunite, pyroxenites, and chromitites, it is convenient to have a single short name for the group and we suggest the word cumulate (Latin *cumulus*, a heap)' (Wager and others, *JPt*, 1960). 'Some problems of the cumulate theory' (Campbell, *L*, 1978).

**Cumulose.** Chiefly for the accumulations in place of the residue of partly decomposed vegetable matter. Peat is the most typical deposit of this kind.

**Cupola.** A relatively small upward projection from a batholite. (Daly, *Igneous Rocks*, 1933.) [Italian f. L. *cupula*, a small tub; adaptation of its architectural usage.]

**Current bedding.** See CROSS-BEDDING, CROSS-STRATIFICATION.

**Curve of erosion.** The theoretical profile curve produced as the consequence of erosion. This curve may differ for different climates and different erosive agents, the curve usually considered being that produced in a temperate region undergoing normal erosion. Exactly what is this theoretical curve is not thoroughly understood, but it appears to be a primary convexity with a concavity superimposed as a result of control by base level, and it also appears likely that the curve is fundamentally the same whether taken along a stream course, down a valley side, in the shapes of hills generally, across a coast, or along a skyline.

**Cut-off.** See MEANDER.

**Cuvette.** '"Cuvette" is a convenient term for a basin in which sedimentation is going on (German *Sammelmulde*), as distinct from a tectonic "basin" due to folding of pre-existing rocks, and not necessarily basin-shaped as far as the present structure is concerned, e.g. the Anglo-Parisian cuvette was the region in which the Lower Cainozoic rocks of Britain and North France accumulated. It is now folded into a number of basins—Paris, Hampshire, and London Basins' (Wills, *Physiographical Evolution of Britain*, 1929). See BASIN. [French, a basin.]

**Cwm.** See CIRQUE. [Welsh, valley, dale.]

**Cycle.** The only truly appropriate use of this term seems to be for a complete round of circumstances, events, records; usually implying that it is one that recurs in the same order, several or many times. One such round, cyclic unit, is a 'cyclothem'. Examples of such cycles are the 'geological cycle', 'metamorphic-plutonic cycle', 'cycle of sedimentation'. The term is also used in the sense of a course of events which runs to completion, the last state being quite different from the first. This inappropriate use is exemplified in the often employed term 'cycle of erosion' or 'geographical cycle'; also in 'igneous cycle'. Hence 'cyclic', 'cyclical'. See CYCLIC SEQUENCE. (Barrell, *BGSA*, 1917.)

**Cycle of erosion.** See GEOGRAPHICAL CYCLE.

**Cycle of sedimentation.** 'We find . . . a similar order of sequence and of mineral composition . . . geological cycles reproducing themselves at distant intervals' (Hull, *QJS*, 1869). 'According to Mr. Hull a natural cycle of sedimentation consists of three phases: 1st, a lower stage of sandstones, shales, and other sedimentary deposits; 2nd, a middle stage, chiefly of limestone, representing general quiescence; 3rd, an upper stage, once more of mechanical sediments indicative of proximity to land' (Geikie, *Textbook*, 1882). 'The oscillation of the water-level in the Anglo-Franco-Belgian Basin enables the beds to be grouped into a series of "cycles of sedimentation", each cycle commencing with the deposits of a marine invasion, followed by shallow then deep water marine beds. These in turn gradually become of shallower-water type as estuarine conditions spread seawards, and so pass up into continental deposits. The latter are cut off abruptly by the marine invasion, which commences the next cycle' (Stamp, *Stratigraphy*, 1923; first expounded, *GM*, 1921). Hence 'cyclic sedimentation'. *Symposium on Cyclic Sedimentation* (Merriam, ed., *BKSGS*, 1964). *Cyclic Sedimentation* (Duff and others, 1967). See quotations under MARINE SAND (Edwards and Stubblefield) and RHYTHMIC SEQUENCE (Shotton). The cycle of sedimentation is a true cycle, in the appropriate sense; a cyclothem.

**Cyclic sequence.** The sequence in a cyclothem. This sequence may regain its starting-point, thus initiating a new unit, either by taking a circuitous route so to speak, or by retracing its steps (swinging like a pendulum). In the first case, the sequence would be of the type ABCDABCDAB, &c., in the second, ABCDCBABCDCBAB, &c. It will be seen that although, in these examples, there are four kinds of element in each, in the first case there are four elements to a unit, in the second case, six. The division between units may be taken at any repeated boundary but would be most appropriately taken at the most marked boundary, that is, the one that seems to mark the episode of most rapid and profound change. The circuitous cycle is probably the commonest and typical sequence, yet some would call this a 'pulsatory' sequence, using 'cyclic' only for the oscillatory type, for which the term, in any case, is less appropriate.

**Cyclothem.** See CYCLE and MESOTHEM. 'The word "cyclothem" is therefore proposed to designate a series of beds deposited during a single sedimentary cycle of the type that prevailed during the Pennsylvanian period' (Weller, *BGSA*, 1932). 'The Wealden cyclothems in the Weald' (Allen, *PT*, 1959). [G. ***kuklos***, a circle, *tithēmi*, deposit.]

**Cylindrical fold.** See FOLD DIRECTION.

**Cymatogeny.** A geomorphological concept named and elaborated by Lester-King (*Morphology of the Earth*, 1962/67). 'By such deformations, in which only vertical displacement is involved, the earth's surface is thrown into gigantic undulations, sometimes measuring hundreds of miles across and with vertical displacement amounting to thousands, or even tens of thousands of feet.' See SWELL. [G. ***kuma***, a condition of swelling, particularly the swell of the sea.]

**Cystidea.** An extinct class of the Echinodermata (subphlyum Pelmatozoa); the test very varied in shape but usually more or less globular. The five-rayed symmetry is often imperfect and the polygonal plates form an irregular mosaic. They range from the Middle Cambrian to the Devonian; rather rare, but least uncommon in the upper part of the Ordovician. (von Buch, 1844.) [G. ***kustis***, a bladder.]

# D

**'D' numbers** See 'F' NUMBERS.

**Dacite.** Quartz-andesite, approximately the fine-grained equivalent of granodiorite. (Stacke, 1863.) [Dacia, the Roman name for a part of Hungary.]

**Dalradian.** The highly folded and metamorphosed series (metamorphic assemblage) of rocks of varying original sedimentary types and some volcanic types, occurring over about the south-east third of the Scottish Highlands. The deformation and metamorphism is that of the Caledonian orogeny (*sensu lato*); the age of the rocks themselves is thought to be partly Precambrian and partly Cambrian and in any case almost certainly younger than the Moine series. Probably the earliest extended notice of the series is that by Williams (*Mineral Kingdom*, 1789). 'It is well known that from the old kingdom of Dalriada, in the north of Ireland, a colony settled in Argyllshire, and gradually acquiring dominion over the whole of Scotland, gave that kingdom its present name. I would therefore propose that the term "Dalradian" might be adopted as an appropriate and useful appellation for the crystalline schists of the north of Ireland and centre and southwest of Scotland. (The adjective ought properly to be "Dalriadan", with the accent on the second syllable; but I feel compelled to alter it into a form more consonant with English habits of pronunciation.)' (Geikie, *QJ*, 1891).

Definition further discussed in Johnstone, *BRG*, *Grampian Highlands*, 1966. Called the 'Dalradian supergroup' in *GSSR* (2), (Cambrian), 1972, and *GSSR* (6), (Precambrian), 1975.

**Danian.** See CRETACEOUS SYSTEM and EOCENE (placed by some authorities in the former and by others in the latter). Absent in Britain, well developed in Denmark.

**Days in a year.** Certain modern shells and corals have corrugations which seem to be annual and, between these, many fine striations which seem to be diurnal. Fossils with a similar ornamentation have been examined and it has been found that these have more than 365 striations between the corrugations, the older they are the more they have; thus those from the Upper Cretaceous have 370, those from the Lower Devonian, 400. These figures (which are approximate) agree with astronomical calculations that there were more days to the year in past geological ages. See Berry and Barker, *N*, 1968, and Panella and others, *S*, 1968.

**Death assemblage.** See THANATOCOENOSE and LIFE ASSEMBLAGE.

**Debris.** 'The fragments of rocks; the ruins of strata; the rubbish, sand, grit, &c., brought down by torrents' (Humble, *Dictionary of Geology and Mineralogy*, 1840). See quotation (Greensmith) under CROSS-BEDDING. [French, *débris*.]

**Decapoda.** See MALACOSTRACA. [G. *deka*-, ten.]

**Decke** (pl. **Decken**). See NAPPE. [German, a cloth.]

**Decollement.** Rupture resulting from folding; particularly that occurring between a relatively rigid block and an overlying cover of more easily folded strata. 'Buxtorf came to the conclusion that the whole anticlinal zone [Jura] had been formed by sliding (*Abschering* or *décollement*) of the upper structures across an unfolded basement with a salty medium in between acting as a lubricant. . . . He imagined that the lubricant effect of the Triassic salt would allow this free development of rootless folds' (Lees, *QJ*, 1952). 'The hypothesis advanced here treats the overall structure of the Lower Palaeozoic pile as one of repeated decollement during accumulation accompanied by creep in the soft sediment and controlled broadly by a slope facing [the present] Cardigan Bay' (Davies and Cave, *SG*, 1976). [French *décollement*, becoming unstuck.]

**Decussate.** A microstructure developed particularly in thermally metamorphosed rocks. 'By an arrangment in which corresponding axes of contiguous crystals lie in diverse directions, the stresses [due to the growth of new crystals] can be made in great measure to cancel one another by mutual accommodation. This is the arrangement typically exhibited in thermally metamorphosed rocks, and the microstructure described, a criss-cross or decussate structure, is for such rocks highly characteristic. The component crystals lie in all directions; not at random, by the operation of a mathematical law of chance, but as part of a definite mechanical expedient for minimizing internal stress' (Harker, *Metamorphism*, 1950). [L. *decusso*, divide crosswise.]

**Dedolomitization.** Elimination of the magnesian carbonate constituent from the mixed magnesian and calcic carbonates of dolomite (mineral or rock-mass). 'Von Morlot (1848) considered that the replacement of dolomite by calcite was a possible process in limestones and coined the term "dedolomitisation". J. J. H. Teall (*GM*, 1903) used the term to describe the metamorphic transformation of the mineral dolomite, and it has since become established in metamorphic petrology' (Shearman and others, *PGA*, 1961).

**Deduction.** Reasoning from the general to the particular; finding implications in, and inferring consequences from, general principles. These inferences can then be checked by observation. It is a mental process not always given its proper priority in considering geological questions. An explanation or hypothesis in accord with deduction must provisionally be preferred to any other. The principle of 'Ockham's razor' demands this. To give two examples from geomorphology: a fairly level hill-top surface in a uniform rock is the logical consequence of the established principle of differential erosion; it cannot be used to suggest or support a hypothesis of the uplift of a peneplain formed at a lower level though in fact it may have been so formed in particular instances. Where along a fault a more resistant rock on one side stands higher than a less resistant rock on the other, the scarp must be provisionally taken to be a fault-line scarp, not a fault scarp, because, as a deduction from the most elementary principles, the universal and perpetual process of erosion will inevitably tend to produce that effect. (See FAULT-LINE SCARP.) Again, in evolutional palaeontology, the ontogeny of a structural feature (such as successive ammonite sutures) cannot be used as evidence of the hypothesis of recapitulation if the ontogeny would in any case have taken that course in the normal process of growth of the individual. The following quotation from W. M. Davis (*Essays*, 1909: *Rivers of New Jersey*) illustrates the distinction between deduction and induction: 'What kinds of rivers are these? Such a question can hardly be answered until we have examined rivers in many parts of the world, gaining material for a general history of rivers by induction from as large a variety of examples as possible; and until we have deduced from our generalizations a series of critical features sufficient to serve for the detection of rivers of different kinds wherever found.' The most famous occasions of the use of deduction in the whole history of geology are those when James Hutton, towards the end of the eighteenth century in Scotland, sought and found (1) unconformities as a result of deducing their existence from his principle of the geological cycle (see UNCONFORMITY) and (2) the penetration by granite of the surrounding rocks as a result of deducing this from his principle of the igneous origin of granite (see GRANITE).

**Deep sea.** Usually means the sea beyond the continental shelf, thus including both the bathyal and the abyssal zones (Wiseman and Ovey, *PGA*, 1950). But as 'deep-sea deposits' it sometimes refers specifically to the abyssal depths (see ABYSSAL DEPOSITS).

**Deep structure.** See UNDERGROUND GEOLOGY. This particular term is used chiefly when the results of geophysical methods, particularly gravity studies, are assessed. 'Deep structure' (Bott in Moseley, ed., *Geology of the Lake District*, 1978). 'The deep structure and dynamics of East Africa' (Maguire and Khan, *PGA*, 1980).

**Deep-seated geology.** See UNDERGROUND GEOLOGY. 'General remarks on the deep-seated geology of the London basin' (Whitaker, *QJ*, 1886).

**Deflation.** In geology, the blowing away of dry incoherent rock material, sand, and dust, by the wind; a form of transport (denudation) chiefly at work in deserts. (Very different from the ordinary meaning of releasing the air from something inflated.) 'We say of the wind that it "sweeps" over the ground; for this word means nothing else than that the wind clears the ground of all loose particles that cover it. Translated into technical geological language, it is called deflation' (Walther, *NGM*, 1893, quoted in *AGI Glossary*, 1957). [L. *flatus*, a blowing.]

**Deformation.** Any change in shape or structure in a rock-unit, on any scale. Also used (technically) to include change in size alone, although the word itself seems specifically to exclude this. 'Mechanisms of crustal deformation' (Turcotte, *JGS*, 1983). 'Eocene deformation on the continental margin SW of the British Isles' (Masson and Parsons, *JGS*, 1983).

**Deformation fabric.** See FABRIC.

**Deformation style.** See TECTONIC STYLE. 'Deformation styles' (Dewey, *QJ*, 1967).

**Deglaciation.** The removal of ice by melting, used particularly for whole regions. 'The deglaciation of Scotland' (Lacaille, *PGA*, 1948/50).

**Degrade.** In general (in geology), to wear down rocks to a lower level. 'It is admitted that most mountains were originally much higher than at present, having been successively degraded and lowered by various subsequent accidents; particularly by disintegration and decomposition. . . . subject to this degradation' (Kirwan, *Geological Essays*, 1799). See EROSION. It is often used, in particular, for the downward-eroding action of a river.

**Dehydration.** The removal of a water-constituent; particularly one from a chemical compound.

**Deltaic.** Particularly for deposits, those forming part of a delta. 'The assemblage of beds which make up the [Millstone Grit] series is such as to leave no reasonable doubt that the material represents the deltaic deposits of some large river comparable in size with such a one as the Mississippi' (Gilligan, *QJ*, 1919). 'Recent advances in the study of deltaic sedimentation' (Van Straaten, *LMGJ*, 1960). 'Sedimentary features of an ancient deltaic complex: the Wealden rocks of South-eastern England' (Taylor, *Sd*, 1963). Hence 'palaeodelta', 'palaeodeltaic'. 'Modern river deltas exhibit considerable morphological variation, primarily in response to the interaction of fluvial and marine processes which define the regime of the depositional area. For example, the morphology of the river dominated Mississippi delta may be contrasted with that of the wave influenced Rhone and Nile deltas, the tide influenced Brahmaputra delta, and the tide and wave influenced Niger delta. It therefore follows that studies of ancient deltaic sediments should aim to elucidate the regime and morphology of the palaeodelta' (Elliott, *JGS*, 1976).

**Denbighshire Grits and Flags.** The representatives of the Salopian series in (roughly) Denbighshire. First referred to, descriptively, by Sedgwick, *PGS*, 1838; and for long a well-known name in regional stratigraphy. Described in detail by Boswell, *Middle Silurian Rocks of North Wales*, 1949.

**Dendrite.** Usually in the plural; arborescent or moss-like growths ('dendritic'), within a mineral or stone, composed (most commonly) of oxide of either manganese or iron; e.g. the Cotham Marble, moss agate. 'Dendrites' (Van Straaten, *JGS*, 1978). See LANDSCAPE MARBLE. [G. *dendron*, a tree.]

**Dendroidea.** One of the two main orders of the class Graptolithina. The rhabdosome typically has many stipes, the thecae being of three kinds arranged in regularly alternating triads along each branch. They range from the Middle Cambrian to the Carboniferous. (Nicholson, 1872.)

**Density.** Mass of unit volume. In geology, usually refers to a mineral, a rock, or an earth-shell. The density of a rock may be that of a dry sample, a sample saturated with water, or the rock-mass as a whole *in situ* ('Density measurements of rocks in South-west Scotland', McLean, *PRSE*, 1961).

**Density current.** In sedimentary geology, the same as 'turbidity current'.

**Denudation.** Strictly, the laying bare of rocks by the removal of material covering them. 'The examination which I made in 1805 and 1806, of the south-eastern parts of England and the discovery of the great Southern Denudation [of the Weald, *PM*, 1810], shewed the necessity of introducing a new term for this geological phaenomenon, which has been explained in Dr. Rees's new Cyclopaedia [part issued *c.* 1808], article "Denudation"' (Farey, *Derbyshire*, 1811). Lyell (*Principles*, 1833) gives: 'The carrying away of a portion of the solid materials of the land, by which the inferior parts are laid bare'. A. Geikie in the first of his many famous books (*Story of a Boulder*, 1858) has: '"Denudation" is a geological term used to denote the removal of rock by the wasting action of water, whereby the underlying mineral masses are "denuded" or laid bare'. The term is sometimes used as synonymous with 'erosion' in its widest sense, that is, to denote the whole of the destructive and removal processes of weathering, transport, and corrasion. The strict distinction between 'denude' and 'erode' is at once obvious from such a sentence as 'longshore drift denudes the shore thus exposing it to erosion'. 'The concept of denudation in seventeenth-century England' (Davies, *JHI*, 1966). See EROSION. [L. *denudo*, lay bare.]

**Deplanation.** See PLANATION. (Cairnes, *AJS*, 1912.)

**Deposition.** The constructive process of the laying down of any kind of loose rock material. Hence 'depositional', as in 'depositional structures', 'depositional fabrics'. The complement of the destructive process of erosion or denudation (in their widest senses).

**Derivative.** (In rock-classification.) See SEDIMENTARY ROCK. 'We have thus three more or less well-marked types of rock, which may be designated Igneous, Derivative, and Metamorphic respectively' (J. Geikie, *Structural and Field Geology*, 1905).

**Derived fossil.** A fossil which was removed (by erosion) from a pre-existing rock or deposit to become embedded in a new deposit. It is thus not representative of the fauna or flora of the time when the new bed was laid down and is to be distinguished from the contemporary, native, fossils of the bed. See ASSEMBLAGE (1). Derived fossils are both out of place and out of time. 'The fossil contents of the Cambridge Greensand are readily divisible into two groups or faunas: the one of these is derivative, the other is *in situ* . . . appended list of derived fossils . . . considering the derived fauna first . . .' (Jukes-Browne, *QJ* 1875). 'The derived Cephalopoda of the Holderness Drift' (Thompson, *QJ*, 1913). 'Derived ammonites [chiefly of Oxford Clay age] from the Lower Greensand of Surrey' (Arkell, *PGA*, 1939). 'Derived Upper Llandovery fossils in Bunter pebbles' (Lamont, *CLG*, 1940).

**Descriptive term.** In proposing, using, interpreting, or adapting terms it is obviously important to distinguish between a purely descriptive term for an observable fact and a genetic term that implies a particular cause, origin, or not directly observable relation. For instance, 'hypabyssal' and 'turbidite' clearly refer to modes of origin, while 'quartz-porphyry' and 'sandstone' as clearly refer solely to texture and composition.

**Desert régime.** 'Throughout the Triassic Period deposition of red sands and red and green mottled muds continued in shallow "dead" seas surrounded by deserts. As in the Permian, evaporation was high enough to cause the precipitation of the calcium sulphates anhydrite and gypsum, and rock salt. . . . The desert regime in northern England was ended by widespread submergence beneath the Rhaetic sea' (Taylor and others, *BRG, Northern England*, 1971).

**Desert varnish.** 'A surface stain or crust of manganese or iron oxide, of brown or black color and usually with a glistening luster, which characterizes many exposed rock surfaces in the desert' (Bryan, *BUSGS*, 1922.)

**Desk-structure.** 'The Mountain Limestone [Carboniferous] country has a character of its own, marked by long scars and gentle slopes, the escarpments and dip slopes of the usually gently inclined strata. This structure is often known as desk-structure, recalling the appearance of a writing-desk on a large scale' (Marr, *Lake District*, 1916).

**Desquamation.** See EXFOLIATION. (Richthofen, 1886.) [L. *desquamo*, scale off.]

**Detrital.** Worn off by breaking or rubbing; for mineral grains, rock particles, stones, or a mass of material (detritus) worn away from pre-existing rocks. In sedimentary petrography, 'detrital' often refers more particularly to the 'accessory' or 'heavy' minerals. See FRAGMENTAL DEPOSITS. [L. *detritus*, worn away.]

**Detritus.** Disintegrated rock material, usually moved from the site of origin. 'The quantity of detritus brought down by the rivers' (Playfair, *Illustrations*, 1802). 'This grit is an aggregate formed from a detritus of granite' (Macculloch, *TGS*, 1811). See quotation under COLLUVIAL. Occasionally, any fragmental material such as pyroclastic (see quotation under EXTRUSION) or organic (shelly).

**Deuteric.** Particularly for minerals and textures in an igneous rock that have arisen secondarily as a result of changing conditions during the cooling of the magma, such as those produced in the reaction process; as distinct, on the one hand, from the primary features of direct crystallization and, on the other, from those later changes more properly classed as pneumatolytic. (Sederholm, 1916.) [G. *deuteros*, second.]

**Devitrification.** The transformation of glass into crystalline matter. 'Devitrification in glassy igneous rocks' (Bonney and Parkinson, *QJ*, 1903).

**Devolatilization.** The loss of volatile constituents; particularly those of a coal, whereby the proportionate carbon content is increased and the rank of the coal becomes higher. Anthracite is a coal of high rank, being largely devolatilized. It is a form of metamorphism due, perhaps, to orogenic pressure or merely to lapse of time, depth of burial, or some other persistent cause. 'The devolatilization of coal seams in South Wales' (Trotter, *QJ*, 1948). Hence 'isovol', a line or surface of equal volatile constituent (that is, everywhere at the same degree of devolatilization).

**Devonian system.** Overlies the Silurian and is succeeded by the Carboniferous. There are (in Britain) two quite distinct facies; the marine Devonian and the Old Red Sandstone. The latter was investigated first, but the normal marine development of the system obviously became the type. 'Devonian', applied to strata and formations, usually implies the marine facies; applied to the system or the period as a whole it includes the Old Red Sandstone. The marine Devonian was first worked out in S. Devon and Cornwall and was found to have a fauna intermediate in character between that of the Silurian and that of the Carboniferous; moreover, in that region it passes up into the Carboniferous, but the base was not seen. It was later found that the strata were most complete, most fossiliferous, and were clearest in their stratigraphy on the Continent, particularly in the Rhineland and the Ardennes. The passage upwards from the Silurian, however, has been most clearly discerned in various parts of the Mediterranean region, eastern Europe, and the Sahara. As to the fauna, the trilobites are less important than in the Lower Palaeozoic systems. Brachiopods and corals are the most abundantly and variously represented groups; of the former the genera *Stringocephalus* and *Uncites* and, of the latter, *Calceola*, *Phillipsastraea*, and *Heliophyllum* are quite or very nearly confined to the system. Mollusca are important; lamellibranchs, gastropods, nautiloids, and, particularly, the ammonoids (the early goniatites and the Clymenidae). Stromatoporoids and crinoids are common. 'The Devonian System in South Devonshire' (Dineley, *FS*, 1961). *A Correlation of Devonian Rocks in the British Isles* (*GSSR* (8), 1977). *The Silurian–Devonian Boundary* (Martinsson, ed., 1977). *The Devonian System* (House and others, eds., *PA*, 1979). See DOWNTONIAN and OLD RED SANDSTONE. (Sedgwick and Murchison, *PGS*, 1839, *TGS*, 1840; Lonsdale, *TGS*, 1840.) The following stages are recognized, particularly on the Continent: Lower—Gedinnian, Siegenian, Emsian; Middle—Eifelian, Givetian; Upper—Frasnian, Famennian.

**Dewatering.** The elimination of water from a rock during consolidation and any subsequent tectonic pressure.

**Dextral fault.** A tear (wrench) fault in which the movement on the far side (as viewed from either side) has been to the right. The reverse case is a sinistral fault. It should be noted that outcrops of dipping beds will be offset to the right or the left whether or not the fault movement had any strike-slip component; the terms dextral and sinistral apply to tear (wrench) faults only. The distinction between a dextral and a sinistral fault has only a very limited structural significance; for instance, one lateral movement of a block between two parallel stationary faults causes one of them to become dextral and the other sinistral. (Anderson *Dynamics of Faulting*, 1942.)

**Diabase.** This name (not now much used in Britain, but extensively used in America) has had a complicated history and still carries somewhat different meanings but, on the whole, it may be taken, as synonymous with 'dolerite'. In British geology it tends to mean an altered Palaeozoic dolerite. [G. *diabasis*, transition.]

**Diachronous.** Cutting across the time-planes; applied particularly to a bed of rock or a stratigraphical formation. 'It is now proposed to introduce the term "diachronous" to describe a bed having such relations to the zonal succession. The word is self-explanatory, and avoids a cumbersome circumlocution' (Wright, 'Stratigraphical diachronism in the Millstone Grit of Lancashire', *BA*, 1926). 'The Dolomitic Conglomerate may pass laterally into rocks varying from Bunter to Upper Keuper in age and is therefore regarded as a diachronous marginal facies' (Welsh and others, *BRG, Bristol and Gloucester*, 1948). Boswell (*TLBS*, 1921) has given diagrams (often reproduced since) showing how lithological divisions will tend to cut across time-planes as a result of deposition on subsiding or rising sea-bottoms.

The term is applicable in other connexions; as in the case where a phase of deformation moves in time as it affects the rocks across a stretch of country: 'Diachronous deformation' (Phillips and others, *JGS*, 1976).

**Diagenesis.** 'Diagenesis may be broadly defined as including the processes of physical and chemical change which take place within a sediment after its deposition and before the onset of either metamorphism or weathering . . . the modification of mineral textures during diagenesis is termed "neomorphism"' (Knox, *Petrology for Students*, 1978). 'Even after sediment reaches its final resting place it may still undergo important changes during its transformation into rock. Walther (Jena, 1894) introduced the term "diagenesis" to embrace all such changes that result from sedimentary processes—but excluding metamorphism. (The term was first used by Guembel in 1888 and applied to metamorphic changes, but Walther specifically excluded these, and his usage has since been generally accepted)' (Dunbar and Rodgers, *Stratigraphy*, 1957). *Diagenesis in Sediments* (Larsen and Chilingar, eds., 1968). 'The birth and development of the concept of diagenesis (1866-1966)' (Dunoyer de Segonzac, *ESR*, 1968). 'Stromatoporoid diagenesis' (Riding, *GM*, 1974). 'Sandstone diagenesis' (symposium papers, *GS*, 1978).

**Diagnostic fossil.** See CHARACTERISTIC FOSSIL.

**Diamictite.** A terrigenous sedimentary rock that contains a wide range of particle sizes. (Flint and others, *BGSA*, 1960.) [G. *mignumi*, mix.]

**Diamond.** The crystalline form of carbon; the hardest known mineral, with brilliant lustre, a very high refractive index, and a high specific gravity. Diamonds occur (1) as rare constituents of ultrabasic igneous rocks, chiefly in the 'diamond pipes' of S. Africa and (2) washed out and redeposited in alluvial gravels (placers), their weight and hardness being favourable to such an occurrence. *Mineralogy of the Diamond* (Orlov, 1977).

**Diaphthoresis.** See RETROGRADE METAMORPHISM. (Becke, 1904.) [G. *diaphthora*, a destruction.]

**Diapir.** 'In 1910 Mrazec, working in the foothills of the Carpathians, coined the term diapir (through-piercing) to describe anticlinal folds in which the underlying older rocks (e.g. salt) pierce upwards through the vault of the anticline. Wegmann (1930) used the term diapir to describe similar piercement-folds where the core of an anticline, undergoing granitization, had moved relatively upwards and, like salt, had pierced the anticlinal arch' (Reynolds, *GM*, 1958). The term is applicable to any kind of anticlinical core, and to any igneous intrusion, so behaving. 'Some of the folds of the Kingswood Anticline are diapiric, i.e., they burst up through the overlying strata' (Welsh and others, *BRG, Bristol and Gloucester*, 1948). 'The Ardara granitic diapir of County Donegal' (Akaad, *QJ*, 1956). Some sedimentary injections (auto-intrusions) have been called 'diapirs' (e.g. Ager and Wallace, *PGA*, 1966). *Diapirism and Diapirs* (Braunstein and O'Brien, eds., 1968). 'Part of the Murdafil diapir in the foothills of the Zagros simply folded belt [SW. Iran].

The diapir consists of gypsums and mudstones, thin limestones and marls . . . intruded into [a formation of] sandstones and mudstones. The diapir is about one mile across at its widest' (pl. in Falcon in Kent and others, eds. *Time and Place in Orogeny*, 1969). [G. *diapeirō*, pierce through.]

**Diastem.** See NON-SEQUENCE and UNCONFORMABLE (table). 'A break represented in other regions, often within the same formation, by a bed or series of beds' (Barrell, *BGSA*, 1917). This definition seems to make the term synonymous with 'non-sequence', and this is a common usage, but Hills (*Structural Geology*, 1963) suggests that Barrell had relatively small intervals of time particularly in mind, and this is a useful restriction. [G. *diastēma*, an interval.]

**Diastrophism.** 'I find it advantageous to follow J. W. Powell in the use of "diatrophism" as a general term for the processes of deformation of the earth's crust. The products of diastrophism are continents, plateaus and mountains, ocean beds and valleys, faults and folds. Its adjective is "diastrophic". It is convenient also to divide diastrophism into orogeny (mountain-making) and epeirogeny (continent-making)' (Gilbert, *Lake Bonneville*, 1890). It is the latter aspect that is sometimes more particularly implied, as in a definition by Powell himself (*Processes*, 1895): 'Regions sink and regions rise and the upheaval and subsidence may be called diastrophism'. 'Diastrophism and mountain building' (Billings, *BGSA*, 1960). [G. *diastrophē*, distortion.]

**Diatom ooze.** See OOZE.

**Diatomaceae.** Diatoms; unicellular algae in which the cell-wall is impregnated with silica forming two halves like a box and its lid. In the genus *Diatoma* itself, and rectangular individuals are connected by their alternate angles to form a zigzag chain. [genus *Diatoma* (G. *diatomas*, cut through).]

**Diatomaceous earth.** See SILICEOUS ORGANIC DEPOSIT.

**Diatomite.** See SILICEOUS ORGANIC DEPOSIT. A diatomaceous earth, hardened and dried artificially for industrial use, also becomes 'diatomite' commercially.

**Diatreme.** A hole (filled up) going right through a rock-structure, produced by igneous activity. 'A small plug-like body is made up of an intrusive breccia contained within a ring of felsite. Attention is drawn to the association of this diatreme with the suite of felsite, hornblende lamprophyre, and hornblendite intrusions clustered around the diapiric pluton of Ardara' (French and Pitcher, *GM*, 1959). [G. *trēma*, a hole.]

**Dibranchiata.** (Dibranchia.) An order of the class Cepalopoda including all the modern cephalopods (e.g. the cuttle fish) except the one genus *Nautilus*. The skeleton, where present, is internal; most elaborate and least unlike the shell of the other groups of cephalopods in the extinct Mesozoic suborder Belemnoidea. [G. *bragchia*, gills.]

**Differential compaction.** In the first place, there are different degrees to which deposits of different kinds become compacted under a given load. Thus, for example, a newly-formed deposit of mud passing laterally into an equally thick deposit of sand will eventually, owing to its greater susceptibility to compaction, form a bed of mudstone thinner than the corresponding bed of sandstone. Secondly, in a deeper part of the sea floor, particularly where there is differential subsidence going on, the greater thickness of sediments accumulating there will result in greater compaction giving, perhaps, an increase to the initial (depositional) dips on the submarine slopes. These two considerations are combined if we visualize deposition of a compactible deposit bordering a relatively upstanding and relatively incompactible mass such as a limestone reef.

**Differential erosion.** Unequal reaction to a uniform process of erosion. Rocks vary in their resistance, so that erosion is selective in its effects and the more resistant rocks stand out while the less resistant are more rapidly worn away. It largely depends on the relative hardness of the rocks, but relatively soft rocks may be resistant if much of the rain sinks in instead of forming erosive streams. 'Now concerning the exaltation of the mountains above the vallies it appeareth to come to pass by the waters in former times, whose property is to wear away by its motion the most loose earth, and to leave the more firm ground and rocky places highest' (Platter, *Subterraneal Treasurers*, 1738, quoted by Green, *Physical Geology*, 1876). The principle was expounded by Hutton in his *Theory of the Earth* (1795). 'The Coalfield itself is sharply delineated by the differential erosion of hard and soft beds' (Pringle and George, *BRG, S. Wales*, 1970). 'To the differential erosion of these hard igneous masses and the softer sediments among which they lie is largely due the varied scenery so characteristic of the region' (Macgregor and Macgregor, *BRG, Midland Valley of Scotland*, 1948). Differential erosion results in the 'topographical expression' of structure and this has been called the 'law of structures' (Lake and Rastall, *Textbook*, 1910/41). 'Differential weathering' is one kind of small-scale differential erosion and is conspicuous (or at least visible) on nearly every exposure of bare rock. See STRUCTURE AND SURFACE.

**Differential structural response.** The structures produced in a rock mass as a result of a particular force depend on the lithological character of the mass (thus slaty cleavage, competent and incompetent folding, fault-structures, etc.). 'The rocks consist of hard and thick bands of grit, and soft shales with thin bands of grit. The hard bands have been bent into a sharp anticline, while the overlying softer strata have been squeezed, upturned, and shifted' (Woodward, description of a *BA* photograph of Culm Measures at Cockington Beach, Devon, *c.* 1903).

**Differential subsidence.** Of the sea floor; with consequent difference in the thickness of sediments. 'There remained a tendency for the old-established Caledonian elevations to persist as "positive" areas or "blocks" where subsidence was less than in the intervening parts of the crust. The result was that sequences of strata laid down in the intervening basins were thicker than the time-equivalents over the "blocks". This tendency for differential subsidence persisted throughout the Namurian, but by Westphalian times it had almost ceased' (Taylor and others, *BRG, Northern England*, 1971).

**Differential weathering.** See DIFFERENTIAL EROSION.

**Differentiation. 1.** In igneous petrogenesis: the separation of a homogeneous magma into chemically unlike portions. This is the most usual sense of the term. 'Whatever view we take as to the nature of silicate-magmas, there can be no doubt that in general the process of consolidation is a process of differentiation' (Teall, *QJ*, 1901). **2.** In sedimentary petrogenesis: the separation of a rock-mass (of one or more petrographical types) undergoing erosion into physically and chemically unlike products which are re-sorted and deposited as sediments over more or less separate areas. 'Sedimentation processes result in a sedimentary differentiation, both physical and chemical, whereby the parent igneous rocks provide four great classes of sediments—the psephites, psammites, pelites and carbonates. Some of you may doubtless regard the first three names with no great affection, but I would remind you that they were introduced more than a century ago by no less figures than Haüy and Naumann' (Read, *PGA*, 1958). ' "The largest and least decomposed part of the material [from a land area undergoing erosion] was deposited near the land, while the smaller and more decomposed part (the finer material) was carried further away" [Vogt]. This constitutes a clear case of "sedimentary differentiation" which the writer would define as progressive change [areally considered] in chemical composition within a well-defined formational unit, the change being in the direction of increasing residual character' (Kennedy, *GM*, 1951). 'If sediment is being distributed by wash it is likely to be separated into fractions determined by ease of transport, and, as the easiest fractions will be carried farthest, horizontal differentiation will be favoured' (Bailey, *TGSG*, 1958).

**3.** In metamorphic petrogenesis: the separation of a rock, containing several minerals, into practically monomineralic masses of these minerals. The rock thus affected may already be a metamorphic rock. (Amin, *GM*, 1952.)

**Dilatation.** (Dilation). Expansion, enlargement; used in various connexions, including the enlargement of a joint, fissure, or bedding-parting by the intrusion of a dyke or sill. It is also used, in a technical sense, conveniently but perhaps rather absurdly, to include the very opposite—compression, then called 'negative dilatation'. [L. *dilato*, make wider.]

**Diluvium.** Applied, during the earlier part of the 19th century, to certain widespread superficial deposits supposed to be due to a universal and catastrophic deluge, with Noah's flood usually particularly in mind. These deposits are now known to be mostly glacial drift. 'The term was first employed by [William] Smith' (Phillips, *Geology*, 1855). Hence the 'diluvial theory' (Conybeare, *PM*, 1831, quoted by Chorley and others, *History of the Study of Landforms*, 1964), and 'diluvialist' for a person advocating it. 'The word diluvium I apply to those extensive and general deposits of superficial loam and gravel, which appear to have been produced by the last great convulsion that has affected our planet' (Buckland, *Reliquiae Diluvianae*, 1823). The term was similarly defined by Conybeare and Phillips (*England and Wales*, 1822). 'A splendid meeting [at the Geological Society] last night. Sedgwick in the chair. Buckland present to defend the "Diluvialists", as Conybeare styles his sect, and us he terms "fluvialists"' (letter from Lyell to Mantell, April 1829, in *Life of Lyell*, 1881). [L. *diluvium*, a deluge.]

**Dimetian.** See PEBIDIAN.

**Dimorphism.** See POLYMORPHISM.

**Dinantian.** A now preferred name for the Lower Carboniferous; being a subsystem of the Carboniferous system. *A Correlation of Dinantian Rocks in the British Isles* (*GSSR* (7), 1976), in which 'a new classification is proposed based on precisely defined stratotyped stages [each named after the location of the boundary stratotype], to provide the framework of a new comprehensive reorganisation of the Dinantian rocks of the British Isles.' These stages are (lowest first): Courceyan, Chadian, Arundian, Holkerian, Asbian, Brigantian. (Munier-Chalmas and de Lapparent, France, 1893.) [Dinant, Belgium.]

**Dinoflagellates.** A group of microscopic unicellular planktonic organisms which seems to straddle the boundary between the animal and plant kingdoms, having, on the whole, the greater affinity with the latter. Living forms show a flagellate phase and a more strongly-walled cyst phase. Dinoflagellate cysts (dinocysts) may be spiny or non-spiny; fossils of both kinds of cyst occur, being abundant from the Middle Jurassic onwards. Certain microscopic fossils, both spiny and non-spiny, were first described, from Cretaceous flints, by Ehrenberg in Germany in 1838; he recognized the non-spiny kinds as belonging to the dinoflagellates but thought the spiny kinds belonged to the desmid group of algae, and for nearly a century they were known as Xanthidia. These spiny kinds were placed as of uncertain affinity, and called Hystrichospheres, by Wetzel in Germany in 1933. The hystrichospheres have since been shown to be the spiny cysts of dinoflagellates, though some may belong to the Chlorophyceae (algae). Spiny dinoflagellate cysts were definitely recognized among the present-day plankton only after they had been fully described from fossil assemblages; the modern hystrichospheres are therefore examples of 'living fossils'. There are certain other organisms, again known chiefly as fossils, somewhat similar to dinoflagellate cysts, but regarded as of uncertain affinity; these have been maned Acritarchs (Evitt, *PNAS*, 1963). These are known mainly from rocks of Palaeozoic and upper Precambrian age; the latter is a fact of especial interest. The following are five among many discussions : 'Dinoflagellates, hystrichospheres and the classification of the acritarchs' (Downie and others, *SU*, 1963); 'Studies on Mesozoic and Cainozoic dinoflagellate cysts' (Davey and others,

*BBM*, 1966); ' "Living fossils" in western Atlantic plankton' (Wall and Dale, *N*, 1966); 'The Xanthidia' (Sarjeant, *MG*, 1967); *Fossil and living Dinoflagellates* (Sarjeant, 1974).

**Dinosaurs.** The collective name for the two orders of Mesozoic reptiles, Saurischia and Ornithischia, which between them comprise the vast majority of those that lived on land in that age and all the large spectacular forms. These two orders, however, do not appear to be a specially nearly related pair so that it is now found to be inappropriate to unite them, in a special systematic group (the Dinosauria), though it is convenient to refer to them as 'the dinosaurs'. Remains of a dinosaur were found by Gideon Mantell in the Wealden beds of Tilgate Forest, north Sussex, in 1822. These were teeth, similar to those of the modern lizard *Iguana* but much larger; Mantell described them and named the herbivorous reptile to which they belonged *Iguanodon* (*PT*, 1825). Later, bones were found which could be definitely associated with the teeth, and the whole animal (together with other types of dinosaur) was reconstructed, under the direction of Richard Owen, as a life-size model in the grounds of the Crystal Palace (1854). Meanwhile William Buckland had described fossils of the large carnivorous dinosaur *Megalosaurus* from the Stonesfield Slate (*TGS*, 1824). It is possible that the actual discovery of the *Megalosaurus* was made before that of the *Iguanodon*. In 1832 Mantell made the first discovery (again in Tilgate Forest) of a member of a strongly armoured group, which he described and called *Hylaeosaurus* (*SE. England*, 1833). Thus the three dinosaurs first to be known, the *Iguanodon*, the *Megalosaurus*, and the *Hylaeosaurus*, belonged each to a quite distinct group, later to be called, respectively, the Ornithopoda, the Theropoda, and the Ankylosauria. Owen proposed the name Dinosauria in 1842 (*BA*). *Dinosaurs: Their Discovery and their World* (Colbert, 1962). *Men and Dinosaurs* (Colbert, 1968). *The Dinosaurs* (Swinton, 1970). 'The earliest discoveries of dinosaurs' (Delair and Sarjeant, *Is*, 1975). [G. *deinos*, terrible, *saura*, a lizard.]

**Diorite.** A coarse-grained intermediate igneous rock composed essentially of plagioclase (oligoclase or andesine) in excess of alkali felspar, and mafic minerals, especially hornblende. (Haüy, 1822.) Plutonic. [G. *diorizō*, distinguish.]

**Diopside.** See PYROXENE. [G. *opsis*, view.]

**Dip.** The inclination of a rock-body or structural rock-surface. Primarily it is a vertical angle measured (downwards) from the horizontal plane in the direction of greatest slope. This may be further specified as 'true dip'. This direction (measured in the horizontal plane) is the 'direction of dip' and is at right angles to the strike. 'Apparent dip' is the dip in any direction, varying from zero in the direction of strike to a maximum in the direction of (true) dip; called 'apparent' because it is the dip seen in a random vertical section. 'Dip' is most commonly used for stating the disposition of strata; but it is also used for fault surfaces, cleavage surfaces, surfaces bounding or within igneous bodies, imaginary surfaces within fold structures, &c. In the case of a curved surface, the dip at any point is the dip of the tangent line at that point. 'This declining or dipping of the coal is sometimes greater and sometimes lesser. All these dipps are to be seen in several places in Lothian and one may see very different declinations, who is curious to observe them' (Sinclair, *Short History of Coal*, 1672). 'On proceeding westwards along the chalk towards Sandown fort, the section lines of the strata on the southern face of the cliff dip to the east. Hence as on the eastern face of the cliff their dip is to the north, the true dip of the planes of the strata must be to the north-east' (Webster in Englefield, *Isle of Wight*, 1816). Lyell (1833) defines 'dip' as the direction; specifying the angle as 'the angle of dip or inclination'. See INITIAL DIP and STRIKE.

**Dip-and-fault structure.** See SUPERFICIAL STRUCTURE.

**Dip-fault.** A fault of which the strike approximates to the direction of dip of the beds it affects.

**Dip-slip fault.** See SLIP.

**Dip-slope.** A slope of the ground that is determined, more or less exactly, by the dip of the beds. It is applied particularly to such a slope that ends upwards along the top of an escarpment and that is then opposed to the, usually steeper, scarp-slope on the other side.

**Dirt.** 'Seat-earth or fossil soil beneath coal seams or forest beds. Especially the Dirt Beds near the base of the Purbeck Beds. Also shaly or clay partings in coal seams and other rocks' (Arkell and Tomkeieff, *Rock Terms*, 1953; with quotations). 'Purbeck Dirt Bed with stones, Bacon Hole [Dorset]' (pl. in Arkell, *MGS*, *Weymouth [etc.]*, 1947.)

**Disconformity.** 'An erosional surface separating two parallel series of beds and indicating an interruption in the stratigraphical succession representative of a considerable period of geological time' (*SA*, 1956). See UNCONFORMABLE. (Grabau, 1905.)

**Discontinuity.** There is physical discontinuity of rock, within a rock-mass, which is particularly likely to occur as weathering proceeds along bedding-planes and joints. But the term usually refers to a surface where the arrangment (bedding, etc.) of the rocks on one side is not continuous with that on the other side. The two main types of structural discontinuity of this kind are unconformity and faulting. For the use of the term in describing the constitution of the earth as a whole see CONSTITUTION OF THE EARTH.

**Discordant. 1.** As applied to bedding: = cross-bedding (Tyrell, *Petrology*, 1929). **2.** As applied to an igneous intrusion: cutting across planes of stratification or schistosity. Contrasted with a concordant relationship.

**Disharmonic folding.** Folding in which the folded beds do not follow a consistent (harmonious) pattern. A close association of competent and incompetent beds tends to produce disharmonic folding (e.g. McIntyre, *GM*, 1951). Consistent folding, whether 'concentric' or 'similar' or a combination of the two, would hardly be referred to as 'harmonic'.

**Displacement. 1.** In faulting. The net relative displacement of points over the fault surface is the measure of the whole net bodily displacement, but 'displacement' usually refers to the effect of fault-movement on bedding surfaces, intrusions, veins, axial surfaces, previous faults, &c., and this depends largely on the disposition of these features. Thus, for example, net relative movement over a fault-surface will cause no displacement of bedding surfaces where (*a*) the fault coincides with the bedding or (*b*) where the movement has been in the direction of the lines of intersection of fault and bedding. **2.** In igneous action. Used particularly in connexion with one method of emplacement of igneous bodies; the country rocks may be displaced, pushed aside, by invading magma. 'We are now fully assured that granite has been made to break, displace and invade the Alpine schistus or primary strata, having been previously forced to flow in the bowels of the earth, and reduced into a state of fusion' (Hutton, *TRSE*, 1790). **3.** In orogeny. 'Although the [amount of absolute] displacement of any one point can never be determined accurately, it is possible to determine the displacement difference [relative displacement] between any two points. The displacement difference of two points on either side of a deformed zone records the "crustal shortening" across that zone' (Ramsay in Kent and others, eds., *Time and Place in Orogeny*, 1969). **4.** Displacement theory. See CONTINENTAL DRIFT.

**Distortion.** Technically, change in shape due to stress. It has a particular application to fossils: 'On slaty cleavage, and the distortion of fossils' (Haughton, *PM*, 1856). The verb 'distort' is naturally often used in its ordinary, general sense: 'Broken, tumbled and distorted strata' (Hutton, *TRSE*, 1788).

**Dittonian.** See OLD RED SANDSTONE. (King, *TWNFC*, 1921.) [Ditton Priors, Salop.]

**Division.** In biological classification: see HIERARCHY and PHYLUM.

**Divisional plane.** The chief kinds of plane (or more or less near-plane surface) that divide a rock-body into distinct parts are three: bedding-planes, joints, faults.

**Dogger.** **1.** Ironstone nodules in the Yorkshire Jurassic. **2.** Any lumpy mass of sandstone in any part of the country. **3.** The basal part of the Inferior Oolite series in Yorkshire; this restricted meaning being developed from (2). (From Arkell and Tomkeieff, *Rock Terms*, 1953.)

**Dolerite.** A medium-grained basic igneous rock composed essentially of the more calcic plagioclase felspars and pyroxene, often with olivine. Usually non-porphyritic and typically with ophitic texture. The commonest rock of minor intrusions, but it may occur as a finer-grained representative of gabbro in a plutonic mass; 'microgabbro' is indeed the better name (Hatch and Wells, *Igneous Rocks*, 1965). (Haüy, 1822; Johannsen, *Petrography*, 1937, gives the complicated history of the term.) [G. *doleros*, deceptive (as being difficult to distinguish from diorite).]

**Dolgelly beds.** The highest of the three divisions of the Lingula flags series; the beds are chiefly dark mudstones. Characteristic fossils are the trilobites *Parabolina* and *Peltura* of the family Olenidae. (Belt, *GM*, 1867.) [Dolgellau, Meirionnydd district of Gwynedd.]

**Dolomite.** A white, or tinged, rather soft mineral, carbonate of calcium and magnesium, $CaCO_3.MgCO_3$ Hexagonal system, often forming curved, saddle-like crystals. The name is also used for the rock, more exactly called 'dolomite-rock'. (Saussure, 1796.) [Dolomieu, French scientist, 1750-1801.]

**Dolomite-mudstone.** See CALCITE-MUDSTONE.

**Dolomite-rock.** A rock composed almost entirely of the mineral dolomite, with little or no admixture of the simple carbonate of calcium.

**Dolomitization.** The conversion, normally by metasomatism due to percolating meteoric waters, of a (calcic) limestone, into a rock with an appreciable amount of magnesium carbonate, or the increase of the magnesian constituent. Complete dolomitization would be the conversion of all the purely calcic carbonate into the mineral dolomite (the double carbonate of calcium and magnesium).

**Dome.** Applies generally to form and, unless qualified, it implies that the form as such actually exists, as, for instance, a dome-shaped hill or region, or a dome-shaped igneous intrusion. In structural geology, however, it refers to the manner of the arrangement of the rocks, whether these now assume, or ever assumed, a real and complete dome-form or not. It is important to keep these conceptions clear and it is better to use 'dome structure' for the latter, the tectonic, meaning. The Harlech Dome, for instance, is both a real, physiographical dome (or somewhat approaching it) and a dome structure; the Weald Anticline is an oval-shaped dome structure but, physiographically, an oval basin with a broad median ridge. Snowdonia has a dome-shaped hill-top surface but is synclinal in structure.

**Dome structure.** See DOME. It is an anticline in which the dips are away from a centre rather than away from a line. An elliptical upfold may be equally well called an elongated dome structure or a short anticline (a 'brachy-anticline', with opposed pitch at each end). It bears the same relation to an anticline as does a basin structure to a syncline. 'Suppose the five formations [of the Cretaceous in the Weald] to lie in horizontal stratification at the bottom of the sea; then let a movement from below press them upwards into the form of a flattened dome, and let the crown of this dome be afterwards cut off, so that the incision should penetrate to the lowest of the five groups. The different beds would then be exposed on the surface in the manner exhibited in the map' (Lyell, *Principles*, 1833). 'Southward of the Snowdon syncline follows the Harlech Dome' (Fearnsides in Monkton & Herries, eds., *Geology in the Field*, 1910).

**Dormant volcano.** See VOLCANO.

**Downthrow.** The effect of a fault in causing the rocks on one side to become 'thrown down' relative to the rocks on the other side. Chiefly used to denote this 'downthrow side'. More correctly, perhaps, it is the side of a fault on which there has been a downward component of net relative movement (but which of the two sides this is may be a matter of assumption rather than certainty in uniformly dipping rocks). The effect on the other side of the fault is 'upthrow', and that is the 'upthrow side'. Downthrow is the effect more usually referred to; the corresponding upthrow is immediately implied. The amount of such movement and the amount of certain effects on the bedding are the 'amount of downthrow' (or the 'amount of upthrow'), but more generally the 'amount of throw'. See THROW.

**Downtonian.** This term was proposed by Lapworth (*AMNH*, 1879) for the uppermost of three divisions of Murchison's Silurian system, the others being Valentian (lowest) and Salopian. It included Murchison's Upper Ludlow, also the Ludlow Bone-bed and the Downton Castle Sandstone both of which Murchison had recognized as passage beds to the Old Red Sandstone but had, on the whole, included in his Silurian. In the 1920's it was claimed that there was a natural stratigraphical series, based particularly on the fish faunas, which extended upwards into what had previously been regarded as unquestionably Old Red Sandstone. For this series the name Downtonian was used, and the base was placed at the base of the Ludlow Bonebed; it was thus bodily and drastically shifted upwards. The term is now used in this latter sense. The series is chiefly one of sandstones with fossils of various kinds, probably peculiar to the conditions which were certainly not typically marine, even in the lower part. In this lower part (Grey Downtonian as distinct from the Red Downtonian above) are such fossils as the brachiopod *Lingula minima*, the lamellibranch *Modiolopsis complanata*, the gastropod *Platyschisma helicites*, the ostracod *Beyrichia kloedeni*, and eurypterids.

The lower half or so of the Downtonian is now considered to be below the base of the Devonian as defined by the boundary stratotype of that system (in Czechoslovakia). Thus while the whole of the Downtonian is called Old Red Sandstone the lower part is taken to be Silurian (post-Ludlow) as now defined. This arrangement is reflected, for instance, in the title of the paper 'A late Silurian flora from the Lower Old Red Sandstone of Dyfed' (Edwards, *P*, 1979). [Downton, Herefordshire.]

**Drag fold.** Drag folds are those resulting from the shearing stress induced in a non-rigid bed or set of beds by the slip of a rigid bed or rock-mass. There are 'drag' structures of various kinds (e.g. due to faulting or thrusting or produced superficially by the movement of an ice mass), but the term 'drag folds' is usually reserved for those produced in an incompetent bed (or set of beds) by bedding-plane slip.

**Dreikanter.** See VENTIFACT. 'The Pleistocene dreikanter in the Vale of York' (Edwards, *SPGS*, 1936). [German, f. *drei*, three, *kante*, edge.]

**Drift.** Murchison (*Silurian System*, 1839) introduced 'drift' for what was then called 'diluvium', preferring it to the latter term (which implied a catastrophic flood, if not The Flood), as he thought all this material had drifted in marine currents and accumulated under the sea (in comparatively recent times). It is now known that practically the whole of Murchison's drift was formed as a result of glacial conditions and the term is now taken to include all deposits referable directly or indirectly to ice, that is, to all glacial and glaciofluvial deposits; for these, 'glacial drift' is more explicit. Glacial drift may be seen in the neighbourhood of existing ice masses, but the term is almost exclusively reserved for the deposits of the Glacial Period in areas no longer glaciated. 'Every observing person must have frequently seen in plains, at a little distance from large and high mountains, a prodigious quantity of coarser and finer gravel, some of it containing large boulders. This gravel is often found spread out with a plane superficies, and as often accumulated into long, round, oval, or semi-circular hillocks of lesser or greater height [drumlins], some of them so large that they may be called little hills' (Williams, *Mineral Kingdom*, 1789, referring particularly to Scotland). 'Drift' is not always and everywhere 'glacial drift'. 'The "Solid" formations end with the Pliocene System. All the later deposits, whether belonging to the Pleistocene System or still later (Recent) are called "Drifts"' (Sherlock, *BRG*, *London*, 1962). The 'drift' editions of many Geological Survey maps include river, and other non-glacial superficial, deposits not shown on the corresponding 'solid' editions (which themselves often show the more extensive spreads of alluvium and blown sand). 'A map of the drift geology of Great Britain and Northern Ireland' (Clayton, *GJ*, 1963). See quotation (Shotton) under ERRATIC.

**Drift theory. 1.** The theory, favoured by some leading geologists during the period from about 1800 to 1840, which attributed the phenomena of erratic blocks and scattered gravels to the action of floating ice. **2.** There is also the 'drift theory' as to the origin of coal, in contrast to the 'growth-in-place theory'.

**Dripstone.** 'The single term dripstone will be used to replace the cumbersome pair, stalactite and stalagmite' (Davis, *BGSA*, 1930).

**Drumlin.** 'Boulder-clay, which as a rule forms plains, gives rise sometimes to the peculiar "drumlin landscape". Typical drumlins are smooth, oval mounds or elongated ridges. The term, from *druim* a mound or rounded hill, a gaelic word extremely common in place-names in Scotland and Ireland, was used by J. Bryce in 1833 and brought into glacial literature by Close in 1866; W. M. Davis introduced it into America' (Charlesworth, *Quaternary Era*, 1957). Drumlins (consisting of boulder-clay) appear to be essentially original, not residual, forms, connected in some way with the character and motion of the ice, but they also appear to have been moulded and streamlined by its overriding action. 'The drumlin topography of S. Donegal' (Wright, *GM*, 1912). 'The shape of drumlins' (Chorley, *JGl*, 1959). 'Drumlins near New Galloway, Kirkcudbrightshire' (pl. in Craig, ed., *Geology of Scotland*, 1965). 'A review of the literature on the formation and location of drumlins' (Menzies, *ESR*, 1979).

**Drusy.** 'Closely covered with minute implanted crystals; example, quartz' (Dana and Ford, *Mineralogy*, 1953). 'Drusy has been adopted from the German term *drusen* for which we have no English word. The surface of a mineral is said to be drusy when composed of very small prominent crystals, nearly equal to each other; it is often seen in iron pyrites' (Phillips, *Mineralogy*, 1819). The term is applied also to the lining of a geode when this is covered with a 'drusy growth' of small crystals; and sometimes to the cavity itself.

**Dry valley.** A valley without a river. Dry valleys are common on porous formations, particularly limestones, having been formed in the ordinary way but now being deserted owing to some change in conditions. A dry valley, often a gorge, will be exposed if erosion cuts down into a deserted underground drainage channel. Glacial overflow channels (of the past) may be said to be another kind of dry valley.

**Ductile deformation.** See BRITTLE DEFORMATION.

**Ductile fault.** A fault in which the rocks adjoining the fracture (on one or both sides) have been stretched, pinched, and curved as a result of the stress which produced the fault. (Fig. in Davis and Coney, *Geoly*, 1979.)

**Dune.** 'A low ridge or hillock of drifted sand, mainly moved by the wind' (Monkhouse, *Dictionary of Geography*, 1965). *Physics of Blown Sand and Desert Dunes* (Bagnold, 1941).

**Dunite.** A peridotite essentially composed of olivine. (Von Hochstetter, 1859.) [Dun Mountain, New Zealand.]

**Durain.** 'The term durain was introduced by M. C. Stopes in 1919 [*PRS*] to designate the macroscopically recognisable dull bands in coals' (*Nomenclature of Coal Petrology*, 1957). [L. *duro*, make hard, -ain (to match 'fusain').]

**Duricrust.** 'The name Duricrust is proposed for the widespread chemically formed capping in Australia, resting on a thoroughly leached substratum, and formed *in situ*, and apparently synchronously in every region' (Woolnough, *GM*, 1930). 'These duricrusts are concentrates, in the upper part of the soil profile of aluminous, siliceous, ferruginous, or calcareous materials [depending on the underlying rock]. They were thought to have been formed under a semi-arid climate. Under the present arid conditions duricrusts are undergoing destruction' (Thornbury, *Geomorphology*, 1954) 'An important family of landforms owes its origin to the dissection of indurated horizons, crusts and layers formed during weathering. Such "duricrusts" are also important as morphostratigraphic horizons (or time markers) in some areas, and are locally significant from a hydrological standpoint' (Twidale, *Analysis of Landforms*, 1976). *Duricrusts in Tropical and Subtropical Landscapes* (Goudie, 1973).

**Durness Limestone.** A formation chiefly of dolomitic limestones, about 450 metres thick, outcropping along a narrow belt in NW. Scotland from the north coast at Durness south-westwards into Skye. It overlies the Lower Cambrian quartzite group and the top of the formation is cut across by the Moine Thrust. The lowest beds contain Lower Cambrian fossils; the upper a fauna, chiefly gastropods and cephalopods, only known elsewhere in the Beekmantown Limestone in N. America which is probably of Ordovician (Arenig) age. The middle beds are unfossiliferous. (Murchison, *QJ*, 1858/59; Peach and Horne, *QJ*, 1888.)

**Dyke.** (Dike.) A wall-like body of rock which, if among stratified rocks, is disposed discordantly to the stratification. The rock is usually igneous, forming an 'igneous dyke', in which case the dyke is an igneous intrusion of sheet form, normally intruded in a more or less vertical position. If intruded into stratified rocks, the property of discordance is essential and would over-ride the question of original attitude. (Dykes and sills may, subsequently to intrusion, be tilted into any attitude.) 'For the study of these manifestations of volcanic energy the British Isles may be regarded as a typical region. It was thence that the word "dyke" passed into geological literature' (Geikie, *Ancient Volcanoes*, 1897). The North Star dyke, a Tertiary dolerite intruding Lower Carboniferous coal measures, seen conspicuously in relief on the shore near Ballycastle, N. Ireland, is a well-known example (about 4 metres across). A dyke composed of sedimentary rock, a 'sedimentary dyke', usually a sandstone ('sandstone dyke') may be a Neptunian dyke or an 'injection dyke'; see SEDIMENTARY INSERTION. A sedimentary dyke may occur in an (extrusive) igneous rock: 'Carbonate dikes in lava (Karadag, Crimea)—the dikes are obviously injections synchronous with their host lavas. They were produced at the floor of a marine basin where the thin sheet of lava was easily cooled and fissured on its contact with water and by seismic oscillations of the ground' (Bondarenko, abstracted by Sokoloff, *IGR*, 1966). Before the modern sense became established, the term 'dyke' was sometimes used for a fault (see quotation from Kirwan under FAULT), and also for a wash-out '. . . what are termed in mining phrase "sand-dykes" or "clay-dykes"' (Dick, *TEGS*, 1870). It has also been used in a general sense for any extensive, relatively thin, sheet-form intrusion: 'dikes which are parallel to the stratification' (Strickland, *TGS*, 1840). Any such wide senses as those above may be said now to be definitely excluded. (Brand, *History of Newcastle*, 1789; Lyell, *Principles*, 1833.) [Middle English *dic*, a ditch or wall.]

**Dyke-swarm.** 'It has always been recognised that ordinary dykes tend to be numerous about a plutonic centre, and that in such a situation they are often parallel; but it is to H. B. Maufe that we owe the clear-cut idea of a centred swarm of parallel dykes. The Devonian northeast swarm of Etive and the Tertiary north-west swarm of Mull, crossing one another in Argyll, provide as fine examples as any one could desire. . . . Some swarms of parallel dykes do not seem to be centred. L. R. Wager and W. A. Deer have described a dyke-swarm in Greenland located by the stretch of a monoclinal fold (1939)' (Bailey, *TGSG*, 1958; see Bailey, *MGS, Mull*, 1924, and Clough and others, *QJ*, 1909).

**Dynamic metamorphism.** Metamorphism due to mechanical stress. The term is usually restricted to that in which there is no significant rise of temperature (pure dynamic metamorphism). (Bonney, *QJ*, 1886, suggested 'pressure metamorphism'.)

**Dynamic equilibrium.** A geomorphological concept which explains erosional topography in terms of a balance and adjustment between erosional process on the one hand and rock composition and structure on the other, rather than in terms of an evolutionary development through time. It is thus largely opposed to the concepts of Davis (see GEOGRAPHICAL CYCLE) and Johnson (*Shore Processes and Shoreline Development*, 1919). 'The concept of dynamic equilibrium' (Hack, 'Interpretation of erosional topography in humid temperate regions', *AJS*, 1960).

**Dynamical geology.** 'Embraces an investigation of the operations which lead to the formation, alteration, and disturbance of rocks' (Geikie, *Textbook*, 1903). It thus deals with the action of the geological processes. 'Mr. Lyell will have reaped the honour of being the first writer in our country to make known a general system of "geological dynamics",—a new province gained by the advance of modern science' (Sedgwick, *PGS*, 1831). 'I may perhaps be allowed to advert to a distinction of the subject into Descriptive Geology and Geological Dynamics; the former science having for its object the description of the strata and other features of the earth's surface as they now exist; and the latter science being employed in examining and reducing to law the causes which may have produced such phenomena' (Whewell, *PGS*, 1838). Also called, concisely, 'geodynamics'.

**Dynamothermal.** Pertaining to force and heat combined. 'In the following pages the various hypotheses relative to the age of the Moine sediments and their dynamothermal metamorphism are presented' (Phemister, *BRG*, *N. Highlands*, 1965).

# E

**Earth.** This, in the most comprehensive, unambiguous, and the most usual sense of the term means the whole planet on which we live—The Earth, Planet Earth. It has also been used in variously restricted, and more or less vague, senses; particularly for the land, or for the softer material of which the surface of the ground is generally composed (soil, mould, clay, etc., and see SILICEOUS ORGANIC DEPOSIT).

The earth comprises several concentric 'zones' or 'spheres'. The first distinction is that between the atmosphere and the rest of the planet. Of this rest of the planet (for which there is no single name) we can next distinguish between the oceans and what is sometimes called the 'solid earth' but better, perhaps, called the 'earth-body'. It is the study of this rocky-metallic earth-body that forms the science of geology. It is curious that both the body and the material ('rock' in an uncomfortably wide sense) with which geology deals evade a satisfactory naming. The earth-body itself comprises the relatively thin crust and the main mass, the endosphere. The endosphere comprises the core and the mantle.

*The Earth* (Reclus, 1872). *The Earth* (Jeffreys, 1924/76). *Planet Earth* (Bates, 1957/64). *The Earth* (Dunbar, 1966). *The Story of the Earth* (Geological Museum, London, booklet, 1972). *Earth* (Press and Siever, 1974). *Planet Earth* (Hallam, ed., 1977). *The Earth: its Origin, Structure and Evolution* (McElhinny, ed., 1979). [Old English *eorthe*.]

**Earth history.** The main object of the pursuit of geology as a pure science may be said to be the deciphering and piecing together of all evidence concerning the history of the earth, 'earth history' in short. This evidence lies in the observable facts of composition and structure and the study of process, and concerns everything as to time, place, activity, geography, and life in the past. The employment of ever-advancing scientific methods and techniques leads to an ever-increasing knowledge and understanding of earth history; at no time has this been more rapid and penetrating than during the last two decades (e.g. seismic survey, radioactivity age-method, palaeomagnetic detection).

The one major part of earth history, throughout geological time that goes largely unrecorded in the rocks is that concerning the physiography of the land during the working of the geological cycles in operation at various stages in the several geographical regions of the earth. We know much, and in great detail, about the marine-deposition phase and the deformation phase of the cycle (also about igneous activity and metamorphism), but very little about the land phase, in which (1) earth-movement and (2) erosion and the resulting denudation are the two factors. See PATTERN OF EARTH HISTORY and GEOMORPHOLOGY.

**Earth Science.** 'Earth Science' is a literal translation of 'Geology', but is used chiefly when some scientific fields are to be covered which are not, or which might not be considered to be, included in geology. It has more latitude of interpretation. 'The co-ordinated group known as earth sciences. These sciences comprise geology in all its aspects, including such borderline fields as physical geodesy, geophysics (including seismology), geochemistry (including isotope geology), oceanography (including marine geology), hydrogeology, glaciology, and pedology (including soil surveys); and applied aspects such as mining geology, engineering geology, and military geology. This group occupies a unique position which cannot be accommodated in either physical or biological science' (*CGS*, 1964). The term is sometimes used to include all aspects of physical geography. The 'geosciences' (a hybrid term) may be said to be the same as the 'earth sciences'. 'Geonomy' was suggested by Lapworth (*QJ*, 1903), analogous to 'astronomy'; but '-nomy', from G. *nemō*, arrange, is inappropriate here. ('-nomy' is hardly appropriate, in association with 'astro-', for the whole science which is called astronomy; but 'astrology' is preoccupied.) 'Geological sciences' means the same thing as 'geology'. That 'earth science' is found convenient is evident from its use in the title of several recent periodicals and books, but it should not be allowed to supersede 'geology' where the latter is indisputably appropriate. *Principles of Earth Science* (Strahler, 1976).

**Earth sculpture.** 'Sculpture' implies carving, shaping by cutting away, not by adding on. Thus 'earth sculpture' is highly appropriate for the production of land forms by erosion, but not so appropriate for the shaping of the land by the combined processes of erosion and deposition, and still less appropriate for the whole of geomorphology. Yet Lake, in his chapter headed 'Earth Sculpture' (*Physical Geography*, 1915), allows the term to include the effects of deposition, and J. Geikie's *Earth Sculpture or the Origin of Land Forms* (1898) even embraces land forms produced directly by earth-movement. 'Glyptogenesis', the origin and development of carved forms (earth's surface implied), is sometimes used. 'The sculpture of mountains by glaciers' (Davis, *SGM*, 1906).

**Earth-body.** See EARTH. 'Solid earth' has been used, as in 'solid earth geophysics', but this is hardly appropriate as we are well aware of liquid igneous magma and the geophysicists themselves tell us that most, if not all, of the earth's core behaves as a fluid.

The 'earth-body' may be said to include the thin outer veneers, (1) vegetation and (2) the constructive works of man, and even (3) animal life. These are not included in the domain of geology.

**Earth-movement.** Movement of the earth's crust due to internal stress; diastrophism in the wide sense.

**Earth-pillar.** A pillar of soft rock or coherent deposit left standing by the erosion of the surrounding material. The most striking earth-pillars are those formed out of such material as moraine and boulder-clay, and coarse conglomerate; each pillar being capped by a temporarily protective piece of hard rock. Well-known examples occur in the Alps and Dolomites (in moraine) and incipient pillars are often seen in boulder-clay where this is being eroded into steep cliffs along a coast (in Britain and elsewhere). 'Excellent examples of rain-eroded columns are to be seen in a group of ravines worn out of the Old Red Sandstone on the right bank of the Spey above Fochabers (fig. 3)' (Geikie, *Scenery of Scotland*, 1887). The less steeply sided, more conical, 'pillars' are appropriately termed 'earth-pyramids' (Cole, *Open-air Studies*, 1895).

**Earthquake.** A vibration of a part of the earth, particularly the vibration at the earth's surface, caused by a shock, usually taken as being due to a sudden slip over a fault-surface (or possibly to a sudden slip in folding); but it has recently been suggested (e.g. Evison, *BSSA*, 1963) that earthquakes cause faulting rather than the reverse and that they themselves are due to sudden volume changes in the earth. A series of shocks, separated by minutes, hours, or days, may sometimes be said to constitute one large earthquake and be classifiable as fore-shocks, principal shock, and after-shocks. 'Earthquake tremor' is often used for an earthquake of no great violence, particularly as 'preliminary tremors' (faint fore-shocks). The 'intensity' of an earthquake refers to any particular point (see SEISMIC INTENSITY); its 'magnitude' is an expression of the total energy released. Local earthquakes may be caused by volcanic action. The slight natural vibrations caused by wind or sea-waves, by landslips or rockfalls, are not usually included in the term 'earthquake', but the artificial vibrations produced by a subsurface explosion in seismic prospecting may legitimately be said to be 'artificial earthquakes'. The connexion between earthquakes and faulting seems to have been first realized by John Michell ('Conjectures concerning . . . earthquakes', *PT*, 1760): 'There is another very remarkable appearance in the structure of the earth, though a very common one; and this is what is usually called by miners, the trapping down [faulting] of the strata, this is, the whole set of strata on one side of a cleft are sunk down below the level of the corresponding strata on the other side . . . it may have a great effect in producing some of the singularites of particular earthquakes'. 'Along certain of these faults slight movements accompanied by earth-tremors and shocks are still taking place. The Highland Boundary fault in the neighbourhood of Comrie, and the Ochil Fault, a powerful dislocation running along the foot of the Ochil Hills, are two of the most notable centres of seismic disturbance in Britain' (Macgregor and Macgregor, *BRG, Midland Valley of Scotland*, 1948). 'Some shallow earthquakes are evidently associated with fault movement; but at depths greater than a few tens of kilometres the pressure is too great to permit fault slippage. The causes of deeper earthquakes are unknown' (*OE*, 1979). *History of British Earthquakes* (Davison, 1924). *Study of Earthquakes* (Matuzawa, 1964).

**Earthquake waves.** See SEISMIC WAVES.

**Earth's crust.** The outermost shell of the earth. In the first place, this is a long-standing and somewhat vague concept, based on the assumption that the composition, physical properties, and structure of the outermost parts of the earth differ, in all probability, from those of the interior. 'It was formerly the prevalent belief that the exterior and interior of the globe differ from each other to such an extent that, while the outer parts are cool and solid, the vastly more enormous inner intensely hot part is more or less completely fluid. Hence the term "crust" was applied to the external rind in the usual sense of that word. This doctrine was reluctantly abandoned by most geologists. Nevertheless, the term "crust" has continued to be used' (Geikie, *Textbook*, 1903). Another concept had been put by Lyell (*Elements*, 1838): 'By the "earth's crust" is meant that small portions of the exterior of our planet which is accessible to human observation. It comprises not merely all of which the structure is laid open in mountain precipices, or in cliffs overhanging a river or the sea, or whatever the miner may reveal in artificial excavations; but the whole of the outer covering of the planet on which we are enabled to reason by observations made at or near the surface'. The surficial deposits on land and the deposits on the sea floor are also to be included if we seek a fully comprehensive definition of the 'earth's crust'.

In order to define the base of the 'crust' we need to find some surface (interface) across which, as we proceed downwards, there is a sharp change, a discontinuity, in some property. Seismology has revealed the presence of such an interface, across which there is an abrupt increase in the velocity of the seismic waves. This indicates that the material of the earth immediately below the interface must be denser than that above. This seismic interface is the 'Mohorovičić discontinuity', discovered (and published in Zagreb) by A. Mohorovičić in 1909 while studying a Balkan earthquake. Its existence was foreshadowed by Milne (*PRS*, 1906) as pointed out by Hawkes (*QJ*, 1958). Below the Mohorovičić discontinuity (often abbreviated to 'M discontinuity' or 'Moho') is the thick mantle zone of the earth and the interface is known also as the 'crust-mantle boundary' (CMB). Beneath the continents the interface (discontinuity) lies at an average depth of about 40 km., reaching about twice that amount under the greatest mountain ranges. Beneath the floors of the oceans it lies at about 6-8 km. There are correspondingly two kinds of crust. The thick 'continental crust' has a complicated structure, the upper half or so being chiefly granitic in composition, with the lower half chiefly basaltic. The comparatively thin 'oceanic crust' has a simple layered structure and a uniform basaltic composition.

*Physics of the Earth's Crust* (Fisher, 1881). *The Earth's Crust* (Heacock and others, eds., 1977). The crust-mantle boundary' (symposium papers, *JGS*, 1977). *Evolution of the Earth's Crust* (Tarling, ed., 1978).

**Earth-slide.** See LANDSLIDE.

**Echelon.** Various geological features may occur primarily, or be subsequently displaced, so as to appear in echelon (or *en echelon*) at outcrop. 'In the neighbourhood of Ilfracombe may be seen parallel interrupted veins of quartz disposed in echellon' (Greenough, *Critical Examination of the First Principles of Geology*, 1819). 'Folds [in Aberystwyth Grits] replace one another en echelon' (Price, *GM*, 1962). 'An "en échelon" model for the Mediterranean' (Kostov, *PGA*, 1978).

**Echinodermata.** (Echinoderma.) A phylum of marine animals which, in the adult, have primarily a radial, usually a five-rayed, symmetry and secondarily a more or less bilateral symmetry. There is a mesodermal skeleton of calcareous plates and rods which form a coherent shell (test). There is a water vascular system. [G. *echinos*, hedgehog, *derma*, *dermat-*, skin.]

**Echinoidea.** Sea-urchins; a class of the phylum Echinodermata (subphylum Eleutherozoa). The test is globular, heart-shaped, or discoidal and is formed of a mosaic of plates, bearing spines, arranged in five narrow radial double-rows (ambulacrals) alternating with similar, but wider, double-rows (interambulacrals), from the apical disc to the peristome. Symmetry is either almost perfectly radial (Regularia) or has a bilateral symmetry superimposed (Irregularia). They range from the Carboniferous (very rarely occurring earlier), where the forms are abnormal, and are important fossils in the Mesozoic and Tertiary. [genus *Echinus* (G. *echinos*).]

**Eclogite.** 'The name is an old one, introduced in 1822 by Haüy for rocks occurring in the Fichtelgebirge. Two components are regarded as essential—red garnet and the bright green pyroxene, omphacite. Eclogite has much the same chemical composition as olivine-gabbro, but contains none of the normal gabbroic minerals. . . . In Britain eclogites occur in northern Scotland in the so-called Lewisian inliers near Glenelg—the first record in this country (Teall, 1891)—in central Sutherland and central Ross-shire. The chief problem of the eclogites concerns their origin. They are believed to be products of "plutonometamorphism"—to lie on the borderland between the igneous and metamorphic' (Hatch and Wells, *Igneous Rocks*,

1952). [G. *eklogē*, a selection, referring to the unusual association of minerals.]

**Economic geology.** 'The applications of geology to practical uses, resulting in the development of the science generally known as Economic Geology. The science is not sharply marked off from the sciences of geology proper; almost any phase of geology may at some time or some place take on its economic aspect' (Leith, *Economic Aspects of Geology*, 1922). *Geology in the Service of Man* (Fearnsides and Bulman, 1944). 'Let us ever remember that our science is not only the interpreter of Nature, but also the servant of Humanity' (Lapworth, *QJ*, 1903). The chief departments of economic geology are Mining geology, Oil geology, Engineering geology, Agricultural geology, and Hydrogeology.

**Ecostratigraphy.** The study of the stratified rocks with respect to the environment in which they were deposited. 'Epeiric sedimentation and sea level: synthetic ecostratigraphy' (Cisne and others, *Let*, 1984).

**Edinburgh Rock structure.** See BARREL FOLD.

**Effusive.** Pouring, or poured, out. Applicable to lavas and *nuées ardentes*. See quotation (Loewinson-Lessing) under HYPABYSSAL. [L. *effundo, effus-*, pour out.]

**Eifelian.** See DEVONIAN SYSTEM. (Dumont, 1848.) [Eifel district, Germany.]

**Ejectamenta.** Material thrust, thrown, or shot out of a volcano, as distinct from the effusive material poured out. 'The cone [in Auvergne], an incoherent heap of scoriae and spongy ejectamenta, stands unmolested' (Lyell, *Antiquity of Man*, 1863). [L. *ejectamentum*, what is cast out.]

**Elbow of capture.** See RIVER CAPTURE.

**Elements of geology.** Any of the distinct component parts of the science: particular facts, generalized facts, explanations; also may be said to include hypotheses, concepts, theories, principles. ('Elementary geology' means a simple or introductory treatment of the subject.) *Elements of Geology* (Lyell, 1838 and later editions).

**Eleutherozoa.** One of the two subphyla of the phylum Echinodermata. The animal is able to move about freely. It includes the classes Asterozoa (subclasses Asteroidea and Ophiuroidea), Echinoidea, and Holothuroidea. [G. *eleutheos*, free.]

**Elutriation.** The separation of the lighter from the heavier particles in a mixture of loose rock-material, such as sand, by a process of washing. [L. *elutrio* from *eluo*, wash out.]

**Eluviation.** A pedological term, concerning the soil profile. 'We may refer to the translocation of material, either mechanically or in solution, as "eluviation", and two main types may be distinguished, (*a*) "mechanical eluviation", in which the finer fractions of the mineral portion of the soil are washed down to lower levels, and (*b*) "chemical eluviation" in which decomposition occurs and certain products thus liberated are translocated in true or colloidal solution, to be deposited in other, "illuvial" [washed in], horizons' (Robinson, *Soils*, 1932).

**Eluvium.** The term is strictly appropriate for a residual deposit left behind in place after removal of soluble material by percolating water, but wind-drifted deposits were included when it was first introduced: 'for [certain] atmospheric accumulations Trautschold has proposed the name "eluvium". They originate *in situ*, or at least only by wind-drift, whereas "alluvium" requires the operation of water and consists of materials brought from a greater or less distance' (Geikie, *Textbook*, 1882). Hence 'eluvial' (washed out).

**Emanation.** Chiefly used for gases and gaseous solutions emitted in volcanic action or from earthquake-fissures, and for those given off from igneous intrusions and permeating the surrounding rocks. Also suggested for plutonic 'rock-transforming gaseous solutions' of whatever origin (Holmes, *Physical Geology*, 1965). 'During a tremendous earthquake which destroyed a great part of St. Domingo (1770), innumerable fissures were caused throughout the island, from which mephitic vapours emanated and produced an epidemic' (Lyell, *Principles*, 1834). 'The emanations which take place from volcanic events' (De la Beche, *Geological Observer*, 1851).

**Emsian.** See DEVONIAN SYSTEM. (Dorlodot, 1900.) [Ems, Germany.]

**Emplacement.** Used particularly in connexion with igneous rock-masses; their establishment in position by whatever means this may be effected. The country rocks may be displaced, pushed aside by invading magma; they may be replaced and incorporated by invading magma, or they may be completely metamorphosed into an igneous rock *in situ*. 'On the emplacement of granites' (Marmo, *AJS*, 1956). 'Volcanic neck emplacement' (Francis, *TRSE*, 1962).

**Enclave.** A rock-body entirely enclosed (until exposed by erosion) by another and larger rock-body. The larger xenoliths are examples. *Granites and their enclaves* (Didier, 1973).

**End-member.** The member at either end of an orderly, graded series, such as an isomorphous series of minerals, a morphological series of fossils, or a series of sedimentary rock-types.

**Endogenetic.** 'Endogenetic' and 'exogenetic' (and the synonyms 'endogenic' 'endogenous' and 'exogenic' 'exogenous'), are two contrasting terms, 'originating within' and 'originating without'. 'The surface features of the earth are commonly split into two categories, the first of which comprises those features that are due to processes occurring inside the solid earth (endogenetic features) and the second those that are due to processes occurring outside the solid earth (exogenetic features). Specifically, the endogenetic features are treated in the science of geodynamics, the exogenetic features in the science of geomorphology' (Scheidegger, *Theoretical Geomorphology*, 1961; and see the same author under ANTAGONISM). We have 'endogenetic' and 'exogenetic' veins and ore deposits (though the termination -genous is more usual here). The terms are perhaps most often used for the classification of all rocks into two mutually exclusive groups, but here there are two different usages based on two different interpretations of the etymology of the words. One usage, suggested by Crook (*MM*, 1914) and adopted by Tyrrell (*Petrology*, 1929), though Haug (1907) had used similar terms in a similar way, divides rocks into those originating or formed within the earth ('endogenetic') and those originating and formed on the earth's surface ('exogenetic'). The former includes the igneous and the thoroughly metamorphic rocks, the latter all the sedimentary rocks (in the wide sense). The other usage, due to Grabau (*AG*, 1904, and *Stratigraphy*, 1913), takes 'endogenetic' to mean 'formed by agents acting from within; i.e. by agents intimately associated with the rock mass forming; they produce rocks by solidification [igneous rocks], precipitation or extraction of mineral matter'. He thus includes chemical deposits and shelly accumulations (bioliths). 'The exogenetic rocks, on the other hand, are those formed by agents acting from without upon already existing rock matter, reducing it to a finer condition while still leaving it in a solid state', thus comprising all the rest of the sedimentary rocks (in the wide sense). [G. *endon*, within.]

**Endosphere.** A useful term (though hardly in current use) for all that part of the earth-body other than (below, within) the earth's crust. 'The endosphere, or inner sphere, comprises the great bulk of the Planet' (Greenly, *The Earth*, 1927).

**End-product.** Something (usually a rock is in mind) resulting from the completion of a process or series of processes. 'The end-product in the sedimentary record depends on the factors of parent material, weathering, erosion, transport, and depositional environment' (Hollingworth, *QJ*, 1962).

**Engineering geology.** Geology applied to engineering. *A Geology for Engineers* (Blyth and De Freitas, 1943/74). *Principles of Engineering Geology* (Attewell and Farmer, 1976). See GEOTECHNICAL.

**Enstatite.** See PYROXENE. 'The mineral from Mount Zdjar in Moravia was described by Kenngott under the name Enstatite in 1855' (Miers, *Mineralogy*, 1929). [G. *enstatitēs*, an opponent, because so refractory before the blowpipe.]

**Entombment.** The act or fact of burial of a rock-body of distinctive form and origin by rocks of a different kind. 'Entombment of tuff-cone at Saline, Fife' (Francis and Ewing, *GM*, 1961).

**Envelope.** This term is widely applicable in geology and is used in a number of special senses of which the following are some. **1.** In a compound fold; see CORE (1). **2.** In granitization; see CORE (2). **3.** In a mineral, an outer part separate in origin from (later than) an inner part. **4.** The hydrosphere and atmosphere are sometimes referred to as the watery and gaseous envelopes of the earth.

**Environmental geology. 1.** 'Environmental geology is not sharply defined but is generally concerned with those aspects of geology that touch on man's environment. Broadly speaking, all aspects of geology influence man's environment, but, more specifically, environmental and urban geology deal in large measure with those aspects of geology that directly influence man's use of the land. These include the stability of sites for building and other civil features, sources of water supply, contamination of waters by sewage and chemical pollutants, and selection of sites for burial of refuse so as to minimize pollution by seepage. Questions of geologic hazards . . . earthquakes and floods . . . broadly relevant to environmental and urban geology. The source of geological building materials, including sand, gravel, and crushed rock is also relevant. Finally, the distribution of naturally occurring toxic or hazardous materials, such as compounds of mercury and certain radioactive materials, has an influence on man's health and therefore is properly within the sphere of environmental geology' (*Encyclopaedia Britannica*, 1974). 'Environmental geology' (Knill, *PGA*, 1970). *Environmental Geoscience* (Strahler and Strahler, 1973). *Geology in Environmental Planning* (Howard and Remson, 1978). **2.** In a quite different sense, the term would apply to the study of the environment in which physical, chemical, and biological processes acted in past geological ages (e.g. 'The environmental history of the Boulonnais, France', Ager and Wallace, *PGA*, 1966; *Sedimentary Environments and Facies*, Reading, ed., 1978).

**Eocambrian.** The period represented by sediments immediately underlying, and intimately connected with, beds with Lower Cambrian fossils; such sediments being best known in the arenaceous Sparagmite formation in Scandinavia. The period might be placed as earliest Cambrian (as the name suggests) or latest Precambrian. (Brögger, *c.* 1900; Holtedahl, *QJ*, 1952; Glaessner, *JGSA*, 1963.)

**Eocene.** The lowest general stratigraphical group of the Tertiary succession, usually ranking as a series; if a Palaeogene system is recognized, the Eocene is the lower of the two series within it. The Cretaceous/Eocene (Mesozoic/Cainozoic) boundary is drawn either at the base of the Montian (see PALAEOGENE) or at the base of the Danian (see CRETACEOUS); for discussion see Eames in Davies (*Tertiary Faunas*, 1975) and *A Correlation of Tertiary Rocks in the British Isles* (*GSSR* (2), 1978). Both Danian and Montian are absent in Britain, the Eocene (sands and clays) being separated from the Cretaceous (chalk) by a widespread discontinuity, with slight unconformity, indicating a retreat and readvance of the sea. However, in S. Europe the Eocene is represented by a great thickness of nummulitic limestones. In Britain, the rocks which were apparently laid down over about the south-eastern half of what is now England are today preserved in the tectonic basins of London and Hampshire. The British Eocene is divided into the following divisions (mentioning some of their characteristic fossils): the 'Lower London Tertiaries' (Thanet Sands and the Woolwich and Reading beds) with the lamellibranchs *Arctica morrisi* and *Ostrea bellovacina;* the London Clay with the annelid *Rotularia bognoriensis*, the lamellibranch *Pinna affinis*, the nautiloid

*'Nautilus' imperialis*, the crustacean *Xanthopsis leachi*, and the fish *Myliobatis striatus;* the Bagshot Sands; the Bracklesham Beds with the foraminifer *Nummulites laevigatus*, the lamellibranch *Cardita planicosta*, and the gastropod *Turritella sulcifera;* the Barton Beds with their very rich molluscan fauna (in Hampshire), the lamellibranchs *Chama squamosa*, *Crassatella sulcata*, the scaphopod *Dentalium striatum*, the gastropods *Clavilithes macrospira*, *Conorbis dormitor*, *Fusinus porrectus*, *Typhis pungens*, *Rimella rimosa*, *Turritella imbricataria*, and *Athleta luctator*. Many well-known Barton fossils were first described by Solander in Brander's *Fossilia Hantoniensia* (1766). The lower part of the Eocene is now often separated from the rest and called the Palaeocene (or Paleocene), but the boundary is placed at different levels: (1) at the base of the Thanetian, (2) at the base of the Sparnacian, (3) at the base of the Ypresian (see PALAEOGENE). For discussion see Curry, *PGA*, 1966 (Lyell, *Principles*, 1833; see PLIOCENE.) [name expressing 'the dawn of recent species'.]

**Eolith.** Eoliths are flints, associated with the Crag formations in eastern England, chipped and shaped in such a way as to suggest human workmanship and which are thus claimed by some to be the earliest signs of man. It is difficult to determine the age of these objects on stratigraphical grounds, but in so far as they are not older than the Red Crag they are, according to the latest terminology and correlation, not older than the Pleistocene. If eoliths are truly artificial they may be either older than, or contemporary with, palaeoliths elsewhere.

**Eon.** See AEON and CHRONOSTRATIGRAPHY.

**Eonathem.** See CHRONOSTRATIGRAPHY.

**Eozoic.** Of the dawn of life; sometimes applied to an earlier part of Precambrian time.

**Epeiric sea.** An extensive shallow sea spread over a continental area; used particularly for such a sea in the geological past. 'The evolution of an epeiric sea' (Sellwood and Jenkyns, *JGS*, 1975).

**Epeirogeny.** 'The displacements of the earth's crust which produce mountain ridges are called "orogenic". For the broader displacements causing continents and plateaus, ocean beds and continental basins, our language affords no terms of equal convenience. Having occasion to contrast the phenomena of the narrower geographic waves with those of the broader swells, I shall take the liberty to apply to the broader movements the adjective "epeirogenic", founding the term on the Greek word ηπειρος, [ēpeiros], continent. The process of mountain formation is "orogeny", the process of continent formation is "epeirogeny", and the two collectively are diastrophism' (Gilbert, *Lake Bonneville*, 1890). 'Epeirogenesis' is synonymous. See DIASTROPHISM.

**Epibole.** The time-rock unit corresponding to the hemera, the time-unit. 'A stratigraphical term to cover deposits accumulated during a hemera' (Trueman, *PGA*, 1923). [G. *epibolē*, a laying on.]

**Epicentre.** The point or small area on the earth's surface vertically above the seismic focus; the projection of the focus on to the surface. If, in any case, the epicentre is to be a point, it is the projection of the focus considered as a point (see SEISMIC FOCUS) or the centre about which isoseismal lines may be said to be grouped, the spot where the shock was most intense. (Either of these definitions would give almost, if not quite, the same point.)

**Epiclastic rock.** A rock formed from the fragments or particles broken away from pre-existing rocks to form an altogether new rock in a new place. All the rocks usually called 'clastic' are epiclastic, except the pyroclastic kind. The term is contrasted with 'autoclastic rock' (in the sense recommended in this Dictionary).

**Epidiorite.** A basic igneous rock that has become metamorphosed so as to be rich in hornblende. (Gümbel, Germany, 1874; Wiseman, *QJ*, 1934.) Well known among the sills in the Grampian Highlands of Scotland.

**Epidote.** A group of minerals, complex basic silicates with similar atomic structure but varying atomic composition. Characteristic of a mild degree of dynamothermal metamorphism. (Haüy, 1801.) [G. *epidosis*, increase, apparently referring to some feature of the crystal form.]

**Epifauna.** A fauna of animal organisms that have grown upon the hard shell of a 'host' animal. Thus we frequently find fossil brachiopods, molluscs, and echinoids with fossil bryozoans, serpulid worms, barnacles, &c. firmly attached. 'The epifauna of a Devonian spiriferid' (Ager, *QJ*, 1961). The organisms themselves are 'epizoans' ('epizoic' organisms).

**Epigene.** Operating or formed on the surface of the earth. 'Epigene or surface action—the changes produced on the superficial parts of the earth, chiefly by the circulation of air and water set in motion by the sun's heat' (Geikie, *Textbook*, 1882). 'Epigene or surface-formed rocks—aqueous (and aeolian), volcanic' (Judd, *Student's Lyell*, 1911). See quotation (Hollingworth) under HYPOGENE.

**Epigenetic.** Usually in the sense 'formed after', particularly for ore-deposits formed later than the enclosing rocks; also for sedimentary structures formed subsequently to the deposition of the sediment. (The opposite of syngenetic.) Occasionally used in the sense 'formed upon'. 'Epigenetic mineralization in Yorkshire' (Dunham, *PYGS*, 1959). (See SUPERIMPOSED DRAINAGE.)

**Epizoan.** See EPIFAUNA. 'Epizoans as a key to ammonoid ecology' (Seilacher, *JP*, 1960).

**Epoch.** In a specialized stratigraphical sense, the time equivalent of a series. See CHRONOSTRATIGRAPHY.

**Equilibrium.** The conception of equilibrium in geology is that a condition or process has produced its total effect or finished its work and thereupon either continues in existence without causing any further change or automatically ceases. Thus in metamorphism a rock tends to be produced which has a mineral composition, texture, and structure appropriate to the temperature and pressure to which it is subjected. If these conditions remain constant, the rock may eventually attain a character which is in perfect equilibrium with them; but if they are changing, the character of the rock will tend to change, probably lagging behind. The insistence of the conditions towards producing an 'equilibrium rock' varies greatly; thus a rise to a high temperature will inevitably cause new minerals to be formed which, however, may remain in existence (theoretically, in a metastable state) when the temperature falls. Again, in geomorphology, equilibrium between form and process is everywhere seen developing or attained. Thus, for instance, in the differential erosion of a coast (with stable sea-level), the irregularity of plan is an expression of the balance between the relative hardness of the different rocks and the vigour of wave-action in the bays and on the headlands.

**Equisetales.** The horsetails, a group of plants belonging to the division Pteridophyta, ranging from the Upper Devonian to the present day but flourishing most abundantly, and to the largest size (as trees), in the Carboniferous period. [genus *Equisetum* (L. *equus*, a horse, *saeta*, a bristle).]

**Era.** In a specialized stratigraphical sense, a prime division of geological time. Palaeozoic, Mesozoic, and Cainozoic are the well-established eras; Pre-Palaeozoic eras may come to be recognized. See CHRONOSTRATIGRAPHY.

**Erathem.** See CHRONOSTRATIGRAPHY.

**Erosion.** The wearing away of rocks; a general term for all the processes of the loosening, breaking up, physical and chemical disintegration, and solution of rocks; a combined term for weathering and corrasion. Often used in a still wider sense to include transport as well, and thus to denote the whole process of the lowering of the land surface; the whole destructive process, as opposed to the constructive process of deposition, synonymous with 'denudation' and 'degradation' in their widest senses. 'M. de Saussure says that the formation of this valley depends upon the mountains themselves and not upon the erosion of the rivers' (Hutton, *Theory of the Earth*, 1795). 'The force of aqueous erosion' (Lyell, *Principles*, 1830). 'Stated in their natural order, the three general divisions of the process of erosion are (*a*) weathering, (*b*) transportation, and (*c*) corrasion. The rocks of the general surface of the land are disintegrated by weathering. The material thus loosened is transported by streams to the ocean or other receptacle. In transit it helps to corrade from the channels of the streams other material, which joins with it to be transported to the same goal' (Gilbert, *AJS*, 1876). 'Erosion may be regarded from several points of view. It lays bare rocks which were before covered and concealed, and is hence called "denudation". It reduces the surfaces of mountains, plateaus, and continents, and is thence called "degradation". It carves new forms of land from those which before existed, and is thence called "land sculpture" ' (Gilbert, *Henry Mountains*, 1877; both extracts quoted by Mather and Mason, *Source Book*, 1939). The following table shows these usages:

| *sensu stricto* | | *sensu lato* |
|---|---|---|
| erosion | weathering | |
| | corrasion | erosion |
| denudation | transport | =denudation |

Weathering is sometimes excluded from 'erosion', but this is not justified on either etymological or historical grounds. [L. *erodo*, gnaw away.]

**Erosion surface.** Usually in a specialized sense for the more or less detached remnants of a surface supposed to have been produced by erosion during a particular phase in the physiographical history of a region. 'Erosion level' is also used. Sometimes these 'surfaces' and 'levels' are correlated with supposed uplifts of the region.

**Erratic. 1.** In a special, and the usual, sense: 'Erratics are fragments of rock which ice or its streams have transported from their parent source. They may be embedded in till or rest on rock or any kind of drift and vary in size from small pieces to huge boulders. The latter, the erratic blocks of Brongniart and De la Beche, early attracted attention because of their dimensions and often impressive difference in composition and colour from the underlying rocks' (Charlesworth, *Quaternary Era*, 1957). The full term is 'erratic block', 'glacial erratic', or 'glacial boulder'. 'The erratic blocks [near Como]' (De la Beche, *Sections and Views*, 1830). Bernhardi and de Charpentier (in the 1830's) were the first to suggest a land-glacial origin for them, and the acceptance of this was part of the acceptance of the glacial theory as a whole during the 1840's (North, *PGA*, 1943). See quotation under GLACIAL THEORY. 'The difficulties which had long attended every attempt to explain the phenomena of the distribution of the Shap Fell boulders, Dr. Buckland considers, are entirely removed by the application of the glacial theory' (*PGS*, 1840, Dec. 2). 'On the distribution of the erratic boulders . . . of South America' (Darwin, *TGS*, 1842). 'There are transported boulders . . . I will confine my notice to the travelled blocks of Shap granite; they have too distinct a mineral structure to be mistaken, in whatever company we may meet with them. . . . By what power were these "erratic blocks" scattered over the north of England, and lodged in positions that seem so utterly strange and anomalous? . . . May we not suppose that some of the highest valleys of England and Scotland were filled with glaciers, and that numberless blocks of stone which had rolled down the mountain sides . . . were then packed up in thick-ribbed ice?' (Sedgwick, letters to Wordsworth, 1842, in Wordsworth and Hudson's *Guide to the Lakes*). Distinctive erratics, such as those of the Shap Granite, that can be traced to a parent source indicate the course of iceflow and are thus 'indicator erratics'. Hutton (*Theory of the Earth*, 1795) has the following: 'Let us now consider the height of the Alps, in general, to have been much greater than it is at present; and this is a supposition of which we have no reason to suspect the fallacy; for, the wasted summits of those mountains attest its truth. There would then have been immense valleys of ice sliding down in all directions towards the lower country, and carrying large blocks of granite to a great distance, where they would be variously deposited, and many of them remain an object of admiration to after ages, conjecturing from whence, or how they came'. An early allusion to the form of erratics, not so called and not attributed to ice, is that by White Watson in his *Strata of Derbyshire* (1811): 'Pieces of granite, limestone, basalt, etc., with their angles rounded off, having been carried and deposited by floods'. Transported shells may also be included: 'The marine shells in the high-level drift-deposits of our islands are "erratics", carried by the ice-sheet which occupied the basin of the Irish Sea' (J. Geikie, *Fragments of Earth Lore*, 1893). The term has been used for boulders dropped in the sea by floating ice, in considering the origin of a boulder-bed in the Dalradian (Sutton and Watson, *GM*, 1954). '[By about the beginning of the 19th century] ice had been invoked to explain the existence of "erratics", as Saussure had called them in 1779. Whalers had brought back tales of icebergs in polar seas carrying large blocks of rock which eventually fell into the sea. No other explanation seemed needed to explain the erratics found on land; but the heights at which they occurred called for a great rise of sea level and substantial drowning of the continents. It was the postulated drifting of the icebergs which gave rise to the term "glacial drifts" which persists illogically to the present day' (Shotton, *OE*, 1980). **2.** 'Erratic' by itself is used in a more general sense to include stray stones foreign to the geological formation in which they are found. 'Erratics in coal-seams' (Lister and Stobbs, *TNSFC*, 1917). 'The erratics of the Cambridge Greensand' (Hawkes, *QJ* 1943). 'Thc erratics of the English Chalk' (Hawkes, *PGA*, 1951). Sarsen stones were called 'erratic blocks' by Mantell (*SE. England*, 1833). Stones transported by human agency might deserve the name 'erratics'. The pebbles in a conglomerate are not called 'erratics'.

**Eruptive.** Usually, for both intrusive and extrusive igneous rocks, implying a breaking-through from lower to higher levels in the earth's crust. 'Eruptive or igneous rocks' (Geikie, *Textbook*, 1903). *Eruptive Rocks* (Shand, 1927/50). Some authorities take 'eruptive' or 'erupted' to imply a breaking-through to the earth's surface (as is implied in 'volcanic eruption'): 'The extrusive rocks are "erupted" to the surface; the intrusive rocks are "irrupted" into the crust' (Tyrell, *Petrology*, 1929). [L. *erumpo*, break out, burst.]

**Escarpment.** 'In common language "escarpment" and "scarp" have the same meaning, any line of cliffs, or abrupt slope breaking the continuity of a surface' (Cotton, *Landscape*, 1948). Distinctions are sometimes attempted between the meaning of the two terms; 'escarpment' sometimes, but 'scarp' never, implying a whole ridge, with both its slopes, especially when this is formed in a resistant stratum, with a dip-slope on one side and a scarp-slope on the other. An escarpment in this wide sense is often conveniently distinguished by the term 'cuesta', it being understood that this Spanish word is used, in English, with this special meaning. A 'fault-line scarp' is one formed along the line of a fault, where a resistant rock is brought against a less-resistant one (the higher ground may be either on the upthrow or the downthrow side), a 'fault scarp' is one formed directly by fault-movement (the higher ground necessarily being the upthrow side). In these latter cases 'escarpment' is seldom used. 'Escarpment' (or 'scarp') has been used for a sea-cliff and any associated bevel above (Arber, *Coast Scenery of N. Devon*, 1911), but is not generally used for the abrupt slopes of the sides of an ordinary (non-structural) valley. 'A thick mass of limestone which terminates suddenly in the steep escarpment of Wenlock Edge' (Buckland, *TGS*, 1821). The association of certain escarpments with regularly dipping strata was early recognized by John Williams: 'I have mentioned more than once that in general the strata dip towards the east and rise towards the west; and I will now observe that it is this disposition of the strata that was the occasion or natural cause of the many abruptions of the mountains to the west in this country' (*Mineral kingdom*, 1789). In Britain, the most prominent escarpments (of the usual dip-and-scarp kind) are those produced in the Silurian limestones of Salop and Herefordshire, the Old Red Sandstone of South Wales, the Millstone Grit of the Pennines, the Jurassic Oolites (particularly the Cotswold Hills), and the Chalk (particularly the Chiltern

Hills and the North and South Downs, with the Lower Greensand ridge in front of the North Downs). [French *escarpment*.]

**Esker.** 'It appears advisable to use "esker" in a wide sense, as a general descriptive term for glacial sands and gravels arranged in mounds or ridges [as original forms], irrespective of their genesis' (Charlesworth, *Quaternary Era*, 1957). The esker then includes the two types, the kame and the os. The name became usual in glacial literature about the middle of the 19th century. (For a discussion and history see Francis in Wright and Moseley, eds. *Ice Ages: Ancient and Modern*, 1975.) [Irish *eiscir*, applied to these ridges.]

**Essential mineral.** See ACCESSORY (1).

**Estuarine.** For the sedimentary and biological environment of estuaries, and the corresponding sediments and forms of life.

**Eucrite.** An olivine-gabbro with various pyroxenes and a felspar at or near the anorthite end of the plagioclase series. Important in some of the Tertiary ring complexes of western Scotland, the Great Eucrite of Ardnamurchan being a typical example. 'Anorthite-augite meteorites were called "eucrites" (G. εὔκριτος [*eukritos*], "easily discerned") by Rose (Berlin, 1864). The first-described terrestrial rock of this group was the "syenite" from Grange Irish, Carlingford, Ireland, which was analyzed and described by Haughton (*QJ*, 1856)' (Johannsen, *Petrography*, 1937).

**Eugeosyncline.** See GEOSYNCLINE.

**Euhedral.** See IDIOMORPHIC and ANHEDRAL.

**Eurypterida.** An extinct subclass (or order) of the class Merostomata, phylum Arthropoda. Large and striking, but not generally common, fossils, ranging from Ordovician to Permian, but especially characteristic of the Old Red Sandstone. They appear to have lived in brackish or fresh water. (Burmeister, 1843.) [genus *Eurypterus* (G. *eurus*, wide, *pteron*, a wing).]

**Eustatic.** Of the land surface: everywhere at rest. The term appears to have been coined by Suess in *Das Antlitz der Erde*, as *eustatische Bewegungen*, in 1888; translated as 'eustatic movements' in the English translation, 1906, and as *mouvements eustatiques* in the French translation, 1921. What was meant by 'movements' was not earth-movements but changes in sea-level, and 'eustatic' meant those occurring uniformly over the whole globe. Changes in sea-level the most surely uniform and world-wide would be those due to causes other than earth-movement of any kind (including isostatic movements), such as deposition of sediment or submarine eruption of lava (if on a large enough scale, causing sea-level to rise) or abstraction of water to form ice-sheets (causing sea-level to fall). But world-wide, uniform changes in sea-level would also be caused by earth-movements, so long as these were confined to the ocean floors, and Suess specifically included these, particularly a sagging causing sea-level to fall (an uprise would cause it to rise). Further confusion may arise over the words 'positive' and 'negative' which were introduced by Suess in connexion with his 'eustatic movements'. A rise in sea-level he called 'positive', a fall 'negative'; but, viewed from the land rather than the sea, 'positive' seems to imply an emergence of the land, 'negative' a submergence (see POSITIVE). 'Eustatic changes in sea level' (Fairbridge, *PCE*, 1961). The term for the principle is 'eustasy'. [G. *statikos*, causing to stand still.]

**Eutaxitic.** This term could apply to any 'well ordered' condition in a rock but is, in practice, reserved for a certain structure or foliation texture in volcanic rocks. Holmes (*Nomenclature of Petrology*, 1920/28) has the following: 'a term describing the streaked or blotched appearance of certain volcanic rocks due to the alteration of bands or elongated lenses of different colour, composition or texture; the bands, etc., having been originally ejected as individual portions of magma which were drawn out together in a viscous state and formed a heterogeneous mass on welding.' When extreme welding (particularly in ignimbrites) has initiated secondary flowage the term 'parataxitic' has been proposed (Beavon and others, *LMGJ*, 1961). (Fritsch and Reiss, 1868.) [G. *taxis*, arrangement.]

**Eutectic.** Literally, melting readily. 'Eutectic mixture.—A discrete mixture (not a compound) of two or more minerals which have crystallized simultaneously from the mutual solution of their constituents, the two or more minerals being in definite proportions' (Holmes, *Nomenclature of Petrology*, 1928). The temperature at which the simultaneous crystallization of these minerals must have occurred is the 'eutectic point'; it is the lowest temperature (at a given pressure) at which the constituents of the minerals can exist together in a liquid state. [G. *tēkō*, melt.]

**Evaporite.** A deposit of precipitated salt, evaporation having caused the necessary concentration; particularly for such a deposit of a former geological age, such as the Permian evaporites of Durham and Yorkshire or the rock-salt among the Triassic rocks of Cheshire. *Salt Deposits: The Origin, Metamorphism, and Deformation of Evaporites* (Borchert and Muir, 1964). *Marine Evaporites* (Kirkland and Evans, eds., 1973).

**Evolution.** The process of evolving, unrolling, opening out; history as a continuous series of conditions or events, one leading to the next. *Earth Evolution and its Facial Expression* (Hobbs, 1921). *The Evolution of Climate* (Brooks, 1922). *Evolution of the Igneous Rocks* (Bowen, 1928, and Yoder, ed., 1979). *Physiographical Evolution of Britain* (Wills, 1929). 'The evolution of a geosyncline' (Jones, *QJ*, 1938). 'Some aspects of magmatic evolution' (Tilley, *QJ*, 1950). 'The geological evolution of Wales' (Jones, *QJ*, 1955). 'The tectonic evolution of the Midland Valley of Scotland' (Kennedy, *TGSG*, 1958). 'The structural and geomorphic evolution of the Dead Sea Rift' (Quennel, *QJ*, 1958). *Evolution of the Earth* (Dott and Batten, 1971). *The Evolving Continents* (Windley, 1977/84). *Geological Evolution of North America* (Stearn and others, 1979). 'The evolution of petrological ideas' (Teall, *QJ*, 1901). Above all other applications it is used in biology, for the theory that new forms of life have developed from pre-existing forms, by descent (through successive generations) with inheritable modification. It is thus directly opposed to the idea that new forms of life have been specially created. Direct evidence as to biological evolution in the past is entirely confined to the palaeontological record. *Patterns of Evolution, as illustrated by the Fossil Record* (Hallam, ed., 1977). Hence the adjectives 'evolutional', 'evolutionary'; which are almost synonymous though the latter carries the particular sense of 'in accordance with' evolution (thus 'evolutional palaeontology' but 'evolutionary series'). [L. *evolvo*, roll forth, unfold.]

**Evolutional palaeontology.** The study of fossils from the evolutionary point of view. 'Evolutional palaeontology in relation to the Lower Palaeozoic rocks' (Elles, *BA*, 1923).

**Evolutionary series.** In palaeontology, a morphological series which, when the time element is added, is found to correspond with time to a significant degree. This graded change may be in the individual fossils as wholes, affecting several features while the other features remain constant; the whole series constituting an evolutionary lineage or evolving plexus, a 'phylogenetic series'. A closely graded series, as so far observed among fossils, usually covers only a small group, a species-group (e.g. *Zaphrentis delanouei*, *sensu lato*, Lower Carboniferous), a genus (e.g. *Micraster*, Chalk), or a few related genera (e.g. *Ostrea* and *Gryphaea*, Lower Lias). A not very closely graded series of species, extending perhaps through a number of genera or even higher groupings, might also form an evolutionary series of, essentially, the same kind. This kind of series (whether closely or loosely graded) is what is usually implied and it is perhaps advisable to restrict the term to it. It is the *Formemreihe* of Waagen; see MUTATION (1). Another kind of series which might be called evolutionary is one selected to show an evolutionary trend in one feature only, the state of the other features being ignored (see TREND).

**Excursion guide.** A book or pamphlet giving itineraries for geological excursions and the location and details of the more interesting geological exposures etc. *en route*. An example of a book is *A Geological Excursion Handbook for the Bristol District* (Reynolds, 1921) and pamphlets are represented by the *Geologists' Association Guides*, in course of publication since 1958. 'Bibliography and index of geological excursion guides and reports for areas in Britain' (Bassett, *WGQ*, 1967-69).

**Exfoliation.** The scaling-off of the outer surface of a rock, either the result of weathering of an exposed rock-mass or due to disintegration of a concretion, &c. The former is the more usual application (also called 'desquamation') and is most strikingly seen in deserts, the result of rapid changes in temperature with, probably, some chemical changes. One early use in geology refers to septarian concretions: 'This stone is subject to perpetual exfoliation' (Playfair, *Illustrations*, 1802). See INSOLATION. [L. *folium*, a leaf.]

**Exhume**. To remove something buried from beneath the ground. Fossils are 'exhumed' when they are unearthed from rock, naturally or artificially. A landscape of a past geological age is said to be 'exhumed' when it is revealed by the removal (denudation) of overlying rocks which were deposited over it, smothering and preserving it. An exhumed surface is a 'fossil' erosion surface. A particular feature, such as a ridge, thus restored to its previous status in existing relief, is said to be 'resurrected'. 'Exhumed paleoplains of the Precambrian Shield of North America' (Ambrose, *AJS*, 1964). 'The most evocative relic of the Trias is a whole fossil landscape, now being exhumed in Charnwood forest. . . . At one time, before the advent of machinery, the great brick pits in the Upper Jurassic clays yielded a fine harvest of complete marine vertebrates. These reptiles are still as common as ever, but it is now seldom possible to exhume them whole' (Sylvester-Bradley in *Geology of the East Midlands*, 1968). See remarks under VOLCANO.

**Exogenetic.** See ENDOGENETIC.

**Exotic.** Introduced from a foreign source. Used occasionally, particularly in the discussion of the source of the material constituting a sedimentary rock. 'Source rocks of the Lower Old Red Sandstone: exotic pebbles from the Brownstones' (Allen, *PGA*, 1974).

**Expanding earth.** The hypothesis that the earth is expanding. 'The expanding earth' (Egyed, *N*, 1963). 'Continental displacement and expansion of the earth' (Owen, *PT*, 1976). *The Expanding Earth* (Carey, 1976).

**Experimental geology**. Research on the working of geological processes by carrying out physical and chemical experiments (1) of general physical, mechanical, and chemical kinds as these might apply to geology, (2) with scaled-down models, (3) with natural geological phenomena. The experiments of Sir James Hall (*TRSE*, 1790/1825) concerning the effects of heat and pressure on rock-substances and of pressure on models of stratification are famous in the history of geological investigation (see Eyles, *E*, 1961). 'The experimental method was being used in petrology before the middle of the eighteenth century by Réamur (1726)' (Loewinson-Lessing, trans. Tomkeieff, *Historical Survey of Petrology*, 1954). 'Report of the experiments made at Holyhead to ascertain the transmit-velocity of waves, anologous to earthquake waves, through the local rock foundations' (Mallet, *BA*, 1861). 'Experimental geology' (Rudler, *PGA*, 1889). 'My object is to apply experimental physics to the study of rocks' (Sorby, *QJ*, 1908). 'Experiments in geology' (Kuenen, *TGSG*, 1958). *Bibliography of Experimental Rock Deformation* (Riecker and others, *U.S. Air Force*, 1965). 'Experimental structural geology' (Currie, *ESR*, 1966).

**Explosion breccia. 1.** A breccia produced by volcanic explosion. See BRECCIA (3). **2.** A breccia produced by the explosive impact of a meteorite. See METEORITE CRATER.

**Explosion structure.** See CRYPTOEXPLOSION STRUCTURE.

**Exposure.** The condition of being exposed at the earth's surface; a place where a rock is exposed to view, naturally or artificially, not being hidden by vegetation, buildings, &c.

**Exsolution.** In crystallogenesis, the process of the rearrangment of the atoms in a homogeneous crystal on its cooling from a high temperature, so that an intimate intergrowth of two separate minerals results. The original crystals may be looked on as a mutual solution of the two products. (Alling, *JG*, 1921.)

**Extended succession.** A stratigraphical succession that is relatively thick because deposition was rapid during the time represented. The opposite of a condensed succession.

**Extinct volcano.** See VOLCANO. *The Extinct Volcanos of Central France* (Scrope, 1826/58).

**Extinction.** In addition to its obvious use in palaeontology, there is a special application of the term in mineralogy and petrology: the darkness that occurs in certain positions as a birefringent crystal is rotated on the stage of a polarizing microscope.

**Extrusion.** Refers almost exclusively to igneous action at the earth's surface resulting in the emission of magma as lava (effusive material) and (or) the thrusting out of more or less solid material. It might be held to include all ejectamenta (as in the quotation below), but hardly gaseous matter. Also for a body of igneous rock so formed. Hence 'extrusive'. 'Extrusive rocks may be classified in two great groups—(*a*) the Lavas, or those which have been poured out in a molten condition at the surface; and (*b*) the Fragmental Materials, including all kinds of pyroclastic detritus discharged from volcanic vents' (Geikie, *Ancient Volcanoes*, 1897).

# F

**'F'numbers.** A series of 'F'numbers (F1, F2, F3, F4) was used by Rast (*QJ*, 1958) to denote successive sets of movements, and the successive sets of folds to which they gave rise, constituting the polyphase orogeny and deformation of the Caledonides of the Scottish Highlands. A similar notation of 'F'numbers has been applied to other regions and other periods of deformation, e.g. to the Hercynian folds of Devon and Cornwall (Hobson, *PGA*, 1976). Polyphase deformation is most clearly evidenced by the folding, but 'S'numbers have been used for successive slaty cleavages (e.g. Phillips and Byrne in Kent and others, eds., *Time and Place in Orogeny*, 1969). 'D'numbers for deformation in general (e.g. Treagus, *JGS*, 1974) and 'M' numbers for metamorphism (Shackleton and others, *JGS*, 1979).

**Fabric.** Of a rock or deposit; the texture, and the manner of arrangement of all the particles and crystallographic elements. A fabric may be primary, formed at the time of deposition or solidification, by the apposition of the particles, or secondary, the result of subsequent deformation. Hence 'micro-fabric'. The term is most often used when there is preferred orientation. 'These rocks [Torridonian and Cambrian] being comparatively unaltered sediments, it is of course possible that they may reveal some kind of sedimentation (apposition) fabric, but such fabrics in rocks of this type are never so well marked as deformation fabrics' (Phillips, *QJ*, 1937). *Depositional Fabrics* (Sander, 1951). 'Pebbles of a conglomerate may show a primary fabric such as an alignment of their long axes in the direction of the operative current' (Read, *PGA*, 1958). 'Fabrics: textures and structures—"fabric" is used here with this meaning following Knopf (*AJS*, 1933) and Fairbairn (*Structural Petrology of Deformed Rocks*, 1942) who thus translated the *Gefüge* of Sander' (Bathurst, *LMGF*, 1958). 'Terminology of crystallization textures and fabrics in sedimentary rocks'

(Friedman, *JSP*, 1965). 'Statistical analysis of till fabric' (Andrews and Smith, *QJ*, 1969). *An Introduction to the Study of Fabrics of Geological Bodies* (Sander, trans., Phillips and Windsor, 1970). For more about the history of the term see Dennis, *Tectonic Dictionary*, 1967. See PETROFABRIC.

**Face.** 'I follow Shrock (*Sequence in Layered Rocks*, 1948) in using the term "to face" rather than "to young", but extend its application from strata to structures. A stratum is deposited face upwards and in any subsequent attitude it faces towards the side that was originally upwards and younger. A fold faces in a direction normal to its axis, along the axial plane, and towards the younger beds. This coincides with the direction towards which the beds face at the hinge. A normal upright fold faces upwards. An anticline closing downwards [i.e. over-turned] faces downwards' (Shackleton, *QJ*, 1957).

**Facies.** Appearance, character. A term that is inherently clear and general but, in geological use, nearly always somewhat restricted and made to carry certain (but varying) implications. The following are some pronouncements and usages.

'Whether these argillaceous masses [Wenlock-Ludlow] be examined in the wilds of Radnor forest and the eastern parts of Montgomery, in the western parts of Shropshire (Long Mountain), or in many tracts of North and South Wales, they present the same "facies" of a thick, yet finely laminated, dark, dull grey shale, in which hard stone of any strength or persistence is the rare exception. Their dominant character is "mudstone"' (Murchison, *Siluria*, 1854). 'The organic remains all have a Lower Silurian "facies"' (Murchison, *Siluria*, 1867). 'The general facies of the Carboniferous vegetation' (Nicholson, *Palaeontology*, 1872). 'The general facies of the mollusca [Pleistocene] indicates marsh conditions' (Kennard and Woodward, *QJ*, 1919). 'Tournaisian facies in Britain' (George, *IGC*, 1948). 'The earlier Silurian deposits show a broad grouping into two main facies: on the one hand, an off-shore, possibly a subbathyal, facies of shales and mudstones, typically stocked with a graptolitic fauna, and on the other hand, a shallower-water facies of sandy and silty beds with a shelly fauna of tribolites, brachiopods, and (in the calcareous layers) corals' (Smith and George, *BRG*, *N. Wales*, 1948). 'Faunal facies' (George, *SP*, 1959). 'An alkali facies of granite at granite-dolomite contacts in Skye' (Tilley, *GM*, 1949). 'Facies has long been a convenient term for the sum of characters (whatever they be) deriving from a particular environment or set of conditions. . . . A proper description of tectonic facies should include at least the three related aspects of form, scale, and composition' (Harland, *GM*, 1956). *Metamorphic Reactions and Metamorphic Facies* (Fyfe and others, *MGSA*, 1958). 'The metamorphic facies concept' (Lambert, *MM*, 1965). (The metamorphic facies is usually taken to be expressed by the mineral composition of the rock, which, naturally, depends in the first place on the chemical composition of the rock as a whole, and also on the degree and kind of metamorphism it has undergone. Sometimes, however, it is defined as being equivalent to metamorphic grade. The term, in connexion with metamorphism, was introduced by Eskola in 1915). 'Origin of the Hastings facies in NW. Europe' (Allen, *PGA*, 1967). 'Facies, in stratigraphy, can mean aspect, nature, or manifestation of character (usually reflecting conditions of origin) of rock strata or specific constituents of rock strata. It is also used as a substantive for a body of rock strata distinctive in aspect, nature, or character. The general term "facies" has been greatly overworked. Rock strata may show differences in facies of various sorts so that one may speak of lithofacies, biofacies, mineralogic facies, marine facies, volcanic facies, boreal facies, and so on. If the term is used, it is desirable to make clear the specific kind of facies to which reference is made' (Hedberg, ed., *International Stratigraphic Guide*, 1976). For the 'facies approach' in the study of palaeoenvironments see Scott, *PGA*, 1979. *Facies Models* (Walker, ed., *GC*, 1979). Dunbar and Rodgers (*Stratigraphy*, 1957) show that Gressly (Switzerland, 1838) used and explained the term *facies* (in French) for the varying lateral modifications in the lithological and fossil aspects of sedimentary rocks depending on the correspondingly varying environments of deposition. [L. *facies*, appearance, figure, shape.]

**False bedding.** See CROSS-BEDDING. This, or the like, is probably the oldest established term for cross-bedding: 'This diagonal arrangement of the layers sometimes called "false stratification"' (Lyell, *Principles*, 1833). It usually implies current bedding in particular.

**False cleavage.** See CLEAVAGE (1*b*).

**Famennian.** See DEVONIAN SYSTEM. (Dumont, 1855.) [Famenne region, Belgium.]

**Family.** Apart from its primary use, which is in connexion with human beings; a unit, usually a fairly small one, in a scheme of classification. In palaeontology (as in biology generally) it is a division of an order and comprises one or more genera. In petrography it may be said to be equivalent to the group now more definitely called the 'clan'. In mineralogy it is sometimes used, but rather loosely; e.g. the 'felspar family'.

**Fan deposit. 1.** A fan-shaped deposit laid down by a river ('alluvial fan'), particularly by a constricted and steep tributary stream as it emerges into a main valley or onto an open plain. If its surface has a high angle of slope it may be called an 'alluvial cone' **2.** See OUTWASH DEPOSIT. **3.** See SOLIFLUXION.

**Fan structure.** An anticlinorium in which the subsidiary folds have their axial planes tending to converge to a centre. 'In mountain chains the rocks have sometimes been so folded and compressed that a double series of isoclinal flexures has been developed; the axes of the folds in each series sloping inwards on both sides of a great central anticlinal fold. This arrangement has been termed "fan structure", and is well exemplified in the Mont Blanc range' (Jukes-Browne, *Physical Geology*, 1884, quoted in Dennis, *Tectonic Dictionary* 1967). 'A further modification of the folded structure is presented by the fan-shaped arrange (*structure en éventail*, *Facher-Falten*) into which highly plicated rocks have been thrown' (Geikie, *Textbook*, 1903). 'Along this northern line [of folding] the axial planes of the parallel flexures dip inwards on the north-west and south-east sides towards a central vertical axis, thus producing a fan-shaped or pseudo-synclinal arrangement of the beds. The southern line commences at St. Abb's Head and extends by way of Hawick and Dumfries to the Mull of Galloway, and from this central vertical axis the axial planes of the parallel folds slope away on opposite sides, and thus form an inverted fan-structure or pseudo-anticlinal' (Pringle, *BRG, S. Scotland*, 1948).

**Farlovian.** See OLD RED SANDSTONE. (King, *QJ*, 1934.) [Farlow, Salop.]

**Fault.** A surface of fracture, often nearly plane, in a rock-body, along which there has been permanent displacement. Hence 'fault surface', 'fault-plane', 'faulting'. 'The uniform course of seams of coal, and of the strata that accompany them, is frequently interrupted by obstructions called slips, dykes, troubles, faults, etc. These never fail to elevate, or depress, the strata beyond them; or rather, the strata on each side of them are found at different heights' (Kirwan, *Geological Essays*, 1799). 'A term derived from the miners, chiefly those working coal, who, when these dislocations are met with, often find themselves "at fault", the amount of the dislocation produced not being always clear. They are also known as "troubles" by the miners' (De la Beche, *Geological Observer*, 1851). Faulting in the Lothian coalfield is described by Sinclair (*Short History of Coal*, 1672): 'There is a gae [fault] encountered with, at which gae the coal is cut off . . . the gae, wearing out towards the west, the two parts of the coal that was separated by it, joynes themselves again'. 'The word "fault" as employed by miners has not necessarily a strict geological significance. It may denote anything which interrupts a coal seam . . . a well-known feature in the South Staffordshire Coalfield known as the "Symon fault" is an unconformity' (Edmunds, *Geology and Ourselves*, 1955). 'In the mining sense the term "fault" is commonly restricted to something which interrupts or displaces the lode, and it is not generally realised that the lodes themselves are true faults, which have produced more or less displacement of the country-rocks' (Jones, *MGS, Mining District of N. Cardiganshire*, 1922). 'It had long been known that the basalt overlies the coal-field [Titterstone Clee Hill] in thick sheets, but it was not ascertained till lately that this basaltic matter had been erupted through the strata of the coal measures before it overflowed them [see quotation under FEEDER]. . . . [This feeder] the miners termed the great Jewstone fault; the basalt being known in this country under the name of "Jewstone"' (Murchison, *Silurian System*, 1839). Faulting, with its slips or 'shifts', was described by Playfair (*Illustrations*, 1802), but the effects of faults on strata and their outcrops were first fully explained (with diagrams) by Farey (*Derbyshire*, 1811). And see EARTHQUAKE (where, in the quotation from Michell, 'trapping' and 'cleft' are further old terms). 'The three fundamental fault types—wrench, normal, and thrust . . .' (Williams, *GM*, 1958). A 'fault set' is a group of parallel faults. A 'fault system' is a group of branching or intersecting faults. 'The San Andreas fault system through time' (Crowell, *JGS*, 1979). For full analysis of kinds of faults, terminology, and history see Dennis, *International Tectonic Dictionary*, 1967.

**Fault block.** See BLOCK-FAULTING.

**Fault coast.** 'The essential feature of a fault coast is a fault scarp separating a higher-standing earth-block which, after faulting, forms the land, from a lower-lying block which, after faulting, is depressed below sea-level' (Cotton, *GR*, 1916, quoted in *AGI Glossary*, 1957).

**Fault gash.** A gash caused by erosion along a fault. They occur particularly in coastal cliffs. 'Huntsman's Leap, Bosherston, Pembrokeshire. A vertical fault gash in a Carboniferous Limestone cliff' (pl. in Steers, *Coastline of England and Wales*, 1946).

**Fault scarp.** See ESCARPMENT and FAULT-LINE SCARP.

**Fault trough.** See RIFT STRUCTURE. 'Fault troughs' (Taber, *JG*, 1927).

**Fault-breccia.** See BRECCIA (6).

**Fault-drag.** The flexing, due to drag, of strata near a fault. If there is drag on one side only of a fault, this strongly suggests that that side moved while the other remained steady. See SHIFT.

**Fault-line scarp.** See ESCARPMENT. 'It must be clearly realized that, throughout Britain and Western Europe and, indeed, over the greater part of the land areas the existing land surface is hundreds, or even thousands, of feet below the surface upon which the original fault-scarp was formed. Any irregularity which occurs at the line of the fault is thus a secondary feature developed during erosion. Where rocks of differing resistance are brought together an inequality of surface must normally result. Such features have been termed "fault-line scarps" by W. M. Davis to distinguish them from the original steps or "fault scarps"' (Wooldridge and Morgan, *Geomorphology*, 1959). 'It is impossible to over-emphasise the importance of distinguishing with meticulous care between fault-line scarps and fault scarps, thus avoiding the possibility of introducing grave errors into interpretations of geological history from surface forms' (Cotton, *Landscape*, 1948). 'Fault scarp or fault-line scarp?' (Challinor, *GM*, 1966). See quotation under TOPOGRAPHICAL EXPRESSION. 'Fault-line scarp of erosion' would be more explicit.

**Fault-line valley of erosion.** A valley formed along the line of a fault, this line being the outcrop of a surface of easily erodable rock.

**Fault-trough valley of erosion.** A valley formed by differential erosion along the strip of a fault trough (rift structure). Note (1): the strip might become a fault-trough ridge by differential erosion, instead of a fault-trough valley; (2) a fault-trough valley of erosion must be carefully distinguished, in principle, from a rift valley (cf. the distinction between fault-line scarp and fault scarp).

**Fauna** (pl. **-ae** or **-as**). **1.** The animal-fossil content of a deposit, bed, formation, or other stratigraphical unit; in this sense it is sometimes used alone when, in fact, 'and/or flora' is implied. A 'microfauna' is a fauna of microfossils. Hence 'faunal'. 'The mammal-fauna of the caves of Cresswell Crags' (Dawkins, *QJ*, 1877). 'The fauna of the Keisley Limestone' (Reed, *QJ*, 1896/7). 'The Lower Ludlow formation and its graptolite-fauna' (Wood, *QJ*, 1900). 'The *Carbonicola* fauna of the Midlothian five foot coal (Leitch, *TGSG*, 1936). 'The faunal succession of the Carboniferous Limestone of the Midland area' (Sibly, *QJ*, 1908). 'The Lower Palaeozoic brachiopod and trilobite faunas of Anglesey' (Bates, *BBM*, 1968). **2.** In reconstructing the life of the past the term carries its ordinary meaning of the collective animal life of a particular region, environment, or time. [L. *Fauna*, sister of *Faunus*, a god of farming.]

**Faunizone.** A term proposed by Buckman (*GM*, 1902) for 'zone', to avoid possible misunderstandings. By implication it includes 'florizone.'

**Faunule.** A small faunal unit.

**Fayalite.** See OLIVINE. The only kind of olivine found in association with free silica, that is, in acid igneous rocks. 'One of the commonest crystalline products of iron slags . . . the name was originally given to a substance from Fayal Island in the Azores, which may, however, be only an iron slag originally brought to that coast as ship's ballast' (Miers, *Mineralogy*, 1929). (Gmelin, 1840).

**Feather-edge.** The fine edge of a bed or igneous instrusion where it thins out (wedges out). See WEDGE. Usually applied to strata; overlapping and offlapping strata will have original feather-edges as will (usually) any locally deposited kind of rock. An early, and appropriate, application of 'feathering' is to the constituent parts (such as the stony beds and fine muds) of an alluvial mass: 'These seldom, or perhaps never, appear stratified in the uniform manner peculiar to regular strata, but the beds thereof, when such are discernible, frequently feather-out, or are wedge-like, and intermix with each other' (Farey, *Derbyshire*, 1811). There is a Feather-edge Coal (seam) at the base of the Coal Measures in N. Staffordshire.

**Feature. 1.** The primary meaning, according to the *OED*, is 'make, form, fashion, shape; proportions'. Thus its use, common in referring to the relief-effect of a geological condition, as in such a phrase as 'the outcrop of the sandstone forms a feature', is near or within this meaning (and if we add, for example, '. . . which is an asymmetrical ridge' we should be strictly within it). **2.** It is most usually employed for 'a distinctive part of anything', as in, for instance, 'stuctural features', 'features of the bedding'; but the *OED* gives this as merely the transferred meaning.

**Feeder.** In igneous geology, the channel or pipe through which magma passes from a general reservoir to feed the growth of a localized intrusion, such as a laccolite, or the accumulation of an extrusive mass of lava. In the latter case, the feeder is a volcanic pipe, in a central eruption, or a feeder-dyke in the case of a linear outpouring. 'A passage must exist somewhere in the vicinity of basaltic rocks through which the basalt was ejected: though this passage may be concealed by the surrounding rock. A remarkable instance of this kind may be observed in the Titterstone Clee Hill in Shropshire. I visited this hill in the autumn of 1811 [and] I found that a vast fissure or dyke, more than 100 yards wide, filled with the same basalt, intersected the hill, and separated the coal fields. It rises from an unknown depth. Where the basalt comes in contact with coal, it has injured its quality, and reduced it to a sooty state' (Bakewell, *Geology*, 1813). 'Basic dykes with a predominant north-north-easterly trend and an average thickness of about 10 ft. are numerous in eastern Iceland, and there are convincing grounds for identifying the dykes as the feeders of the basalt lavas' (Walker, *QJ*, 1963). See SUB-VOLCANIC.

**Felsic.** A portmanteau term for the (light coloured) minerals, felspar (etc.), and quartz (silica) in igneous rocks; also applied to rocks largely composed of these minerals. The other minerals, in this terminology, are the mafic minerals.

**Felsite.** A very fine grained igneous rock composed predominantly of quartz and felspar or, if porphyritic, with a ground mass so composed. The individual crystals are in any case not visible to the naked eye; but the name is sometimes restricted to rocks which are cryptocrystalline, the crystals not visible even under the microscope. With phenocrysts of quartz the rock is a quartz-felsite (a quartz-porphyry). Hence 'felsitic'. *The Felsitic Lavas of England and Wales* (Rutley, *MGS*, 1885). 'The word was first applied in 1814 by Gerhard, an early geologist, to the fine ground-masses of porphyries' (Rice, *Dictionary of Geological Terms*, 1951). 'Felstone', though sometimes separately defined, may be said to be synonymous, but is not now in use.

**Felspar.** (Feldspar.) The most important family of the rock-forming minerals, silicates of aluminium and, variously, potassium, sodium, calcium, and (rarely) barium. They differ in symmetry and twinning habit but occur as hard, white, light grey or pink block-like, tabular, or lath-shaped crystals with a pearly lustre. Ignoring the barium felspar there are two subgroups. (1). The alkali-felspars, with potassium and/or sodium. The commonest potassium felspar is orthoclase while the sodium felspar is albite. Intimate mixtures of these two felspars are perthite, in which the potassium and sodium may be present in various proportions, but the series is not strictly isomorphous. (2). The isomorphous series of the plagioclase felspars, ranging from albite (with sodium) to anorthite (with calcium). Thus albite is common to both the alkali felspars and the plagioclase felspars. Frequently the felspar content of a rock is given as a ratio between alkali felspar and plagioclase, even though sodium felspar be present, so that the terminology is obviously ambiguous ; but it is usually implied that the sodium felspar is being put solely among the alkali felspars while plagioclase means 'plagioclase excluding albite'. 'The word feldspar was used by Wallerius in his *Mineralogy* of 1747 in the Swedish form *felt-spat*, meaning fieldspar. It did not originate with him probably, but may have been a popular name in his time. Da Costa used it in 1757 (*Natural History of Fossils*) in the German form *feldspath*, and this form was current until 1794, when we find in Kirwan's *Mineralogy* the following note: "This name seems to be derived from *fels*, a rock, it being commonly found in granites, and not from *feld*, a field; and hence I write it thus, felspar". This assumption of Kirwan has been taken for fact by all English writers, and the corrupt form is in very general use' (Chester, *Names of Minerals*, 1896). Notwithstanding Kirwan's apparent misreading of the history of the derivation of the name, his interpretation is, in fact, the more appropriate. Apart from that, 'fel' seems a legitimate contraction and is less awkward in English than the unaltered 'feld'; 'feldspar' is a hybrid name neither the original *feldspath* nor the completely anglicized 'fieldspar' being in use in English. For these reasons we prefer 'felspar' to 'feldspar'. Both forms of the name are about equally common today. Aikin (*Mineralogy*, 1814) gives 'felspar' as the English equivalent of the Continental 'Feldspath'. 'Feldspath is an old German or Swedish name of doubtful signification, which was in use as early as 1750. Several different varieties of the mineral were distinguished first by their crystallographic, then by their chemical characters. The name orthoclase was used by Breithaupt (1823) to denote felspar with a rectangular cleavage (i.e. monoclinic felspar), and the name plagioclase was subsequently [Breithaupt, 1847] applied to the anorthic felspars whose cleavage-angle is not 90°. He also introduced [1826] the name oligoclase for the soda-felspar from Stockholm, which had been distinguished as a new mineral by Berzelius, and possessed an ill-defined cleavage. These names are universally employed at the present time' (Miers, *Mineralogy*, 1929). *The Feldspars* (Mackensie and Zussman, 1974).

**Felspathization.** The formation of felspar in a rock usually as a result of metamorphism towards granitization. Material for the felspar may be contributed both by the country rock and by introduced solutions, magmatic or other.

**Felspathoid.** (Feldspathoid.) A family of rock-forming minerals, silicates of aluminium and, variously, potassium, sodium, and calcium. They thus contain the same elements as the felspars, but are less rich in silica, and calcium is less common. Rocks rich in alkalis and poor in silica usually express this composition by the presence of one or more of the felspathoid minerals, of which the chief are leucite (potassium) and nepheline (sodium).

**Felstone.** See FELSITE.

**Femic.** One of the two main groups into which normative minerals are divided (the other being the salic group). It includes chiefly ferromagnesian minerals, the term being a mnemonic term to express this. It is used only when these minerals are being considered as components of the norm of the igneous rock, not for the minerals themselves (the 'mafic' minerals).

**Fenite.** A metasomatized igneous country rock surrounding a carbonatite intrusive complex. There are peculiar mineralogical and textural features, the general chemical changes being loss of silica and gain of alkalis. Hence 'fenitization', the form in which the name is most commonly employed. (Brögger, Norway, 1921.) [Fen, Norway.]

**Ferricrete.** See CALCRETE.

**Ferro-magnesian minerals.** A convenient name for the rock-forming silicates of iron and (or) magnesium. Nearly all the dark essential minerals of igneous rocks belong to this group, so that the phanerocrystalline constituents of these rocks may generally be visualized as falling into three groups: (1) quartz (if present), (2) felspars, (3) ferro-magnesian minerals. The ferro-magnesian minerals comprise the amphiboles, the pyroxenes, the olivines, and dark mica (biotite). (The presence of some other important mineral, such as muscovite, would spoil the comprehensiveness of this three-fold grouping.)

**Ferruginous.** Literally, rusty or rust coloured. Used for rocks in the more general sense of containing a considerable amount of a compound of iron. But see SIDEROSE. [L. *ferrugo, ferrugin-*, rust.]

**Ffestiniog beds.** The middle division of the three composing the Lingula flags. Characteristic fossils are species of the trilobite *Olenus* and an abundance of the brachiopod *Lingulella davisi*. (Sedgwick, *QJ*, 1847/52; Belt, *GM*, 1867.) [Ffestiniog, Meirionnydd district of Gwynedd.]

**Field.** As 'field geology', 'field work', 'field meeting', &c. 'The term Field-Geology points to practical work in the open field, as distinguished from the researches which may be carried on in the library or laboratory' (Geikie, *Field-Geology*, 1892). 'The role of the field geologist' (Harrison, *AdS*, 1965). 'A code for geological field work' (*Geologists' Association*, 1975). *Field Geology* (Bates and Kirkaldy, 1976). (Occurs also in such combinations as coal-field, oil-field, &c.)

**Figure of the earth.** See GEOID.

**Filicales.** The ferns, a group of plants belonging to the division Pteridophyta, ranging from the Upper Devonian to today. Fern-like fronds are common fossils in the Upper Devonian to Permian, but for many of the genera it is not known whether they are true ferns or pteridosperms (the majority appear to be pteridosperms). [L. *filix, filic-*, a fern.]

**Fining.** In graded bedding, the direction, upwards or downwards (usually the former), in which the bedding becomes finer. 'Fining-upwards cycles in alluvial successions' (Allen, *Geol J*, 1965).

**Fireclay.** A refractory clay. Fireclays occur chiefly in the Coal Measures, most of them as underclays.

**Fissile.** Capable of being easily split. Applicable to shales, flags, slates, schists, &c. [L. *fissilis*, split, capable of being split.]

**Fissure eruption.** Volcanic eruption from a long fissure, usually with the emission of flows of basic (basaltic) lava. 'Corresponding with the two principal categories of crust movements, and the two characteristic tectonic types which result from them, we have to recognise two contrasted modes in which volcanic action manifests itself. We shall distinguish these as "fissure-eruptions" and "central eruptions"; and they are the expressions, as regards igneous action at the surface, of plateau-building and mountain-building forces respectively'. (Harker, *Igneous Rocks*, 1909.)

**Fjord.** (Fiord.) 'The words fjord in Norway [and equivalents in Sweden, Iceland and the Faeroe Islands, Denmark and in Germany] are all derived from the same root. J. D. Dana (1849), who was the first to recognise the fjord as a special coastal type, separated the first from the others as a technical term' (Charlesworth, *Quaternary Era*, 1957). Fjords are essentially large-scale U-valleys submerged beneath the sea. 'That glaciation has played an appreciable role in modelling the fjord is overwhelmingly clear and generally acknowledged but the problem of the fjord still awaits its complete solution' (Charlesworth, *ibid.*).

**Flag, Flagstone.** 'Sandstone or sandy limestone rocks, usually more or less micaceous, which are fissile along the bedding-planes, splitting into slabs' (Arkell and Tomkeieff, *Rock Terms*, 1953). 'Flagstone: used like slate for the covering of houses, particularly at Bath' (Woodward, *Method of Fossils*, 1728). 'The grey flaggy strata of Caithness . . . this stone is strong and tough, and all over this country it is disposed in thin, regular strata. In several parts of the country, these flags are so thin and regular, and are raised so light and broad, that three or four of them cover the side of a small house' (Williams, *Mineral Kingdom*, 1789, describing the Caithness Flags of the Old Red Sandstone). 'The "Eastern Schists" [Moine] have had various names applied to them . . . such as "gneissose flagstones", "quartzose schists", "flaggy schists"' (Horne and Teall, *MGS*, *NW. Highlands*, 1907). See quotation under SLATESTONE. [Old Norse *flaga*, a stone slab.]

**Flexure.** In a general way, the same as 'fold', as in Busk's *Earth Flexures*, 1929 (and see quotation from Dana under GEANTICLINE); but often implying a gentle, rather than an acute, bend in the rocks and sometimes implying, in particular, a not very acute monocline. Also the verb 'to flex' (but usually 'flexuring' and 'flexured' refer to the condition, not the action).

**Flint.** A special variety of chert, in characteristically irregular nodules, occurring in the Chalk. Formerly used to cover chert in general, and locally used for any hard rock. 'Flint is most commonly found in nodules: but 'tis sometimes found in thin strata, when 'tis called chert' (Woodward, *Method of Fossils*, 1728). 'During the Stone Age flint was the undisputed key mineral of the land' (Bailey, *JRSA*, 1944). 'The nature and origin of flint' (Oakley, *SP*, 1946). 'Flint is impure cryptocrystalline silica formed, after the Chalk was deposited, by the separation of the silica contained therein (derived from the skeletons of sponges) into masses, sometimes around sponges as nuclei' (Chatwin, *BRG*, *East Anglia*, 1964.) [Old English.]

**Flint-meal.** Fine, mealy material occurring in the enclosed cavities of some flints. 'One of the larger of these flints [from the Chalk near Norwich] had been splintered and presented, instead of a solid mass of stone, a central cavity, which contained a quantity of material resembling fine flour in appearance and feel, and of a creamy-white tint. An examination with a hand-lens showed that this floury material abounded with minute fossils, more particularly sponge spicules. The cavity appeared to have been hermetically sealed up in its interior and preserved unharmed from mechanical injury a small portion of the mud of the Cretaceous ocean . . . this "flint meal", if thus it may be termed . . .' (Hinde, *Sponge Spicules*, 1880).

**Floetz.** An old name for the whole range of strata from the Devonian to Tertiary (excluding Pleistocene) which were supposed to lie comparatively flat as distinct from the 'Primary' and 'Transition' series; equivalent to the 'Secondary' of the contemporary classification. Used by Werner

(1787) and his followers, particularly, in Britain, by Jameson (as expounded in his *Geognosy*, 1808). [German *flötz*, a flat layer.]

**Flood basalt.** See PLATEAU BASALT. 'Flood basalts and fissure eruption' (Tyrrell, *BV*, 1937).

**Floor.** Particularly, in geology, for that of (**1**) a coal-seam and (**2**) an igneous intrusion. See ROOF. ('Floor' is not used as often as 'roof'.) Also (**3**) see COMPOUND STRUCTURE. 'The Palaeozoic floor north of the Thames' (Davies, *QJ*, 1913); 'The rocks belong to the "floor" over which the Tertiary basalt lavas were extruded' (Richey, *BRG*, *Tertiary Volcanic Districts*, 1961).

**Flora** (pl. **-ae** or **-as**). **1.** The (fossil) plant remains of a stratigraphical unit, geological or geographical region, &c. *Fossil Flora of Great Britain* (Lindley and Hutton, 1831/7). 'The fossil flora of the Cumberland Coalfield' (Arber, *QJ*, 1903). 'Wealden floras' (Seward, *QJ*, 1913). 'The flora of the upper portion of the Coal Measures of North Staffordshire' (Dix, *QJ*, 1931). **2.** In reconstructing the life of the past the term carries its ordinary meaning of the collective plant life of a particular region, environment, or time. [L. *Flora*, the goddess of flowers.]

**Florizone.** A zone defined by an assemblage of plant (not animal) fossils.

**Flow.** Water (carrying sediment) flows either as a laminar flow or a turbulent flow. Molten magma flows simply as a viscous fluid; sedimentary rocks may undergo 'plastic flow', a complicated combination of different behaviours, under the influence of stress which produces deformation in excess of the elastic deformation but not resulting in a noteable loss of cohesion (rupture).

**Flow-banding.** See FLOW-STRUCTURE (1).

**Flow-breccia.** See BRECCIA (3b).

**Flow-fold.** Flow-structure, igneous or sedimentary, in the form of a fold.

**Flow-structure. 1.** Igneous. 'Admirable examples of this structure may often be seen in old lavas, as well as in dykes and sills, the streaky lines of flow being marked as distinctly as the lines of foam that curve round the boulders projecting from the surface of a mountain-brook' (Geikie, *Ancient Volcanoes*, 1897). 'These lavas [Uriconian] include nodular and beautifully flow-banded examples of rhyolite' (Pocock and Whitehead, *BRG*, *Welsh Borderland*, 1948). 'Flow structures [in the Main Donegal Granite] are shown by mineral orientation and a perfect banding' (Pitcher and Read, *QJ*, 1958). Also called 'fluxion-structure'. **2.** Sedimentary. A structure resulting from subaqueous flow. 'Flow structures in sedimentary rocks' (Prentice, *JG*, 1960).

**Fluidization.** 'Fluidization is an industrial process in which gas is passed through a bed of fine-grained solid particles in order to facilitate mixing and chemical reaction. At a particular rate of gas flow the bed expands and the individual particles become free to move. With increase in the rate of gas flow a bubble phase forms and travels upwards through the expanded bed in which the particles are violently agitated; the bed is now said to be fluidized. With continued increase in the rate of gas flow more and more of the gas travels as bubbles containing suspended solids, until ultimately the solid particles become entirely entrained and transported by the gas' (Reynolds, 'Fluidization as a geological process and its bearings on the problem of intrusive granites', *AJS*, 1954). 'Examples of tuff and agglomerate necks are of widespread occurrence, but it is only in recent years that their full significance has been realised, primarily as a result of the detailed work of Hans Cloos on the Tertiary tuff pipes of Swabia, east of the Black forest. . . . The Swabian pipes illustrate the operation of a process known to industry as "fluidization"; a process that has far-reaching geological applications, as Doris Reynolds discovered from her work on . . . Slieve Gullion [Ireland]' (Holmes, *Physical Geology*, 1965). 'Minor intrusions from west Cork are divided into two groups, the younger of which is shown to have originated by a process of fluidization. A review of fluidization literature is followed by a description of the west Cork bodies, in which emphasis is placed on those features shown to be characteristic of fluidization' (Coe, *QJ*, 1966; and see Reynolds in discussion). 'Fluidization as a volcanological agent' (Reynolds, *PGS*, 1969).

**Fluorine dating.** A test for the age of Pleistocene bones. 'Founded on the fact that fluorine, which occurs as a trace in ground water, combines with the calcium phosphate of bones, the fluorine content increasing with the time that the bones lie under the influence of this water: the average fluorine content of lower Pleistocene bones is c.2%, of middle Pleistocene bones 1%, and of upper Pleistocene bones 0.5%' (Charlesworth, *Quaternary Era*, 1957).

**Fluorspar.** (Fluorite, Fluor.) A mineral, fluoride of calcium, $CaF_2$, sometimes colourless but often blue or purple or other colours; almost always crystallizing in cubes. Occurs chiefly as a veinstone. [L. *fluo*, flow; because it melts easily.]

**Flute-cast.** A sole-mark; the form is elongate, bulbous at one end, and is the infilling of a lobate incision, the cutting of such incisions being 'fluting'. (Crowell, *BGSA*, 1955; Prentice, *GM*, 1956.)

**Fluvial, Fluviatile.** Of or pertaining to rivers. Fowler (*Modern English Usage*) remarks that 'there is no difference in meaning, and no reason why both should exist'. Nevertheless, geologists tend to speak of 'fluvial action' but of 'fluviatile faunas' and 'fluviatile environment'. For deposits, both terms are about equally common. [L. *fluvius*, a river.]

**Fluvialist.** A name used in the earlier part of the 19th century for one who advocated the view that the widespread superficial deposits, now known to be mostly glacial drift, were due to the ordinary action of rivers; as opposed to the diluvialist view of the time. See quotation (Lyell) under DILUVIUM.

**Fluvioglacial.** See GLACIOFLUVIAL. 'Glacial deposits are of two kinds. One of these is . . . Boulder Clay. The other consists of gravel, sand and clay, in which some part of the materials are ice-borne, but the depositing agent was water. To these the term fluvioglacial may be conveniently applied' (Pocock, *MGS, Macclesfield* [&c.], 1906).

**Fluvio-marine.** For a series of beds some of which are of fluvial, and others of marine, origin. *Tertiary Fluvio-marine Formations of the Isle of Wight* (Forbes, *MGS*, 1856, also *QJ*, 1853).

**Fluxion-structure.** See FLOW-STRUCTURE. 'The structures of igneous rocks may be modified in consequence of differential movements in the magma during the process of consolidation. Here belong the familiar "fluxion-structures"' (Harker, *Igneous Rocks*, 1909).

**Flysch.** Originally indicated the strata of late Cretaceous and early Tertiary age, in the Alpine region, which are largely fissile sandy and calcareous shales (all marine). When it was found that this facies was bound to that particular phase of the Alpine orogeny which just preceded the main paroxysmal phase, the term took on the expression of this relationship, in contrast to the molasse facies, a post-orogenic facies. 'This has become such a widespread usage that Scandinavian geologists, for instance, discussing Precambrian geology, tried until recently to outline their otherwise undated orogenic periods by establishing the chronological sequence of the flysch and molasse facies' (de Sitter, *Structural Geology*, 1964). *Sedimentology of some Flysch Deposits* (Bouma, 1962). *Sedimentary Features of Flysch and Greywackes* (Dżulyński and Walton, 1965). 'Attempt at classification of flysch and molasse' (Contescu, trans. King, *IGR*, 1966). (Studer, 1827.) [Swiss dialect.]

**Fold.** A verb, 'to fold' (hence 'folded', 'folding') and a noun, 'a fold'. A fold is a pronounced bend in stratified rocks, even a very slight bend if that results in reversal of the direction of dip. The term is, from its ordinary meaning, most appropriate for strata that are bent so that one part 'lies reversed over or alongside another' (*OED*);

in older works it was used only in this sense, while for gentler cases such terms as bending, curvature, inflexion were used. However, we find it used to cover the general case in Geikie's *Textbook* (1882): 'From the abundance of inclined strata all over the world we may readily perceive that the normal structure of the visible part of the earth's crust is one of innumerable foldings of the rocks'. 'We were no less gratified in our views with respect to the mineral operations by which soft strata, regularly formed in horizontal planes at the bottom of the sea, had been hardened and displaced. Fig. 4 [showing a sharp anticline and syncline] represents one of those examples; it was drawn by Sir James Hall from a perfect section in the perpendicular cliff at Lumesden burn. Here is a fine example of the bendings of the strata' (Hutton, *Theory of the Earth*, 1795). A less diagrammatic drawing by Hall of the same scene is reproduced in Lyell's *Elements* (1838 and later). Sinclair (1672) recognized the broad folding of strata; see quotation under COALFIELD. An early section across a large extent of folded structure is that by White Watson across Derbyshire (1811). Lapworth (*GM*, 1883) takes one fold to comprise a complete undulation of anticline and complementary syncline, but usually such an anticline and syncline would be taken as two folds (with the middle limb common to both). 'The description of folds' (Fleuty, *PGA*, 1964). *Structural Geology of Folded Rocks* (Whitten, 1966). *Folding and Fracturing of Rocks* (Ramsay, 1967).

**Fold direction.** Where a folded bedding surface is cylindrical, that is, has a degree of regularity such that it may be considered as being generated by a line moving in space parallel to itself, the direction of this line (one way and the opposite) is the fold direction of the surface. As this will be very nearly, if not quite, the same for all the surfaces within the fold, the fold direction is a general element of the fold structure as a whole. The direction is one in three-dimensional space; but the plan projection of its gives the bearing in the horizontal plan and the vertical angle between it and the horizontal is the 'pitch' or 'plunge'. The term 'fold direction' is not very satisfactory. It is here defined as the 'b' tectonic axis, but it might possibly be taken to imply the 'a' tectonic axis, the direction of movement (at right angles to the b-axis), particularly where there is overfolding. 'Axis of folding' would probably be preferable were it not for the many senses of 'axis' in folding.

**Fold-fault.** 'An old word for a fault formed in close causal connexion with folding. At first I employed the word "slip" in this sense, but on Lapworth's suggestion exchanged it for "slide". Now that reversed and unreversed limbs are often distinguishable the word "slide" is less necessary, since it can often be replaced by "thrust" (a slide replacing an inverted limb) or "lag" (a slide replacing a normal limb)' (Bailey, *QJ*, 1934).

**Fold-system.** The folds, collectively, produced in one district or region during one period or phase of orogeny. 'The three fold-systems in the metamorphic rocks of Upper Glen Orrin' (Fleuty, *QJ*, 1961).

**Foliation.** A layered arrangement of minerals, usually implying a super-induced structural character of a metamorphic rock. The layers may be more or less plane, gently undulating or crumpled, and are usually lenticular, the whole suggesting a mass of leaves. Foliation may be parallel to original bedding or to axial planes or be not obviously related to any other structural direction; it may also have its own structures and tectonic style. It is most characteristically represented in schists—schistosity is typical foliation—but the term is sometimes inappropriately restricted to the coarser and much less leaf-like structure seen in gneisses, where there is an aggregation of particular minerals into lenticles, streaks, or inconstant bands (gneissic structure). 'The increase of mica creates a foliated structure by the pallets being placed in a parallel position, and then the rock passes into gneiss, or foliated granite' (Scrope, *Considerations on Volcanos*, 1825). 'The term "foliated" expresses very well the peculiar structure of mica schist' (Sedgwick, *TGS*, 1835). 'I cannot doubt that in most cases foliation and cleavage are parts of the same process' (Darwin, *South American Geology*, 1846).

'On the arrangement of the foliation and cleavage in the rocks of Scotland' (Sharpe, *PT*, 1852). 'Coincidence between foliation and stratification' (Murchison and Geikie, *QJ*, 1861). 'Stratification-foliation' and 'Cleavage-foliation' (Sorby, *QJ*, 1880). 'The term "foliation" is also used to describe parallel structures in igneous rocks, due to flow during cooling' (Rastall, *Textbook*, 1941). Further discussion and history in Dennis, *Tectonic Dictionary*, 1967. [L. *folium*, a leaf.]

**Folklore of fossils.** Old ideas that have been associated with fossils. 'Folklore of fossils' (Oakley, *Ant*, 1965). 'Folklore and fossils (Bassett, *Amg*, 1971).

**Foot wall.** The lower side of a fault or vein, opposite to the hanging wall. See quotation under HANGING WALL.

**Footprint.** See TRACK. 'An account of the tracks and footprints of animals found impressed on sandstone in . . . Dumfriesshire' (Duncan *TRSE*, 1831). 'A history and bibliography of the study of fossil vertebrate footprints in the British Isles' (Sarjeant, *PPP*, 1974).

**Foraminifera.** Protozoa possessing chambered shells, usually calcareous, sometimes arenaceous (sand grains or other particles cemented together), often built into an elaborate structure. They are mostly about the size of a pin-head but some disk-shaped forms may be an inch or so across. They occur throughout the Palaeozoic and in great abundance and variety in many Mesozoic, and particularly Tertiary, formations. Representative assemblages are often present in small samples, such as may be obtained from the cores of borings. Mostly marine. Micropalaeontology is largely concerned with this group of small fossils. [L. *foramen*, an opening, *fero*, bear.]

**Forceful intrusion.** 'As regards the manner of emplacement of igneous intrusions there are the "forceful intrusions" which forcibly make room for themselves by shouldering aside the country rocks and the "permitted intrusions", formed more passively by magma rising to occupy spaces left by the subsidence of detached masses of the country rock' (Read and Watson, *Introduction to Geology*, 1962).

**Foredeep.** The deep part of a geosyncline. 'During the Cambrian period North Wales occupied a position (the foredeep) in the geosyncline where the geosynclinal floor was relatively weak and subsided to great depths' (Smith and George, *BRG, N. Wales*, 1948).

**Foreland.** In orogeny, the resistant block towards which the sediments of a geosyncline move when subjected to compression by the movement of the hinterland behind them. 'The Moine thrust carries the Moine series over the Cambrian and Pre-Cambrian rocks of the foreland' (Giletti and others, *QJ*, 1961).

**Foreland folding.** 'The foreland, north and west of the northern Appalachian Mountains, is characterized by gentle folds . . . these foreland folds . . .' (Sherrill, *JG*, 1934). 'Under the title of foreland folding, I shall describe some structural problems in the transition zone between a strongly deformed mountain zone and its adjacent rigid shield' (Lees, *QJ*, 1952).

**Foreset beds.** Inclined beds built forward, in a deltaic deposit, each above and in front of the previous one, in much the same way (but in a normal case with reduced vertical scale) as an embankment may be built forward by tipping over the end.

**Forest Marble.** See GREAT OOLITE SERIES. (Smith, *MS*, 1799.) [Forest of Wychwood, Oxfordshire.]

**Formation.** See LITHOSTRATIGRAPHY. 'The formation is the primary local unit. It should possess some degree of internal lithological homogeneity, or distinctive lithological features that constitute a unity by comparison with adjacent strata. Above all, a formation is mappable at the earth's surface and/or is traceable in the subsurface. Thickness is not a determining feature in the establishment of a formation. In any case the thickness will normally vary from a feather edge to a maximum development' (*Stratigraphical Guide*,

*GSSR* (11), 1978). 'The term "formation" expresses in geology any assemblage of rocks which have some character in common, whether of origin, age, or composition. Thus we speak of stratified and unstratified, fresh water and marine, aqueous and volcanic, ancient and modern, metalliferous and non-metalliferous formations' (Lyell, *Elements*, 1838). 'A formation, which is the unit of geological stratigraphy, is a rock sheet composed of many strata possessing common lithological characters. The formation may be simple, like the Chalk, or compound, like the New Red Sandstone, but, simple or compound, local or regional, it must be always recognizable, geographically and geologically, as a lithological individual' (Lapworth, *BA*, 1892). 'A formation is a convenient lithological rock unit in a particular area and the basis of geological mapping' (Arkell, *Jurassic Geology*, 1956).

**Formenreihe.** An evolutionary series or lineage. See MUTATION. [German.]

**Form-genus, Form-species.** See PARATAXON. Chiefly in palaeobotany; taxonomic categories for use in the classification of those plant-fossils that are detached parts of the whole plant and that almost always occur quite separate from the other parts. Thus in many geological formations, and for many types of plant, separately occurring stems, leaves, seeds, &c. may be discriminated, classified, and named in the same way as whole individuals, but it is not possible to say which 'species' of stem, which leaf, which seed really belong to the same whole-plant species. Parkinson (*Organic Remains of a Former World*, 1804) remarked on the difficulty of determining a whole plant from 'little more than the outlines of a leaf', and even the famous botanist, Sir J. E. Smith, to whom he applied for help, had to admit that such were 'a sort of botanical riddles'. Occasionally, fuller knowledge allows some of the form-species to be linked as belonging to the same plant. Thus the stem *Lyginopteris oldhamia*, the frond *Sphenopteris hoeninghausi*, the seed *Trigonocarpus lomaxi*, and the male fructification *Crossotheca hoeninghausi* are all found to belong to the same pteridosperm plant, which is still nameless as a whole species though some would say that the name first given to any part of it is the valid name for the whole. For animal fossils, 'form-genus' and 'form-species' (in so far as these terms have been used) may now be said to be superseded by two corresponding categories of parataxa. The term 'form-genus' is sometimes used in a rather different sense, for leaves, for instance, which are much alike but which are suspected of belonging to quite different kinds of plants. In America, 'form-genus' has been used for a genus *sensu lato*, or for a genus which comprises a large number of species of the same general form but which, it is suspected, belong to diverse lines of descent (e.g. *Monograptus*).

**Forsterite.** See OLIVINE. This end-member of the olivine family 'stands in a category by itself: it is a characteristic product of the thermal metamorphism of magnesian limestones and dolomites which contained the necessary silica in the form of detrital quartz grains, sponge spicules or tests of radiolaria. The crystals and grains of forsterite are embedded in a matrix of crystalline calcite, and the rock is termed a forsterite marble' (Hatch and Wells, *Igneous Rocks*, 1965). 'Probably named in honor of Prof. J. R. Forster' (Chester, *Names of Minerals*, 1896). (Levy, 1824.)

**Fossil.** The remains, representation, impression, or trace, in the rocks, of parts of an animal or plant. 'A fossil' is a substantive; 'fossil' is also an adjective. [L. *fossilis*, dug up.]

The term 'fossil' was, as the word suggests, originally given to anything extracted from the earth or the rocks. It included minerals, all kinds of stony objects, and pieces of the rock itself, as well as the remains of organisms. 'Fossilia' in the wide sense and not, in fact, including organic remains, was used by Agricola in 1546 (*De Natura Fossilium*). The first illustrated work on fossils, including organic remains, was Gesner's *De Rerum Fossilium et Gemmarum Figuris*, 1565. In Britain organic fossils were called 'petrified shells' (Hooke, 1665), 'formed stones' (Plot, 1677), or 'figured stones' (Lhwyd, 1699). Lhwyd (1695) used the term 'fossil-shells' and Ray (1721) 'marine fossils', 'fossil fish teeth', &c. In 1728 Woodward wrote a work on *Fossils of all Kinds digested into a Method* and classified them into 'native' (minerals, &c.) and 'extraneous' (fossil shells, &c.). Da Costa's *Natural History of Fossils* (1757) is an excellent early treatise on petrography; but Brander and Solander's *Fossilia Hantoniensia* (1766) deals entirely with fossil shells. Owing, no doubt, to these various confusing usages, the term 'fossil' dropped out for a time, 'petrifaction' largely taking its place (e.g. Walcott's *Petrifactions found near Bath,* 1779, and Martin's *Petrifactions collected in Derbyshire,* 1793). The always appropriate 'organic remains' then became popular, as in Parkinson's *Organic Remains of a Former World,* 1804/11, and was being used much later, as in the *Figures and Descriptions of British Organic Remains,* 1849 and following years, of the Geological Survey. Meanwhile 'fossil' was again coming into use, but now for organic remains only, though usually with, or as, a qualifying adjective; e.g. *Strata Identified by Organized Fossils* (Smith, 1816) and *Fossil Organic Remains* (Parkinson, 1822). Already, however, the word by itself was beginning to be used. Parkinson (1804) remarks that 'in the common language of those most conversant with these substances' their nature 'is conveyed by the substantive ("fossil") alone'. Lamarck in France seems to have been the first definitely to restrict the term (*Systeme des Animaux sans Vertèbres,* 1801, and *Hydrøgéologie,* 1802). The substantive 'fossil', alone and exclusively for organic remains, became thoroughly established some twenty years later (e.g. Mantell's *Fossils of the South Downs,* 1822). It is recorded that in the N. Staffordshire Coalfield a miners' term for a fossil is a 'miracle' (Stobbs, *TNSFC,* 1916).

Camden (*Britannia,* 1586) quotes 'Ralphe the Monke of Coggeshall, who wrote 350 years ago' as recording that 'at a village called Eadulphnesse [in Essex], were found two teeth of a certaine Giant, of a huge bignesse' (Holland's translation, 1610). Gibson, in his edition of the *Britannia* (1695) remarks that these 'are supposed to be the bones [teeth] of elephants', a supposition since confirmed. This must be one of the earliest records (*c.* 1236) of a find of fossils in Britain. Fossils in the British rocks (ammonites and lamellibranchs from the Jurassic) are mentioned by Leland (*Itinerary, c.* 1538) and by Camden (1586) but Childrey (*Natural Rarities,* 1661) seems to have been the first to specify and localize various kinds of British fossils. They began to be carefully described and figured during the latter part of the 17th century, particularly by Hooke, Plot, and Lhwyd. *The Meaning of Fossils* (Rudwick, 1972).

A 'fossil' need not have been actually extracted (though the etymology suggests that it should have been), nor even seen; the term naturally being extended to cover any occurrence of the kind, even if for ever hidden. The term occasionally breaks out, urged by the sense of preservation it carries, into use for a 'fossil landscape', 'fossil glacier' (rock ingredients preserved), 'fossil rain-drops', &c.; but when we find it as 'fossil axes of rotation of the earth' (*N,* 1963) it seems to have strayed too far. Fossils are normally the remains of the harder, skeletal, parts of animals and of the cell wall structures of plants; including the casts, moulds, and impressions of these.

**Fossil association.** See ASSEMBLAGE (1).

**Fossil botany.** Palaeobotany. *Studies in Fossil Botany* (Scott, 1900/23).

**Fossil forest.** A more or less extensive occurrence, in a bed of rock, of abundant fossil trees (stumps) in position of growth. The 'Fossil Forests' at Partick, near Glasgow (Carboniferous) and near Lulworth, Dorset (Jurassic) are well-known British exposures.

**Fossil fuel.** 'A combustible solid, liquid, or gas formed in past geological ages. Principal fossil fuels are coals (solids which are largely organic), oil shales (solids which are largely inorganic), petroleum (largely liquid hydrocarbons), tar sands, and natural gas (largely gaseous hydrocarbons). Each of these varies widely in composition and properties. No single theory seems to account for the formation of all types of coal or even for all deposits of a single type, and the same thing is true for all other fossil fuels. The deposits . . . all contain carbon and hydrogen. Peats resemble some fossil fuels but are relatively young' (*McGraw-Hill Encyclopedia of Science and Technology*, 1960).

**Fossil record.** That part, of the general 'geological record', of which the 'documents' are fossils. Also called the 'palaeontological record'. It is thus a record, preserved in the rocks, of life in the geological past. This record is an incomplete representation of that life, due to the restricted conditions, and the natural accidents, of fossilization. There is also the 'fossil record' meaning the actual specimens, photographs, delineations, and descriptions, together with statements of stratigraphical occurrence, at present available to the palaeontologist in museums, collections, monographs, and papers. This again is, of course, partial: 'The fact that new species of fossil organisms are discovered daily confirms the incompleteness of present palaeontological knowledge' (Brouwer, trans. Kaye, *General Palaeontology*, 1967). Finally, there are the works in which certain kinds of facts, as at present known, are compiled, selected, and integrated; as in the work published by the Geological Society, *The Fossil Record* (Harland and others, eds., 1967), which thoroughly and exactly summarizes, with full documentation, our knowledge of the ranges of all fossil groups (families and higher taxa).

**Fossil zoology.** Palaeozoology. 'Fossil zoology' (Murchison, *PGS*, 1833; Lyell, *PGS*, 1836; Whewell, *PGS*, 1838). 'Their rock structure and fossil zoology are replete with subjects of extensive and varied interest' (Lycett, *Cotteswold Hills*, 1857).

**Fossiliferous.** Fossil-bearing. Chiefly applied to rocks. The term was used by Lyell (*Elements*, 1838). 'These organic-remain bearing, or "fossiliferous rocks", as they have been termed' (De la Beche, *Geological Observer*, 1851). 'A fossiliferous band at the top of the Lower Greensand near Leighton Buzzard' (Lamplugh and Walker, *QJ*, 1903). Also applied to exposures and places: *Directory of British Fossiliferous Localities* (1954).

**Fossilize.** To become, or cause to become, fossil. Chiefly occurs in the form 'fossilized', or as the derivative 'fossilization'. The ways in which shells may become fossilized were clearly explained by Hooke in his *Micrographia*, 1665. There are innumerable modern expositions of the matter, e.g. by Müller in the Introductory volume of the *Treatise on Invertebrate Paleontology*, 1979. See TAPHONOMY. 'Bones that are fossil' (*PT*, 1794). 'To fossilize' may also mean 'to search for fossils': 'I fossilized for three days very diligently at Shell Bluff' (Lyell, *Travels in N. America*, 1845).

**Fossilology.** 'Fossilology' and 'Fossilogy' are in the *OED* (the former 'rare', the latter 'obsolete') but these combinations of Latin and Greek are not in serious use by geologists. 'From that hour [in 1824] the acquisitions I had made in "fossilogy", as we then termed the magnificent branch of study now known as Palaeontology, brought me perpetual engagements in Yorkshire' (Phillips, 1874, quot. *QJ*, 1875).

**Fractional crystallization.** Separation of a cooling magma into successive 'fractions' of different minerals. (Becker, *AJS*, 1897). The term is sometimes used with 'fractions' meaning the two phases, crystals and liquid. 'Fractionation' is another form of the term.

**Fracture.** The breaking of a rock; usually a more or less clean-cut break is implied, otherwise 'rupture' or 'crushing' or some other term would more likely be used. Also the break itself; a fault is a definite case of fracture in a rock-mass. The 'fracture' of a mineral or rock is also used for the appearance of the fresh surface in a broken piece.

**Fracture cleavage.** See CLEAVAGE (1*b*). (Leith, *BUSGS*, 1905.)

**Fracture zone.** This is, in the first place, a simple descriptive statement, but the term itself has largely come into use in connexion with plate tectonics, for such zones along plate margins, where they are, in effect, transform fault systems. See papers in *JGS*, 1979.

**Fragmental deposits.** 'Sediments such as gravels, sands and clays, which consist predominantly of the solid fragments formed from the waste of pre-existing rocks, are grouped together as "fragmental" or "clastic deposits", a term which, for our purposes, may be regarded as synonymous with those used above' (Hatch and Rastall, *Petrology of the Sedimentary Rocks*, 1971).

**Frasnian.** See DEVONIAN SYSTEM. (Gosselet, 1879). [Frasne, Belgium.]

**Freestone.** 'Any fine-grained sandstone or limestone that can be sawn easily' (Arkell and Tomkeieff, *Rock Terms*, 1953).

**Freeze.** The ordinary use is for the conversion of water to ice; but it is also legitimately used for the solidification of a magma on cooling (see FUGATIVE CONSTITUENT).

**Freshwater.** Chiefly as applied to formations which were evidently laid down under such conditions, and particularly to those of the Oligocene. 'The freshwater formations of the Isle of Wight' (Webster, *TGS*, 1814).

**Front.** In geology this ordinary word is used particularly for a process that is advancing; deposition, deformation, metamorphism, metasomatism, weathering, igneous action, &c. Thus, the advancing front of the Millstone Grit delta, structural front (e.g. Appalachians, Rodgers, *KGS*, 1953; Scotland, Kennedy, *QJ*, 1954), weathering front (Mabbutt, *PGA*, 1961). The term is used particularly in connexion with ideas as to migmatization and granitization. The migmatizing process is visualized as moving forwards, with regional metamorphism preceding it and granitization following it, there being two fronts, one at each of the two zones between the three advancing belts corresponding to these three processes (in so far as these belts are separable). It is the front of the completely granitized rock that appears to be the more definite and it is often characterized particularly by an enrichment of mafic minerals in the adjoining belt, as if these constituents were being driven out of rock that was becoming completely granitized; hence the expression 'basic front'. 'The junction is a true "front" of metamorphism, exactly comparable to "fronts" of granitization, migmatite fronts, and basic fronts' (Reynolds, *GM*, 1947).

**Fucoid.** A body within a rock, or a marking on a rock surface, of a form suggesting that it might be the remains or impression of a sea-weed. Fucoids are problematica particularly common in the Lower Palaeozoic rocks of Wales whence they were described by Murchison (*Silurian System*, 1839) and Keeping (*GM*, 1882). 'Black shales and thin limestones [Carboniferous, Mixon, N. Staffs.]—Branched fucoids' (Green, *MGS*, *Stockport* [&c.], 1866). '"Fucoids" on Hambleton Oolite [Jurassic], Filey Carr Naze' (Wilson, *PGA*, 1949). [L. *fucus*, G. *phukos*, sea-weed.]

**Fugitive constituent.** 'By fugitive constituents we mean substances which were present in the magma before freezing set in, but were for the greater part lost during the process of crystallization, so that they do not commonly appear as rock constituents. These substances may be gases and vapours which escape from the magma when the external pressure is reduced, or soluble salts which are carried away in solution by magmatic or even meteoric water' (Shand, *Eruptive Rocks*, 1927).

**Fulgurite.** 'One effect of lightning is to produce in loose sand or more compact rock, tubes termed fulgurites, which range up to 2½ inches in diameter. These descend vertically but sometimes obliquely from the surface, occasionally branch, and rapidly lessen in dimensions till they disappear. They are formed by the actual fusion of the particles of the sand or rock surrounding the pathway of the electric spark' (Geikie, *Textbook*, 1903). 'An unusual ring-like structure was observed to have weathered out from a bedding surface [sandstone, probably Permian, in Arran] . . . later recognized as a "fulgurite", i.e., a tube formed in sand as the result of fusion by a lightning strike. . . . More fulgurites were discovered. As far as we know this is the first record of an ancient lightning strike i.e. a palaeo-fulgurite, re-exposed at the present land surface, as distinct from the many fulgurites of relatively recent origin already described from recent and ancient rocks exposed at the present surface' (Harland and Hacker, *AdS*, 1966). 'The formation of fulgurites' (Williams and Johnson, *GM*, 1980). 'In the hillocks of drifted sand near to Drigg, in Cumberland, hollow tubes of a vitreous substance were discovered rising above the surface perpendicularly through the sand. . . . It appears that the tubes have all the marks of fusion: lightning seems to be the only agent that could at once supply the heat and the force requisite to make them' (Anon., *TGS*, 1814). The *OED* gives an early use of the term itself in 1834. [L. *fulgur*, lightning.]

**Fuller's earth.** 'Fuller's earth is the name given to a particular form of clay composed substantially of montmorillonite, plus anorthoclase felspar (containing lime), the whole possessing low plasticity and strong decolorizing and degreasing properties' (Milner, *Sedimentary Petrography*, 1962). 'Fuller's earths particularly rich in montmorillonite approach the rock bentonite in nature and may be derived from windswept clouds of volcanic ash, deposited on the surface of shallow waters' (Hallam and Sellwood, *N*, 1968). In Britain, occurs particularly in the Lower Greensand of the Weald and Bedfordshire and the Middle Jurassic (the Fuller's Earth) of western England. 'Fullers' earth, from Wooburn in Bedfordshire. Another sample, from Detling, near Maidstone, Kent; this lay 33 foot deep, the stratum of it is about a foot thick' (Woodward, *Catalogue*, 1729).

**Fuller's Earth.** See GREAT OOLITE SERIES. (Smith, *MS*, 1797/9.)

**Fumarole.** A vent through which smoke or (usually hot) vapour issues from the ground, chiefly one in a region of (usually waning) volcanic activity. 'A more proper name for these ignited hills and spots would be fumaroles' (Pinkerton, *Petralogy*, 1811). 'Fumaroles or small crevices in the cone through which hot vapours are disengaged' (Lyell, *Principles*, 1830). 'At Holworth Cliff [Dorset] . . . are many small fumaroles that exhale bituminous and sulphureous vapours' (Buckland and De la Beche, *TGS*, 1835; and see Cole *GM*, 1974). 'Tertiary fumaroles on the island of Raasay, Inverness-shire' (Selley, *GM*, 1966). [L. *fumus*, smoke.]

**Fundamental complex.** A 'basement complex' but referring more particularly to the Precambrian 'shields'. A rather old-fashioned term.

**Fusain.** 'The term fusain was introduced by Grand 'Eury in 1882 to designate the black silky, lustrous bands, recognised macroscopically in coal' (*Nomenclature of Coal Petrology*, 1957). 'The French name, adopted into English by J. J. Stephenson (1911-13) and Stopes and Wheeler (1918) to replace our native un-wieldly names "mother of coal" and "mineral charcoal" . . . based on the Latin *fusus*, and its application came about in a circuitous way' (Stopes, *PRS*, 1919).

**Fusion.** The idea of rocks having been in a state of fusion was one of the dominant ideas of Hutton's theory of the earth. He carried it too far, but was right as regards the igneous rocks. 'We are now fully assured that granite has been forced to flow in the bowels of the earth and reduced into a state of fusion' (Hutton, *TRSE*, 1790; see also quotation under GRAPHIC).

# G

**Gabbro.** A coarse-grained basic igneous rock composed essentially of the more calcic plagioclase felspars and pyroxene or olivine or both and occurring usually as a plutonic body. 'The term is derived from an old Florentine name for various rocks, including rocks which we now call gabbro. Leopold Von Buch (1810) was the first to introduce this local name into petrographic literature in his description of the rocks of Upper Wallis' (Johannsen, *Petrography*, 1937).

**Gala beds.** The beds, some 1000-1200 m. thick, forming an extended, sparsely fossiliferous, succession of the four graptolite zones, *Monograptus turriculatus* to *M. crenulatus* (the top zones of the Llandovery series), immediately following the highly condensed and fossiliferous Birkhill Shales of S. Scotland. (Lapworth, *GM*, 1870.) [Gala Water, S. Scotland.]

**Galena.** A grey mineral, lead sulphide, PbS, with a shining metallic lustre, relatively soft and very heavy, often crystallizing in nearly perfect cubes. The most important ore of lead; usually occurring in veins. [L; name used by Pliny.]

**Gangue.** See VEINSTONE. (French f. German *gang*, a mineral vein.]

**Ganister.** (Gannister.) A hard, even grained highly siliceous rock occurring as beds (usually as underclays) in the Coal Measures. The term is, and has long been, used particularly in Yorkshire, Lancashire, and Derbyshire. (*MGS*, *Mineral Resources*, vol. 6, 1920.) [origin obscure.]

**Ganoid.** See OSTEICHTHYES. [G. *ganos*, brightness.]

**Gap.** The chief purely geological use of this term is in stratigraphy. See NON-SEQUENCE.

**Garnet.** A family of minerals, silicates of aluminium, iron, manganese, chromium, calcium, and magnesium (the several members having these elements variously); red or green in colour, with a high refractive index and crystallizing in the cubic system. They occur as accessory minerals in a wide range of igneous rocks, but most commonly, and as the finest crystals, in the metamorphic rock, garnet-mica-schist. [L. *granatum*, pomegranate, 'from its resemblance in colour to the pulp of the fruit' (*OED*).]

**Gash breccia.** See BRECCIA (4).

**Gastrolith.** A stomach-stone; either one that has grown in the animal body or one picked up from a pebbly beach by an animal and swallowed. Either kind might fall on the death of the animal and become part of a deposit on the sea floor; but it is the latter kind, apparently picked up by the Mesozoic reptiles (?plesiosaurs), that is usually considered by geologists. [G. *gaster*, *gastro-*, the belly.]

**Gastropoda.** (Gasteropoda.) One of the three chief classes of the phylum Mollusca, having a shell all in one piece and nearly always coiled into an asymmetrical spiral of a helicoid (like the garden snail, *Helix*) or turreted form (or in some intermediate or exaggerated form). This readily distinguishes nearly all gastropods from nearly all cephalopods (coiled in a symmetrical spiral) but the well-known limpet (*Patella*) with its tent-like form, and the well-known fossil *Bellerophon*, coiled in a plane spiral, are among the exceptions. A constant difference is the absence of regular partitions (septa) in the gastropod shell. They live mostly in water, but some on dry land. They range from the Cambrian and become steadily more abundant and varied up to the present day.

**Gault.** The clay formation between the Chalk (or Upper Greensand where present) and Lower Greensand. The typical representative of the Albian series of the Lower Cretaceous. Its outcrop forms a narrow strip, but in E. England, northwards from Cambridge, it becomes a thin condensed succession and tends to lose its characteristic lithology. In SE. England it is about 60 m. thick. It is highly fossiliferous, the most distinctive fossils being the ammonites, particularly the family Hoplitidae. Some other characteristic fossils are the annelid *Rotularia concava*, the brachiopod *'Terebratula' biplicata*, and the lamellibranchs *Inoceramus concentricus*, *I. sulcatus*, and *Nucula pectinata*. Before becoming rerestricted to a particular stratigraphical meaning it was used for various stiff clays. (In stratigraphy: Michell, *MS*, 1788, 'Golt'; Greenough, *Map*, 1819, 'Gault'.) [origin obscure.]

**Geanticline.** 'In the movements of the earth's crust, there would necessarily be upward as well as downward flexures—that is, "geanticlinals" as well as geosynclinals' (Dana, *Manual*, 1875). Nevertheless, the geosyncline is a feature *sui generis* and does not have the geanticline as a necessary complement in the same way as the syncline has its complement in the anticline. Judd (*Volcanoes*, 1881) has the following: 'To the mass of folded, crumpled, and altered strata, formed from a geosynclinal by lateral pressure, geologists have given the name of a "geanticlinal". The formation of the Alpine geanticlinal . . .'. This is Dana's 'synclinorium'. 'Geanticline' now tends to be used in a restricted sense, for an upbulge developing within a geosyncline as this becomes laterally compressed.

**Gedinnian.** See DEVONIAN SYSTEM. (Dumont, 1848.) [Gedinne, Belgium.]

**Genetic term.** See DESCRIPTIVE TERM.

**Gens** (pl. **gentes**). A group of species and varieties supposed to be especially nearly related. They would usually be covered by one generic name. A term now largely fallen out of use, 'species-group', or a specific name *sensu lato* being preferred. (Vaughan, *QJ*, 1905.) [L. *gens*, a clan.]

**Genus** (pl. **genera**). In the classification of animals and plants, recent or fossil, a genus is a division of a family and itself comprises one or more species. These species are those that are sufficiently alike (when all their characters are considered) to be deemed worthy of being all grouped together in one small group. Consequently, a genus will normally comprise species that form one 'twig' of the evolutionary tree; but the nearness of approximation between degree of likeness and actual phylogenetic relationship cannot be known with certainty. The name of a particular genus is a Latin noun or a word treated as such. [L. *genius*, a kind.]

**Geobiology.** In effect, palaeobiology in the widest sense. 'The earth sciences include not only geology, but the hybrid sciences geophysics, geochemistry and geobiology' (Sylvester-Bradley, *JGS*, 1972).

**Geobleme.** A cryptoexplosion structure caused from within the earth. See CRYPTOEXPLOSION STRUCTURE.

**Geobotany.** The application of a knowledge of the character and details of the vegetation cover towards a knowledge of the physical state and chemical composition of the underlying soils and rocks, and the mode of occurrence of groundwater. *Plant Indicators of Soils, Rocks, and Subsurface Waters: Proceedings of a Conference on Indicational Geobotany* (Chikishev, ed., 1965).

**Geochemistry.** The chemistry of the earth; the chemical composition and chemical processes of the earth and its parts, and of rocks and minerals. The chemistry of the hydrosphere and the atmosphere have sometimes been included. From the chemical, rather than the geological point of view, it is the study of the natural occurrence of chemical elements and compounds. *Data of Geochemistry* (Clarke, *BUSGS*, 1908/24). 'On geochemical prospecting' (Horvitz, *Gp*, 1939). 'The geochemistry of some Caledonian plutonic rocks: a study in the relationship between the major and trace elements of igneous rocks and their minerals' (Nockolds and Mitchell, *TRSE*, 1948). 'The copious work of Bischof (1847-1855) provides a veritable "Data of Geochemistry" of his period, as he gave all the important analyses of gases, waters, minerals and rocks and also discussed the physico-chemical processes of the earth' (Leowinson-Lessing, trans. Tomkeieff, *Historical Survey of Petrology*, 1954). *Principles of Geochemistry* (Mason, 1952/67). *Geochemistry* (Goldschmidt, in translation, 1954). *Researches in Geochemistry* (Abelson, ed., 1959). 'The geochemistry of scapolite' (Shaw, *JPt*, 1960). 'The geochemistry of petroleum' (*GCA*, 1960). 'A geochemical survey of the Nottinghamshire oil-fields' (Evans and others, *QJ*, 1962). *Physical Geochemistry* (Smith, 1963). *Advances in Organic Geochemistry* (Colombo and Hobson, eds., 1964). *Geochemistry of Sediments* (Degens, 1965). *Introduction to Geochemistry* (Krauskopf, 1967/79). *The Wolfson Geochemical Atlas of England and Wales* (Webb, ed., 1978). *Handbook of Geochemistry* (Wedepohl, ed., 1978). *Geochemistry* (Brownlow, 1979). 'The word "geochemistry" was first used, in 1838, by the German chemist C. F. Schönbein, professor at the University of Basel, and the discoverer of ozone' (Manten, *CGe*, 1966, in an article that discusses the history and various aspects of geochemistry).

**Geochronology.** See CHRONOLOGY. 'It covers human prehistory as well as the whole of the geological past' (Zeuner, *Dating the Past: an Introduction to Geochronology*, 1958). 'A geochronology of the last 12,000 years' (de Geer, *IGC*, 1912). 'A geochronological study of the metamorphic complexes of the Scottish Highlands' (Giletti and others, *QJ*, 1961). 'Coral growth and geochronology' (Wells, *N*, 1963). *Applied Geochronology* (Hamilton, 1965). *The Geochronologic Time Scale* (*AAPG*, 1978). *Geochronology of N. America* (Wetherill and others, 1965). The measurement of geological time is sometimes more specifically called 'geochronometry'.

**Geochronometry.** See GEOCHRONOLOGY. 'Geochronometry is the science which attempts to measure the age of rocks in years' (*Stratigraphical Guide*, *GSSR*, 1978).

**Geocosm.** See EARTH.

**Geocryology.** The study of the effect of frost action at or below the earth's surface. From the etymology it is a more general term than either 'cryopedology' or 'cryoturbation' and it specifically refers to frost action whereas the term 'periglacial' (by itself) does not. (See reference to Washburn, 1980, under PERIGLACIAL.) [G. *kruos*, frost.]

**Geode.** A hollow potato-shaped nodule, often lined inside with crystals projecting into the cavity, occurring chiefly in calcareous rocks. 'A Geodes, of a ferruginous colour, with several cavities. 'Tis broke, and in two of the caverns are several small cylinders' (Woodward, *Catalogue*, 1729). See DRUSY. [G. *geōdēs*, earthy.]

**Geodesy.** The determination of the figures and areas (and relative positions) of large portions of the earth's surface and the figure of the earth as a whole. It is the same as the old meaning of the word 'geometry': the measurement of the earth. 'The literal meaning of "Geodesy" is "dividing the earth", and its first object is to provide an accurate framework for the control of topographical surveys' (Bomford, *Geodesy*, 1952). 'Geodesy is both theoretical and practical. Its theoretical function is to determine the size and shape of the earth. . . . Its practical function is to perform the measurements and computations that will give the coordinates of selected control points on the earth's surface; i.e. to fix their positions' (Heiskanen and Meinesz, *Earth and its Gravity Field*, 1958; who give a historical survey). *Physical Geodesy* (Heiskanen and Moritz, 1967). [G. *daio*, divide.]

**Geodynamics.** See DYNAMICAL GEOLOGY. *Principles of Geodynamics* (Scheidegger, 1958). *Geodynamics Today* (The Royal Society, 1975).

**Geofault.** A large, deep-seated fault. 'Geofracture' is synonymous. A geofault of the wrench-fault type is sometimes called a 'geosuture'.

**Geoflexure.** A regional monoclinal flexure.

**Geofracture.** See GEOFAULT.

**Geogeny.** Has been used for various aspects of geology, such as the origin and mode of formation of the earth as a whole, or of the earth's crust; mineralogical changes; historical geology. Now seldom used at all.

**Geognosy.** Knowledge of the earth. A term much in use 160 years or so ago in a general way for geology when, indeed, little was known. It is particularly assocoated with the teaching of Werner (1749–1817) and his followers: *Elements of Geognosy*, title of Robert Jameson's book (1808). 'Werner dismissed the word geology, then only just introduced [it came in gradually during the last two decades of the 18th century], as one to be used only for the speculative theorising of earlier writers, and substituted the term "geognosy" to embrace his own doctrines which, he claimed, were entirely based on ascertained facts' (Eyles, *HS*, 1964). 'The term "well educated geognost" as used by some writers denotes a perfect disciple of Werner, who has lost the use of his own eyes by constantly looking through the eyes of his master' (Bakewell, *Geology*, 1813). It continued to be restricted to absolute knowledge (e.g. by Lyell), as distinct from the reasoning and speculation of 'geology', and survived in this sense, or as the study of the actual materials of the earth, into this century (e.g. Geikie's *Textbook*, 1903). It is now obsolete. [G. *gnosis*, knowledge.]

**Geogony.** The origin and mode of formation of the earth, particularly a speculative discussion of this. Seldom used. [G. *gignomai*, become.]

**Geogram.** Used by Marr (*QJ*, 1905) for a geological column, including a wide column in which lateral variations over an area of any chosen extent could be shown. Marr expressly did not advocate its adoption, but it seems to be a useful term. (For an example see Krumbein and Sloss, *Stratigraphy and Sedimentation*, 1951, fig. 10-2.)

**Geographical cycle.** 'The wearing down of highlands to featureless lowlands: all historic time is hardly more than a negligible fraction of so long a duration. The best that can be done at present is to give a convenient name to this unmeasured part of eternity, and for this purpose nothing seems more appropriate than "a geographical cycle"' (W. M. Davis, *GJ*, 1899, reprinted in *Essays*, 1909; *AGI Glossary*, 1957, gives first use by Davis in *NGM*, 1893). 'Cycle' is, however, hardly appropriate, as the conception is of nothing circular or cyclical, but of a gradual sequential passage from one extreme state to another; it is the erosional phase of the whole geological cycle. (See CYCLE.) Nor is it a period of time as implied in Davis's definition. The 'geographical cycle', or 'cycle of erosion', has been divided into three stages: youth, maturity, and old age. The validity of this concept is very much open to question. See DYNAMIC EQUILIBRIUM.

**Geography.** Essentially, the science that describes the earth's surface; its form, character (land or sea), physical features, and climates. This is 'physical geography'. Also in the sense of regional ('geographical') occurrence and distribution of phenomena, such as volcanoes, and of peoples ('human geography', 'anthropogeography'), animal and plant life ('biogeography'), &c. 'The concept and scope of geography have undergone considerable change, and it is highly unlikely that any definition would satisfy everyone' (Monkhouse, *Dictionary of Geography*, 1965).

**Geohydrology.** See HYDROGEOLOGY.

**Geoid.** The generalized shape of the earth as a whole; that is, of the mean sea-level of the oceans and their imaginary continuations through the continents. 'The word "geoid" is used to designate the actual figure of the surface of the waters of the earth . . . it is an irregular figure peculiar to our planet' (Merriman, *Figure of the Earth*, 1881). The conception was clearly defined by Playfair (*Illustrations*, 1802): 'The expressions "figure of the earth" and "surface of the earth" are each of them occasionally taken in two different senses. The surface of the earth, in its most obvious sense, is that which bounds the whole earth, and includes all its inequalities; it is a surface extremely irregular. This may be called the "actual" surface, and the figure bounded by it, the "actual" figure of the earth. The surface of the earth, in another sense, is one that is everywhere horizontal, and is the same which water assumes when at rest; it is the surface marked out by levelling, and may be supposed to be continued from the sea, through the interior of the land, till it meets the sea again. The figure bounded by this horizontal surface may properly be called the "statical" figure of the earth' (A few years later Playfair published a paper 'On the figure of the earth', *TRSE*, 1805.) *The Earth's Shape and Gravity* (Garland, 1965)

**Geo-isotherm.** See ISOTHERM.

**Geological agent.** A force or instrument by means of which a geological process operates. See quotation (Playfair) under WEATHERING. 'The chief agents of transport are: gravity, wind, and especially water both in the liquid form and when solidified into snow and ice' (Rastall, *Textbook*, 1941). *Man as a Geological Agent* (Sherlock, 1922).

**Geological ages.** This term is used when we want to particularize or emphasize those long-past periods of time with which geology specifically deals, as distinct from the present day or recent times. As to when past ages cease to be 'geological', this depends on the subject, the context, and the writer. Thus, in the history of Man, the ages since the close of the Neolithic period (that is, since about 2000 B.C.) definitely cease to be geological; for marine sedimentation, the geological ages would probably be taken to end with the exposed sedimentary rocks which, for most regions, means about the early part of the Pleistocene: in considering the evolution of an existing landscape by the process of normal subaerial (and marine) erosion, the geological ages grade into recent times as there is no cessation of a characteristically 'geological' record; but the Ice Age is certainly a 'geological' age.

**Geological cartography.** See GEOLOGICAL MAP.

**Geological column.** A geological, or stratigraphical, column shows a rock sequence, local or general, on a large or a small scale, expressed as a diagram; the beds, &c., being placed horizontally in a vertical column, the thicknesses shown to scale. Features of the beds are indicated; conventional markings for lithology, names of fossils written at the side at the appropriate levels, &c. The geological column is the most concise and graphic method of recording a sequence, and comparative columns show correlations and lateral variations. Early examples are found at the end of the 18th century, and almost every recent stratigraphical work provides examples, as do the *BRG* volumes. One of the earliest must be that through the Coal Measures near Newcastle-upon-Tyne, shown in the translation (1799) of Faujas Saint Fond's *Voyage* (1797) and in Erasmus Darwin's *Botanic Garden* (1799, reproduced by Arkell and Tomkeieff). A much more detailed column, adding up to a length of 1·7m, is given in Westgarth Forster's *Treatise on a Section of the Strata from Newcastle-upon-Tyne to the Mountain of Cross Fell* (1809). These illustrate the 18th century use of vernacular rock-names by miners; such as girdles, post, thill, hazle, plate, tuft. *Guide to the Geological Column* (Sherlock, *GS*, 1938).

**Geological conservation.** 'Geological conservation can be defined as the maintenance of those geological localities in Britain which form the indispensable bases of geological education and research' (*Geological Conservation, Nature Conservancy*, 1968).

**Geological cycle.** The general cyclic sequence of the major geological processes and conditions throughout a region of the earth's crust: (*a*) deposition and (at least partial) consolidation in a geosyncline, (*b*) diastrophism, with profound deformation, uplift, metamorphism (probably) of the down-squeezed region of the geosyncline, together with (probably) igneous intrusion, (*c*) erosion, (*d*) submergence, (*e*) deposition of a further set of deposits (unconformable to the first). The regions of successive geosynclines will normally not be the same, and a region in the erosion stage may well be supplying material for a neighbouring region in the deposition stage. The idea of the geological cycle is inherent in the Huttonian theory (1788), and the term is used by Playfair (*Illustrations*, 1802): 'The great geological cycle, by which the waste and reproduction of entire continents is circumscribed'. The term 'geological cycle' as meaning just this is not very firmly fixed, and 'geostrophic cycle' has been proposed. See CYCLE OF SEDIMENTATION.

In visualizing the concept of the geological cycle there is a tendency to over-simplify the matter and to see the successive phases as being rather sharply marked off in time, one from the next. It seems reasonable, however, to suppose that they largely overlap.

**Geological history.** The history of the earth and its inhabitants throughout the geological ages, that is, in its widest application, the same as 'earth history'. Geological history may be taken for certain areas, over certain periods of time, and in certain aspects: 'The history of volcanic action in the geological past' (Geikie, *QJ*, 1891); *History of Devonshire Scenery* (Clayden, 1906); 'Fossil vertebrates and their geological history' (Trueman, *Introduction to Geology*, 1938); 'The geological history of the Lake District' (Mitchell, *PYGS*, 1956). 'The tectonic, plutonic and metamorphic histories [of Dalradian and associated rocks] in north-west Donegal' (Knill and Knill, *QJ*, 1961). *The Geological History of the British Isles* (Bennison and Wright, 1969).

**Geological map.** A geological map is one on which are recorded the geological features of a region, particularly the delineation of outcrops, the occurrence of faults, mineral veins, fossil localities, &c. and, by conventional signs, the directions and quantities of dip, cleavage, &c. The geological map shows a whole three-dimensional structure of rocks truncated at the earth's present surface. 'An ingenious proposal for a new sort of maps of countrys. . . a soile or mineral map as I may call it' (Lister, *PT*, 1684). 'I have often wished for a mappe, coloured according to the colours of the earth, with marks of the fossiles and minerals' (Aubrey, *Wiltshire*, 1685). Amongst the earliest purely geological maps of any part of Britain are Maton's of the southwestern counties (1797) and the much more satisfactory maps of Derbyshire by Farey (1811) and of the Isle of Wight by Webster (1816); but the most famous of all geological maps are William Smith's (1799 onwards), particularly his great map of the whole of England and Wales (1815). This latter map was closely followed by the more accurate map edited by Greenough (1819). The earliest geological maps of Scotland are those of Necker (*MS*, 1808, facsimile in *TEGS*, 1939) and Boué (*Essai gelogique sur l'Écosse*, 1820); and of Ireland, Griffith (1815). The history of geological mapping ('geological cartography') throughout the world is given by Greenly and Williams (*Methods in Geological Surveying*, 1930); see also Butcher, *The History and Development of Geological Cartography* (University of Reading pamphlet, 1967). Owen's description of the outcrop of the 'vaynes' of limestone in S. Wales (1603) is the first attempt to map, if only in words, a British geological formation. Wales, although some areas still remain virtually unmapped, is well covered bibliographically: *Geological Maps—their History and Development with special reference to Wales* (North, 1928); *A Source-book of Geological, Geomorphological and Soil Maps for Wales and the Welsh Borders* (Bassett, 1967).

**Geological mineralogy.** 'Is it not time for mineralogy to venture forth from the laboratory and find out something of the origin of minerals through field studies? The natural history of mineral deposits is geological mineralogy' (Landes, *AM*, 1946).

**Geological model.** The representation of a geological structure in three dimensions; either a generalized type of structure or the actual structure of a particular piece of country. The beautiful little 'Sopwith's models', of polished wood, are among the earliest of the former kind; drawings of them have appeared in several textbooks, beginning with Lyell's *Manual of Elementary Geology*, 1851. Smithson's models published by Murby & Co. (*c.* 1928) form a well-known later set. Models of particular pieces of country may be conspicuous exhibits in geological museums; some have been described in the *Memoirs of the Geological Survey*, for instance, *Guide to the Geological Model of the Isle of Purbeck* (1906/32), . . . *Ingleborough and District* (1910), . . . *Ardnamurchan* (1932). Block diagrams are representations of imaginary models.

**Geological processes.** The physical and chemical processes concerned with the formation, deformation, and destruction of rocks and with changes in the constitution and structure of the earth (mainly the earth's crust). The chief of such processes are erosion and deposition, stresses and strains, igneous action, and metamorphism. The biological process of evolution may be said to come also within the field of the geological processes.

**Geological record.** The rocks and their contents constitute in themselves the 'geological record'. 'The imperfection of the geological record' (Darwin, *Origin of Species*, 1859). 'The Geological Record, as the accessible solid part of the globe is called' (Geikie, *Textbook*, 1903). *Records of the Rocks* (Symonds, 1872).

**Geological sciences.** See EARTH SCIENCE. 'Geology has developed such strong links with certain of the other sciences that virtually autonomous sub-divisions have come into being, and the general term geological sciences is now used to cover the total field of interest to the geological community. Hence, the International Union of Geological Sciences' (Harrison, *PGS*, 1968). Hence also the Institute of Geological Sciences, in London, which was formed in 1966 by the incorporation of the Geological Survey of Great Britain and the Museum of Practical Geology with Overseas Geological Surveys.

**Geological section.** See SECTION (1-5).

**Geological survey.** A general view, an inspection, or a detailed examination and record, of the geological facts of an area; particularly the making of a geological map.

**Geological Survey.** An official body with the function of carrying out the geological survey of a nation, state, or other (usually political) regional entity. *Contributions to a History of American State Geological and Natural History Surveys* (Merrill, 1920). *The First Hundred Years of the Geological Survey of Great Britain* (Flett, 1937). 'The Geological Survey of India, 1846-1947' (Fox, *N*, 1947). 'Geological Surveys and their influence' (Edmunds, *S-ENA*, 1949). 'The first national Geological Survey' (Eyles, *GM*, 1950). *Geological Survey of Great Britain* (Bailey, 1952).

**Geological time.** The time during which the earth has been a separate planetary body. 'Towards the end of the eighteenth century it began to be realized that there was such a thing as "geological time" ' (in Bell, ed., *Darwin's Biological Work*, 1959). 'A revised geological time scale' (Holmes, *TEGS*, 1959; repeated in *Physical Geology*, 1965). 'Geologic time scale' (Kulp, *S*, 1961). *Phanerozoic Time-scale* (Harland and others, eds., 1964). *Geologic Time* (Eicher, 1976). *Geological Time* (Kirkaldy, 1971). *A Geologic Time Scale* (Harland and others, 1982). Geology is the study of processes working through time: 'The mere passing of time obviously cannot change anything' (Geikie, *Classbook*, 1890). See GEOCHRONOLOGY.

**Geologize.** To practise geology, particularly in the field.

**Geology.** The science of the earth; its composition, structure, processes, and history. Hence 'geological', 'geologist'. Lyell gives two definitions, each repeated unchanged through all the editions of each work. That in the *Elements* (1838 and later) may be taken as the authoritative statement on the matter and is more satisfactory than that in the *Principles* (1830 and later). It runs as follows: 'Of what materials is the earth composed, and in what manner are these materials arranged? These are the inquiries with which Geology is occupied, a science which derives its name from the Greek, γη, *gē*, the earth, and λογος, *logos*, a discourse. Such investigations appear, at first sight, to relate exclusively to the mineral kingdom, and to the various rocks, soils, and metals, which occur upon the surface of the earth, or at various depths beneath it. But, in pursuing these researches, we soon find ourselves led on to consider the successive changes which have taken place in the former state of the earth's surface and interior, and the causes which have given rise to these changes; and, what is still more singular and unexpected, we soon become engaged in researches into the history of the animate creation, and of the various tribes of animals and plants which have, at different periods of the past, inhabited the globe'. Of definitions in later textbooks, one of the best is in the first edition of Lake and Rastall's (1910): 'The main object of the science of geology is the study of the structure and history of the earth'; but this, so simple and appropriate, is reworded, elaborated, and obscured in later editions. Geology may also be defined (though rather restrictively) as 'the study of rocks', or, to be more precise, 'the study of rocks and naturally-occurring rock-materials; what they are, what they contain, and what they teach'. 'Definitions of geology' (*WGQ*, 1965/6). 'The definition of geology' (Challinor, *MG*, 1977). See EARTH, EARTH SCIENCE, and ROCK.

'The mediaeval Latin word *Geologia* covered the study of anything, such as law, which was earthy rather than divine' (Edwards, *Early History of Palaentology*, 1931). It eventually came to be used approximately in its present sense of 'geology'. 'So far as is known at the present time, Lovell's work *Pammineralogicon or an Universal History of Minerals*, published in 1661, is the first work in the English language in which the word *geologia* appears. The work of Escholt [*Geologia Norvegica*], written in the Danish language but published at Christiania in 1657, is the first printed work in which the word occurs. But Aldrovandus employed the word, in some manuscript notes and in his will, essentially in the modern sense, at least as early as 1605, the year in which he died' (Adams, *Birth and Development of the Geological Sciences*, 1938. See also *BGSA*, 1932/3). The term then begins to appear in non-classical form. 'To De Saussure, so far as I have been able to discover, we owe the first adoption of the terms "geology" and "geologist". This science had formed a part of mineralogy, and subsequently of physical geography. The earliest writer who dignified it with the name it now bears was the first great explorer of the Alps. In the year 1778 there appeared at the Hague the first imperfect edition of De Luc's *Lettres Physiques et Morales sur les Montagnes*, in the introduction to which the author states that for the science that treats of the knowledge of the earth he employs the designation of Cosmology. The proper word, he admits, should have been Geology, but he "could not venture to adopt it because it was not a word in use". In the completed edition of his work, published the next year, he repeats his statement as to the use of the term Cosmology, yet he uses Geology in his text notwithstanding. In the same year (1779), De Saussure employs the term Geology in his first volume without any explanation or apology, and alludes to the geologist as if he were a well-known species of natural philosopher' (Geikie, *Founders of Geology*, 1905). 'Geology was used in manuscript by John Walker at least as early as 1779 as the subject of a series of lectures and as one of the major divisions of natural history, deserving to stand alone as a separate science' (Scott, *Lectures on Geology by John Walker*, 1966): 'We come now to another branch of our subject vizt.—Geology or The Natural History of the Earth'. The term seems to have been firmly in use in Britain in 1795: 'I hope fully to refute the geological, as well as mineralogical notions with regard to that body granite' and '. . . a person who has formed his notions of geology from the vague opinion of others and not from what he has seen' and 'When, in forming a theory of the earth, a geologist shall indulge his fancy in framing, without evidence, that which had preceded the present order of things, he then either misleads himself, or writes a fable for the amusement of his reader' (Hutton, *Theory of the Earth*, 1795). 'Without calling it "Geology", Steno gives in the little treatise of 1667 the first outline of a scientific history of the earth arrived at through exact studies of Nature and through inductive reasoning. Geology as a science was born. Two years later he gave in *De Solido* (1669) a more detailed "geology"' (Garboe, *Earliest Geological Treatise*

(*1667*), 1958). Williams's *Natural History of the Mineral Kingdom* (1789) may perhaps be said to be the first text-book of geology in English (he does not use the term itself), especially if we ignore the speculative parts of the two volumes; but the first formal treatise, at least in Britain, appears to be Bakewell's *Introduction to Geology, illustrative of the general structure of the Earth; comprising . . . an outline of geology and mineral geography of England*, 1813. See GEOGNOSY.

Geology is one of the major sciences comprising a number of branches or component sciences, of which the chief are Mineralogy, Petrology, Structural Geology, Stratigraphy, Geomorphology, and (?) Palaeontology. The study of any branch of geology naturally involves the study of the relevant Geological Processes and the reading of the relevant aspects and details of Geological History. Knowledge from any branch of geology may of course be 'applied' to any other science or activity. 'Geology' is used also in the sense of the geological features of a district or country (e.g. *The Geology of North Wales*, *MGS*, 1866/81); and in the sense of a treatise on the subject (e.g. *A Geology for Engineers*, Blyth, 1943/74). See EARTH SCIENCE.

The adjectives 'geological', 'geologic', are often treated as synonymous, particularly in Britain where, in fact, 'geologic' is unusual. A distinction is, however, sometimes made, particularly in America, when 'geologic' refers to things forming parts of the subject-matter of geology while 'geological' refers to things concerning geology: 'Geologic history of the Red Sea area', 'Geological notes' (titles in one number of the *BAAPG*, 1960) and 'Geological Society'.

'During the last few years frontiers of geology have widened to encompass not only the Moon but Mars and Mercury' (Guest, *CGA*, 1975).

**Geomagnetism.** The earth's magnetic phenomena. The study of the earth's magnetic field, its origin and its practical significance in the reading of geological history, is a branch of geophysics. See PALAEOMAGNETISM.

**Geomathematics.** See MATHEMATICAL GEOLOGY. *Geomathematics* (Agteberg, 1974). *Recent Advances in Geomathematics* (Merriam, ed., 1976).

**Geomicrobiology.** See PALAEOBIOLOGY. 'Frontiers in geology: geo-microbiology' (Moore, *AdS*, 1966).

**Geomorphic.** Concerning the relief of the land. As an example of the (not very rigid) distinction between the adjectives 'geomorphic' and 'geomorphological', 'geomorphic features' and 'geomorphological terms' would probably be the preferred alternative in each case.

**Geomorphology.** The science which deals with the surface features of the land and ocean floors; their relief, form, character, origin, and evolution. 'A genetic classification of geologic phenomena . . . will apply equally to geography, whether observational or of the more philosophic nature which Powell has called Geomorphology' (McGee, *NGM*, 1888, quoted by Chorley and others, *History of the Study of Landforms*, 1964). Nevertheless the study of this branch began as soon as geology as a whole began to be studied; the foundations were laid by James Hutton in his *Theory of the Earth* (1795). If we separate the two aspects of geomorphology, purely descriptive character on the one hand and origin and evolution on the other, there is 'geomorphogeny' available for the latter, and we might use 'geomorphography' for the former. 'Geomorphology' has been restricted to the treatment of features produced only by erosion and deposition (see quotation from Scheidegger under ENDOGENETIC); Sparks (*Geomorphology*, 1960) also narrows the meaning, emphasizing erosion, and Peel remarks (*AdS*, 1967): 'Whatever its name might logically be held to imply, geomorphology has in fact become very much the study of terrestrial denudation and of the forms and features that result from it'. *The morphology of the Earth* (King, 1967). 'Early British geomorphology, 1578-1705' (Davies, *GJ*, 1966).

Of all the earth-sciences, geomorphology is the one where the essential evidence has been virtually obliterated from the geological record. We can see today the results on the landscape of weathering, the action of rivers, the waves of the sea, glaciers, and the wind; and we can detect evidence of recent crustal instability, such as that provided by raised beaches—and there are earthquakes. From all this we can infer and predict to some extent the changes that have occurred in the past and will occur in the future for a period of time very brief on the geological scale (measured in tens of thousands of years perhaps), a period far too short for us to draw from it any conclusions of a general nature concerning the balance of earth-movement and denudation during the whole land phase of a geological cycle. See EARTH HISTORY and RATE OF PROCESS.

**Geonomy.** See EARTH SCIENCE.

**Geophysics. 1.** The physics of the earth in all its various parts, from the core to the outermost atmospheric fringe. Geophysics in past geological ages is 'palaeogeophysics'. **2.** The study of the earth by physical methods. These are chiefly exact quantitative methods; taking advantage of natural occurrences, particularly earthquakes, but usually by artificial experiments, such as making explosions to observe the wave effects ('seismic survey'); and measuring variations in properties such as gravity ('gravity survey'), magnetism ('magnetic survey'), electrical conductivity, &c. It is the name of a science, therefore a singular noun, though plural in origin and form. *Elements of Geophysics* (Ambronn, trans. Cobb, 1928). *Contributions in Geophysics in Honor of Beno Gutenberg* (Benioff and others, eds., 1958). *International Dictionary of Geophysics* (Runcorn and others, eds., 1967). *Theoretical Geophysics* (Officer, 1974). *Principles of Applied Geophysics* (Parasnis, 1979).

**Geoplatform.** See PLATFORM.

**Geosciences.** See EARTH SCIENCE.

**Geosphere.** See LITHOSPHERE. It has been proposed (Rodgers and Adams, *Fundamentals of Geology*, 1966) that 'the three subdivisions of the solid earth [core, mantle, and crust], plus the atmosphere, hydrosphere, and biosphere' be termed 'geospheres'.

**Geostrophic cycle.** 'A commemorative article on Hutton may be a suitable time and place to suggest an accurate title for the process which he was the first to define. I should like here to propose the term "geostrophic cycle" (from the Greek *strophe*, a turning, or *strepho*, to change) for the major cycle of changes in the earth' (Tomkeieff, *PGA*, 1946; and see Tomkeieff, *TEGS*, 1948). See GEOLOGICAL CYCLE.

**Geostructure. 1.** A regional structure, such as a geofault or geoflexure. **2.** The whole structure of a large part of the earth's crust, such as the Pacific or Antarctic regions.

**Geosuture.** See GEOFAULT.

**Geosyncline.** An elongated downwarp in the earth's crust, the bottom subsiding deeply beneath accumulating sediments. 'The making of the Alleghany range was carried forward at first through a long-continued subsidence—a "geosynclinal" (not a true synclinal, since the rocks of the bending crust may have had in them many true or simple synclinals as well as anticlinals)—and a consequent accumulation of sediments, which occupied the whole of Paleozoic time; and it was completed, finally, in great breakings, faultings and foldings or plications of the strata, along with other results of disturbance' (Dana, *AJS*, 1873). 'Hall had demonstrated [in 1859] that an enormous thickness of sediments had accumulated in the Appalachian area during the Palaeozoic period, and that this great thickness of sediments occurred within the region which was subsequently severely folded and faulted to form the Appalachian chain. Since the sediments bore evidence that shallow-water conditions had prevailed during their deposition, he inferred that the area of deposition had been steadily depressed, and further that the depression was caused by the weight of the accumulated deposits. Dana, however, contended that the accumulation of a great thickness of sediments was a consequence and not a cause of the subsidence; but he accepted the evidence adduced by Hall as to the relation between the existence of thick masses of sediments and a broad depression of the crust, and the term "geosynclinal" was introduced by him to indicate this relationship' (Jones, *QJ*, 1938; his address 'On the evolution of a geosyncline' being concerned with the Lower Palaeozoic geosyncline of the British area). The form 'geosyncline' is used in Geikie's *Textbook* (1903). *North American Geosynclines* (Kay, *MGSA*, 1951). *Geosynclines* (Aubouin, 1965). 'On geosynclinal nomenclature' (Kay, *GM*, 1967). An elaborate terminology has been proposed for geosynclines, based on shape, size, occurrence in oceanic or continental regions, proximity to shelf-seas or land areas, association with volcanism, subsequent deformation history, &c. The typical deep oceanic geosyncline is then a 'eugeosyncline'. Of the other terms 'miogeosyncline' (etymologically, 'lesser geosyncline') is in current use, but it is variously defined, for instance as (*a*) a marginal part of a eugeosyncline, (*b*) any geosyncline lacking active volcanism, (*c*) any minor geosyncline. 'The status of geosynclinal nomenclature' (Cobbing, *JGS*, 1978). See GEANTICLINE.

Many of the features included in the 'classical' concept of a geosyncline are now explained in terms of sediment accumulation either at continental margins, particularly where these margins are also plate margins, or beside island arc systems.

**Geotechnical.** Though seeming, from the construction of the word, to be widely applicable, it usally means: relating to the geological aspects of engineering. Apart from the facts of geological occurrence and structure, the aspects of importance are the physical and bulk properties of the soils, superficial deposits, and 'solid' rocks encountered by the engineer, or with which he is concerned, in any particular case. Hence 'geotechnology' (*Geotechnology*, Roberts, 1977). See SOIL MECHANICS.

**Geotectonic.** 'Geotectonic, or Structural Geology—the Architecture of the Earth—the mode of arrangment of the various materials composing the crust of the earth' (Geikie, *Textbook*, 1882). 'Naumann, in his *Lehrbuch der Geognosie* (1850) was the first author who devoted a special chapter to Geo-tectonics" ' (Zittel, *History*, 1901). Whereas 'tectonic' tends to imply secondary deformation structures only, 'geotectonic' is used in a comprehensive way, or for large areas, and includes the character, mode of formation, and original disposition of the various kinds of rock-bodies: strata, igneous intrusions, lava-flows, &c. *Basic Problems in Geotectonics* (Belousov, 1962). 'Recent trends in geotectonics' (Rast, *ESR*, 1966).

**Geothermal.** Of, or pertaining to, the heat of the earth itself; that is, the heat of the earth's interior, excluding the heat superficially derived from the sun. 'The earth's heat and internal temperatures' (Sass in Gass and others, eds., *Understanding the Earth*, 1972). 'Geothermal exploration' (Lovering, *TSME*, 1965). Hence 'geothermal heat flow', 'geothermal energy'.

**Geothermal gradient.** Difference in temperature (within the earth's crust) per unit of depth, pressure, distance, or time. 'Penetration of the earth's crust by boreholes and mines shows that the temperature increases with depth. The rate of temperature increase or, in other words, the "geothermal gradient", varies considerably from place to place.' (Holmes, *Physical Geology*, 1978). In regional metamorphism it is depth and pressure that are concerned: 'A geothermal gradient of about 40°C/kilometre' and 'A thermal gradient of slope greater than 50°C per kilobar' (Rutland and Chinner, *PGS*, 1962). In contact metamorphism the 'gradient' might refer to distance from the intrusion; in the crystallization of a magma it might refer to time.

**Geyser.** A hot spring, being a manifestation of a late phase of volcanic activity, from which a gushing column of hot water and steam is explosively discharged at intervals. Usually the orifice is in the centre of a mound of siliceous sinter, geyserite. (The correct pronunciation seems to be gay-, possibly guy-, not gē-.) [Icelandic *geysir*, a roarer; particularly for the Great Geyser in Iceland.]

**Geyserite.** See GEYSER and SINTER.

**Ghost-stratigraphy.** The lingering relict traces, in highly metamorphosed strata, of the original lithology and stratification.

**Gio.** (Geo.) 'Another singular feature of these northern coasts [of Scotland] is the number of "gios", or narrow steep-walled gullies, or inlets, by which the sea-cliffs are indented. Here again we trace the dominant influence of the joints' (Geikie, *GM*, 1878, and *Geological Sketches*, 1882). [Norse, *gya*, a creek.]

**Givetian.** See DEVONIAN SYSTEM. (Gosselet, 1879.) [Givet, France.]

**Glacial.** Characterized by the presence of ice; produced by, or concerning, glaciers or ice-sheets. [L. *glacies*, ice.]

**Glacial boulder.** See ERRATIC (1).

**Glacial drag.** In particular, terminal curvature produced by the force of moving ice. Examples are illustrated from Rishton, near Blackburn (Ranson, *PGA*, 1928) and from Aberystwyth (Challinor, *GM*, 1947).

**Glacial drainage channel.** A channel cut by the drainage from a glacier; includes overflow channels and meltwater channels. 'Newtondale. This direct glacial drainage channel crosses the North Yorkshire Moors' (pl. in Rayner and Hemingway, eds., *Geology of Yorkshire*, 1974).

**Glacial drift.** See DRIFT.

**Glacial erratic.** See ERRATIC (1).

**Glacial geology.** The study of the geological effects of glaciers and ice-sheets and the reading of the glacial history of a region. *Glacial Geology of Great Britain and Ireland* (Carvill Lewis, 1894). 'The glacial geology of the Derbyshire Dome' (Jowett and Charlesworth, *QJ*, 1929). 'The drift or glacial deposits of Ayrshire (Smith, *TGSG*, 1898). 'Glacial erosion in North Wales' (Davis, *QJ*, 1909). 'The glacial geology of Spitsbergen' (Garwood and Gregory, *QJ*, 1898). 'Initiation and growth of the East Antarctic ice sheet' (Drewry, *JGS*, 1975). *Glacial geomorphology* (Embleton and King, 1975). *Glaciers and Landscape* (Sugden and John, 1976). 'Glacial history of the Spitsbergen archipelago' (Boulton, *BO*, 1979).

**Glacial period.** See ICE AGE. 'Professor Agassiz writes, in his essay entitled "A period in the history of our planet" (*Edinburgh New Philosophical Journal*, vol. xxxv) as follows: "A crust of ice covered the superficies of the earth . . . a period appeared in which all life was annihilated . . . the Glacial period".' Forbes realizes the less universal and catastrophic conception: 'throughout this essay I have used the epithet "glacial" for want of a better, always intending to express by the phrase "Glacial epoch" that section of geological time which was typically distinguished by the prevalence of severe climatal conditions through a great part of the northern hemisphere' (*MGS*, vol. 1, 1846).

**Glacial theory.** The theory now universally accepted that such phenomena as striated rock-pavements, erratic boulders, and the drift are due to the action of glaciers and ice-sheets; first propounded, separately, by the Swiss geologists de Charpentier and L. Agassiz about 1840. 'Having established his theory as completely as he could by repeated investigations of Switzerland and the adjacent portions of France and Germany, M. Agassiz became desirous of investigating a country in which glaciers no longer exist, but in which traces of them might be found. This opportunity he has recently enjoyed by examining a considerable part of Scotland, the north of England, and the north, centre, west and south-west of Ireland; and he has arrived at the conclusion, that great masses of ice, and subsequently glaciers, existed in these portions of the United Kingdom at a period immediately preceding the present condition of the globe, founding his belief upon the characters of the superficial gravels and erratic blocks, and on the polished and striated appearance of the rocks *in situ*' (*PGS*, 1840, Nov. 4). See ERRATIC. For the history of the theory see De la Beche, *Geological Observer*, 1851; North, *PGA*, 1943; Bailey, *Charles Lyell*, 1962; Chorley and others, *History of the Study of Landforms*, 1964; Davies, 'The tour of the British Isles made by Louis Agassiz in 1840' (*AS*, 1968); Shotton, *OE* 1, 1980.

**Glaciation.** Glacial action. 'The glaciation of the Black Combe district, Cumberland' (Smith, *QJ*, 1912). Also, a minor period of glacial conditions or a glacial episode.

**Glacier lake.** A lake dammed up in a valley, or some other depression of the ground, by the edge or front of a glacier or ice-sheet. Also called 'glacial lake', 'ice-dammed lake', 'pro-glacial lake'. 'A system of glacier lakes in the Cleveland Hills' (Kendall, *QJ*, 1902). 'Glacial Lake Lapworth came into being when the lake around Newport was united with Glacial Lake Buildwas by the withdrawal of the ice-front from the foot of the Wrekin massif' (Wills, *QJ*, 1924). 'The glacier-lakes of Eskdale, Miterdale, and Wasdale' (Smith, *QJ*, 1932). 'Glacial lakes and spillways in the vicinity of Madeley, North Staffordshire' (Yates and Moseley, *QJ*, 1957).

**Glacigene.** Of glacial origin. 'Upper Proterozoic glacigene rocks' (Bowes, *PGA*, 1970). Also in the form 'glaciogenic'.

**Glaciofluvial.** Of or pertaining to streams which are meltwaters from a body of ice; particularly for the rock material worked over and deposited by water within, underneath, or in the neighbourhood of a glacier or ice sheet. Englacial deposits of this nature may be let down on the floor of the ice as it melts. Glaciofluvial deposits are fluvial as regards their construction and glacial as regards their immediate derivation. 'Glaciofluvial' seems preferable to the synonymous 'fluvioglacial'.

**Glaciolacustrine.** Of or pertaining to a lake caused by the presence of a mass of ice in the immediate vicinity, particularly for deposits laid down in such a lake through the intermediary action of a stream emerging from the ice into the lake. Glaciolacustrine deposits may be deltas, and thus be considered as also glaciofluvial; or they may be deposits on the floor of the lake when they will be evenly stratified sediments and may show seasonal (varved) bedding.

**Glaciology.** The science of glaciers and ice-sheets; their distribution, character, action, and effects. 'Speculation and research in Alpine glaciology: an historical review' (Garwood, *QJ*, 1932).

**Glaciomarine.** Due to glacial and marine action combined; particularly for deposits, laid down in the sea, having been transported there by a glacier or ice sheet, or its streams; they would usually be of the nature of a delta. Rock material dropped on the sea floor by icebergs would also be included. 'Ordovician-Silurian glaciomarine deposit from Newfoundland' (McCann and Kennedy, *GM*, 1974).

**Glass.** Hard, brittle, lustrous amorphous mineral material, particularly (in geology) that produced by the sudden cooling of magma. Glass is liable to devitrification and practically no pre-Carboniferous igneous rocks are known which are now quite glassy. 'On rock glass, and the solid and liquid states' (Hawkes, *GM*, 1930).

**Glauconite.** Essentially a hydrous silicate of iron and potassium; amorphous, and some shade of green in colour. Occurs extensively in small grains in the Chalk Marl and Upper Greensand; also in other strata, e.g. the Comley Sandstone (Cambrian). 'Geology of glauconite' (Galliher, *BAAPG*, 1935). '"Glauconite" pellets: their mineral nature and applications to stratigraphic interpretation' (Burst, *BAAPG*, 1958). 'Glauconite is abundant in many Palaeogene rocks, and it's contained potassium can be used for radiometric dating' (Odin and others, *JGS*, 1978). (Keferstein, 1828, *glauconit.*) [G. *glaucos*, gleaming blue-green.]

**Glaucophane.** See AMPHIBOLE. Occurs particularly in the metamorphic 'glaucophane-schist' ('blueschist'), which represents either igneous rocks modified by regional metamorphism or a metasomatic transformation of a siliceous sediment (Hausmann, 1845). [G. *glaucos*, gleaming blue-green, *phaneros*, plainly visible.]

**Glenkiln Shales.** The strata, only about 6 m. thick, in S. Scotland, representing, in a condensed succession, the lower two graptolite zones of the graptolitic facies of the Caradoc series (zones which used to be placed in the Llandeilo series). The zone fossils *Nemagraptus gracilis* (below) and *Climacograptus peltifer* (above) occur, with their respective associates. (Lapworth, *QJ*, 1878.) [Glenkiln Burn, Dumfriesshire.]

**Global tectonics.** Used chiefly as 'the new global tectonics' meaning, especially, continental drift and plate tectonics.

**Globigerina ooze.** See OOZE.

**Glossopteris flora.** The fossil flora, characterized by the genus *Glossopteris* and its allies (Pteridospermae?), which is the chief distinguishing feature of the Gondwana series (in the wide sense).

**Glyptogenesis.** See EARTH SCULPTURE. (Grabau, *Stratigraphy*, 1913.) [G. *gluptos*, carved.]

**Gneiss.** A coarse-grained banded crystalline rock, the product of high grade regional metamorphism. Gneisses may have any mineralogical composition but are commonly granitic. 'Gneiss is a venerable rock name, derived long ago from a slavonic word meaning "nest" and applied by the medieval miners of Central Europe to the foliated rocks in which the metalliferous veins occur; the rock, so to speak, being the "nest" of the ores' (Holmes, *Physical Geology*, 1978). The term was adopted by Werner and used by him and his followers with its present meaning. Whereas most of the Wernerian terminology is obsolete, 'gneiss' is still very much alive.

**Gneissic structure.** The coarse foliation or banding characteristic of gneisses. 'There is a curious variety of stone, found in the Highlands, which may be called streaked or stripped, the rock being composed of strips of different colours, commonly white and black alternately, sometimes different shades of a lighter and darker grey. Some of these streaks are not above the eighth part of an inch thick, and some are up to half an inch' (Williams, *Mineral Kingdom*, 1789).

**Golden spike.** See BOUNDARY STRATOTYPE.

**Gondwana series.** 'The marine older and middle mesozoic, and probably the upper palaeozoic formations of other countries are represented in the Peninsula of India by a great system of beds, chiefly composed of sandstones and shales, which appear, with the exception of the rocks just noticed, to have been entirely deposited in fresh water, and probably by rivers. Remains of animals are very rare in these rocks, and the few which have hitherto been found belong chiefly to the lower vertebrate classes of reptiles, amphibians, and fishes. Plant remains are more common, and evidence of several successive floras has been detected. The Geological Survey [of India] has now adopted the term Gondwana for the whole of this great plant-bearing series. This term is derived from the old name for the countries south of the Narbada valley, which were formerly Gond kingdoms, and now form the Jabulpur, Nagpur, and Chhatisgarh divisions of the Central Provinces. In this region of Gondwana the most complete sequence of the formations constituting the present rock system is to be found' (Medlicott and Blanford, *Geology of India*, 1879). The Gondwana series was later recognized over a much wider area and is taken, in a wide sense, to include its equivalents outside India, particularly the Karroo series of Africa. It is characterized, above all, by the fossils of the *Glossopteris* flora. 'The name Gondwana was introduced by H. B. Medlicott in 1872 in a manuscript report, but appeared for the first time in print in a paper by O. Feistmantel published in 1876 . . . though the Gondwanas are generally referred to as a System, their extent and magnitude in space and time entitles them to be considered as a major group. They span the time from the Upper Carboniferous to the Jurassic or Middle Cretaceous and comprise strata whose total thickness is from 20,000 to 30,000 ft. in different parts of the world' (Krishnan, *India and Burma*, 1956).

**Gondwanaland.** The continental mass which, as stratigraphy and palaeontology give us reason to believe, existed at the time, and in the places, of the accumulation of the Gondwana series (in the widest sense). Of this continent we see detached parts in several regions which today have nothing in common: Brazil, nearly all Africa, Arabia, peninsular India, Australia, and Antarctica. 'We call this mass Gondwana-Land after the ancient Gondwana flora [the *Glossopteris* flora] which is common to all its parts' (Suess, *Das Antlitz der Erde*, 1885, English trans., 1904; Suess did not envisage quite such a large area). 'The Indian Ocean and the status of Gondwanaland' (Fairbridge in Sears, ed., *Progress in Oceanography*, 1965).

**Goniatite.** There is the well-known genus *Goniatites*, but 'the goniatites' is a familiar and useful term for all similar members of the order Ammonoidea (class Cephalopoda, phylum Mollusca). It is used in wider and narrower senses but always for normally coiled forms in which the septal suture is simply folded, not denticulated or frilled, and these are nearly all Palaeozoic. A stricter sense would confine the name to those with somewhat angular, 'goniatitic', sutures and a still stricter sense to the smooth forms of these. 'Goniatites in carbonaceous shales show that black mud was not ill suited to them' (De la Beche, *MGS*, vol. i, 1846). [genus *Goniatites* (G. *gōnia*, an angle).]

**Gossan.** The weathered or otherwise decomposed upper part of a lode. 'Samples out of various veins all near Redruth. The miners call them "gossens". They show a variety of mixed matter' (Woodward, *Catalogue*, 1728). [A Cornish term.]

**Gothlandian.** In use on the Continent for the Silurian system in the restricted sense, as used in Britain and America. See SILURIAN. [Island of Gothland in the Baltic Sea.]

**Graben.** Usually regarded as synonynous with 'rift valley' but might refer to a purely structural feature. Used unqualified (as it almost always is) it is thus a dangerously ambiguous term (as is the unqualified 'rift'). See TAPHROGENESIS and quotations under AGE AND ORIGIN (Tromp) and RIFT VALLEY (Cotton). *Graben Problems* (Illies and Mueller, eds., 1970). [German, a ditch.]

**Grade. 1.** A supposed condition of balance achieved during the progress of river action, the conception of which was formally introduced (though Gilbert had previously suggested the term) by W. M. Davis (*JG*, 1902; reprinted in *Essays*, 1909). Davis admits that 'this condition cannot be understood without rather careful thinking'. The present writer confesses that he can get no clear mental picture of 'grade' either from his or from anyone else's discussion; he remains unconvinced that any special condition is reached at any particular stage. 'The concept of grade' (Dury in Dury, ed., *Essays in Geomorphology*, 1966). **2.** In metamorphism, the degree to which the rocks have been subjected to the metamorphic agents, temperature and pressure. Within one general rock-type this corresponds to the actual degree of alteration, but for two different rock-types the same degree of temperature and pressure (same metamorphic grade, 'isograde') may produce different degrees of actual alteration. There are thus 'high grade' and 'low grade' metamorphic rocks. **3.** Of a coal, refers to the amount of inorganic matter it contains, the ash left behind after burning. The less the ash, the higher the grade. **4.** The size of the constituent particles of a clastic sedimentary rock (grain size); e.g. clay-grade, sand-grade, pebble-grade.

**Graded bedding.** Bedding characterized by an upward gradation in grain size. Normally, relatively coarse material below grades into fine material above, the next graded bed beginning abruptly with the coarse material. 'In what may be called "graded bedding" a bed of sandstone, ranging from a fraction of an inch to two or three feet in thickness, starts at the bottom with relatively coarse material and grades upwards to relatively fine material. There are wide ranges of texture. Some graded beds are coarse grit at the bottom and pass upwards to fine grit or sandstone at the top. Others are fine sandstone at the bottom and pass to sandy mudstone above' (Bailey, *GM*, 1930). 'Significant features of graded bedding' (Kuenen, *BAAPG*, 1953). 'Graded bedding, with observations on Lower Palaeozoic rocks of Britain' (Kuenen, Netherlands, 1953). 'Textural studies in graded bedding' (Scheidegger and Potter, *Sd*, 1965). See RIBBON BANDING (George, 1970).

**Grain. 1.** A grain is a mineral particle, particularly a small, hard, more or less rounded, particle such as a sand-grain. **2.** The grain of a rock is its texture as regards fineness or coarseness (granularity). **3.** The lineation of a texture or structure. **4.** We speak of the 'grain of the country' meaning the direction of strike of beds or the fold direction; that is, the 'structural grain'. 'The imprint of structural grain on the micro-relief of the Welsh uplands' (Thomas, *GeolJ*, 1970).

**Grain size.** The general size of the grains in a sediment or rock or of the grains of a particular mineral in the sediment or rock.

**Granite.** In the wide sense, a coarse-grained acid igneous rock, sometimes porphyritic, consisting essentially of quartz, felspar, and dark (mafic) minerals which usually include micas or hornblende or both. Subdivision into 'true' granite, adamellite, and granodiorite is based on the felspars; in true granite, granite in the restricted sense, alkali felspar is predominant. Granite (in the wide sense) is by far the most abundant of all plutonic rocks. From the point of view of petrogenesis it seems likely that there are 'granites and granites', some formed directly by crystallization from a magma, others by the ultrametamorphism of pre-existing rock (granitization). Hence arises the 'granite problem'. 'The continuing controversy regarding the origin of granite can be reduced in part to Bowen's purposely simplified statement (1948): "The real question is, then: how much granite is magmatic and how much metamorphic"' (Tuttle and Keith, *GM*, 1954). 'I now propose to close this sketch with a brief account of the history of the granite problem as it has fared in Britain' (Bailey, *TGSG*, 1958). From the point of view of relationship with orogenic movements, granites have been classified into syntectonic (emplaced during such movements), post-tectonic, and atectonic (unrelated to orogeny) (Bott, *QJ*, 1956).

Granite was a word used at least as early as the 16th century, but it acquired its present scientific meaning chiefly through its use by Werner (in Germany) in the latter part of the 18th century. That granite is an intrusive igneous rock, not an aqueous crystallization as the Wernerians contended, was established by James Hutton and James Hall, particularly in papers 'On granite' in 1790 (*TRSE*). There is the celebrated account of the geological relations to be seen in Glen Tilt, Perthshire, given in the third volume of Hutton's *Theory of the Earth*, written in 1795 but not published till 1899, of which the following are a few sentences: 'The present question regards the granite, how far it is to be considered as a primary mass in relation to the alpine schistus [here the Dalradian]; in that case, fragments of the granite might be found included in the schistus, but none of the schistus in the granite. But besides this point to be considered, I had in a preceding part of this work drawn a very probable conclusion concerning the natural history of granite, so far as those masses might be considered as analogous to basaltes, or subterraneous lava, in having been made to flow. We have both those points now perfectly decided; the granite is here found breaking and displacing the strata in every conceivable manner, including the fragments of the broken strata, and interjected in every possible direction among the strata which appear. This is to be seen, not in one place only of the valley, but in many places, where the rocks appear, or where the river has laid bare the strata . . . the granite must be considered as posterior to those strata, notwithstanding that these are found superincumbent on the granite.' 'Meditations on granite' (Read, *PGA*, 1943/4). 'Natural history of granite' (Holmes, *N*, 1945). *Geology of Granite* (Raguin, trans. Kranck and Eakins, 1965). *Granite Petrology and the Granite Problem* (Marmo, 1971). 'The origin of granite magmas' (Leake and others, *JGS*, 1980).

In commercial usage any other very hard rock suitable for roadstone may be called 'granite'; this is usually some kind of igneous rock but is occasionally a tough sedimentary rock such as a quartzite. [L. *granum*, a grain.]

**Granite-porphyry.** The medium-grained equivalent of a porphyritic granite, that is, a porphyritic microgranite. Usually occurs as minor intrusions.

**Granitic.** Having the irregularly macro-crystalline texture typically and most commonly seen in a non-porphyritic granite. When used for the first time (?) by Judd (*QJ*, 1886) it was being applied to a gabbro.

**Granitization.** The making of granite out of some pre-existing rock; that is, the conversion (transformation) of a rock *in situ*, by metamorphism, into granite (see TRANSFORMISM). In a narrower sense it is the final stage of such a transformation. 'It will be well to be quite clear as to what is meant by this word granitization. Here is Professor Read's synthetic definition derived from analysis of many sources: Granitization means the process by which solid rocks are converted to rocks of granitic character without passing through a magmatic stage' (Rastall, *GM*, 1945). 'I believe that all the granitic rocks I have seen are simply the result of the extreme of metamorphism brought about by great heat with presence of water' (Ramsay, *Physical Geography and Geology of Great Britain*, 1872). (Read, *PGA*, 1944, and elsewhere; Reynolds, *QJ*, 1946, *GM*, 1947, 1958; Backlund, *GM*, 1946; Marmo, *ESR*, 1967.)

**Granoblastic.** See CRYSTALLOBLASTIC.

**Granodiorite.** A granite, in the wide sense, in which plagioclase (excluding albite) considerably exceeds alkali felspar; intermediate in composition between a granite and a diorite.

**Granophyre.** 'A fine-grained granitic rock having a micrographic texture, or a granite- or quartz-porphyry having a micrographic groundmass' (Holmes, *Nomenclature of Petrology*, 1928). (Rosenbusch, 1872.) [L. *granum*, a grain, G. *phurō*, mix.]

**Granular.** Consisting of grains. In the widest sense, consisting of mineral particles of any kind; in a narrower sense, consisting of small hard somewhat rounded mineral particles, as in a sandstone. It is sometimes used in a special sense for a texture, in any kind of rock, in which the mineral particles are all of about the same size (equigranular). [L. *granulum*, a little grain.]

**Granularity. 1.** The property of being granular, in one sense or another. **2.** The grain size of the particles in a rock (degree of granularity).

**Granulite.** 'According to British usage a granulite is any rather fine-grained quartzo-feldspathic metamorphic rock without conspicuous schistosity (e.g. the Moine "granulites")' (Turner and Verhoogen, *Igneous and Metamorphic Petrology*, 1960). Typical granulites (in the above sense) have the granoblastic texture.

**Granulitic. 1.** See GRANULOSE. **2.** 'In the [basic] rocks exhibiting the "granulitic structure" . . . the most notable distinction is found in the character of the pyroxene and olivine grains, which, as seen in section, assume more or less rounded outlines, and are imbedded in a plexus of lath-shaped crystals of felspar. In masses of molten rock where internal movements accompany the crystallizing process, a "granulation" of the pyroxene and olivine is the result' (Judd, *QJ*, 1886).

**Graphic.** Intergrown so as to resemble writing. Applied to a particular, and common, eutectic texture of a quartz-felspar rock in which the interpenetration of the two minerals result in the quartz appearing, on a surface, like hieroglyphics against a background of the felspar. Most conspicuously seen in a granite-pegmatite. Hence 'micrographic', this texture with a microcrystalline condition, and 'cryptographic' with a cryptocrystalline condition. 'We shall now consider one particular species of granite; and if this shall appear to have been in a fluid state of fusion, we may be allowed to extend this property to all the kind. The species now to be examined comes from the north country, about four to five miles west from Portsoy, on the road to Huntly. The singularity of this specimen consists, not in the nature or proportions of its constituent parts, but in the uniformity of the sparry ground, and the regular shape of the quartz mixture. [This quartz has] not only separately the forms of certain typographical characters, but collectively gives the regular lineal appearance of types set in writing. It is evident that the sparry and siliceous substances had been mixed together in a fluid state; and that the crystallization of the sparry substance, which is rhombic, had determined the regular structure of the quartz' (Hutton, *TRSE*, 1788). 'A granite described by M. Patrin in the *Journal de Physique* for 1791, under the name of "pierre graphique", seemed to Dr. Hutton to have so great a resemblance to the granite of Portsoy, that he ventured to consider them both as the same stone' (Playfair, *Illustrations*, 1802).

**Graphical presentation.** The presentation of geological facts (known, inferred, or suggested) in the form of a map, section, or any kind of diagram. Geological 'reconstructions' would usually be excluded. 'Graphic representation' (another form of the term) is fully analysed in Weller's *Stratigraphic Principles and Practice*, 1960.

**Graphite.** A black mineral, one of the dimorphic states of carbon (the other being diamond); very soft with a grey streak, hence its use for writing. Occurs in veins or as lenticular masses in metamorphic rocks. (Werner, 1789.)

**Graptolite.** The familiar name for any member of the class Graptolithina. By itself it sometimes implies a member of the order Graptoloidea (Graptolitoidea), the members of the other order, the Dendroidea, being then specified as the 'dendroid graptolites. Hence 'graptolitic'. 'A specimen of black slate, with graptolites, from Haverfordwest' (De la Beche, *MGS*, vol, i, 1846). *Monograph of British Graptolites* (Elles and Wood, *MPS*. 1901/18).

**Graptolite shale.** A distinctive Lower Palaeozoic rock-type, characterized by a combination of lithology—a black or dark grey colour, due partly to finely disseminated iron sulphide and partly to carbon—and faunal content—graptolites almost exclusively. This suggests that the surface waters were favourable to life but the bottom layers were unfavourable. 'A totally distinct group of fossils, and one hitherto regarded as of little geological significance, occurs in certain beds of black carbonaceous shales and mudstones. The striking mineralogical features of these black bands, their prolific fauna, and their great longitudinal extent, where for thousands of square miles no other continuous stratum, separable either by lithological or palaeontological characters, relieves the wearisome monotony of the interminable greywackes, soon convince the geologist that it is by their aid alone that he can ever hope successfully to unravel the more than ordinary complexity of the South Scottish succession. It is to the consideration of the numerous facts recently made out regarding the black graptolitic shales that the present paper will be devoted' (Lapworth, *QJ*, 1878).

**Graptolithina.** A group of extinct colonial organisms (Cambrian to Carboniferous, but chiefly Ordovician and Silurian) having an external skeleton of chitin. This skeleton is in the form of rows of cups or tubes (thecae) along one or more branches (stipes), the whole colony (rhabdosome) originating in a conical cell (sicula). The stipes commonly present the appearance of a minute saw. Their biological affinities are uncertain; they were for long placed with the Hydrozoa in the phylum Coelenterata, but are now placed provisionally as a class within the Stomochorda (Hemichordata) which itself, but again only provisionally, is placed as a subphylum of the phylum Chordata, right at the other end of the invertebrate classification (Kozlowski, *JP*, 1966). (Brown, 1846). Some neontologists use the group-name 'Graptolita'. [G. *graptos*, delineated or written, *lithinos*, stony.]

**Graptolitic facies.** A facies of (Lower Palaeozoic) sedimentary rocks characterized by graptolites in a shaly lithology. See POURED-IN DEPOSIT and quotation (Smith and George) under FACIES.

**Graptoloidea.** (Graptolitoidea.) One of the two main orders into which the class Graptolithina is divided, and much the more abundant. It ranges from the Tremadocian into the Gedinnian (lowest Devonian). The stipes are generally few; where there is only one, the thecae may be arranged in one, two or, rarely, four rows. There is only one type of theca. For a member of this group in particular, 'graptoloid' is to be preferred to 'graptolite' because of the wide sense of the latter term. (Lapworth, 1875.)

**Gravel.** Loose detrital material composed chiefly of small pebbles usually mixed with sand. [Celtic.]

**Gravel train.** A term used in glaciology. 'Glaciation of the London Basin. . . . The fundamental basis upon which the work has proceeded is that of treating all well-marked spreads of fluviatile or fluvio-glacial gravel as marking the low ground of their time and occuring in gravel-trains which fall in level downstream' (Wooldridge, *QJ*, 1938). 'Pleistocene gravel trains of the River Thames' (Green and McGregor, *PGA*, 1978). 'Valley train' is a related term; more general as regards composition, more specific as regards associated topography.

**Gravelstone.** As sandstone is consolidated sand, mudstone, consolidated mud, so gravelstone is consolidated gravel; but the term is very seldom used. A 'gravel stone' is a pebble constituent of a gravel.

**Gravitational compaction.** See COMPACTION.

**Gravitational sorting.** The sorting produced by the readier sinking of the heavier particles. If the medium is water, the sediment tends to be 'graded' (see GRADED BEDDING); if the medium is molten magma, in process of cooling, the resulting igneous rock tends to be layered (see LAYERED INTRUSION), or phenocrysts may be sorted (Blake, *GM*, 1968).

**Gravity survey.** See GEOPHYSICS and BOUGUER ANOMALY. 'A Gravity Survey seems clearly to be called for' (Strahan, *QJ*, 1913).

**Gravity tectonics.** The role of gravity in tectonics. Deformation structures may be produced within a rock-mass by the force of gravity, as distinct from being the result of a general lateral compression; the chief of these are the gravity-collapse structures. Another category comprises those movements and displacements due to instability resulting from heavier rock-masses overlying lighter ones, as in cases of gravity-controlled diapirism and (supposed) gravity-induced doming of a basement gneiss. 'Gravitational tectonics' (Hills, *Structural Geology*, 1963). 'Gravity tectonics' (Dennis, *Tectonic Dictionary*, 1967). *Gravity and Tectonics* (De Jong and Scholten, eds., 1973).

**Gravity-collapse structure.** A structure, in stratified rocks, produced on the limb of a simple fold as a result of collapse under the force of gravity. 'We were at length forced to explain some of the structures we encountered [in Persia] as being produced by collapse of great sheets of limestone after removal of their original supporting cover accompanied by unsticking from their basement. Gravity alone provided the motive force' (Harrison and Falcon, *GM*, 1934). 'Gravity collapse structures' (Harrison and Falcon, *QJ*, 1936).

**Great Oolite.** See GREAT OOLITE SERIES. (Smith, *MS*, 1797/9, 'Upper bed of freestone'; Smith in Warner, *Guide through Bath*, 1811, 'Great Oolite'.)

**Great Oolite series.** The British representative, together with the Lower Cornbrash, of the general Bathonian series (upper part of the Middle Jurassic). It includes the following members (where most fully and typically developed), with characteristic fossils (lowest first): Fuller's Earth beds (*Ostrea acuminata*, lamellibranch), Stonesfield Slate (*Trigonia impressa*, lamellibranch), Great Oolite (*'Terebratula' maxillata*, brachiopod, *Capulus tessoni*, gastropod), Bradford Clay (*Apiocrinus parkinsoni*, crinoid, *'Rhynchonella' boueti*, brachiopod), and Forest Marble (*'Pecten' vagans*, lamellibranch).

**Green mud.** A terrigenous mud owing its colour to the mineral glauconite.

**Green schist.** See CHLORITE.

**Greensand.** A sandstone containing the green mineral glauconite; usually weathering to an orange or yellow colour, as the mineral is iron-bearing. In stratigraphy, either or both of the Greensands (Lower and Upper) of the Cretaceous system. Also the Cambridge Greensand, of Lower Chalk age.

**Greenstone. 1.** An old term for all those varieties of dark, greenish igneous rocks which, on a finer discrimination, and in more modern terminology, would mostly resolve themselves into diorites, dolerites (particularly), and somewhat altered basalts. 'The prevailing variety of the trap-dyke near Hereford is a highly crystalline greenstone, made up of hornblende, olivine, and felspar' (Murchison, *Silurian System*, 1839). 'Through the whole length of the upland valley that lies between Cader Idris and Mynydd-gader, the ashes are penetrated by numerous injections of greenstone' (Ramsay, *MGS*, *N. Wales*, 1866). The term is still occasionally used: 'The Cornish greenstones' (Hawkes, *PGS*, 1969). **2.** 'Freshly quarried stone containing "quarry water" (Watson, *Building Stones*, 1911)' (Arkell and Tomkeieff, *Rock Terms*, 1953).

**Greisen.** A light-coloured rock composed of quartz and minerals rich in fluorine and lithium (particularly a variety of white mica, topaz, and fluorspar), produced by the pneumatolysis of granite by fluids containing those elements. Hence 'greisening', 'greisenization'. [German, *greis*, hoary.]

**Greywacke.** Anglicized from the German *grauwacke* (grey rock) and introduced by Jameson (1805/8) from the writings of Werner. In its wider sense, it included practically all the 'transition' (Upper Precambrian and Lower Palaeozoic) sedimentary rocks other than limestone ('Transition limestone'), that is, the hard conglomerates, sandstones, and siltstones ('greywacke proper') and the hard shaly and slaty mudstones ('greywacke slate'). Thus even in the narrower ('proper') sense the term was still wide and vague. 'Some will find greywacke in one of these divisions, some in another and some in all; while others ridicule the name as one invented to supply the defect of a better' (Otley, *English Lakes*, 1825).

In stratigraphy, the name makes its first appearance in the British tables in that of Buckland in Phillips's *England and Wales*, 1818. Here we have 'Porphyry Greenstone', 'Greywacke Slate', 'Greywacke', and 'Limestone' (from below upwards) as constituting the 'Greywacke formation' or 'Transition Rocks'. Such terms as 'Greywacke series', 'Greywacke group', continued to be so used by De la Beche in his various works published between 1830 and 1839, and by Lyell who, in the 1834 edition of the *Principles*, included the rocks from the 'Longmynd' to the 'Ludlow' (names which had just been given by Murchison). But Murchison later remarks (*Silurian System*, 1839): 'It has already been amply shown that this word should cease to be used in geological nomenclature, and I shall in the following pages give further proofs that it is mineralogically valueless.'

Since those early times, and particularly during the last thirty years (the term is popular at present), various characters have been assigned to the rock-type 'greywacke' (of any age); characters of colour, hardness, mineralogical make-up, coarseness, sorting, angularity of grain, and bedding. 'The graded, bedded sandstones which Bailey (*GM*, 1930) has suggested should be named "greywacke" are essentially deposits of the geosyncline and are characteristically found near one or other of its margins. Such sandstones are so frequently associated with grey or greenish mudstones of similar constitution that they might be referred together to a greywacke suite' (Jones, *QJ*, 1938). See quotation (Murchison, 1839) under SILURIAN SYSTEM. 'The term graywacke' (Boswell and Pettijohn, *JSP*, 1960). 'The greywacke problem' (Cummins, *LMGJ*, 1962). Also spelt 'graywacke'.

**Greywether.** See SARSEN.

**Grike.** See CLINT. [N. Country dialect.]

**Grit.** A wide, vague term for any sedimentary rock that looks or feels gritty. Typical grits are the coarse rough sandstones of the Millstone Grit (Carboniferous). Other grits occur among the Lower Palaeozoic 'greywackes' and are rocks that are on the whole fine-grained but contain larger, somewhat angular fragments (such as those in the Aberystwyth Grits formation). Morton (*Northamptonshire*, 1712) refers to oolite as 'spherical grit', and we have the Trigonia Grit (Jurassic). 'Stone, soft, and of a pretty small gritt [i.e. fine grained]. . . . A gritty stone, of a reddish brown colour' (Woodward, *Catalogue*, 1729). 'The common grit, or sandstone . . .' (Playfair, *Illustrations*, 1802). 'When sandstone is coarse-grained, it is usually called "grit"' (Lyell, *Elements*, 1838). Sometimes the term is used for any hard sandstone bed to distinguish it, by a conveniently short word, from other kinds of beds in a sequence. 'Gritstone' is synonymous with 'grit', as a lithological term, but 'grit' is the only form in use in a stratigraphical sense.

**Groove-cast.** A sole-mark; the infilling of a long straight furrow. (Shrock, *Sequence in Layered Rocks*, 1948.)

**Ground-ice.** Analogous to 'groundwater'; ice enclosed in the frozen ground (permafrost). (Also has the meaning: ice formed on the bed of a river, lake, or shallow sea.)

**Groundmass.** Used most frequently as a term in igneous petrography; see PHENOCRYST. What might be called 'groundmass' in sedimentary petrography (in a conglomerate, for instance) is usually called 'matrix'.

**Ground-water.** Water contained within the soil, subsoil, and underlying rocks. It is usually implied that subterranean streams (and, perhaps, vadose water) are excluded; these are, however, included in the more general expression 'underground water'. 'The thirsty ground' (Seneca, A.D. 3-65, quoted by Holmes). 'Groundwater' (*JIWE*, 1963).

**Group.** An ordinary term for collections or assemblages of any kind. In stratigraphy it has been restricted and specialized to indicate a particular grade in the lithostratigraphical hierarchy (see LITHOSTRATIGRAPHY): 'A group consists of two or more adjacent and naturally related or associated formations. It consists wholly of units defined as formations. Formations need not necessarily be gathered into groups' (*Stratigraphic Guide*, *GSSR* (11), 1978).

**Growth-in-place theory.** As to the origin of coal. See DRIFT THEORY (2) and UNDERCLAY. 'As regards the natural history of coal, I accept, on the evidence of rootbeds, that most of our British coal seams must be attributed to vegetation that grew on the sites now occupied by the coal itself' (Bailey, *JRSA*, 1944).

**Guano.** A phosphatic deposit, the excrement of sea-fowl. (A valuable agricultural fertilizer.) 'The principal deposits of guano are formed on rocky islands frequented by sea-birds, the greatest accumulations being found in the dry trade-wind belts, as in the West Indies, and in the islands of the eastern Pacific Ocean' (Hatch and Rastall, *Sedimentary Rocks*, 1965). [Spanish, *huanu*.]

**Guide fossil.** A species (or genus) of fossil useful as a guide to stratigraphical horizon 'In the chronologic correlation of the stratified rocks most dependence is put upon a few species, known as "guide fossils", together with the collateral evidence of associated forms' (Schuchert and Barrell, *AJS*, 1914). '*Tylonautilus nodiferus:* a Carboniferous guide fossil' (Pringle and Jackson, *Nst*, 1924).

**Guilielmite.** A shining discoidal or conical body, a centimetre or two across, found chiefly in coal-measure shales. At first these 'guilielmites' were thought to be plant-remains and were given the generic name *Guilielmites*, but they are now thought to be slip structures formed round a collapsing shell or fragment of shell debris. (Wood, *GM*, 1935, *JSP*, 1965.)

**Gull.** 'The word "gull" is an old term meaning chasm or gully, and it has long been applied by quarrymen to widened fissures with or without an infilling' (Hollingworth and others, *QJ*, 1944). 'This above described, and all others the like wide and deep perpendicular fissures, the quarriers here call gulls, gullies, or gulfe-joints' (Morton, *Northamptonshire*, 1712). See SUPERFICIAL STRUCTURE.

**Gutenberg discontinuity.** The discontinuity between the earth's core and mantle. Other names associated with this discontinuity are Oldham and Weichert. See CORE (3). [B. Gutenberg, American geophysicist, 1889–1960, who discovered it in 1914.]

**Guyot.** See SEAMOUNT. (Also called 'table-mount'.) (Hess, *AJS*, 1946; Hamilton, *MGSA*, 1956.) [Arnold Guyot, 19th century French geographer.]

**Gymnospermae.** One of the main divisions of the plant kingdom including the Coniferales, Ginkgoales, Cycadales, and the extinct Bennettitales (Mesozoic), and the Cordaitales (Carboniferous and Permian). [G. *gumnos*, naked, *sperma*, seed.]

**Gypsum.** A white, rather soft mineral, hydrated sulphate of calcium, $CaSO_4.2H_2O$. It crystallizes in the monoclinic system (selenite is the fully crystallized form) and occurs chiefly as an evaporite. Alabaster is a very fine-grained and compact, massive variety. Anhydrite is calcium sulphate without the water of crystallization, also occurring as a salt deposit. 'The gypsum bed upwards of 20ft. thick with a middle band of anhydrite [in New Red Sandstone]' (Trotter and Hollingworth, *MGS, Brampton*, 1932). [G. *gupsos*, chalk or burnt gypsum, plaster.]

# H

**Habit.** A special usage is that in crystallography, to describe the shape of a crystal depending on the variation in the degree of development of its several crystal faces.

**Hackly.** Rough or jagged; for this type of fracture in rocks and minerals.

**Hade.** The angle between a fault plane, or the plane of a mineral vein, and the vertical; thus the complement of the angle of dip. 'Hade. This is the deflection or deviation of the fissure from its perpendicular line, as it is followed in depth like the slope of the roof of a house' (Pryce, *Mineralogia Cornubiensis*, 1778). [a miners' term.]

**Haematite.** A grey, reddish-black, or iron-black mineral, iron oxide, $Fe_2O_3$, with a metallic lustre and red streak. Commonly forms fibrous aggregates of crystals, often kidney-shaped (kidney ore). Usually occurs in pockets, replacing limestone. 'The name haematite is as old as Theophrastus and Pliny, and denotes the blood-red colour of some of the ordinary massive varieties' (Miers, *Mineralogy*, 1929). [G. *haima, haimat-*, blood.]

**Half-life period.** A measure of the rate of transmutation of a radioactive substance. It is the (constant) time which must elapse, starting at any instant, for the radioactivity to decay to half the value it has at that instant. Theoretically, therefore, the activity, while decreasing all the time, never falls quite to zero. See RADIOACTIVITY AGE-METHOD.

**Halite.** See ROCK-SALT. [G. *hals, halo-*, salt.]

**Halokinesis.** The movement of a saltmass. 'The Mesozoic strata [in the North Sea area] are strongly folded as a result of halokinesis within evaporites of the English Zechstein Basin' (Dingle, *JGS*, 1971). [G. *kinēsis*, movement.]

**Hammer.** The geologist's hammer, with its firmly fixed head of hardened steel, is his chief instrument and his recognized symbol. Hooke (*Micrographia*, 1665) remarks, as to the internal moulds of fossils and the matrix, that 'they may be easily separated by a knock or two of a hammer'. See also quotation under AGGLUTINATE.

**Hand specimen.** A specimen of rock of a size convenient for holding in the hand

and for showing those characters that are (1) not so small or of such a nature that they can only be seen under the microscope and/or in thin section, and (2) not so large that they can only be seen in a mass of the rock at least some feet across. 'Structure to be observed . . . in hand specimens' (Jameson, *Geognosy*, 1808). 'Hand-specimens of slate' (Sedgwick, *TGS*, 1835).

**Hanging valley.** A valley that hangs, as it were, on the side of a steep slope. Inland, a side valley, particularly a small one, may emerge some way up the steep slope of the main valley side. Such a valley will be commonly caused by the vigorous downcutting of a main river with which the downcutting of a relatively small side tributary cannot keep pace; the tributary ('hanging stream') emerging from its own valley and cascading down the sides of the main valley. Glacial erosion tends to produce striking hanging valleys with special features, the side valleys being hollowed out into the bowl-like form peculiar to glacial action and the descent to the main, U-shaped, valley being sudden and steep. Hanging valleys on a small scale are common features along a coast where cliffs are being rapidly worn back. (Davis, *PBSNH*, 1900, from *AGI Glossary*, 1957.)

**Hanging wall.** The upper side of a fault or vein, opposite to the foot wall. 'Thus the fissure, in which the ore lies, is all the way environed and bounded by two walls of stone, which are generally parallel to each other, and include the breadth of the vein or lode; so that when the miners dig down or along in a large lode, then the roof, i.e. the upper, the hanging wall, or incumbent wall of the lode or fissure, is (in a certain proportion according to its inclination or underlie) over their heads; and the lower, or other wall or rind, is under their feet' (Pryce, *Mineralogia Cornubiensis*, 1778).

**Haphazard folding.** 'The extensive exposures of the Aberystwyth Grits along the shore are characterized by closed, haphazard folds well seen in plan from the cliffs above, in which whale-back anticlines and their counterpart synclines, with wavelengths and amplitudes up to 10m, rarely exceed 200m in length. Their axial traces are often so curved that it is difficult to suggest a common regional compression' (Davies and Cave, *SG*, 1976). This haphazardness is important evidence concerning the question as to whether the folding was penecontemporaneous with deposition or due to subsequent tectonic forces.

**Hard pan.** See PAN.

**Hardground.** A part of the sea-floor composed of indurated sediment, usually on the continental shelf and often perforated by burrowing organisms. Currents sweep the hardground clear and prevent the accumulation of fresh sediment. In the geological record hardgrounds are defined as 'horizons of early-diagenetic cementation beneath intraformational erosion surfaces' (Bromley, 'Burrows of thalassinidian Crustacea in chalk hardgrounds', *QJ*, 1967).

**Hardness. 1.** 'The hardness of a mineral is measured by the resistance which a smooth surface offers to abrasion. The degree of hardness is determined by observing the comparative ease or difficulty with which one mineral is scratched by another, or by a file or knife. In minerals there are all degrees of hardness, from that of talc, impressible by the finger nail, to that of diamond. To give precision to the use of this character, a scale of hardness was introduced by Mohs. It is as follows:—1. Talc, 2. Gypsum, 3. Calcite, 4. Fluorite, 5. Apatite, 6. Orthoclase, 7. Quartz, 8. Topaz, 9. Corundum, 10. Diamond' (Dana and Ford, *Mineralogy*, 1932). **2.** The hardness of a rock may be said to depend on the hardnesses of the component mineral materials and the degree of their cohesion, but there are various ways in which 'hardness' may be considered as being expressed. The following are two authoritative statements. 'The hardness or softness of a rock—in other words, its induration, friability, or the degree of aggregation of its particles' (Geikie, *Textbook*, 1882/1903). '"Hardness" and "softness" are terms indicating the resistance of rocks to weathering' (de Sitter, *Structural Geology*, 1964).

**Harlech Grits.** A name for the almost unfossiliferous series of sandstones and shales forming the central part of the Harlech Dome in N. Wales and underlying the Menevian Shales of upper Middle Cambrian age. They extend downwards visibly for about 1,500 m., the base being concealed at an unknown depth. It is not known at what horizon within them the division between Lower and Middle Cambrian should be drawn. (Sedgwick, *QJ*, 1852, *British Palaeozoic Rocks*, 1855.)

**Hartfell Shales.** The typical development, though in a very condensed form, of the graptolitic facies of the Caradocian and Ashgillian series, occurring in the Southern Uplands of Scotland. (The lower two zones of the Caradocian are placed in the Glenkiln Shales.) The whole formation is only about 30 m. thick. (Lapworth, *QJ*, 1878.) [Hartfell, Dumfriesshire.]

**Hastings Sands.** See WEALDEN. Clays alternate with the sands. (Fitton, *AP*, 1824.) [Hastings, East Sussex.]

**Hawaiian type.** See VOLCANICITY. 'In what we may conveniently call the "Hawaiian type", represented by the giant volcanoes of the Sandwich Isles, the lavas, of basaltic composition, are exceptionally fluid. The violently explosive element is lacking in the eruptions, and there is consequently an absence of fragmental products. The volcanic mountain itself, built up entirely by successive outflows of very fluid lava, presents very gently inclined slopes, and, as illustrated by Kilauea and Mauna Loa may attain enormous dimensions' (Harker, *Igneous Rocks*, 1909). Such volcanoes have been called 'shield volcanoes' (described by Dana, *Characteristics of Volcanoes*, 1890, who called them 'basalt-volcanoes').

**Head.** A Pleistocene deposit. 'The head, which De la Beche early described (*Cornwall* [&c.], 1839) is an accumulation of local rocks, sharp and angular, which is coarse close to the hills but farther away becomes smaller and mixed with fine material. It spreads irregularly over low ground, sheeting the sides and floors of valleys, as in Pembroke and Cornwall, and in the west Midlands and those parts of west Yorkshire which were ice-free during the later stages. Its subaerial origin under cold solifluxion conditions was early recognized and is proved by the angularity of its constituents and by distortions in the underlying strata' (Charlesworth, *Quaternary Era*, 1957). In a wide sense, 'head' includes 'coombe rock'. (Dines and others, *GM*, 1940.) The origin of the term is obscure.

**Heave.** Now restricted to a horizontal measurement in a faulted structure, analogous to the vertical measurement, 'throw', but the same vagueness and variety of meaning surround each term. The verb 'heave' was formerly used in a general sense, in connexion with faulting, as equivalent to 'displace': 'Not only the strata are shifted, but veins, which intersect one another, are also shifted themselves . . . sometimes in a horizontal, sometimes in a oblique direction. They are "heaved", as it is called in the significant language of the miners, and forced out of their direction' (Playfair, *Illustrations*, 1802).

**Heavy mineral.** See ACCESSORY. As a matter of practice, 'heavy minerals' are those with a specific gravity greater than that of bromoform (2.85). 'The heavy minerals of the plutonic rocks of the Channel Islands' (Groves, *GM*, 1927). 'The heavy minerals of the Keele, Enville, "Permian" and Lower Triassic rocks of the Midlands' (Fleet, *PGA*, 1927).

**Height of the land.** The mean height above sea-level at any time over any area, large or small, is the net resultant of the earth-movement and the denudation over that area from the earliest period up to that time.

**Hemera** (pl. **-ai, -as,** or **-ae**). Hemerai are small successive geological time units based on the conception of the successive rise and fall of particular organic species, each hemera being called by the name of the species of fossil on which it is based. The total, world-wide ranges in time of the commoner species would hardly allow successive time-intervals to be defined in this way; even if there were no overlapping of ranges or gaps between them, a hemera could be recognized in the strata (by its epibole) only where the species happened to live and become fossilized. Nor is it likely that the period of acme of a species in one region would be the same as its period of acme in another. Thus the hemera is an abstract conception based on an unsatisfactory criterion, but the examination of the validity of the conception is not unimportant. Hemerai, derived from the occurrence of particular ammonites, have been much used in connexion with Jurassic chronology, but hardly, if at all, outside that system or with other groups. 'The term "hemera" is intended to mark the acme of development of one or more species. It is designed as a chronological division and will not replace the term "zone" or be a subdivision of it, for that term is strictly a stratigraphical one' (Buckman, *QJ*, 1893; also *GM*, 1902). [G. *hemera*, day. It seems that the plural should be either the Greek -ai or the English -as; the commonly used -ae, used by Buckman himself, might possibly be justified as being modern Latin derived from the Greek.]

**Hemicrystalline.** 'A term applied to igneous rocks to denote that they consist partly of crystals and partly of glass or devitrified glass' (Holmes, *Nomenclature of Petrology*, 1928).

**Hercynian.** Originally introduced by von Buch in Germany for 'a very important set of more or less north-westerly fractures and parallel folds that characterize much of Czechoslovakia, Germany, and Scania. The dates of these fractures are various: Cretaceous rocks are often affected and sometimes Tertiary. Bertrand reintroduced the term in 1887. For Bertrand and his heirs Hercynian, as applied to tectonics, means Late Palaeozoic, in any part of the world, and carries no general directional significance. Among German-speaking tectonists Hercynian is still commonly employed in von Buch's sense, but Hercynian in Bertrand's sense has established itself apparently for all time in the literature of France, Switzerland, Britain and America' (Bailey, *Tectonic Essays*, 1935; quotations rearranged). In Bertrand's sense the term thus includes Suess's Armorican and Variscan orogenies. Another review of the history of the name is given by Evans (*QJ*, 1926). See MOUNTAIN SYSTEM.

**Hercynides.** The fold-structure mountains produced by the Hercynian orogeny; in a more comprehensive sense, the whole Hercynian orogenic belt. When 'Hercynian' is taken to mean the same as 'Variscan', the 'Hercynides' are the same as the 'Variscides'.

**Heteromorphism.** Applied to the phenomenon whereby a magma having and keeping a particular chemical composition may crystallize into several different mineral assemblages as a result of different cooling histories. 'Although a certain mineral assemblage uniquely determines the chemical composition of a rock, the opposite is not true; the mineral assemblage is not a one-valued function of the chemical composition (heteromorphism of rocks)' (Barth, *Theoretical Petrology*, 1962). (Lacroix, 1917.)

**Hettangian.** See JURASSIC SYSTEM. [Hettange, Lorraine.]

**Hexacoralla.** See SCLERACTINIA. [G. *hex*, six, *korallion*, coral.]

**Hierarchy.** In classification, the system of ranks of the several taxonomic categories. Chiefly in connexion with the organic world. Fossils are classified in the same way as, and together with, the animals and plants of today. The taxonomic categories (taxa) form the following hierarchy, from above downwards:

Kingdom (Regnum)
Phylum (phylum) for animals, Division (Divisio) for plants
Class (Classis)
Order (Ordo)
Family (Familia)
Genus (Genus)
Species (Species).

Intermediate categories may be intercalated, (*a*) for animals: by using the prefixes Sub and Super and, occasionally, by placing Cohort between Class and Order, and Tribe between Family and Genus, (b) for plants: by using the prefix Sub and, occasionally, by placing Tribe between Family and Genus and Section between Genus and Species. There are also hierarchies in stratigraphy; see CHRONOSTRATIGRAPHY and LITHOSTRATIGRAPHY.

**High-level shelly drift.** Rare but significant marine-shell-bearing glacial gravels and sands at rather high levels (about 300 metres, or more), presumably dredged up by ice moving through the sea and depositing the material after it has moved onwards and upwards over the neighbouring land. The two well-known occurrences in Britain are to the S. and SE. of the Irish Sea: at Moel Tryfan, Caernarvonshire (Trimmer, *PGS*, 1831; Jeffreys, *QJ*, 1880) and on the hills E. of Macclesfield, Cheshire ('Prestwich's Patch': Darbishire, *GM*, 1865; Sainter, *Rambles round Macclesfield*, 1878; Pocock, *MGS*, *Macclesfield*, 1906).

**Hill-top surface.** An imaginary surface, as smooth as possible, touching the tops of the hills of a region. There would probably be a number of such surfaces imaginable for one region, according to whether only the highest or many or all the hills were touched. It might be geomorphologically significant if the highest hills all reached about the same summit level or if their hill-top surface formed an even slope or curve, but if a particular hill-top surface is to be interpreted as representing an uplifted peneplain (in the widest sense), it has to be borne in mind that any such peneplain will doubtless, in most cases, have been considerably higher and have been itself completely 'lost' by the erosion of the hill-tops themselves. 'Summit-level', 'summit-plain', 'accordant summits', 'accordant summit level(s)', are terms sometimes used for a flat hill-top surface. 'The hill-top surface of North Cardiganshire' (Challinor, *G*, 1930). The term also has its literal meaning of a real surface; see quotation under STACK.

**Hilt's law.** 'Hilt, after working on the coals of the Pas de Calais field, came to the conclusion that in a vertical succession at any point in the coal field the rank of the coals [carbon content] increased with the depth. This observation has come to be known as Hilt's law' (Raistrick and Marshall, *Coal and Coal Seams*, 1939). 'Hilt's Law, according to which, in any particular vertical succession encountered in mining, deeper seams usually retain less of their original volatiles than shallower seams' (Bailey, *JRSA*, 1944). Also called 'Hilt's rule'. (C. Hilt, Liège, 1873).

**Hinge.** There is usually a line, or narrow strip, along a folded bedding surface, where the curvature is at a maximum This line, or the median line of the strip, is given various names; of these 'hinge' (or 'hinge-line') seems to be the best.

**Hinge fault.** A fault in which the blocks are hinged about an axis at right angles to the fault line, the throw increasing away from the hinge.

**Hinterland. 1.** The district behind the coast. **2.** In orogeny, the moving block which compresses the sediments of a geosyncline and forces them towards the foreland. (Also called 'backland'.) [German *hinter*, behind.]

**Histogram.** A diagram in which columns represent frequencies (actual or proportional) of ranges of value of a quantity, such as grain-size in a sedimentary rock (e.g. Greensmith, *Petrology of the Sedimentary Rocks*, 1978). [G. *histos*, mast.]

**Historical geology.** The study of geological history, as a whole or in its various parts and aspects. *Outline of Historical Geology* (Wells and Kirkaldy, 1937/66). See EARTH HISTORY.

**History of Geology.** 'History of the progress of geology' (chapters 2, 3, and 4 of Lyell's *Principles of Geology*, 1830/75; practically unchanged in all editions). 'Notes on the history of English geology' (Fitton, *PM*, in four parts, 1832/3; his anonymous article in the *ER*, 1818, expanded to twice its length but carrying the story no further in time). 'History of geology' (book 18 of Whewell's *History of the Inductive Sciences*, 1837/57; a philosphical review based on the accounts of Conybeare, Lyell, and Fitton). *Passages in the History of Geology* (two lectures by Ramsay, 1848/9). *History of Geology and Palaeontology* (translation and abridgement by Maria Ogilvie-Gordon, 1901, of Zittel's *Geschichte*, 1899). *History of Geology* (Woodward, 1911; a masterly short book). *Geologists and the History of Geology* (Sarjeant, 1980; a five-volume bibliography to 1978). Also (not having 'history' in the title): Conybeare's Introduction to his and Phillips's *Outlines of the Geology of England and Wales*, 1822; Geikie's *Founders of Geology*, 1897/1905, one of the 'classics' of geological literature; Adams's *Birth and Development of the Geological Sciences*, 1938, the standard work on the history of geology down to the early part of the 19th century; Mather and Mason's *Source Book in Geology*, 1939, and Mather's additional volume (1967) covering the period 1900-1950, contain valuable selections of original writings; Bromehead's 'Geology in embryo', *PGA*, 1945, a detailed review, with many references, to A.D. 1600; Gillispie's *Genesis and Geology* (1951 and 1959), on the impact of geology during the decades before Darwin (1790–1850). Loewinson-Lessing's *Historical Survey of Petrology* (trans. Tomkeieff, 1954) is particularly notable among the more specialized histories; and Parkinson's discussion of views on the nature of fossils in his *Organic Remains of a Former World* (vol. i, 1804) is probably the first historical résumé, in Britain at least, of any branch of geology. 'Reference books for history of geology' (White, *S-HBN*, 1963). 'Milestones in the history of geology' (LaRocque, ed., *JGE*, 1974). Much history of geology is contained in the biographies of geologists; Wells and White have compiled a useful list (*OJS*, 1958) of those in book form. There are also histories of the progress of geology in particular regions and histories of geological institutions. 'The history of geology: suggestions for further research' (Eyles, *HS*, 1966).

**Hogback.** (Hog's back.) A narrow-crested symmetrical ridge, particularly an escarpment (cuesta), formed in a relatively hard stratum, in which the dip-slope and scarp-slope are equal (or nearly so). The Hog's Back in Surrey is a gentle-sloped example; steep-sloped examples are those in the Dakota Sandstone (Cretaceous) in Colorado. Similarly shaped ridges may occur at the coast where marine erosion is working into land sloping landwards. 'Little Hangman, Combe Martin, Devon. Typical "hog's-back" cliffs in Hangman Grits' (pl. in Steers, *Coastline of England and Wales*, 1946).

**Holkerian.** See DINANTIAN. [Holker Hall, Levens estuary, Cumbria.]

**Holocene.** See RECENT. 'The term Holocene has officially replaced Recent in nomenclature usage of the U.S. Geological Survey. Holocene neans 'wholly recent' and refers to the percentage of living organisms. Both vertebrate and invertebrate faunas reflect marked changes between Pleistocene [in the restricted sense] and Holocene epochs and the archaeological record provides means for subdividing Holocene deposits' (*Geot*, 1968).

**Holocrystalline.** Wholly crystalline; almost exclusively for igneous rocks that are entirely made up of crystals, not being in any part glassy.

**Holohyaline.** See HYALINE.

**Holotype.** See TYPE (3).

**Hominoidea.** See MAN and PRIMATES. 'New palaeontological evidence bearing on the evolution of the Hominoidea' (Le Gros Clark, *QJ*, 1949).

**Homocline.** 'The term "homocline" (introduced by R. A. Daly in 1916) replaces the ambiguous "monocline" where it is used to signify a succession of strata dipping continuously in one direction, and has been very widely adopted' (Cotton, *Landscape*, 1948). The term is used in other senses; but this sense seems to be the most satisfactory. Hence 'homoclinal'.

**Homoeomorphy.** 'The occurrence of similarities of form in organisms of different ancestry' (George, *CGA*, 1958, *PGA*, 1962). '"Homoeomorphy" is one of a family of terms, erupted during the burst of theorising in the decades after the Darwinian revolution of 1859, that denote in one way or another the occurrence and the significance of structural similarities and differences between organisms. It was first introduced by Buckman (*QJ*, 1895) in a manner, as "homoeomorphous" and "homoeomorph", that suggests he was informally coining or adapting a word merely in descriptive allusion to fossils whose similarities of form were deceptive, and whose differences were revealed on closer inspection . . . When Buckman ("Homoeomorphy among Jurassic brachiopods", *PCNFC*, 1901) gave some formality to the term his definition again was simply descriptive, homoeomorphy being "the phenomenon of species nearly alike so far as superficial appearance is concerned, but unlike when particular structural details are examined . . . the phenomenon of similarity in general with dissimilarity in details". . . . After 1859 theory was adequate to define "affinity" on a historical basis; and homoeomorphy became recognisable when likenesses were revealed as independent in evolutionary origin. . . . The homoeomorphy emerges only when the morphologist evaluates the conflicting similarities and imposes his evolutionary scheme on the passive fossils. . . . Combining its descriptive with its morphogenetic elements, Lang ("Homoeomorphy in fossil corals", *PGA*, 1917) summarised it as "similarity of form in organisms of different ancestry"' (George, 'The concept of homoeomorphy', *PGA*, 1962).

**Homologous.** Corresponding. Implies more than correlation. In stratigraphy, though the term does not seem to be in use, strata in separated regions would (presumably) be said to be homologous if they were not only contemporary but were of the same general lithological and faunal facies. 'The Morar and Loch Hourn anticlines are strictly homologous in that they share an envelope in common, occupy analogous structural positions along strike, and consist of similar rock types' (Kennedy, *QJ*, 1954). The term has been used for faults, thus the (postulated) Malvern fault and the (postulated) Red Rock fault on the east side of the Cheshire basin have been called 'homologous' (Kent, in discussion, *QJ*, 1955). See quotation (Charlesworth, second) under CIRQUE. [G. *homologos*, agreeing.]

**Homotaxis.** Similarity of serial arrangement; for groups of strata, in different localities, showing similar vertical sequences of faunas. The term was proposed by Huxley (*QJ*, 1862) to distinguish this condition from contemporaneity, which can only be more or less uncertainly inferred: 'The mischief of confounding that "homotaxis" or "similarity of arrangement", which *can* be demonstrated, with "synchrony" or "identity of date", for which there is not a shadow of proof'. See CHRONOTAXIS. Hence 'homotaxial', 'having been deposited during the same relative period in the general progress of life in each region' (Geikie, *Textbook*, 1882). 'Abbey Shales . . . . and the homotaxial Lower Alum Shales of Sweden' (Illing, *QJ*, 1915). [G. *taxis*, an arranging.]

**Honeycomb weathering.** Weathering which produces the superficial appearance of an enlarged honeycomb. 'Many a sandstone, well weathered, as it is termed, will exhibit a honeycombed and irregular appearance, arising from the different character of parts of the cementing substance' (De la Beche, *Geological Observer*, 1851). 'A peculiar and very striking phenomenon known as honeycomb weathering takes place even in apparently homogeneous rocks, the surface being weathered into innumerable small pits averaging about one inch in diameter and of the same depth' (King, *S. African Scenery*, 1951). 'The Sand Rock [Tunbridge Wells Sand] shows excellent examples of honeycomb weathering. The customary discussions on the origin of this weathering took place' (Sweeting, *PGA*, 1945).

**Horizon.** 'Geol:—A plane of stratification assumed to have been once horizontal and continuous' (*OED*). This is the only observable feature to which the term can be given; but when we are referring to a wide area, we have very much in mind the idea of contemporaneity, and an horizon then may mean an imaginary isochronal surface (a 'time-plane') within the strata. Though strictly applicable to a surface only, it is often used for a thin bed with some distinct character as regards its fossils. 'The coal-beds of oolitic series in Yorkshire have been long known as occurring on a "geological horizon", to adopt the term of

Humboldt, with limestones, and clays, replete with marine organic remains, on the south of England' (De la Beche, *Geological Observer*, 1851). 'Faunal horizons in the Bristol Coalfield' (Bolton, *QJ*, 1911). See quotation (Hudson) under FAUNA (1). The use of the term in pedology is shown in the quotation under SOIL PROFILE.

**Horizontal surface.** At any point, a plane surface tangent to the generalized earth-surface, the geoid, considered as drawn through that point. Horizontality thus varies from point to point and horizontal surfaces, (topographical, structural), if integrated, form a curved surface. (A vertical surface is one at right angles to a horizontal surface at the same point.) See GEOID.

**Hornblende.** See AMPHIBOLE. Occurs as dark horny-looking crystals and is characteristic of many igneous rocks of intermediate and acid composition; it is also very common in metamorphic rocks. 'An old German name for any dark-colored, prismatic crystals found with metallic ores, but containing no valuable metal, blende meaning a deceiver' (Chester, *Names of Minerals*, 1896).

**Hornfels.** A hard, tough, fine-grained rock produced by the thermal metamorphism of an argillaceous or calcareo-argillaceous rock. It is composed of quartz, felspar, mica, and usually one or more of the metamorphic minerals such as andalusite, sillimanite, cordierite. The texture is typically decussate. [German, horn-rock.]

**Hornstone. 1.**=hornfels. **2.**=chert. **3.** A compact flinty variety of chalcedony.

**Horse. 1.** 'Commonly applied by vein miners to any large detached mass of rock occurring in a vein, or lying between two branches of a vein' (Jukes, *Manual*, 1862). 'It frequently happens in very large and also rich lodes, there appears a kind of stone about the middle of the vein, of the same nature of the ground or stratum nigh the lode, being not at all of a veiny quality, though it is in the body of the lode. The Cornish miners call it a "horse"' (Pryce, *Mineralogia Cornubiensis*, 1778). **2.** See WASH-OUT.

**Horst.** 'If the outer borders of two fields of subsidence approach each other so that a ridge is left between them, on both sides of which the two areas of depression descend more or less in the form of steps, then we have what we shall distinguish, making use again of a common mining word, as a "horst", in this case a "horst of the first order", as opposed to the subsidiary horsts which occur here and there between networks of fracture. As horsts of the first order we may mention for example the Schwarzwald, the Vosges, Morvan, and the Kaibab Plateau in Colorado' (Suess, *Das Antlitz der Erde*, 1885, English trans., 1904). The term has, however, acquired various connotations, as may be illustrated by the following quotations. 'Certain portions of the crust have acquired more or less immobility, and have served as buttresses against which surrounding areas have been pressed and dislocated by subsequent movements. Suess has pointed out various areas of the earth's surface, named by him "horsts", which seem to have served this purpose in the general rupture and subsidence of the terrestrial crust' (Geikie, *Textbook*, 1903). 'A study of differential movements has shown that certain well-defined blocks tend to stand up as hard and immovable masses, while the areas around them sink: these fixed masses are now generally designated by the German term *Horst*, for which there is no satisfactory equivalent in English (the literal meaning of the word is eagle's nest or eyrie). The Highlands of Scotland, north of the Grampian fault, and the Southern Uplands are good examples of horsts, while between them occurs the depressed area of the Central Valley' (Rastall, *Textbook*, 1941). 'Differential elevation of a block relative to its surroundings may produce a "plateau" or a "block-mountain". If denudation has proceeded far, and if the rocks brought into contact with one another by the fault are more or less equally resistant to erosion, the boundary of the block may be marked by no topographical feature. Where erosion has laid bare in the elevated block far older and harder rocks than those forming the surface in the adjacent depressed areas, the block will on account of its hardness form hills or mountains projecting above their surroundings. It is then termed a "horst"' (Wills, *Physiographical Evolution of Britain*, 1929). 'Regions which are divided by faults into a number of differentially elevated or depressed blocks are said to exhibit "block faulting". If such a region has not been subjected to prolonged erosion, the movements of the blocks will be directly reflected in the topography. Upstanding blocks, either plateaux or ridges, are termed "horsts"' (Hills, *Structural Geology*, 1953). For further discussion see Stamp, ed., *Glossary of Geographical Terms*, 1961. See STABLE BLOCK.

**Host rock.** A rock-body in which are lodged other, smaller, and not indigenous bodies; such as a pluton containing xenoliths, or any rock enclosing an epigenetic mineral deposit.

**Hot spot.** See MANTLE PLUMES.

**Hot spring.** 'Ground-water that has circulated to great depths in deeply folded rocks becomes heated, and if a sufficiently rapid ascent to the surface should be locally possible, it will emerge as a warm spring. Such conditions are rare, however, and really hot springs generally occur in regions of active or geologically recent vulcanism, where they owe their high temperature to superheated steam and emanations which rise from subterranean sources and mingle with the meteoric circulation at higher levels. There are three volcanic regions where hot springs and geysers occur on an imposing scale: Iceland, Yellowstone Park, and the North Island of New Zealand' (Holmes, *Physical Geology*, 1944). 'There is only one hot spring in England, at Bath, but that is a remarkable one. The actual temperature of the water is about 120°F. and the yield about 400,000 gallons a day. The water is saline and distinctly radio-active. Since Bath is far removed from any volcanic region, the high temperature is suggestive of a very deep-seated origin, and the same is true of some high-temperature springs in the Worcester district of Cape Colony. Most of the famous hot springs of Europe are in or near regions where vulcanicity has been active in comparatively recent geological times' (Rastall, *Textbook*, 1941).

**Humidhesion.** 'This new term is coined for the fixation of sand by wetness in the aeolian sedimentary environment. Humidhesion results in relative firmness of the sand, mimicking cementation' (Vortisch and Lindstrom, *GM*, 1980).

**Humolith.** A term to include peat, lignite, and coal; but it would exclude the sapropelic coals.

**Humus.** A vegetable mould. 'H. Potonie recognised as long ago as 1908 that a useful distinction can be drawn between the products of the decomposition of land plants and associated organisms under strongly acid conditions, and those resulting from decay of aquatic plants and the organisms associated with them in neutral or nearly neutral water [sapropel]. For the unconsolidated product of the first process I shall use the term "humus", this usage being consistent with that of Waksman (*Soil Microbiology*, 1952) though not with that of authors who refer to marine humus. Humus is a dark brown to black substance, insoluble in water, representing a natural organic system in dynamic equilibrium. As understood here, it is a terrestrial accumulation, reaching its maximum development in swamp conditions; but it is often eroded and transported to lakes or to the sea. The essential constituents are cellulose, hemicelluloses and the more resistant lignins and proteins. . . . Starting from humic constituents, slow progressive elimination of oxygen and hydrogen gives rise to the "coal series"' (Dunham, *AS*, 1961). [L.]

**Huronian.** See PRECAMBRIAN and PROTEROZOIC. [Lake Huron.]

**Huttonian theory.** The theory expounded by the Scotsman, James Hutton (1726-1797), published in his 'Theory of the Earth; or an investigation of the laws observable in the composition, dissolution and restoration of land upon the Globe' (*TRSE*, 1788; read 1785), reprinted with much additional material in his two-volume book *Theory of the Earth, with Proofs and Illustrations* (1795; part of a third volume was rescued and edited by Sir Archibald Geikie for the Geological Society of London in 1899). The theory was further clarified and illustrated by John Playfair in *Illustrations of the Huttonian Theory of the Earth* (1802). In it the two main principles of geology were formulated: the geological cycle, by the demonstration in the field of igneous intrusion and stratigraphical unconformity, and uniformitarianism, by the insistence of continuity of process and limitless time. Whereas previous geological knowledge had been a matter of scattered observations, and theorizing had been mostly wild fantasy, here at last was the first coordination and rationalization. The theory was faulty in some respects, particularly in that it ascribed the consolidation of sediments to heat instead of to pressure and mineral cementation. See PLUTONIC (2), PLUTONISM, PLUTONIST, UNIFORMITARIANISM. (Discussions in Geikie, *Founders of Geology*, 1905; Adams, *Birth and Development of the Geological Sciences*, 1938; Commemorative essays by several authors, *PRSE*, 1950; Gillispie, *Genesis and Geology*, 1951; Bailey, *James Hutton: the Founder of Modern Geology*, 1967).

**Hyaline.** Glassy. A wholly glassy igneous rock is 'holohyaline'. [G. *hualos*, glass.]

**Hyaloclastite.** 'Deposits formed from the comminuted glassy shells of growing pillows [of submarine lava] could perhaps be called "hyaloclastic deposits" or simply "hyaloclastites"' (Rittmann, trans. Vincent, *Volcanoes and their Activity*, 1962). 'Fragmental rocks, called "hyaloclastites" by Rittmann, produced during submarine and sub-lacustrine basalt eruptions by the intense chilling and brecciation of lava against water or ice' (Fitch in discussion of Walker and Blake, *QJ*, 1966; quotation re-arranged). 'The pillow lavas and hyaloclastite breccias of King Island, Australia' (Solomon, *QJ*, 1968).

**Hybrid.** Its ordinary use is in biology, thus applicable in palaeontology; but hybrids are hardly to be certainly recognized as such among fossils. In geology, its chief use is figuratively in petrogenesis where it means a rock produced either by the melting together (syntexis) of two different kinds of rock, or the mixing of two magmas of diverse composition, or the reaction between a magma and rocks incorporated and melted in it. 'Hybridism in igneous rocks' (Harker, *Igneous Rocks*, 1909). 'The Pre-Cambrian Malvernian gneisses originally seem to have been, in some cases, "hybrid" rocks such as might result from the mixture of granite and gabbro' (Pocock and Whitehead, *BRG*, *Welsh Borderland*, 1948).

**Hydatogenetic mineral.** See PYROGENETIC MINERAL. The specialized usage is unfortunate as the term, from its etymology, can obviously refer, and has been used to refer, to any mineral formed in, from, or under the influence of water.

**Hydraulic action.** The mechanical action of water 'through pipes'; in geology such action chiefly occurs when sea-waves rush into caves and tunnels and, combined with the consequent compression of the air, cause a loosening of the rocks.

**Hydroclastic volcanic rock.** A fragmental volcanic rock broken during chilling under water or ice. (Fitch and authors in discussion, Walker and Blake, *PGS*, 1965.)

**Hydrogeology.** The geology of underground water. Also called 'geohydrology', 'groundwater hydrology' (but see quotation, Davis and De Wiest, below). 'The hydrogeology of the Lower Greensands of Surrey and Hampshire' (Lucas, *PICE*, 1879/80). 'A hydrogeological study of the permeability of the chalk' (Ineson, *JIWE*, 1962). *Hydrogeology*, 'a companion volume to *Geohydrology* [De Wiest, 1965]' (Davis and De Wiest, 1966): 'In *Hydrogeology* the emphasis has been placed on the geologic aspects of ground water, while in *Geohydrology*, as the name suggests, the emphasis is on the hydrologic or fluid-flow aspects of ground water'. *Dictionary of Hydrogeology* (Pfannkuch, 1969). *Hydrogeological Map of England and Wales* (*IGSc*, 1977). Lamarck's *Hydrogéologie* (1802, trans. Carozzi, *Hydrogeology*, 1964) is an important early treatise dealing with matters concerning a general 'theory of the earth'. (For some historical facts see Buchan, *AdS*, 1968.)

**Hydrological cycle.** 'The endless interchange of water between the sea, air and land. . . . "All the rivers run into the sea, yet the sea is not full; unto the place from whence the rivers come, thither they return again" (*Ecclesiastes*, i, 7)' (Monkhouse, *Dictionary of Geography*, 1965). 'John Keill's view of the hydrologic cycle, 1698' (White, *WRR*, 1968).

**Hydrology.** The science of water: its properties, distribution, and circulation; and its exploitation for the needs of mankind. It deals with water on the surface of the land, in the soil, and in underlying rocks, and with related aspects of water in the atmosphere (evaporation and precipitation). By general usage, oceans and seas are excluded. Hydrology comes into geology as 'hydrogeology', and also as being concerned with the principles of stream-action.

**Hydrosphere.** A collective name for all the waters (and ice) on the face of the earth, and usually including those within the earth's crust; rain and water vapour in the atmosphere are also sometimes included. [G. *hudōr*, water.]

**Hydrothermal.** Of or by emanations rich in hot water (or steam), and the consequent processes and products. These emanations will in most cases be of magmatic origin. 'The kaolin deposits of Cornwall and Devon were formed by the hydrothermal alteration of the granite masses emplaced during or just after the Hercynian orogenic phase' (Bristow, *IGC*, 1968). *Geochemistry of Hydrothermal Ore Deposits* (Barnes, ed., 1969/79).

**Hypabyssal.** Of or at no great depth; applied to shallow-seated intrusive igneous rock-bodies, such as dykes and sills, so formed. 'Plutonic and volcanic modes; a third and intermediate mode, the hypabyssal, is recognised by many' (Tyrrell, *Petrology*, 1929). 'It was Rosenbusch who proposed the terms "intrusive" rocks and "effusive" rocks (1877). To these two classes he added a third, "dyke rocks" (*Ganggesteine*), which did not, however, meet with universal approval. Originally this class embraced only rocks found in the form of dykes but this limited interpretation was later replaced by one which worked on a chemical and genetic basis. In this connotation the term "dyke-rocks" of Rosenbusch was losing its significance and was eventually supplanted by the term "hypabyssal" proposed by Brögger in 1886 in order to remove the ambiguity' (Loewinson-Lessing, trans. Tomkeieff, *Historical Survey of Petrology*, 1954). [G. *abussos*, the bottomless pit.]

**Hypersolvus.** 'A classification of salic rocks (granites, syenites and nepheline syenites) based on the nature of the alkali feldspar is proposed. This has two major divisions: (*a*) subsolvus, and (*b*) hypersolvus, depending on the whereabouts of the soda feldspar. In the hypersolvus rocks all the soda feldspar is or was in solid solution in the potash feldspar (thus forming perthite) whereas in the subsolvus rocks the soda feldspar is present as discrete grains' (Tuttle and Bowen, *MGSA*, 1958, quotation composite and adjusted).

**Hypersthene.** See PYROXENE. Schillerization lustre is characteristic. 'R. J. Haüy, 1806, probably f. G. ὑπερ [*huper*] and σθένοζ [*sthenos*], strong above others indicating that it possesses certain qualities to a very high degree because it is superior in lustre and hardness to amphibole, with which it had been confounded' (Chester, *Names of Minerals*, 1896).

**Hypidiomorphic.** Having its own proper shape partly developed; for a mineral, in an igneous rock, bounded partly by its proper crystal faces and partly accommodating itself to the crystal facies of adjacent minerals. Also called 'subhedral'. The term is also used for an igneous rocktexture in which most of the essential minerals are more or less hypidiomorphic or for one in which some of the minerals are idiomorphic, some hypidiomorphic, and some allotriomorphic. (Most granites, for instance, are hypidiomorphic in one or the other of these senses.)

**Hypocentre.** See SEISMIC FOCUS.

**Hypogene.** Generated or formed beneath the earth's surface. 'Granite and gneiss (the plutonic as well as the altered rocks) belong to one great natural division of mineral masses . . . the class to which they belong should receive some common name . . . we propose the term "hypogene" for this purpose, implying that granite and gneiss are both nether-formed rocks, or rocks which have not assumed their present form and structure at the surface' (Lyell, *Principles*, 1833). 'Hypogene or plutonic action—the changes within the earth caused by original internal heat and by chemical action' (Geikie, *Textbook*, 1882). 'To an increasing degree geological research is becoming concerned with process, including in that term (as Holmes expressed it) "the whole machinery of the earth": "epigene" in the development of land-forms, both erosional and constructional, and "hypogene" in the structural geology of both the observable crust and the sub-crustal material and the operation of sub-crustal processes' (Hollingworth, *QJ*, 1962). (The opposite of epigene.) The term is also used in a special sense for the production or enrichment of ore minerals by aqueous solutions percolating upwards.

**Hypozoic.** At one time used in a rather vague way (implying a general application) for those rocks underlying the fossil-bearing stratigraphical groups; that is, as roughly equivalent to Cryptozoic or Precambrian. (Sedgwick, *QJ*, 1847.)

**Hystrichospheres.** See DINOFLAGELLATES. [G. *hustrix*, the porcupine.]

# I

**Iapetus ocean.** The ocean inferred as having occupied roughly the site of the present North Atlantic during the period from some time probably in the later Precambrian to some time in the later Lower Palaeozoic era. 'The postulated ocean that preceded the Caledonian orogeny appears not to have separated Africa and South America but was restricted to the area between the Eurasian and North American cratons. It had closed by late Silurian-early Devonian time (say 400 m.y.) at the latest, at least 200 m.y. earlier than the opening stages of the central Proto-Atlantic, so that for 200 m.y. or more there was no ocean. Moreover while the Proto-Atlantic proper is virtually certain, the earlier ocean is hypothetical. It is a concept that needs a distinctive name and we suggest "Iapetus" who in Greek mythology was the son of Earth (Ge) and Heaven (Uranus), brother of Okeanus and Tethys and father of Atlas, from whose name the word Atlantic is derived' (Harland and Gayer, *GM*, 1972). 'In summary the conclusions now emerging are that an ocean (Iapetus) separated the continents for the greater part of the Ordovician, virtually closing towards the end of the period, but with true continental collision and the orogeny delayed until the end of the Silurian, when practically all traces of Iapetus were obliterated. The most probable line for the continental suture is along the Solway where, according to Phillips et al. (*JGS*, 1976), there was also more than 1000 km. of dextral displacement, essentially taking place during Silurian times' (Moseley, *Geology of the Lake District*, 1978). See PROTO-ATLANTIC OCEAN.

**Ice age.** In the widest sense, a period in world or regional earth-history during which a large area is covered by, or under the influence of, ice (ice-sheet, ice-caps, valley-glaciers); being under 'glacial conditions', undergoing 'glaciation'. 'At this moment the Antarctic is surely experiencing an "ice age" but equally clearly the Sahara is not' (Shotton in Wright and Moseley, eds., *Ice Ages: Ancient and Modern*, 1975). More definitely, it is a period in world history during which large parts of the earth's surface, normally enjoying a temperate or warm climate, come under glacial conditions. 'The Ice Age' (or 'The Glacial Period') refers to that of the Quaternary; this is the ice age of which we see such striking evidences to-day in Europe and N. America. The Ice Age is made up of six (?) cold periods ('glacial periods' in a restricted sense, usually called 'glacials') alternating with warmer periods ('interglacial periods', 'interglacials'). Minor interglacial periods are 'interstadial periods', 'interstadials'. 'Post-glacial' means since the latest glacial ended about 10,000 years ago, sometimes appearing to be used on the supposition that the Ice Age is altogether over, but 'we do not know whether we live in a post-glacial period, although we call it that, or an interglacial period. The former would be a comforting belief; the latter more likely to be correct' (Sparks and West, *The Ice Age in Britain*, 1972). *The Great Ice Age* (J. Geikie, 1874).

There is evidence that ice ages occurred at several periods in the geological past, particularly one at the end of Carboniferous times. The causes of ice ages have for long been matters of vigorous discussion, and still are. We do not know whether glacial conditions in polar regions are, or are not, normal as viewed against the general world climate throughout the geological ages.

**Ice wedge.** A wedge-shaped mass of ice in the ground, tapering downwards, which exerts lateral pressure on the surrounding rock by the action of freeze-and-thaw. Ice wedges thus tend to shatter the rocks.

**Ichnofossil.** A trace-fossil, particularly an animal track. [G. *ichnos*, a track, footstep.]

**Ichnology.** The study of those trace-fossils that are animal tracks: footprints or the trail of a crawling invertebrate such as a gastropod, worm, or trilobite. Hence 'ichnofossil'. 'Term proposed by Buckland prior to 1844' (*AGI Dictionary*, 1960). [G. *ichnos*, a track, footstep.]

**Ichthyodont.** A fossil tooth of a fish. (*PT*, 1708.) [G. *ichthus, ichthuos*, a fish.]

**Ichthyodorulite.** (Ichthyodorylite.) A fossil fish-spine. 'Dorsal spines of fishes . . . have been named "ichthyodorulites" ' (Buckland, *Geology*, 1836). [G. *doru*, a spear.]

**Ichthyolite.** A fossil fish. 'In the arrangment of animal reliquia the genera may conveniently be founded on the Linnean classes of the recent subjects. On these principles we found eight genera of animal reliquia . . . 4. *Ichthyolithus*, of fish' (Martin, *Extraneous Fossils*, 1809). 'Ichthyolites, or the fossil remains of fishes' (Mantell, *Fossils of the South Downs*, 1822). The term was used by Murchison in *Siluria* (1854/72). This is one of those terms which have a definite meaning and are useful, but which have dropped out of fashion and not been replaced.

**Ichthyosauria.** One of the two orders of Mesozoic swimming reptiles. The body was shaped like that of a typical fish, with paddle-like limbs. [genus *Ichthyosaurus* (G. *ichthus, ichthuos*, a fish, *saura*, a lizard.)]

**Idioblastic.** See CRYSTALLOBLASTIC. [G. *idios*, one's own.]

**Idiomorphic.** Having it's own natural shape; for a mineral, in an igneous rock, that on crystallization was on the whole able to form it's own proper crystal faces. Also called 'euhedral'. Allotriomorphic (anhedral) is the opposite condition. Although it would be etymologically allowable, the term is seldom used for crystals in metamorphic rocks. As Harker says (*Metamorphism*, 1950): 'since the terms "idiomorphic" and "xenomorphic" carry a connotation which is proper to igneous but alien to metamorphosed rocks, we shall adopt in their stead the terms used by Becke, primarily in connexion with the crystalline schists, "idioblastic" and "xenoblastic" '.

**Igneous.** In the strict sense, pertaining to, or derived directly from, magma (molten rock material). But for long it has occasionally been used in a wider sense: 'Igneous in the sense of extreme metamorphism' (Ramsay, in discussion, *QJ*, 1877); and see IGNEOUS ROCK. Etymologically it is not a very appropriate term. [L. *ignis*, fire.]

**Igneous complex.** A close association of igneous rock-bodies differing in form and/or petrographical type. It may form an 'intrusive centre'. 'The Cairnsmore of Carsphairn igneous complex' (Deer, *QJ*, 1935). 'The series of Tertiary intrusive complexes from Skye to Arran' (Richey,

*BRG, Tertiary Volcanic Districts*, 1961). 'Evolution of the British Tertiary intrusive centres' (Walker, *JGS*, 1975). 'The plutonic complex of Central Anglesey' (Callaway, *QJ*, 1902). 'The Etive granite complex' (Anderson, *QJ*, 1937). 'The Tertiary ring complex of Slieve Gullion' (Richey, *QJ*, 1932). 'The volcanic complex of Calton Hill (Derbyshire)' (Tomkeieff, *QJ*, 1928). 'The magmatic complex of Kingscross Point, Isle of Arran' (Tomkeieff and Longstaff, *TEGS*, 1961).

**Igneous cycle.** The sequence of phases in the history of one main period of igneous activity in a region. This is normally: volcanic, major intrusions, minor intrusions. (Not a 'cycle' in the appropriate sense of the term.) 'Cycles of igneous activity' (Harker, *Natural History of Igneous Rocks*, 1909).

**Igneous dyke.** See DYKE.

**Igneous intrusion.** See INTRUSION.

**Igneous rock.** Although, in a strict sense, an igneous rock would be one derived directly from magma, there is the great petrographically distinctive class of rocks, the 'igneous rocks', probably not all of which have been formed simply by the cooling of a magma. We can see today that modern solidified lava is so formed and is thus demonstrably truly igneous. 'Igneous rock' in a conveniently wide sense comprises the crystalline or glassy rocks which, belonging neither to the sedimentary nor clearly metamorphic classes, look as if they might have cooled from a magma. 'Whinstone is not of volcanic, nor of aqueous, but certainly of igneous origin' (Playfair, *Illustrations*, 1802). *Natural History of Igneous Rocks* (Harker, 1909). *Petrology of the Igneous Rocks* (Hatch and Wells, 1891/1976). *Igneous Rocks* (Iddings, 1909). *Igneous Rocks and their Origin* (Daly, 1914). *The Evolution of the Igneous Rocks* (Bowen, 1928, and Yoder, ed., 1979). 'The classification of igneous rocks: an historical approach' (Tomkeieff, *GM*, 1939). *The Interpretation of Igneous Rocks* (Cox and others, 1979).

**Igneous rock name.** Every localized occurrence of an igneous rock has some unique petrographical feature but nevertheless igneous rocks in general conform to 'types', of a wide category, such as 'granite', 'basalt', or a narrow category, such as 'hornblende-biotite-granite', 'olivine-basalt'. Place-names are commonly used in naming rocks. These may be 'types' if the local occurrence is found to be representative of a general kind of rock, or they may be apparently unique; in the latter case the name specifies merely the kind of rock found at that particular place. This may be illustrated by two rocks of Tertiary age named by Harker in Skye (*MGS*, 1904). Of a rock found in the mountain of Marsco: 'Since this is unlike any type included in systematic classifications and nomenclature, we shall for convenience refer to it under the provisional name "marscoite". This is done merely to avoid repeated paraphrases and it is not intended to establish a new rock-type; the rock indeed is certainly a hybrid one, and therefore not entitled to systematic rank or formal designation'. And of one found at the village of Mugeary: 'The chief mineralogical peculiarities of "mugearite" result from its unusual chemical composition and go to characterise it as a special rock type.' This same rock-type, mugearite, was subsequently found among the Carboniferous lavas of the Midland Valley of Scotland and in the recent Hawaiian volcanic province.

**Igneous tectonics.** See TECTONIC-MAGMATIC COMPLEX. 'Tertiary igneous tectonics of Rhum' (Bailey, *QJ*, 1944).

**Ignimbrite.** An ash-flow pyroclastic volcanic rock; one formed directly from a *nuée ardente*. 'Some ignimbrites lithify after deposition by recrystallization due to the activities of escaping hot gases and fluids. They can then be given the more precise name "sillars". Other ignimbrites are so hot on deposition that the glass shards fuse, collapse and weld together to give the rock a laminated structure. This rock is called a "welded tuff" ' (Hatch and Rastall, revised Greensmith, *Sedimentary Rocks*, 1965).

The term was introduced by Marshall for certain late Tertiary occurrences in New Zealand in 1932 and 1935. Recently there have been many papers on British ignimbrites (in particular) owing to their widespread recognition: e.g. Rast and others (*N*, 1958; *PGS*, 1959; *LMGJ*, 1961, 1962); Fitch (*PGS*, 1966; *BV*, 1967); Dearnley (*BGS*, 1966); Sparks (*Sd*, 1976); Sparks and Wilson (*JGS*, 1976); Ray (*GM*, 1960, describing an intrusive occurrence). See PYROTURBIDITE. [L. *ignis*, fire, *imber*, a rainstorm, shower.]

**Illite.** A name covering the mica-like clay minerals. Important in the study of palaeo-salinity as it has the property of absorbing the element boron, the amount taken in being related to the salinity of the water at the time of its deposition. 'Boron in Holocene illites of the Dovey estuary' (Adams and others, *Sd*, 1965). (Brown, *MnS*, 1956.) [Illinois.]

**Illuvial.** See ELUVIATION.

**Ilmenite.** A black, opaque mineral, oxide of iron and titanium, $FeO.TiO_2$ (when pure), occurring as a common accessory in basic igneous rocks, particularly gabbro and norite. (Kupffer, 1827, *ilmenit.*) [Ilmen Mts., S. Urals.]

**Imbricate structure. 1.** 'The structure resulting from the piling upon one another of wedges or sheets of rock, like the partly-overlapping tiles of a roof' (Boswell, *Nappe Theory*, 1929). It is a series of thrust-slices. Hence 'imbrication'. 'Without incipient folding, the strata are repeated by a series of minor thrusts or reversed faults, which lie at an oblique angle to more important dislocations, termed by us major thrust-planes. They are likewise inclined to the ESE.—the direction from which the pressure proceeded. (Imbricate structure, *Schuppen Struktur*.)' (Peach and Horne, *MGS*, *NW. Highlands*, 1907.) 'An area of imbrication in the Ludlow rocks of the Denbighshire Moors' (Boswell, *PLGS*, 1931). Lees (*QJ*, 1952) uses the form 'imbric' and has, e.g., 'imbric zone', 'seven thrust imbrics in twelve miles'. **2.** "A structure [fabric] characterized by overlapping of tabular fragments or pebbles which display up-current dip. Described by Jamieson (*QJ*, 1860) and others' (Pettijohn and Potter, *Primary Sedimentary Structures*, 1964).

**Impactite.** Glassy or finely crystalline material produced where a meteorite has struck the ground. See METEORITE CRATER.

**Impact structure.** See ASTROBLEME. "Terrestrial impact structures—a bibliography' (Freeberg, *BUSGS*, 1966).

**Impression.** Finds its chief application as one of the normal states of fossil occurrence, where the impression of some part of the organism is retained on the matrix. Usually it is that of a relatively hard structure, such as a shell or the strengthened surface of a leaf, but soft bodies, such as jelly fish, may occasionally leave impressions on the mud (which later becomes hardened). See CAST (1). Tracks and trails, though impressions in a sense, are usually considered as separate states of fossil occurrence (trace fossils). 'These shells have in tract of time rotted and mouldered away, and only left their impressions, both on the containing and contained substances' (Hooke, *Micrographia*, 1665).

**In situ.** In place. A rock *in situ* means that, in an exposure, the visible rock is a part of the rock-body outcropping there, not a detached piece carried from some distance away. Loose weathered material may be said to be *in situ* if it has been derived from the rock immediately underneath (autochthonous). "When a stratum, or rock, is in its natural position, it is said to be *in situ*' (Phillips' 'Outline of the geology of England and Wales', *Mineralogy and Geology*, 1816). 'Bowlder-stones are evidently not *in situ*' (Greenough, *Geology*, 1819). 'It is not so easy as may at first appear to draw a clear line of distinction between the "fixed" rocks (rocks *in situ*, or "in place"), and their covering of travelled materials' (Lyell, *Elements*, 1838). Other applications of the expression are, for instance, to crystallization and replacement. [L.]

**Inarticulata.** One of the two classes of the phylum Brachiopoda, the shells having no tooth-and-socket apparatus. The shell is usually chitinous. (Huxley, 1869.)

**Incertae sedis.** Of uncertain placing. Applied to extinct groups of fossils of which the biological affinities cannot be satisfactorily inferred. [L.]

**Incise.** As a purely descriptive term in geomorphology this is applicable to any steeply-cut erosive feature such as a river gorge or coastal gio, but in physiographical interpretation it nearly always refers to a feature of fluvial erosion which has been produced as a result of earth-movement (relative to base-level) acting 'against' the downward-cutting action of the stream. Thus 'incised meander' usually implies that there is here a rejuvenation so caused, and 'incision' may refer to the cutting induced on the surface of a tectonic structure rising across an antecedent drainage system.

**Inclination.** Whereas dip signifies a downward slope measured from the horizontal, inclination is a wider term. The slope may be considered either in the upward or downward direction and be measured from either the vertical or the horizontal; or it may be the angle which two planes or lines, in any position, make with one another. But, notwithstanding this wider application, 'inclination' usually means, in fact, the same as dip and is sometimes preferred to 'dip' for slopes which are not those of bedding-planes.

**Included fragments.** This or a similar phrase is used when a stratigraphical formation is determined as being younger than another formation or rock-body by reason of its including recognizable fragments of that formation or rock-body. Thus, for instance, the sedimentary Longmyndian must be younger than the Uriconian because it contains pieces of the distinctive Uriconian volcanic rock; the age relations of these two unfossiliferous Precambrian formations is otherwise obscure because they are separated by a fault.

**Inclusions.** In a special petrographical sense: small accessory or accidental minerals enclosed in larger ones, or small particles of glass, or globules of liquid or gas, so enclosed. Crystals of one essential mineral ubiquitously enclosed in larger crystals of another give a poikilitic texture and are not spoken of as inclusions.

**Incompetent bed.** A bed that, in a particular case of folding, has yielded to the lateral pressure by plastic adjustment and flow. This may result in the bedding being thrown into complex structures or in the development of more regular internal structures, particularly drag folds and fracture cleavage. The bed tends to thicken towards the hinges, and to thin in the limbs, of the folds. Hence 'incompetent tectonics': 'Between these two competent beds the incompetent strata, from Purbeck to Lower Chalk inclusive, had to adjust themselves as best they could. There resulted "incompetent tectonics" of extraordinary variety and interest: thrusting, overturning, crumpling, drag-folding, brecciation and reversed faulting with production of discordant dips' (Arkell, *QJ*, 1938). 'The deformation of confined, incompetent layers in folding. The terms "competent" and "incompetent" refer to the structural character of the rocks, and are used in such a way that a weak rock which has undergone considerable deformation is called "incompetent". The terms are relative, and are invaluable in discussing the behaviour of beds during folding' (Williams, *GM*, 1961). See COMPETENT BED.

**Incretion.** A concretion (in the wide sense), a secretion, with the definite specification that growth has been from without inwards. 'Incretionary nodules have been formed by the deposition of mineral matter from percolating water in hollow spaces in rocks: the first coat was laid down on the walls of the cavity, upon the inner surface of this another coat was deposited, and so the growth of the nodule has gone on in the direction just mentioned. Agates are a common instance of this class of nodules' (Green, *Physical Geology*, 1876). Grabau (*Stratigraphy*, 1913, following Todd, *BGSA*, 1903) used the term in a special sense not implied in the word itself ('a cylindrical concretion with a hollow core').

**Incrop.** The (hidden) area over which a rock-body abuts against an overlying formation (or formations) or is restricted by overlying conditions. The term tends to be used only in special cases; incrops in a more general sense would be expressed by a 'subcrop map'. 'In north-east Warwickshire and in south Leicestershire the superficial deposits are thick and the Keuper Marl "outcrop" is more correctly a sub-Pleistocene incrop' (Shotton, *PGS*, 1968). 'Being soluble, salt cannot crop out at surface in Britain. Instead it has an "incrop" against the base of the zone of ground-water circulation' (Hains and Horton, *BRG, Central England*, 1969).

**Incrustation. 1.** In a general way, for instance: 'Calcareous waters "incrust" the rocky surfaces over which they flow with a coating [incrustation] of calc-tuff' (Page, *Geological Terms*, 1859). **2.** In a special sense, for a condition in which a fossil plant may occur. 'An incrustation may be described as an external mould of a plant usually in some incompressible material such as sandstone, ironstone or tufa which undergoes very little subsequent compression. As a rule the plant substance has disappeared and a cavity has been left which has the form of the original plant. Sometimes the cavity is subsequently filled up with mineral matter which thus forms a "cast" of the original plant. The surrounding material, the mould, forms the "incrustation"' (Walton, *Fossil Plants*, 1940).

**Index fossil.** Specially, 'zonal index fossil'. A species selected from among the assemblage of a zone to give a name to the zone. The species would be selected in the first place as seeming to be eminently characteristic of the zone (a guide fossil), but even if this turned out as a result of further research not to be so the zone (defined by its characteristic assemblage) would probably still continue to be known by its original name. More loosely, as in: 'The present discovery of regularity in the strata, to which fossil shells are now become the indices' (Smith, *TSA*, 1815). See ZONAL DESIGNATE.

**Index mineral.** Usually applied to a mineral which is an index of the grade of metamorphism undergone by the rock in which it occurs, requiring a certain minimum temperature or stress for its formation. Such a mineral is probably absolutely stable only under the condition of temperature or stress, or both combined, under which it was formed; but it has persisted ever since in a metastable state. 'Metamorphic zones characterized by the presence of critical index minerals' (Tilley, *QJ*, 1925). See quotation (Kennedy) under THERMOMETRY.

**Indicator erratic.** See ERRATIC and LARVIKITE. 'Map showing the distribution of indicator erratics from the principal centres of dispersion' (Eastwood, *BRG, Northern England*, 1946).

**Indicator fossil.** A fossil useful as indicating the conditions under which the animal or plant lived. The term has also been used with the meaning of 'guide fossil'. '*Ophiomorpha*: a marine indicator?' (Stewart, *PGA*, 1978).

**Induction.** 'That process of collecting general truths from the examination of particular facts' (Whewell, *Inductive Sciences*, 1857). See DEDUCTION.

**Indurate.** Commonly as 'indurated', 'induration'. To harden; in geology, often as meaning 'to make harder still', to harden further an already more or less consolidated rock, by continued gravity-compaction, tectonic pressure, metamorphism, or strong cementation. 'Through the middle of the isle of Wight there runs a ridge of hills of indurated chalk' (Hutton, *TRSE*, 1788). 'Proofs of the igneous formation of whinstone are derived from the induration of the contiguous strata' (Playfair, *Illustrations*, 1802).

**Industrial geology.** All those aspects of applied geology which are relevant to industry. *Industrial Geology* (Knill, ed., 1978).

**Inferior Oolite series.** The British representative of the general Bajocian series (lower part of the Middle Jurassic), predominantly oolitic limestone, outcropping across England from Yorkshire to Dorset and usually eroded into an escarpment, best developed in the Cotswold Hills, facing north-westward overlooking the Lias-Trias plain. In Yorkshire the series is represented by deltaic sands and clays, with ironstones. Fossils are abundant and include the following well-known species:

*Clypeus ploti* (echinoid), *'Rhynchonella' cynocephala*, *'Terebratula' fimbria*, *'T.' globata* (brachiopods), *Pholadomya fidicula* (lamellibranch), *Nerinea cingenda* (gastropod), *Parkinsonia parkinsoni* (ammonite), *Equisetites columnaris* (pteridophyte). (Michell, *MS*, 1788, 'Northampton Limes'; Smith, *MS*, 1797/9, 'Lower Bed of Freestone'; Smith in Townsend, *Character of Moses*, 1813, 'Inferior Oolite'.)

**Infiltration.** This word finds its commonest application in geology in connexion with water percolating through pores and interstices in the rocks (loose or coherent), particularly when such water is depositing mineral matter from solution either as cementing material or as veins. See quotation under CONSOLIDATION.

**Infra-Cambrian.** Whereas 'Precambrian' refers to age, 'Infra-Cambrian' refers to stratigraphical position and is thus the more direct name for rocks below the Cambrian. It usually refers to a rock-series immediately below what is taken to be the base of the Cambrian. 'The Infra-Cambrian Polarisbreen Series [Spitsbergen]' (Wilson and Harland, *GM*, 1964). (Menchikoff, 1949.)

**Ingletonian.** 'The name Ingletonian Series was given by Rastall [*PYGS*, 1906] to a group of unfossiliferous slates and grits occurring in the West Riding of Yorkshire at Ingleton and Horton-in-Ribblesdale. The formation is unconformably overlain by the Carboniferous Limestone, and on tbe south is faulted against Lower Palaeozoic rocks. The age of the series is uncertain . . . but it is in all probability of Pre-Cambrian age' (Leedal and Walker, *GM*, 1950). Recent isotopic evidence, however, has led to an opinion that 'the Ingletonian series was most likely deposited in the Cambrian or early Ordovician' (O'Nions and others, *JGS*, 1973).

**Inherited. 1.** Minerals in a sedimentary rock that, being chemically stable, have remained unaltered ever since they were constituents of the parent rock. Quartz is the commonest of such minerals. A sediment usually contains varying amounts of new minerals and of inherited minerals, and diagenesis will tend to increase the proportion of the former. **2.** Features that owe their character to conditions or events of a former geological age. Thus an exhumed landscape has a form that must be largely inherited; and a direction of folding, if and in so far as it is determined by a fold-direction of a former orogenic period, is inherited. '[Lower Carboniferous] rocks occupying one of several broad downwarps of inherited caledonoid trend . . .' (Dixon, *JGS*, 1972).

**Initial dip.** The dip of the beds, as originally deposited. Unless there are some observed facts or known conditions to suggest otherwise, it is assumed that beds were originally (initially) laid down nearly horizontally and that the present dip has been subsequently imposed by tectonic forces. Also called 'original dip', 'depositional dip.'

**Injection.** (A.) In igneous and metamorphic petrology. **1.** The intrusion of magma. **2.** The penetration of an already altered rock by magma as a last (migmatitic) phase in the total metamorphism, the injections remaining distinct in the final rock (as in 'injection complex', 'injection gneiss'). **3.** The permeation of country rock by residual fluids, with consequent pneumatolysis. (B.) In the structural geology of sedimentary rocks. **4.** The forcing under pressure of sandy or other material (upwards, downwards, or sideways) into a pre-existing deposit or rock, either irregularly or along some plane of parting, crack, or fissure. Also the body of rock so formed, a 'sedimentary injection'; see SEDIMENTARY INSERTION.

**Injection dyke.** See DYKE.

**Inlier.** Certain cases of an outcrop of a lower (normally an older) rock surrounded by upper (normally newer) rocks. Whereas the term 'outlier' implies a main outcrop, probably not far away, the term 'inlier' does not necessarily, perhaps not usually, imply that. Nevertheless there is one kind of inlier that is the counterpart of an outlier: where a patch of outcrop of a formation is exposed by deep erosion and is separated from its main outcrop by overlying beds. A clearly marked inlier may be produced if erosion cuts down through an unconformity to expose a patch of an underlying series (an unconformity-window); one will be produced where a formation has been pushed up between faults so that, after erosion, its outcrop is surrounded by outcrops of newer rocks. A tectonic 'window' is another case of an inlier. The central part of the outcrop in a dome-structure would be called an inlier only if it were in some way specially distinctive, as in the Silurian inliers emerging from beneath the Old Red Sandstone in the Welsh Borderland. 'The regular structure of the plain of [New Red] sandstone is interrupted near Appleby by projecting masses of slate and greenstone. These rocks form an insulated group extending nearly north and south along the base of the escarpment of the great [Carboniferous] limestone series of Cross Fell' (Buckland, *TGS*, 1817; and, for another instance described by Buckland, see quotation under UNCONFORMITY). 'The Cross Fell inlier' (Nicholson and Marr, *QJ*, 1891). 'The faulted inlier of Carboniferous Limestone at Upper Vobster, Somerset' (Sibly, *QJ*, 1912). 'The Silurian inlier of Woolhope, Herefordshire' (Gardiner, *QJ*, 1927).

**Inorganic.** See ORGANIC.

**Insect bed.** A bed characterized by fossil insects. In Britain examples are known from the Lower Lias (Brodie, *History of the Fossil Insects in the Secondary Rocks of Britain*, 1845) and the Purbeck (Brodie, *QJ*, 1854). See quotation (Shrock and Twenhofel) under THANATOCOENOSE.

**Insecta.** A class of the phylum Arthropoda; perhaps the most important of all invertebrate classes to the neontologist but one of the least important to the palaeontologist because of the exceptional conditions necessary for fossilization. Nevertheless, of the thirty-four orders ten are extinct, chiefly Upper Carboniferous, and of the twenty-four living orders all but two are known in the fossil state. [L. pl., insects.]

**Inselberg.** 'As long ago as 1904 Passarge [Berlin], in his studies of arid South Africa, noted the existence of vast flat surfaces, studded with sharply rising residual hills, termed in German literature "Inselberge"' (Wooldridge and Morgan, *Geomorphology*, 1959). Now used in a general sense, equivalent to 'monadnock'. [German, island-mount.]

**Insolation.** Exposure to the sun, hence 'the geological action of the sun's heat upon rocks at the surface' (J. Geikie, *Earth Sculpture*, 1898).

**Interbedded.** Laid down in sequence, between one bed and another; not formed or intruded between these beds at a subsequent time. Applied particularly to a lava which flowed into a region of deposition and, while not being a sedimentary bed itself, has taken its place in the whole rock-succession. See INTERSTRATIFIED and quotation under CONTEMPORANEOUS.

**Intercalate.** There is perhaps some ambiguity in the use of this term in geology. A rock 'intercalated' in a bedded series, or a formation 'intercalated' among others may imply that it has come to take up this relation (by, for instance, intrusion or thrusting) at some time after those beds or formations were formed. On the other hand, it sometimes refers merely to the fact that, for instance, beds of sandstone occur at several levels within a predominantly shaly series, with no implication they they are not in the normal order of deposition.

**Inter-depositional.** A syndepositional process in which the deposition is in stages rather than being continuous. '. . . inter-depositional normal faults downthrowing to the south' (Horne, *BGSI*, 1977).

**Interface.** The surface between two contiguous bodies or two contiguous parts of a body. On any scale; thus between two rock-bodies or two parts of the solid earth-body; and specially in sedimentation for the surface separating the top of the last layer of sediment and the medium (usually water) in which the sedimentation is taking place. 'The kind of contact, or "interface", between till sheets . . .' (White, *IGC*, 1972). It might be useful, as 'bedding interface', for the concept of the imaginary surface between and separating two bedding surfaces, the reference marker points of boundary stratotype sections lie within bedding interfaces. The adjective 'interfacial' is most commonly used for the angle between two faces, as in crystallography. The special application of the term 'interface' in physics does not concern us. See CONTACT.

**Interformational.** Between one stratigraphical formation and another.

**Interglacial period.** See ICE AGE. 'We have found that there is abundant proof to show that the accumulation of a *moraine profonde* by one great ice-sheet was interrupted several times; that the ice-sheet vanished from the low-grounds, and even from many of the upland valleys, and that rivers and lakes then appeared where before all had been ice and snow. We have also learned that during such mild inter-glacial periods, oxen, deer, horses, mammoths, reindeer, and no doubt other animals besides these, occupied the land' (J. Geikie, *Great Ice Age*, 1874). 'Interglacial and interstadial periods' (Lüttig) and 'The definition of interglacial" ' (Suggate, both in *JG*, 1965).

**Interjection.** (Or as parts of the verb 'interject'.) An intrusion emplaced between, and by the separation of, strata; that is, normally, a sill. 'The strata are not broken in order to have the whinstone introduced, they are separated, and the whinstone is interjected in form of strata' (Hutton, *TRSE*, 1788).

**Intermediate.** In petrology, for igneous rocks falling between the acid and the basic groups; that is, for those having a silica percentage between about half and two-thirds of the total mass.

**Intermontane.** Situated between mountainous or upstanding regions. Applied particularly to intermontane basins where there is deposition of the material eroded from the surrounding mountains. 'Intermontane troughs' has been used for subsiding regions of the ocean lying between stable or uprising regions.

**Interstadial.** See INTERGLACIAL.

**Interpretation.** Geologically significant observations and experiments are 'interpreted' when they have provided material for the 'geological record'; the facts of the geological record are 'interpreted' when they have provided evidence of force, process, environment, age, history. 'Interpretation of geophysical observations between the Orkney and Shetland Islands' (Bott and Browitt, *JGS*, 1975). 'Interpretation of facts' (Watts, *Geology for Beginners*, 1912). 'Interpretation of stratigraphical ages: orogenic belts' (Harland in Kent and others, eds., *Time and Place in Orogeny*, 1969). Allen (*ESR*, 1974) gives a penetrating discussion of the physical 'interpretation' of sedimentary rocks.

**Interstadial period.** See ICE AGE.

**Interstratified.** May be said to be synonymous with 'interbedded', but whereas the latter is more usual for a contemporaneous lava flow, 'interstratified' is rather more commonly used to describe a lithological character of a stratified series, such as 'limestone and shale interstratified'.

**Intraclast.** Material created by penecontemporaneous erosion within a basin of deposition.

**Intraformational, Intrastratal.** Within one stratigraphical formation, within one stratum. Used in various connexions. 'Intraformational contorted rocks in the Upper Carboniferous of the southern Pennines. . . . The contorted shales are considered to be of tectonic origin and to be an expression of adjustment through bedding-plane slip' (Cope, *QJ*, 1945). 'When a slumped mass slides down on to undisturbed sediments it may later be covered by younger deposits. Severely disturbed beds will then be found between undisturbed strata, an arrangment that is known as intraformational contortion or corrugation' (Hills, *Structural Geology*, 1953). 'Intra-stratal flow and convolute folding' (Williams, *GM*, 1960). 'The anomalous attenuation does not seem to be due to intraformational non-sequences' (Smith and George, *BRG, N. Wales*, 1948).

**Intraformational conglomerate.** See INTRAFORMATIONAL and CONGLOMERATE.

'Walcott (*BGSA*, 1894) introduced the term intraformational conglomerate, defining it as "one formed within a geologic formation of material derived from and deposited within that formation" ' (Allen, 'Intraformational conglomerates in the Lower Old Red Sandstone of the Anglo-Welsh cuvette', *LMGJ*, 1962). 'Intraformational breccia' is analogous.

**Intraglacial.** Within an ice sheet or glacier. 'Intraglacial volcanoes of the Laugarvato region, south-west Iceland' (Jones, *QJ*, 1968).

**Intratelluric.** See TELLURIC.

**Intrusion.** The action whereby rock material or rock is forced, thrusts itself, or finds its way among other rocks. Also for the body of rock so emplaced. Applies chiefly to such movement of igneous magma within the earth's crust and the resulting body of igneous rock ('igneous intrusion'). Hence 'intruded', 'intrusive', &c. Bodies of rock material such as clay, chalk, common salt, gypsum, may become plastic and be intruded into other rocks ('sedimentary intrusion'). See CARBONATITE. 'Intrusive limestones in . . . Northern Rhodesia' (Bailey, *QJ*, 1961). Small intrusions of one kind of sedimentary material into another are termed 'injections'. Whitehurst (*Strata in Derbyshire*, 1778), in his consideration of the Derbyshire toadstones, was one of the first (perhaps quite the first) to realize the possibility of igneous intrusion. The idea of this process was elaborated by Hutton (*TRSE*, 1788): '[In the Midland Valley of Scotland] the strata consist, in general, of sandstone, coal, limestone or marble, ironstone, and marl or argillaceous strata . . . but through all this space, there are interspersed immense quantities of whinstone. The origin of this form, in which the trap or whinstone appears, is most evident to inspection, when we consider that this solid body had been in a fluid state, and introduced, in that state, among strata, which preserved their proper form.' 'Rocks which have been "intruded" amid the strata . . . greenstones, porphyries . . .' (Murchison, *Silurian System*, 1839). See quotation under HYPABYSSAL.

**Intrusion breccia.** See BRECCIA (5).

**Inversion.** A turning upside down; an overfolding of stratified rocks so that the original upward sequence (newer beds following older) is reversed. 'These strata are inverted; the Lower Silurian overlying the Devonian or Old Red rocks' (Murchison, *Siluria*, 1854).

**Invertebrata.** All animals other than the Vertebrata; that is, all the invertebrate phyla, together with the invertebrate groups (subphyla) of the phylum Chordata (Stomochorda, to which the graptolites are at present provisonally assigned, Urochorda, and Cephalochorda).

**Invertebrate palaeontology.** The palaeontology (palaeozoology) of the invertebrate animals is studied more by scientists who are primarily geologists than by those who are primarily zoologists. This is because zoologists are much more interested in the soft parts than in the hard parts of invertebrates and also because the careful discrimination of invertebrate fossils is of paramount importance in purely geological (stratigraphical) researches. Moreover it is the geologist who inevitably finds these fossils in the course of his ordinary work. *Palaeontology: Invertebrate* (Woods, 1893/1946). *Treatise on Invertebrate Paleontology* (Moore, ed., 1953-).

**Iron pan.** See PAN.

**Ironshot.** 'Ores in which the ooliths are essentially composed of limonite, the "iron-shot" rocks' (Hallimond, *MGS*, *Bedded Iron Ores*, 1925). 'Sandy ferruginous beds, and hard limestones with limonitic ooliths ("ironshots") are the typical Dundry rock-types [Inferior Oolite]' (Welch and others, *BRG, Bristol and Gloucester*, 1948). 'The Fredville ironstone (Kellaways Rock) is an ironshot sand, with limonite ooliths forming about one-half of the rock' (Hatch and Rastall, *Sedimentary Rocks*, 1938).

**Ironstone.** A rock largely made up of an iron compound. It may be a segregation in an igneous intrusion; but it is most commonly a sedimentary rock, either deposited directly as a ferruginous sediment or resulting from chemical replacement.

**Irruptive.** Sometimes used for intrusive rocks as contrasted with extrusive ('eruptive' in the narrow sense). See ERUPTIVE. 'The acid irruptives mark the sources from which the lavas were poured out' (Harker, *Bala Volcanic Series of Caernarvonshire*, 1889). [L. *irrumpo*, break into.]

**Island arc.** This descriptive geographical term carries with it, in geological usage, certain connotations. 'An island arc is a tectonic belt of high seismic activity characterized by a high heat-flow arc with active volcanoes bordered by a submarine trench. It forms where a plate of oceanic lithosphere collides with, and is subducted beneath, another oceanic or continental plate along a Benioff or subduction zone' (Windley, *The Evolving Continents*, 1977). 'On island arcs' (Coleman, *ESR*, 1975).

**Isobar.** A line or surface of equal pressure. In geology this has an application to a surface within the earth's crust where, at any one time during folding or metamorphism (or both together), the pressure was the same. 'Isobars and isotherms in north-eastern Scotland' (Chinner, 'Distribution of pressure and temperature during Dalradian metamorphism', *QJ*, 1966). [G. *barus*, heavy.]

**Isochemical.** Having or keeping the same chemical composition; for a process, a set or series of minerals or rocks, a case of metamorphism, &c. A rock changing its mineral composition isochemically remains a 'closed system'.

**Isochronal.** Formed or occurring at the same time. In stratigraphy, applied to a surface imagined as being drawn through points, in a more or less widespread body of strata, where deposits were formed at the same time. The surface (a 'time-plane') may be further imagined as extending from one body of strata to another, or, indeed, as being world wide.

**Isoclinal.** Of equal dip; but applied exclusively to a fold, or (and usually) a series of folds, in which the limbs have become parallel or nearly so. Hence 'isocline' for such a fold which, unless quite upright, must be an overfold.

**Isodietic.** 'It was found that the peculiar distribution of certain genera and species of British Carboniferous lamellibranchs at widely different horizons in different areas was very striking. As evidence accumulated, it became certain that this distribution depended on similarity of conditions of deposition. It seems to us that by taking some fossil, or group of fossils, whose habits can be compared with some living representatives, and ascertaining at what horizons they occur in different localities, it may be possible to construct isobathymetric lines in the series of rocks or, at least, to draw lines in them connecting points of similar conditions of deposition; such lines we propose to call "isodietic"' (Hind and Howe, *QJ*, 1901). Considered spatially instead of along a line these conditions would be represented by isodietic surfaces. [G. *diaita*, way of living.]

**Isogeotherm.** See ISOTHERM.

**Isogonal.** Of equal angle. 'An "isogonal line" is a line which joins all points of equal dip on a surface, while an "isogonal surface" is a surface which joins all points of equal dip and strike within a three-dimensional body' (Phillips and Byrne in Kent and others, eds., *Time and Place in Orogeny*, 1969). [G. *gonia*, an angle.]

**Isograd.** See ISOGRADE.

**Isograde.** See GRADE (2). '"Isograde" rocks are those which have originated under closely similar physical conditions of temperature and pressure. "Isograd" is a term which may be used in the same sense as isotherm or isobar, a line joining points of similar pressure-temperature values; it is the intersection of an inclined isograd surface with the earth's surface' (Tilley, *GM*, 1924). To make the definition of isograd purely one of observed fact, leaving the temperature and pressure involved to be inferred, Carmichael (*AJS*, 1978) suggests the following: 'the trace of a surface across which a specific change in metamorphic mineralogy takes place'. 'Kyanite isograds of Grampian metamorphism' (Chinner, *JGS*, 1980). [L. *gradus*, a step, a stage.]

**Isomorphism.** 'It is found that certain minerals of analogous composition crystalise in forms showing close relation one with another. Such minerals have their atoms arranged on similar plans. This phenomenon is called "isomorphism" . . . the important group of minerals known as the plagioclase felspars constitutes an excellent example of a series showing "isomorphous mixture"' (Read, *Rutley's Mineralogy*, 1948).

**Isopachyte.** An isopachyte line, isopachyte, or isopach, is an imaginary line over the surface of the ground (or a corresponding real line on a map) below which a certain bed or formation has everywhere the same true thickness. 'The data are at present hardly sufficient to enable us to draw accurate isopachytes (lines joining points where equal thicknesses of beds occur) for either the London Clay or the Lower London Tertiaries' (Boswell, *QJ*, 1915). 'When Gerard de Geer introduced the word "isopachyte" towards the end of the last century and applied it in its English form during the visit to Stockholm in 1910 of the International Geological Congress, he used it for recording actual thicknesses of Pleistocene deposits. Since the publication of the Report in 1912 (after which date I adopted the term in connexion with the plotting of the present-day thicknesses of British formations of different ages) it has been used by many geologists' (Boswell, *LMGJ*, 1961). [G. *pachos*, thickness.]

**Isopleth.** A line, surface, or zone along, within, or related to which some attribute has an equal value, such as isobar, isotherm, isograd, isopachyte line, isoseismal line. 'Isopleth distribution' (Chinner, *QJ*, 1966). [G. *plethron*, a measure.]

**Isoseismal.** An isoseismal line, or isoseismal, is an imaginary line (or a corresponding real line on a map) through all points where an earthquake shock had the same intensity. (Mallet, *BA*, 1858.) [G. *seismos*, a shaking.]

**Isostasy.** 'Theoretical balance of all large portions of the earth's crust as though they were floating on a denser, underlying layer; thus areas of less dense crustal material rise topographically above areas of more dense material' (*AGI Glossary*, 1960). The concept originated as a result of the measurements of the gravitational attraction of the Himalayas on the plumb line. These, and the consequent calculations, were made by Airy and Pratt and published in a series of papers (*PT*), beginning in 1855. The term 'isostasy' was proposed by Dutton (*BPSLW*, 1889, quoted in several works): 'If the earth were composed of homogeneous matter its normal figure of equilibrium without strain would be a true spheroid of revolution; but if heterogeneous, if some parts were denser or lighter than others, its normal figure would no longer be spheroidal. Where the lighter matter was accumulated there would be a tendency to bulge, and where the denser matter existed there would be a tendency to flatten or depress the surface. For this condition of equilibrium of figure to which gravitation tends to reduce a planetary body, irrespective of whether it be homogeneous or not, I propose the name "isostasy" (from the Greek *isostasios*, meaning "in equipoise with", compare *isos*, equal, and *statikos*, stable). I would have preferred the word isobary, but it is preoccupied. We may also use the corresponding adjective, "isostatic".' 'As early as about 1500 Leonardo da Vinci recognized that change of load causes movements of the earth's crust. The earliest recognition in America of what we now call isostatic adjustment appears to have been in 1743 by Lewis Evans, colonial surveyor, cartographer, and geological observer' (White, *S*, 1951). For discussion and historical review see (to mention a few among many works): Fisher, *Physics of the Earth's Crust*, 1889; Joly, *Surface History of the Earth*, 1925/30; Jeffreys, *The Earth*, 1924/59; Heiskanen and Meinesz, *Earth and its Gravity Field*, 1958; Howell, *Geophysics*, 1959. 'Isostasy' (Heiskanen in Runcorn and others, eds., *International Dictionary of Geophysics*, 1967).

**Isotherm.** A line or surface of equal temperature. In geology this has an application to a surface within the earth's crust, sometimes called an 'isogeotherm' or 'geoisotherm'. Such isotherms will, in particular, be related to regions and times of igneous intrusion and metamorphism. See quotation under ISOBAR. [G. *thermos*, warm.]

**Isotope.** See RADIOACTIVITY AGE-METHOD. 'One of two or more atoms having the same atomic number but different mass number. The nuclei of isotopes contain identical numbers of protons, but different numbers of neutrons. Thus although they differ in mass, isotopes belong to the same chemical element. For most elements, both stable and radioactive isotopes are known . . . [but only] approximately a dozen radioisotopes are found in nature in appreciable amounts' (*McGraw-Hill Encyclopedia of Science and Technology*, 1960). 'The term isotopes was coined by F. Soddy in 1913 from the Greek *isos*, "equal", and *topos*, "place", to indicate substances occupying the same place in the periodic table of the elements; i.e. having exactly the same chemical properties, but distinguishable from each other in some other way, such as mode of radioactive decay' (*Encyclopaedia Britannica*, 1962). *Principles of Isotope Geology* (Faure, 1977). Hence 'isotopic' (see RADIOACTIVITY AGE-METHOD).

**Isotropic.** Having physical properties which are the same in all directions; in contrast to anisotropic, where the properties depend, for their development or intensity, on the direction. In sedimentary rocks, 'where the orientation of the fabric elements is random, the fabric is isotropic, where preferred orientation is present, the fabric is anisotropic' (Pettijohn, *Sedimentary Rocks*, 1957). Hence 'isotropy', 'anisotropy'. The term is used particularly in mineralogy; all crystals have directional properties, are anisotropic, but a glass is isotropic. In crystals in the cubic system, the properties that are directional are equal in the several particular directions, and properties depending on the transmission of light are equal in all directions; such crystals are optically isotropic or, simply, isotropic, as in mineralogy 'isotropic' usually means optically isotropic. [G. *tropos*, a direction.]

**Isovol.** See DEVOLATILIZATION. 'The isovol map shows the volatile contents of the Nine Feet Seam in various parts of the [S. Wales] coalfield. Isovols of some other seams are also shown' (Jones, *QJ*, 1951).

# J

**Jargon.** Twittering, chattering. Terminology invented or used where ordinary words would do as well. Technical terminology, geological terminology, however uncouth the terms, is not jargon if it fixes meanings shortly and precisely. ('Geology without jargon', subtitle of Shand's *Earth Lore*, 1933.)

**Jasper.** See CHALCEDONY.

**Jet.** A hard black coaly substance, taking a brilliant polish. In Britain it occurs as lenticles (each appearing to represent a separate piece of waterlogged driftwood) in the Upper Lias of Yorkshire. 'A piece of jet, very fine and black, with impressions of ammonitae upon it. From Whitby, Yorkshire' (Woodward, *Catalogue*, 1729). [Old French *jaiet* f. G. *gagatēs* f. *Gagae*, a town in Asia Minor.]

**Joint.** A surface (usually approximately a plane) within a rock of actual or potential fracture or parting along which there has been practically no displacement. In sedimentary rocks, joints usually occur in several parallel series, one dominant, a second somewhat less conspicuous and perhaps a third or even more series more feebly developed. A 'joint set' is a group of parallel joints. A 'joint system' is a group of two or more intersecting joint sets. The planes are most commonly at right angles to the bedding, and if there are two conspicuous sets of joints these are commonly at right angles to each other. In regions of folding, joints tend to occur particularly in the plane containing the *a* and *c* tectonic axes (cross joints) and in that containing the *b* and *c* tectonic axes (longitudinal joints). (Wilson, *PGA*, 1952). Over regions of simple tilting these are dip joints and strike joints: 'When joints nearly coincide with the strike of the beds they may be called "strike joints"; and when they are nearly transverse to the strike they may be called "dip joints"' (Sedgwick, letters to Wordsworth, 1842, in Wordsworth and Hudson's *Guide to the Lakes*). 'Joints in igneous rocks. . . . The joints in granite are often wonderfully regular—two sets of vertical or steeply inclined division-planes intersecting at various angles, which often do not depart widely from right angles. The rock thus tends to be divided into columnar masses. In addition to its vertical division-planes granite often exhibits a set of cross-joints, arranged at approximately right angles to the others. These cross-joints may be horizontal or inclined, and often give the rock a kind of bedded appearance. . . . Many basalts are jointed so symmetrically that the rock looks like an organised aggregate of prismatic columns. When this structure is fully developed, as in the well-known rocks of Fingal's Cave [Staffa] and the Giant's Causeway, the columns tend to assume hexagonal forms. But although six-sided columns are common enough, yet the faces of the prisms are seldom equally developed, while many columns may show fewer or more faces than six, so that trigonal, tetragonal, pentagonal, and polygonal forms are often associated. . . . They are usually intersected at more or less regular intervals by transverse or cross-joints, which in some few cases show a ball-and-socket arrangment—the convex surface of one segment fitting into the concave surface of the next overlying or underlying block' (J. Geikie, *Structural and Field Geology*, 1953). See COLUMNAR JOINTING. Joints are emphasized on an exposed surface by being etched by weathering; weathering along joints may also proceed below the ground-surface. 'A single joint may determine the position of a great erosion trough or valley hundreds of feet wide' (Leith, *Structural Geology*, 1923). '"Master joints", or those large planes of division which run regularly parallel to each other over large distances' (Jukes, *Manual*, 1862). See quotation under PERPENDICULAR. Morton (*Northamptonshire*, 1712) refers to joints given various names by the quarrymen ('upright-joints', 'thorough-joints', 'gulfe-joints').

'Grey granite or moor-stone, so-called in Cornwall . . . in some places it is disposed into thick irregular unwieldly beds, which are commonly broken transversely into huge masses or blocks of various sizes and shapes' (Williams, *Mineral Kingdom*, 1789). 'Granite is not stratified, but is sometimes separated into tabular masses, which have been mistaken for strata. It is more frequently divided into large masses or blocks, which have a tendency to assume a rhomboidal form. Granite also exists in round masses, which are composed of concentric spherical layers, separated by granite of a less compact kind, and inclosing a harder central nucleus. These globular masses are three or four yards or more in diameter, and are sometimes found detached, and sometimes imbedded in granite of a softer kind: probably the detached globes of granite were also once imbedded in a similar rock, which has been decomposed and worn away' (Bakewell, *Geology*, 1813). 'Many rocks, both stratified and unstratified, are divided into solids of greater or less regularity, by parallel systems of fissures or joints. This gives rise to a jointed structure, and is quite distinct from slaty cleavage. For the joints are at definite distances from each other, and a mass of the rock between them has, generally speaking, no tendency to cleave in a direction parallel to them. The structure in question seems in most cases to have been produced mechanically, either by a strain upon the rock from external force providing, more or less, regular sets of cracks and fissures, or by a mechanical tension on the mass (produced probably by contraction) during its passage from a fluid or semi-fluid, into a solid state' (Sedgwick, *TGS*, 1835). 'Jointing in the Great Scar Limestone of Craven and its relation to the tectonics of the area' (Wager, *QJ*,

1931). 'Mechanics of jointing in rocks' (Price, *GM*, 1959). 'Classification of structures on joint surfaces' (Hodgson, *AJS*, 1961). 'Joints and faults' (Hancock, *PGA*, 1968). 'Joints are the commonest of secondary structures in rocks, and although apparently the simplest, being merely cracks, they include a variety of types of diverse origin. The word "joint" was first used by coal-miners for smooth, straight cracks separating blocks of rock that appeared to them to have been "joined" as in brickwork, the "joints" being at right angles to the bedding planes. This type of joint, this is to say, a fracture which is normal or nearly normal to the bedding, and along which there has been no visible displacement, is the most typical, but also the least understood, of all structures currently classified as "joints"' (Hills, *Structural Geology*, 1963).

**Joint intensity.** 'Joint surface area per unit volume of rock' (Wheeler and Dixon, *Geoly*, 1980). 'Joint spacing', 'joint frequency' refer to the same quality.

**Junction.** Sometimes used for the contact between two rock bodies, especially when there is an upward sequence of stratified groups. Such a junction may be a perfectly conformable one (with or without non-sequence), a disconformity, an unconformity, or be of a tectonic nature (such as a thrust). There is the 'Junction Bed' in the Lias; see CONDENSED SUCCESSION.

**Jurassic system.** Overlies the Triassic system and is overlain by the Cretaceous; it is the central member of the three Mesozoic systems. The base of the system is drawn at the base of the Hettangian stage of the Lias (base of the *Psiloceras planorbis* zone). The top is defined by the base of the Cretaceous system.

In Britain the Jurassic system outcrops mainly in a broad band running diagonally across England and S. Wales from the Yorkshire coast to the Dorset coast. There are outlying patches in Cheshire, Cumberland, and NW. and NE. Scotland. *Jurassic Rocks of Britain*, 5 vols. (Fox-Strangeways and Woodward, *MGS*, 1892/95). *Jurassic System in Great Britain* (Arkell, 1933). *Jurassic Geology of the World* (Arkell, 1956). 'A Jurassic time scale' (Hinte, *BAAPG*, 1976). *A Correlation of the Jurassic rocks in the British Isles*, *GSSR* (14 & 15), 1980. The system is classified as follows (at east in NW. Europe):

| *General stages* | *British formations* |
|---|---|
| Portlandian | Purbeck beds<br>Portland beds |
| Kimmeridgian | Kimmeridge Clay |
| Oxfordian | Corallian<br>Upper Oxford Clay |
| Callovian | Lr. & Md. Oxford Clay<br>Kellaways beds<br>Upper Cornbrash |
| Bathonian | Lower Cornbrash<br>Great Oolite series |
| Bajocian<br>Aalenian | Inferior Oolite series |
| Toarcian<br>Pliensbachian<br>Sinemurian<br>Hettangian | Lias |

The rocks of the system are extremely important both as source rocks and traps for oil and gas in the NW European continental shelf. The fossils which characterize the Jurassic system are particular families, genera, and species from among the groups which make up the Mesozoic fauna and flora as a whole. The ammonites provide most of the families and genera that are confined to the system, and the restricted genera of the terebratulid and rhynchonellid brachiopods are also thus confined; such families and genera, together with particular species of those that range into the Cretaceous or persist from the Trias, are mostly further restricted to particular formations or stages. (Von. Humbolt, Germany, 1795; Brongniart, France, 1829.) [Jura Mts.]

**Juvenile.** Fresh in origin; applied to water, steam, and gases given off from a magma either within the earth's crust or in a volcanic eruption. Juvenile waters are thus distinct from meteoric waters.

# K

**Kame.** One of the two types of esker (in the wide sense; the other being the os). Kames are typically irregular mounds, or ridges, of glacial sands and gravels; they are essentially marginal or terminal moraines, but laid down in water instead of dumped down by the ice itself. In the case of kames that are ridges, the only difference between them and the shorter varieties of osar would be (on the distinctions adopted here, following Charlesworth) that the origin of the kame as a moraine is clear. See ESKER. [Celtic *cam* or *kaim*, crooked, winding.]

**Kaolin.** The clay produced by the decomposition (by weathering or, and particularly, pneumatolysis) *in situ* of the felspars (chiefly) of a highly felspathic rock, such as are the Devon and Cornish granites, where much kaolin occurs. It is composed essentially of the particular hydrated silicate of alumina, kaolinite. It is the natural (unwashed) china-clay. Hence 'kaolinization'. 'The famous china clays and china stones of Cornwall are generally attributed to destructive activities of certain members of the great series of emanations that produced the mineral veins of the district. According to this view, particular emanations traversing already consolidated portions of the granites have attacked the feldspars, removing alkalies and leaving water in exchange, combined with residual aluminium silicate' (Bailey, *JRSA*, 1944). See quotation (Bristow) under HYDROTHERMAL. [Chinese *Kaoling* (meaning 'high hill'), a mountain in N. China, whence the clay was first obtained.]

**Karroo series.** The representative in S. and Equatorial Africa of the Gondwana series. It ranges from the Upper Carboniferous to the Lias inclusive. In addition to the *Glossopteris* flora, it has many characteristic vertebrate fossils, particularly reptiles. '"The Table Mountain Sandstone", wrote F. v. Hochstetter many years ago [1866], "forms to a certain extent the edge of the great continental plateau which consists of the formations of the great Karroo"' (Suess, *Das Antlitz der Erde*, 1885, English trans., 1904). [Great Karroo region, S. Africa.]

**Karst topography.** The topography characteristic of a limestone country, with its underground drainage and effects of surface solution. 'The Great Scar Limestone [Carboniferous, Yorkshire] is commonly a well-bedded, massive, pale-grey rock, and its shelf-like outcrops form the finest areas of "karst" landscape in Britain' (Edwards and Trotter, *BRG*, *Pennines*, 1975). 'Fossil karst in Derbyshire' (Ford, *PBSA*, 1964). [Karst district of the Adriatic coast.]

**Katamorphism.** Change in rocks of a kind in which chemically simple minerals are formed from more complex ones. Such changes occur most commonly in processes, particularly weathering, that are not usually included under the head of metamorphism.

**Kellaways beds.** A Jurassic formation, some 80ft thick, of clays and calcareous sandstones, stretching from Yorkshire to Dorset; the British representative of the zones of *Proplanulites koenigi* and *Sigaloceras calloviense* of the Callovian stage. (Smith in Townsend, *Character of Moses*, 1813, 'Kelloway Rock'.) [Kellaways, Wiltshire.]

**Kenspeckle.** 'More than one member of the Arvonian, in particular the granite, is a most kenspeckle rock and in the Cambrian has been recognized as pebbles. [Footnote] Kenspeckle: a very expressive Scottish word signifying something which we shall know when we see it again' (Greenly, *GM*, 1946). It would certainly be useful to have a word for the recognizability of (especially) a particular rock when it occurs as quite disconnected pieces, perhaps far removed from the visible parent body.

**Kentallenite.** A coarse-grained igneous rock, near monzonite in character, but with abundant mafic minerals which are olivine, augite, and biotite rather than hornblende. (Hill and Kynaston, *QJ*, 1900.) [Kentallen, Argyllshire.]

**Kerogen.** The complex organic matter present in certain carbonaceous shales and oil shales. 'Kerogenite' has been suggested for these (and any other) rocks rich in kerogen (Hatch and Rastall, *Sedimentary Rocks*, 1965).

**Keuper.** The upper half (in Britain) and upper third (in Germany, where first named) of the New Red Sandstone facies of the Triassic system. The rocks are sandstones, clays, and marls, with salt deposits. Fossils are uncommon except for the little crustacean, *Estheria;* but plants and vertebrates are found and various invertebrates, including scorpions. [German, red marl.]

**Keweenawan.** See PROTEROZOIC. [Keweenaw peninsula, Lake Superior.]

**Kidney iron ore.** See HAEMATITE.

**Killas.** 'The slate formations of Cornwall; a term from the extinct Cornish language' (Arkell and Tomkeieff, *Rock Terms*, 1953). 'Slate, near Fowey. The miners and people call this "killas"' (Woodward, *Catalogue*, 1729). Mostly Devonian in age.

**Kimmeridge Clay.** (Kimeridge Clay.) The rather thick Upper Jurassic formation, the British representative of the Kimmeridgian stage, consistently of clay lithology, extending along a low-lying outcrop from the Yorkshire coast at Filey to the Dorset coast at Kimmeridge (Kimeridge) in the Isle of Purbeck. Characteristic fossils are the brachiopod *'Rhynchonella' inconstans*, the ostreid lamellibranchs *Exogyra virgula* and *Ostrea delta*, and the ammonite *Pavlovia rotunda*. For the question of spelling see Arkell, *Jurassic System in Great Britain*, 1933 (but not followed here). (Webster, *TGS*, 1814, 'Clay . . . containing the Kimeridge coal' and in Englefield, *Isle of Wight*, 1816, 'Kimmeridge strata'.)

**Kimmeridgian.** See JURASSIC SYSTEM.

**Kink-band.** See KINK-PLANE. 'Kink-bands and related geological structures' (Anderson, *N*, 1964). 'Nature and origin of kink-bands' (Dewey, *Ip*, 1965).

**Kink-plane.** 'The sharp bending of beds about certain planes is called in German *Knickung* and the planes, *Knickungsebene*. In English the planes are best termed "kink planes"' (Hills, *Structural Geology*, 1963). A (structurally-defined) band of rock between two parallel kink-planes is a 'kink-band'.

**Klippe.** A nappe or pile of nappes detached by erosion or gravity gliding from the parental mass, of which it is now only a remnant, is known as a nappe-outlier or "klippe"' (Holmes, *Physical Geology*, 1965). Whether by erosion or by gravity gliding, the front portion is 'clipped off', as it were, from the main part of the nappe. This seems the most reasonable derivation of the term, and the *OED* gives the etymology of the verb 'clip' as probably from Old Norse *klippa*. However, the term, originating with the Alpine geologists, is often said to be derived from the German *klippe*, a cliff, though a klippe in the structural (geological) sense is not necessarily bounded by cliffs. The above is the usual meaning; for discussion of other meanings see Dennis, *Tectonic Dictionary*, 1967, and Andrasov and Scheibner, *IGC*, 1968.

**Knick point.** (Nick point.) '(German, *knick-punkt*). A break of slope, particularly in a river-profile' (Stamp, ed., *Geographical Terms*, 1961). It is sometimes used specially for a rejuvenation effect, but to the present writer it seems likely that nearly all knick points are due to some commoner and more compelling cause, such as the outcropping of a hard bed or glacial erosion, and that, in any particular instance, such causes must logically be taken first and rejuvenation invoked for consideration only when none of those are found to fit the case.

**Knoll-reef.** A reef that formed as a knoll (small hill) or a chain of knolls. Quaquaversal initial dips will tend to be shown in the material forming the original knolls. See REEF-KNOLL.

**Komatiite. 1.** Ultramafic lava. (Viljoen and Viljoen, 1969). **2.** An igneous suite analagous to the tholeitic, calc-alkaline and alkaline suites, distinguished by the presence of ultramafic lavas. (Arndt and others, 1977.) *Komatiites* (Arndt and Nisbet, 1982). [Komati River, Barberton Mountain Land, Transvaal, South Africa.]

**Kratogen.** See CRATON.

**Kyanite.** (Cyanite.) See ANDALUSITE. (Werner, 1789.) [G. *kuanos*, blue.]

# L

**Labradorite.** See PLAGIOCLASE. Often shows a rich play of blue-green colours on cleavage faces. (Werner, 1780.) [Labrador.]

**Labyrinthodontia.** A superorder of the class Amphibia, named from the character of the teeth; also known as the Stegocephalia from the bony character of the skull. Carboniferous to Trias. [genus *Labyrinthodon* (G. *laburinthos*, a labyrinth.)]

**Laccolite.** (Laccolith.) An igneous intrusion in the form of a low dome with (known or assumed) flat base, intruded among sedimentary rocks which themselves become domed by the force of the intruded magma. It is usually supposed to have a pipe-like feeder below. Typically intruded into flat or gently dipping strata. The term has also been used for a similar intrusion of sedimentary rock: 'Description of a laccolithic sedimentary intrusion in the Miocene near Campina, Rumania' (De Raaf, *QJ*, 1945). 'Laccolite' is much to be preferred to 'laccolith' as being not only the original form of the term but also more euphonious. They are equally appropriate etymologically, both '-lite' and '-lith' being derived from the Greek *lithos*, a stone. For kinds of rock-bodies, '-lite' or '-lithe' is the French suffix, '-lite' the usual English one, and '-lith' or '-lithen' the German (with silent h). The substitution of '-lith' for '-lite' in English was made by American geologists (first, apparently, by Merrill, in 1879) owing partly (it seems) to a confusion between the terminations '-lite' and '-ite', which led to the erroneous idea that '-lith' was etymologically the more correct for a body of rock. British geologists largely followed the American usage in order 'to distinguish such names from mineral and rock names' (Holmes, *Nomenclature of Petrology*, 1928), apart from any question of correctness of etymology. This seems unnecessary and unfortunate. Harker (among the later authorities) always used the original, correct, unambiguous, euphonious 'laccolite'. 'The lava of the Henry Mountains, instead of rising through all the beds of the earth's crust, stopped at a lower horizon, insinuating itself between two strata, and opening for itself a chamber by lifting all the superior beds. In this chamber it congealed, forming a massive body of trap. For this body the name "laccolite" (λάκκος [lakkos], cistern, and "λίθος" [lithos], stone) will be used.' (Gilbert, *Henry Mountains*, 1877). 'The Eildon Hills are regarded as representing the denuded remains of a composite laccolite, and the intrusions have been made sheet by sheet, giving rise to the appearance of stratification' (Pringle *BRG, S. Scotland*, 1948). 'The laccolithic series' (Jones and Pugh, *AJS*, 1949). A nomenclatural question of more general importance has arisen in connexion with the use of the term. It was defined in two ways which were assumed to coincide: (*a*) by description of form and manner of emplacement and (*b*) by naming a type. If Gilbert's type 'laccolite' turns out to be, as a result of further research, really of the form and manner of Harker's 'phacolite', are 'phacolites' to be termed 'laccolites' ? (Watts, discussing Blyth, *QJ*, 1943.) But 'types' outside palaeontological nomenclature (in so far as they are employed at all) have not the rigid significance they have inside, and it seems that a term should always continue to carry the descriptive meaning given to it by its author even if at the same time he cites an instance as the type which is later found to be something different. (Compare also the case of the mineral-name Bytownite.)

**Lacustrine.** Of or pertaining to lakes, particularly as applied to deposits. See LIMNIC. [L. *lacus*, a lake.]

**Lag-fault.** 'A low-angle fault formed during a forward movement of the rocks—the beds above it having lagged behind those below it' (Trotter, *QJ*, 1947). (Marr. *PGA*, 1900.)

**Lagoon phase.** The stratigraphical facies of 'a group of rocks, the characters and development of which show that it has been deposited in a coastal area, of wide extent both parallel and at right angles to the coast, but so extremely shallow as to have been, in effect, isolated from the neighbouring though deeper parts of the sea, and thus to have become the site of peculiar types of sediment and fauna' (Dixon, *QJ*, 1911).

**Lahar.** A mudflow of volcanic material; ash and coarser products mixed with water. There are 'rain lahars', the rain being either the condensation of volcanic vapours or ordinary rain, and 'crater-lake lahars', usually hot and often very acid, the result of the overflow and evacuation of crater-lakes. (Malay.)

**Lamellibranchiata.** (Lamellibranchia.) One of the three chief classes of the phylum Mollusca, having a shell of two hinged valves, with beaks (first-formed parts) more or less prominent. The shell is typically equivalve and inequilateral, and nearly always the valves articulate by means of teeth and sockets on each valve; it is thus distinguished from the shell of a brachiopod, especially as there is no pedicle opening or associated structures in the lamellibranch shell. Aquatic, mostly marine; ranging from the Cambrian and becoming steadily more varied and abundant up to the present day. Pelecypoda is another name for the class. (L. *lamella*, a little plate, *branchiae*, gills.]

**Lamina** (pl. **-ae**). In geology the term is now largely reserved for a thin separable layer in stratified rocks, representing a layer of original deposition. A bed or stratum is thus often made up of laminae, as the thickness of a book is made up of leaves. Hence 'laminated', 'lamination'. A shale is the most perfectly laminated type of rock. It may, however, be used in its ordinary general sense of any 'thin plate, scale, layer, or flake' (*OED*), and in former times it was used in connexion with slaty cleavage (see quotations from Conybeare and Phillips, and Otley, under CLEAVAGE (1); and De la Beche, *Cornwall* [&c.], 1839).

**Lamprophyre.** A basic igneous rock rich in porphyritic crystals of biotite or other mafic minerals, typically in a fine-grained groundmass of alkali felspar. (Von Gumbel, 1879.) [G. *lampros*, glistening, *phurō*, mix.]

**Landenian.** See PALAEOGENE. [Landen, Belgium.]

**Landform.** This short simple term itself defines exactly what it denotes and it has no synonym. 'Landform' seems to be for the geomorphologist what 'fossil' is for the palaeontologist or 'rock' for the petrologist, yet the term is hardly in constant and universal use (it is not in the *OED*, though it is in *Webster*.) 'The systematic description of land forms' (Davis, *GJ*, 1909). *Geomorphology: Introduction to the Study of Landforms* (Cotton, 1942). *History of the Study of Landforms: or the Development of Geomorphology* (Chorley and others, 1964/73. *Analysis of Landforms* (Twidale, 1976).

**Landscape marble.** A rock showing, on a smooth surface more or less perpendicular to the bedding, the effect of a landscape with trees. A typical occurrence is that of the Landscape Marble ('known as a geological curiosity for over two centuries', Hamilton), the chief variety of the Cotham

Marble, itself a distinctive band in the Ur. Rhaetic Cotham beds of the W. of England. 'The Cotham Marble is . . . a highly ornamental stone of striking appearance. It is a buff, exceedingly fine textured limestone through which solutions have worked their way, depositing dendritic manganese oxide in such a way as to simulate trees and bushes and to give the general effect of a landscape painting in sepia' (Wells and Kirkaldy, *Historical Geology*, 1959). 'Cotham Stone' was first described by Owen (*Observations*, 1754; his description being discussed by Smith, *PBNS*, 1945). Owen attributed the formation of the landscape to rising gas bubbles, an explanation generally accepted (e.g. *BRG*, 1948), but a detailed description and discussion has recently been given by Hamilton (*P*, 1961; also summarized, *PGS*, 1960) who states that 'the landscape is an association of algal growth forms occurring in convex biohermal masses'. 'We may notice another marble, in which there appear many blackish figures which imitate little trees and herbs . . . it must be carefully considered, in order that the true interpretation of this effect may be arrived at, whether the images of this kind of little plants penetrate within the stones: for if they do actually penetrate, without question these appearances cannot arise from a moss that has stuck on but from veins running through the substance of the stone' (Aldrovandus, 1648, quoted by Bromehead, 'A seventeenth century geological museum', *QJ*, 1947). 'In the joints of the Lias-stones, growing over beds of clay, we often meet with a great plenty of elegant landskips' (Beaumont, *PT*, 1676, quoted by Arkell and Tomkeieff). 'Landscape marble' (Woodward, *GM*, 1892; Thompson, *QJ*, 1894).

**Landslide.** A landslide or landslip is the sliding (slipping) downhill, on the land, of a mass of rock material on a more or less steep slope or cliff-face. The rock material may be a superficial deposit or a large mass of strata (or other rock); in the latter case, if the mass retains its character without breaking up it is sometimes distinguished as an 'earthslide'. *Landslides and Related Phenomena* (Sharpe, 1938). 'The coastal landslips of south-east Devon and west Dorset' (Arber, *PGA*, 1940/41). 'The Folkestone Warren landslides' (Hutchinson and others, *QJEG*, 1980).

**Lapidify.** 'Lapidifying process. Conversion to stone' (Lyell, *Principles*, 1833). The same as lithify, but derived from the Latin instead of the Greek; though etymologically more consistent, it is not so commonly used as 'lithify'. 'Modes of lapidification' (Kirwan, *Geological Essays*, 1799). [L. *lapis, lapid-*, a stone.]

**Lapilli.** Small stony pieces of lava, about the size of peas or walnuts, falling as pyroclastic material, having been blown into the air in a volcanic eruption. [L. *lapillus*, a little stone.]

**Laramian.** See LARAMIDE.

**Laramide.** 'The Laramide orogeny (late Cretaceous to early Tertiary) transformed a long belt from the Yukon Plateau and the Rockies of Canada to the Sierra Madre Oriental of Mexico into an irregular chequer-board of great uplifts and deep intermontane basins with thick thrust-sheets of sedimentary strata that travelled outwards from the higher uplifts' (Holmes, *Physical Geology*, 1965). 'The first mention of the Laramide orogeny by name appears to be by Dana (*Manual*, 1896)' (Dennis, *Tectonic Dictionary*, 1967). This orogeny has been called the 'Laramian' orogeny (e.g. Kay and Colbert, *Stratigraphy and Life History*, 1965) with the name 'Laramides' applied to the corresponding mountain ranges, thus making the nomenclature uniform with, e.g., 'Caledonian' and 'Caledonides'. [Laramie Range, Wyoming, U.S.A.]

**Larvikite.** 'A most distinctive type [of syenite] is the very handsome Norwegian rock, widely used for ornamental purposes and termed "larvikite" (or "lauvigite") by Brögger (1890). It is characterized by its coarse grain and distinctive feldspars, which, especially on polished surfaces, exhibit a beautiful blue schillerization' (Hatch and Wells, *Igneous Rocks*, 1965). 'Lauvikite is the delightful South Norwegian rock which serves the double purpose of a facing to public houses, and a very useful indicator boulder' (Trechman, 'The Scandinavian Drift on the Durham coast', *PGA*, 1931). [Larvik, Norway.]

**Lateral.** In stratigraphy, often used in the sense of 'in directions of extension of the strata' (at right angles to the stratigraphically vertical direction). See VERTICAL.

**Laterite.** A weathering product developed over wide areas in wet tropical regions as a result of the deep decomposition of, particularly, igneous rocks. Hence 'laterization'. 'Although there is a lack of detailed information concerning the exact chemical process occurring during laterization, there appears to be some justification for defining it, in a general sense, as the leaching out and elimination of silica, alkali, and alkaline earths and the concentration, in their hydrated form, of iron and aluminium oxide, the latter compound being partly combined with silica' (Mohr and Van Baren, *Tropical Soils*, 1954). At the surface it is hard, and any soil forming over it is infertile and easily removed by rain so that large bare exposures are formed; but underground it is soft and can be cut into blocks which harden on being exposed. Bricks are thus made from it. 'The word "laterite" was originally suggested by Buchanan (*A Journey from Madras*, 1807) as a name for a highly ferruginous deposit first observed in Malabar during his journey through the countries of Mysore, Canara and Malabar in 1800-1' (Prescott and Pendleton, *Laterite and Lateritic Soils*, 1952, who quote and discuss the original definition). *Laterite and Landscape* (McFarlane, 1976). [L. *later*, a brick.]

**Laurentian.** See PRECAMBRIAN. (St. Lawrence river.]

**Lava.** Extruded igneous magma and the rock formed as the result of its solidification. 'The pillars of lava-like matter at Fingal's Cave' (Banks in Pennant, *Tour in Scotland*, 1774). See also quotations under EXTRUSION and TOADSTONE. The term is often extended to cover the magma, and corresponding rock, which, though not actually extruded, is in direct connexion with the outpoured material, such as that in a volcanic neck; nor is it wrong to say, with Playfair (1802), that the magma of a minor intrusion is 'unerupted lava' (and see quotation under PHENOCRYST). [Italian f. L. *lavo*, wash.]

**Lava toe.** See PAHOEHOE.

**Law of structures.** See DIFFERENTIAL EROSION.

**Law of superposition.** See SUPERPOSITION. 'The law of superposition. . . . This is the fundamental principle of stratigraphy' (Lake and Rastall, *Textbook*, 1910/41).

**Laxfordian.** See SCOURIAN.

**Layer.** 'Something laid; a thickness of matter spread over a surface; especially one of a series; a stratum, course, or bed' (*OED*). Of the four parts of this definition, the last may be said to be an exemplification. It is implied that the layer is more or less distinctly limited above and below.

**Layered intrusion.** An igneous intrusion showing layering (banding, stratification), which might be of several possible modes of origin (flowage, successive accumulations, multiple intrusion, composite intrusion, gravitational differentiation, incorporation of strata). (Wager and Deer, 'Skaergaard intrusion', Greenland, 1939; Wells, *QJ*, 1953; 'The layered ultrabasic rocks of Rhum', Brown, *PT*, 1956; 'Layered intrusions', Bailey, *TGSG*, 1958; 'Layered series of the Skaergaard intrusion', Wager, *JPt*, 1960; *Layered Igneous Rocks*, Wager and Brown, 1967.)

**Leach.** To remove the soluble constituents of a soil, rock, ore, &c., by the action of percolating waters.

**Ledian.** See PALAEOGENE. [Lede, Belgium.]

**Lenticular.** For a lens-shaped piece of rock substance, of any size; particularly for a mass or bed of rock thinning out in all directions.

**Lepidoblastic.** 'Different types of schistosity, determined mainly by the mineralogical composition of the rocks, have received appropriate names. Of chief importance are the "lepidoblastic" or flaky, due to an abundance of minerals like micas and chlorites with a general parallel arrangement, and the "nematoblastic" or fibrous, seen in rocks composed largely of such minerals as glaucophane and actinolite' (Harker, *Metamorphism*, 1950). (Becke, 1903.) [G. *lepis*, a scale.]

**Leucite.** See FELSPATHOID. 'Best developed in certain intermediate and basic lavas from the Roman volcanic province in Italy and from the Leucite Hills in Wyoming' (Hatch and Wells, *Igneous Rocks*, 1952). (Werner, 1791.)

**Leucocratic.** Light-coloured; for an igneous rock very rich in the felsic (light-coloured) minerals. 'Leuco-', as a prefix to a rock-name, means that it is a variety of a lighter colour than the normal. See quotation under MELANOCRATIC. [G. *luekos*, light-coloured, *krateo*, predominate.]

**Leucosome.** The light-coloured part of a rock-body, particularly of a metamorphic one. See MELANOSOME.

**Lewisian.** The lower of the two main Precambrian groups in the NW. Highlands and Islands of Scotland. 'To the north-west of the Moine Thrust belt, in the North-west Highlands and Hebrides, lies the north-western foreland of the Caledonides which remained stable throughout the orogeny. Here, a basement of crystalline gneisses, the Lewisian, is overlain with strong unconformity by an unmetamorphosed cover of Torridonian arkoses followed with a second unconformity by Cambro-Ordovician orthoquartzites and dolomites' (Watson, *PGA*, 1963). '[The rocks] form nearby the whole of the Outer Hebrides, and occupy a variable belt of the western parts of the counties of Sutherland and Ross. Murchison proposed to term them the fundamental or Lewisian Gneiss from the Isle of Lewis, the chief of the Hebrides [*QJ*, 1860]. Afterwards he called them Laurentian' (Geikie, *Textbook*, 1882).

**Lias. 1.** Any compact, argillaceous limestones or cementstones, usually banded with beds of shale or clay; especially those quarried in Somerset and other parts of the West of England, but also similar limestones anywhere. [*OED* gives 1404.] **2.** 'The formation name for the Lower Jurassic, consisting mainly of clays and shales, but with the typical "lias limestones" near the base in certain districts. Adopted from Old French *liois*, modern French *liais*, a compact kind of limestone' (Arkell and Tomkeieff, *Rock Terms*, 1953). (Strachey, *PT*, 1719, 'Lyas'; Michell, *MS*, 1788, 'Lyas strata'.)

The British Lias outcrops along a belt crossing England from the Yorkshire to the Dorset coast, and extending into S. Wales. It is famous for its fossils, particularly its ammonites and reptiles (though the latter are relatively uncommon). There are three divisions and some of the characteristic fossils of each are the following: Lower Lias, *Extracrinus briareus* (crinoid), *Spiriferina walcotti* (brachiopod), *Gryphaea incurva*, *Ostrea liassica*, *Plagiostoma giganteum* (lamellibranchs), *Pleurotomaria anglica* (gastropod), *Arietites bucklandi*, *Asteroceras obtusum*, *Oxynoticeras oxynotum*, *Psiloceras planorbis* (ammonites), and *Ichthyosaurus communis* (reptile); Middle Lias, *'Terebratula' punctata*, *'Rhynchonella' tetrahedra* (brachiopods), *Amaltheus margaritatus* (ammonite); Upper Lias, *Lingula beani* (brachiopod), *Nuculana ovum* (lamellibranch), *Dactylioceras commune* and *Hildoceras bifrons* (ammonites).

The Lias, characterized by 'the Gryphites oyster' was traced from S. Wales across the English midland counties by John Strange (*A*, 1782). 'Blue Lias' and 'White Lias' are names that have long been in use. The former refers to the alternations of bluish clay or shale and pale grey argillaceous limestone which makes up the lower part of the Lower Lias in S. England. The White Lias is a name given to a division near the top of the Rhaetic series, underlying the Lias, composed of pale-coloured limestones and thin marls. 'Most of the cottages are built with a bluish limestone, called "lyas" . . . there is also the "White lyas"' (Maton, *Western Counties*, 1797). ('Blue Lias' and 'White Lias' were tabulated by Smith, *MS*, 1797/9.)

**Liesegang rings.** A small-scale, more or less concentric, chemical-reaction effect, seen as discrete, but largely contiguous, structures throughout a part of a rock-body. Each structure shows, on an exposed surface, ring-shaped bands of distinctive colour. The effect is probably caused by rhythmic precipitation in a gel (a jelly-like colloidal solution), perhaps a kind of weathering process. A striking example is seen in Silurian rocks in the Ingleborough district of Yorkshire, particularly at Moughton Whetstone Hole (Crummack Dale), at places just below the Carboniferous Limestone unconformity; it seems that the alternating red and green bands here must be the result of pre-Carboniferous 'weathering'. [R. E. Liesegang, *Geologische Diffusionien*, 1913.]

**Life assemblage.** See BIOCOENOSE. 'Assemblage' here can refer either to all the forms of life living together in the particular area or only to those belonging to some specified organic group or groups. 'It is here suggested that the terms "life assemblage" and "death assemblage", respectively, be employed as the palaeontological equivalents of biocoenosis and thanatocoenosis' ('Life and death assemblages among fossils', Boucot, *AJS*, 1953). 'Brachiopod life assemblages from the Marlstone Rock-bed' (Hallam, *P*, 1962). 'Life or death assemblages? The first problem which arises in any consideration of palaeoecology is that of determining whether the fossils concerned are preserved where they lived or have been moved after death by currents or other agencies, since upon this depend any further deductions concerning conditions of life of the animals' (Middlemiss, *P*, 1962). 'A quiet water deposit containing a life assemblage of *Mya arenaria* . . . in the Red Crag of Essex' (McManus *GM*, 1964).

**Lignite.** A carbonaceous rock, a kind of coal, intermediate in character (rank) between modern peat and bituminous coal. Also intermediate in age, occurring in the late Cretaceous and Tertiary coalfields. The only important occurrence of lignite in Britain is that among the Tertiary deposits of the Bovey basin, Devon. (L. *lignum*, wood.]

**Limb.** A fold is usually said to be made up of two 'limbs', one on each side of the hinge. Thus in a succession of folds, each limb is common to two adjacent folds; but sometimes the parts of adjacent anticlines and synclines are divided into 'arch limbs', 'middle limbs', 'trough limbs'.

**Lime-silicate rock.** A rock resulting from the thermal metamorphism of an impure limestone. The silicate minerals of the impurities have combined with lime to produce new lime-silicate minerals, the calcium carbonate left over recrystallizing as calcite.

**Limestone.** A rock consisting chiefly of carbonate of lime ($CaCO_3$). In the widest sense it includes dolomite-rock (the mineral dolomite being half 'lime' carbonate). See quotation (Owen) under VEIN. 'Structure and origin of limestones' (Sorby, *QJ*, 1879). *Limestones* (North, 1930). 'Terminology of limestone and related rocks' (Rodgers, *JSP*, 1954). 'Practical petrographic classification of limestones' (Folk, *BAAPG*, 1959). 'The classification of Avonian limestones' (George, *JGS*, 1972).

**Limestone knoll.** See REEF-KNOLL. 'The Lower Carboniferous limestone knolls of Clitheroe, Lancashire' (Parkinson, *GM*, 1967/8).

**Limestone pavement.** See CLINT. The limestone pavements of the north crop of the South Wales coalfield' (*Thomas*, *TIBG*, 1970).

**Limnic.** Although 'limnology' has been coined for 'the science of lakes', lake deposits are 'lacustrine' to the geologist; while 'limnic' is usually reserved for deposits laid down in shallow pools and swamps. Nearness to the sea is implied, but the strata are entirely non-marine (cf. 'paralic'). Limnic deposits, in this sense, are well exemplified in the typical strata of the Coal Measures. This distinction between 'limnic' and 'lacustrine' seems to correspond to a slight difference between the meaning of G. *limnē*, a marshy lake in particular, and L. *lacus*, any kind of lake.

**Limonite.** Hydrated iron oxide ($2Fe_2O_3.3H_2O$). Results from the alteration of other iron minerals. 'The name limonite (λειμών [*leimōn*], a meadow) really refers to "bog-iron ore", which is a brownish-yellow deposit common in marshes and derived from the ferrous carbonate and sulphate dissolved in their waters' (Miers, *Mineralogy*, 1929). (Hausmann, 1813.)

**Lineage.** A line of descent. Used more or less synonymously with 'evolutionary series', but the latter is more general, 'lineage' usually referring to one particular line of descent within the plexus. (Not necessarily confined to the organic world.)

**Lineament.** A more or less linear expression of structure, particularly a physiographical expression and one connected with faulting. Hobbs, who first used the term ('Lineaments of the Atlantic Border region', *BGSA*, 1904), has: 'Significant lines of landscapes which reveal the hidden architecture of the rock basement are described as "lineaments". They are character lines of the earth's physiognomy' (*Earth Features and their Meaning*, 1912). 'Several attempts have been made to recognize a worldwide pattern in features such as faults, fractures, and major relief forms, for example continental margins and submarine ridges. Such structures are termed in general "lineaments"' (Hills, *Structural Geology*, 1953; and see quotation under MORPHOTECTONICS). Johnson (*New England-Acadian Shoreline*, 1925) gives some smaller-scale examples. 'These lineaments [in the Outer Hebrides] appear to mark steep zones of deep-seated dislocation along which adjacent gneiss blocks moved relative to each other' (Watson *PGA*, 1977).

**Linear.** Arranged in a line or lines, strung out; to be distinguished from a planar arrangement.

**Lineation.** 'A descriptive and non-genetic term for any kind of linear structure within or on a rock. It includes striae on slickensides, fold axes, flow lines, stretching, elongate pebbles or ooids, wrinkles, streaks, intersection of planes, linear parallelism of minerals or components, or any other kind of linear structure of megascopic, microscopic, or regional dimensions' (Cloos, *Lineation*, *MGSA*, 1946). 'On lineation and petrofabric structure' (Anderson, *QJ*, 1948).

**Lingula flags.** The rocks in Wales which represent the general Olenus series of the Cambrian. They are well seen in N. Wales where they are some 2,000 m. thick, comprising the three divisions the Maentwrog, the Ffestiniog, and the Dolgelly. The rocks are typically flaggy, but by no means everywhere so. The commonest fossil is the brachiopod *Lingulella davisi*. This was formerly placed in the genus *Lingula* (Sedgwick, *QJ*, 1847/52, *British Palaeozoic Rocks*, 1855.)

**Listric surface.** A curvilinear, usually concave upward surface of fracture that curves, at first gently and then more steeply, from a horizontal position. Hence listric fault. [G. *listron*, shovel.]

**Lithic tuff.** See TUFF.

**Lithify.** Turn to stone. Used for the consolidation and induration of sediments. See LAPIDIFY.

**Lithification.** Turning to stone. Used for the consolidation, particularly the hard induration, of sediments. See LAPIDIFICATION.

**Lithofacies.** See FACIES. The general aspect of the lithology of a bed, formation, &c., usually considered particularly as the expression of the local conditions of deposition. *Lithofacies Maps* (Sloss and others, 1960).

**Lithogenesis.** The origin and formation of rocks, particularly sedimentary rocks. *Principles of Lithogenesis* (Strakhov, trans., Fitzsimmons, ed. Tomkeieff and Hemingway, 1967).

**Lithography.** The only current sense is that of drawing or writing on stone (of a certain kind) for the purpose of producing impressions. The *OED*, however, gives 'A description of stones or rocks' as an obsolete sense (quoting an example from *PT*, 1708), a sense that would be useful geologically as there is at present no suitable term for the description of the lithological character of rocks.

**Lithology.** In the widest sense, the general character of a rock, particularly as seen in field-exposures and hand-specimens; that is, its mineral composition, texture, primary structures, and the smaller-scale secondary structures. (Literally, it is the study of this character.) Hence 'lithological'. Jukes (*Manual*, 1862) defines it as 'the study of the mineral structure of rocks'. and includes the igneous and metamorphic; but it more commonly implies the naked-eye mineral composition and texture of a sedimentary rock. Thus Lyell (*Principles*, 1833): 'Lithological. A term expressing the stony structure or character of a mineral mass. We speak of the lithological character of a stratum as distinguished from its zoological character'. 'The lithological succession of the Carboniferous Limestone of the Avon section' (Reynolds, *QJ*, 1921). 'Lithology' was also used, in early times, for the character of the rocks of a district: 'The mountains and the lithology of the environs of Oban' (Faujas Saint Fond, *Voyage en Angleterre* [&c.], 1797, trans. 1799). 'Lithological survey of Schehallien' (Playfair, *PT* 1811). Another application has been suggested by Tyrrell (*Petrology*, 1929): 'Etymologically, it means the science of stones; and, accordingly, there is a tendency, which should be encouraged, to use the term to indicate the study of stones in engineering, architecture, and in other fields of applied geology'. In the 'Concise guide to stratigraphical procedure' (*JGS*, 1972) the term is used in a very comprehensive sense: '. . . lithological characters (structural, petrological, mineralogical, palaeontological, geochemical and geophysical)'.

**Lithosphere.** This term has been used in several senses, particularly as synonymous with 'crust' in the former vague application of that term. It is now used strictly for the crust (in the modern sense) together with a certain upper part of the mantle. 'As long ago as 1926, the celebrated Californian geologist, Beno Gutenberg, suspected that there was a relatively low-velocity seismic zone enclosed within the upper mantle. Thirty years after Gutenberg's suggestion the reality of the zone was established by world-wide records of the blasts from underground nuclear explosions. . . . The low-velocity zone in the upper mantle is thus a kind of lubricated zone, making relative movements between the overlying layers and the interior possible. As a result, it has become usual to refer to the crust and the upper part of the mantle as the "lithosphere", and the low-velocity zone as the "asthenosphere"' (Holmes, *Physical Geology*, 1978, slightly adapted). There is thus a 'sub-crustal' part of the lithosphere. The base of the lithosphere extends well beneath both continental and oceanic crusts to depths (below the earth's surface) of the order of 100 km.

**Lithostratigraphy.** 'Lithostratigraphy (rock stratigraphy) is the description of local rock successions in terms of named and described units (primarily formations) and their spatial relationships. These lithostratigraphical units are described in terms of observable and recognizable petrological, mineralogical, geochemical, or general palaeontological characters.

Lithostratigraphical units are localised in occurrence and readily observable and recognisable by their lithological characters. The boundaries of each unit are placed at positions of lithological change. These may be at sharp contrasts or points fixed arbitrarily in gradational rock sequences. The lower, upper, and lateral boundaries should be based on those criteria that provide the greatest unity and practical utility. Boundaries of lithostratigraphical units are not defined to imply isochronous surfaces but to provide an observable framework of rock for reference in subsequent correlation and interpretation.

The use of fossils in lithostratigraphy is clear in those cases where they form part of the grossly recognisable lithology of the rock: as in coral beds, coquinas, plant beds, etc. There is, however, a less clear area closer to biostratigraphy, where lithostratigraphical units are defined, partly at least, by the identification of fossils. As such these units must remain readily recognisable and, in general, mappable' (*Stratigraphical Guide*, *GSSR* (11), 1978).

The following is the lithological hierarchy:

supergroup
group
formation
member
bed

**Lithotype.** A particular lithological character or category; as in, for instance, 'the dull and bright lithotypes of coal' (Smith, *GM*, 1957).

**Lit-par-lit.** An injection or diffusion of innumerable, more or less parallel, tongues of magma, into a sedimentary rock formation, a foliated gneiss thus resulting from such intrusion and the accompanying metamorphism. This conception is now incorporated in the more general one of migmatization. [French, bed-by-bed.]

**Littoral.** Of or pertaining to the shore (between tide marks), particularly as applied to deposits. 'Vermicular markings, sometimes clearly of organic origin often cover the surface, and indicate "littoral deposition"' (Phillips, *Yorkshire*, 1836). [L. *litus*, the shore.]

**Living fossil.** This usually means an animal (or plant), known as a fossil and thought to be long extinct, which has later appeared as a living member of the world of today. A striking example is the coelacanth fish, belonging to a distinct group of fishes, which all the evidence indicated had died out at the end of the Cretaceous, but of which a few specimens (*Latimeria*) have now been caught off the coast of East Africa. ('A living fossil', J. L. B. Smith, *CN*, 1939, reprinted in *Crust of the Earth*, Rapport and Wright, eds., 1955.) *Neopilina*, discovered as a living animal in 1952, was found to belong to a group of archaic molluscs, related to the annelids, which had come to light, as fossils, in 1940 and had been thought to have been extinct since Palaeozoic times. *Ceratodus*, the Lung fish or Barramunda of the Queensland rivers, Australia, was first known from fossils, those found in the Rhaetic Bone-bed of Aust Cliff, near Bristol, though it is now known from the Upper Carboniferous to the Tertiary. '"Living fossils" in western Atlantic plankton' (Wall and Dale, *N*, 1966). 'A family of living fossil spiders' (Bristowe, *E*, 1975). A recently identified modern member of the crustacean family Glypheoidea, hitherto believed to have become extinct in the early Tertiary (*S*, 1976). Much commoner are 'living fossils' which are the lingering survivors of a race which flourished in the past, such as the lamellibranch family Trigoniidae, of which a few species still live in Australian waters; and the well-known *Nautilus*.

**Llanberis Slates.** The purple, blue, and green slates which, stratigraphically, form the upper half (about 600 m.) of the Lower Cambrian series on the north side of the Snowdon syncline in Caernarvonshire. *Pseudatops viola*, a trilobite considered to indicate a Lower Cambrian age, has been found in the uppermost beds. See quotation (*BRG*) under SLATE. (Sedgwick, *QJ*, 1852, *British Palaeozoic Rocks*, 1855.)

**Llandeilo series.** (Llandeilian.) The series in the middle of the Ordovician succession, between the Llanvirn (or Arenig) below and the Caradoc (=lower Bala) above. It is defined as the general series including all the time equivalents of the Llandeilo formation of Llandeilo (Carmarthenshire). This comprises limestones and flags some 800 m. thick with characteristic trilobites, for instance *Ogygiocarella debuchi*, *'Ampyx' nudus*, and *Trinucleus fimbriatus*. The equivalent in the graptolitic succession is now taken to be only the zone of *Glyptograptus teretiusculus;* the graptolitic fauna, with that of the Llanvirn, constituting a subfauna of the Diplograptid fauna (Bulman, *P*, 1958). The name has been used in wider senses; to include beds now placed in the Caradocian, and the upper of the two zones of the Llanvirnian. (Murchison, *PGS*, 1833/4, *Silurian System*, 1839.) [Llandeilo, Dinefwr district of Dyfed.]

**Llandovery series.** (Llandoverian.) The lowest of the three Silurian series; also known as the Valentian. It has a characteristic brachiopod fauna, of which *Pentamerus oblongus* and *Stricklandia lens* are prominent species, while *Phacops elegans* is a characteristic trilobite. The lowest beds contain the latest subfauna of the Diplograptid fauna, but otherwise the series is characterized by certain subfaunas of the Monograptid fauna (Bulman, *P*, 1958). The shelly facies is typically developed in the Llandovery and Haverfordwest districts of S. Wales. The graptolitic facies of the Llandovery series has its classic exposure in Dobbs Linn, NE. of Moffat, Dumfriesshire, where most of it is there represented by the Birkhill series. In somewhat less condensed succession it is also well developed in Central Wales, for instance at Ponterwyd and Machynlleth, and in the Lake District (Skelgill beds). (Geological Survey maps, 1857/8; Murchison, *Siluria*, 1859; for history of the name see *C*, 1951.) [Llandovery, Dinefwr district of Dyfed.]

**Llanvirn series.** (Llanvirnian.) The second series of the Ordovician system, but sometimes incorporated, the lower part in the Arenig, and the upper part in the Llandeilo, series. In the graptolitic succession it carries, with the Llandeilo, a subfauna of the Diplograptid fauna (Bulman, *P*, 1958). The volcanic rocks of the Cader Idris range belong, for the most part, to this series. (Hicks, *PSR* and *PGA*, 1881.) [Llanvirn farm, Preseli district of Dyfed.]

**Load. 1.** Particularly as 'load of a stream'; the rock material that is being transported, in solution and as particles in suspension, or being moved by traction, rolling, saltation, &c. In many discussions the context makes it clear that only the mechanical, visible load (the rock particles) is meant. The load of a stream cannot be defined as the amount of material it can carry, for a stream is (theoretically) always fully loaded, in the sense that it can carry no more load without loss of velocity and at the same time it is (theoretically) never fully loaded, in the sense that it can always carry more load if the load is composed of still finer particles. **2.** Particularly as 'loading', the weight resulting from the accumulation of sediments, by which those already deposited are compressed.

**Load-cast.** A sole-mark. 'Loadcasts, one of the most common basal characteristics of sandstone beds overlying shales or marls, are due to the loading of a mud deposit with a sandbed. The sand is able to sink into the mud forming basin-shaped hollows' (De Sitter, *Structural Geology*, 1956). (Kuenen, Netherlands, 1953, and *BAAPG*, 1953; Kelling and Walton, *GM*, 1957.)

**Loam.** 'A rich soil composed chiefly of clay and sand with an admixture of decomposed vegetable matter' (*OED*). (See Arkell and Tomkeieff, *Rock Terms*, 1953.) [Old English *lam*.]

**Local geology.** The geology of a restricted district, such as that within a few miles of a particular town. *Local Geology* (Davies, 1923/7). See EXCURSION GUIDE. Compare REGIONAL GEOLOGY.

**Lode.** A mineral vein, especially a vein of metalliferous ore. A term particularly associated with the Cornish mining area. 'From the fissures let us proceed to that which they contain, and whatever fills them, whether clay, stone, mineral or metal, we call in Cornwall a "lode"' (Borlase, *Cornwall*, 1758). 'The lode which separates and goes through another must have been formed subsequent to that which it divides and passes between' (Pryce, *Mineralogia Cornubiensis*, 1778) ['lode', 'load', and 'to lead' are associated words from Old English *lād*, way, course. The derivation of 'lode' in the mining sense was considered by Borlase and, particularly, by William Phillips (*TGS*, 1814); it may signify either the 'leading' or the 'load' aspect of the vein.]

**Loess.** A fine-grained, light-coloured, unindurated but coherent deposit, apparently wind-borne, occurring over wide areas of the land. 'Basin of Vienna. . . . Alluvial loam called Loss, with terrestrial shells, of existing species' (Murchison, *TGS*, 1832). 'Loess. The substance to which this name is applied is a sandy calcareous loam of a yellowish brown colour, slightly coherent, and absorbing water with great avidity. It is a deposit which has been generally considered to be peculiar to the Rhine valley. . . . Loess is specially noticed by Leonhard in his *Charakteristic der Felsarten*, published in 1824, who has adopted this trivial name, given to it in the neighbourhood of Basle, but a fuller account of it has more recently been given by Professor Bronn (1830), in his description of the environs of Heidelberg' (Horner, *TGS*, 1836, read 1833). 'In the *Edinburgh Philosophical Journal* for July 1834, there is a paper by Mr. Lyell on this peculiar deposit: he has treated of it very fully in his *Principles of Geology*, 1834; he read some additional observations upon it at the Geological Society in December 1835, an account of which is given in the Society's Proceedings' (Horner, *TGS*, 1836, written 1836). 'The brickearth which forms the upper part of the sea cliff near Pegwell, SW. of Ramsgate, clearly shows that this particular deposit is a true loess; further, this loess rests unconformably on rocks which have been disturbed by a process of freezing and thawing' (Pitcher and others, *GM*, 1954). 'Loess and coversands' (Catt in Shotton, ed., *British Quaternary Studies*, 1977). 'Distribution of loess in Britain (including 'Loess Commission definitions') '(Lill and Smalley, *IPGA*, 1978; and discussion by Catt, *IPGA*, 1979). 'Loess, by far the most important periglacial accumulation, was first recognised in the valley of the Rhine. Here A. Braun in 1842 accurately described it—the name (German *loss*) belongs to the language of the peasants and brickworkers of this region. It was found afterwards elsewhere in Europe and in 1846 along the banks of the Mississippi by C. Lyell and A. Binney who declared it to be analogous with the Rhine loess, though its vast expanse in North America was not appreciated until much later (1873). China's extensive loess was discovered by R. Pumpelly (1866) and described in detail by Richthofen (1877-1883). Deposits comparable to the recent loess have been found among rocks of former geological age; Devonian, Trias, Oligocene, Miocene' (Charlesworth, *Quaternary Era*, 1957).

**Log.** A record of facts concerning a well or bore-hole, particularly a record of the kinds and vertical thicknesses of the rocks passed through.

**Logan stone.** 'Rocking stone. A block of stone so finely balanced on its base that it will easily "log" or "rock" from side to side. Cornish *log*, to rock' (Arkell and Tomkeieff, *Rock Terms*, 1953). The best-known examples are the large pieces of Cornish granite, weathered along horizontal joints so as to have become loose and poised. 'That stones, which lie on other stones, may, by wearing, be brought very near an equilibrium, is proved by what are called rocking-stones, or in Cornwall "loganstones"' (Playfair, *Illustrations*, 1802).

**London Clay.** The main formation of the Eocene succession in the London Basin, immediately underlying most of London itself. It is a bluish clay, weathering yellowish brown; nearly 180 m. thick near Southend to about 90 m. at Windsor. Fossils are rather sparse. See EOCENE. (Smith in Farey, *Derbyshire*, 1811.)

**Longitudinal.** 'Longitudinal' and 'transverse' are used particularly for topographical features that run parallel to, or at right angles to, the general strike of the strata. 'The direction of vallies relatively to that of the chains of hills among which they range: considered under this relation they may be divided into two classes, commonly distinguished as "longitudinal" and "transverse" vallies. Since the direction of the chain of hills is very generally the same with the direction of their constituent strata, it is unnecessary to add that the longitudinal vallies will in such cases be parallel to, and the transverse cut across, the strata among which they range' (Conybeare and Phillips, *England and Wales*, 1822).

**Longmyndian.** The thick series of unfossiliferous mudstones and sandstones in S. Shropshire (Salop) whose main outcrop forms the Longmynd upland. At first thought to be Cambrian it has for long been classed as Precambrian; but recent isotopic evidence suggests that it may be of Cambrian age after all (Bath, *JGS*, 1974). (Murchison, *Silurian System*, 1839, 'Stratified rocks of the Longmynd'; Callaway, *TSANHS*, 1888, 'Longmyndian'.)

**Lopolith.** 'The lopolith; an igneous form exemplified by the Duluth gabbro . . . a large, lenticular, centrally sunken, generally concordant, intrusive mass, with its thickness approximately one-tenth to one-twentieth of its width or diameter' (Grout, *AJS*, 1918). 'Form of the Sudbury lopolith' (Hamilton, *CM*, 1960). [G. *lopas*, a flat dish.]

**Lower Greensand.** A stratigraphical formation of highly ferruginous, glauconitic sandstone (yellow when weathered), the British representative of the general Aptian series of the Lower Cretaceous. It usually forms an escarpment in front of the Chalk escarpment, the outcrop-band running NE.-SW. across the SE. Midlands and also forming an inner rim round the Weald, where it is particularly conspicuous in the northern part. Among the characteristic fossils are the sponge *Raphidonema farringdonense*, the coral *Holocystis elegans*, the brachiopod *'Terebratula' sella*, and the lamellibranch *Exogyra sinuata*. (Michell, *MS*, 1788, 'Sand of Bedfordshire'; Smith in Farey, *Derbyshire*, 1811, 'Woburn Sand'; Mantell, *South Downs*, 1822, and Fitton, *AP*, Nov. 1824, 'Greensand'; Webster, *AP*, Dec. 1824, 'Lower green sand'.)

**Lower London Tertiaries.** A name given by Prestwich (*QJ*, 1852) to the strata, investigated and named by him, between the Chalk and the main London Clay in south-east England: the Thanet Sands, the Woolwich and Reading beds, and the basement beds of the London Clay (later called the Blackheath and Oldhaven beds). See PALAEOGENE. (Smith in Farey, *Derbyshire*, 1811, 'Black-Heath Sand'; Webster, *TGS*, 1814, 'Plastic Clay'.)

**Lower Palaeozoic.** See PALAEOZOIC. 'Wales is classic ground for the study of Lower Palaeozoic rocks. The system names, Cambrian, Ordovician and Silurian commemorate Wales (Cambria) and the names of two ancient Welsh tribes, the Silures and the Ordovices. In geological circles the names of small market towns like Bala, Llandeilo and Llandovery, a wild and lonely mountain, Arenig, even an isolated farm house in Pembrokeshire, Llanvirn, have gained international currency as stage names in the Lower Palaeozoic' (Wood, ed., *The Pre-Cambrian and Lower Palaeozoic Rocks of Wales*, 1969).

**Ludian.** See PALAEOGENE. [Ludes, France.]

**Ludlow series.** (Ludlovian.) The uppermost of the three Silurian series, typically developed at and near Ludlow, Shropshire. The succession is: Lower Ludlow Shale, Aymestry Limestone, and Upper Ludlow Shale, the last being highly calcareous. The Lower Ludlow Shale sees the last of the Graptoloidea, which constitute a subfauna of the Monograptid fauna (Bulman, *P*, 1958). The characteristic fossil of the Aymestry Limestone (where typically developed) is the brachiopod *Conchidium knighti*, and among those of the Upper Ludlow are *Camarotoechia nucula*, *Dalmanella lunata*, *Orbiculoidea rugata*, and *Wilsonia wilsoni* (all brachiopods). (Murchison, *PGS*, 1833/4, *Silurian System*, 1839.) [Ludlow, Salop.]

**Lunar geology.** See SELENOLOGY.

**Lustre.** Of a mineral; the quality and intensity of the light reflected from its surface, including the surface appearance. Thus for quality and appearance: metallic, resinous, vitreous, pearly, silky; for intensity: splendent (mirror-like), shining, dull.

**Lutetian.** See PALAEOGENE. [L. *Lutetia*, Paris.]

**Lutite.** An argillaceous rock, a mudstone, a pelite. [L. *lutum*, mud.]

**Luxullianite.** (Luxulyanite.) A pneumatolytic modification of granite, particularly of the Devon and Cornwall granites, in which tourmalinization has reached an arrested stage so that porphyritic crystals of pink felspar are strikingly contrasted with the dark-blue groundmass of tourmaline, with quartz, the tourmaline in sheaves of needle-like crystals. (Pisani, 1864.) [Luxulyan, Cornwall.]

**Lycopodiales.** The club-mosses, a group of plants belonging to the division Pteridophyta, ranging from the Upper Devonian to the present day, but flourishing most abundantly, and growing as trees, in the Carboniferous period. [genus *Lycopodium* (G. *lukos*, a wolf, *pous*, *pod-*, from the shape of the root).]

**Lyellian doctrine.** See UNIFORMITARIANISM and PAST AND PRESENT. [Charles Lyell, 1797-1875].

# M

**'M'numbers.** See 'F'NUMBERS.

**Ma.** Symbol for 'million years'; the most often required of several such symbols. See TIME UNIT.

**Maar.** A hollow excavated by a volcanic explosion provides a natural well if deep enough to intersect the water table. Such a hollow, when occupied by a lake, is termed a "maar", the name being taken from examples in the Eifel district of Germany' (Cotton, *Volcanoes*, 1944).

**Macrocrystalline.** See CRYSTALLINE.

**Macrofossil.** Any fossil other than a microfossil.

**Macroscopic.** Applied to rock characters (in particular)—component particles, texture, structure—that are to be seen by the naked eye (or with a simple lens) in hand-specimens. Distinguished from microscopic characters, on the one hand, and from field characters and geotectonic arrangments on the other. 'Megascopic' is synonymous.

**Maentwrog** beds. The lowest of the three divisions of the Lingula Flags. The characteristic fossils are species of the trilobite genus *Olenus*. (Belt, *GM*, 1867.) [Maentwrog, Meirionnydd district of Gwynedd.]

**Mafic.** A portmanteau term for the (dark-coloured) ferromagnesian and other non-felsic minerals in igneous rocks; also applied to rocks largely composed of these minerals. Proposed by Cross and others (1912) to take the place of 'femag' suggested by Johannsen (1911), this latter being thought liable to confusion with femic and also not being euphonious. 'Mafic' grades into 'ultramafic'.

**Magma.** Rock material, completely or to a large extent molten and mobile, within the earth's crust or poured out on the earth's surface; a melt-solution of mineral substances. 'The Greek word "magma" has the meaning of a plastic mass, or a paste of solid and liquid matter. The word has long been used in pharmacy and it passed in the 18th century into chemistry, being defined as "a pasty or semifluid mixture". From chemistry this word passed into petrology, replacing the older term "subterraneous lava"—G. Poulett Scrope wrote in 1862 that "lava in not a homogeneous molecular liquid, but . . . a 'magma', or composed of crystalline or granular particles to which a certain mobility is given by an interstitial fluid". With the development of petrology the emphasis has shifted from the presence of crystalline particles to the liquid. Since some lavas are completely free from crystals, and others contain abundant crystals while they are still flowing, we might say that "rock-magma is the partly or wholly liquid material from which the eruptive [igneous] rocks are formed"' (Shand, *Eruptive Rocks*, 1950). 'Magmas' (Bowen, *BGSA*, 1947). 'Present-day lavas are the most accessible and easily-studied representatives of magmas' (Tyrrell, *Petrology*, 1929). But magmas within the crust, being confined, very hot, and under great pressure, are charged with volatiles and so differ from the extrusive lavas; thus lavas are sometimes excluded from 'magma'.

**Magma chamber.** A chamber (reservoir) of magma within the earth's crust. 'Magma chambers' (Rittmann, *Volcanoes*, 1962).

**Magma type.** A hypothetical magma, having a certain general chemical composition, from which particular petrographical types of igneous rocks may be produced according to the conditions of cooling. A magma type is postulated on the evidence of such particular petrographical types commonly occurring in association. (Bailey and others, *MGS*, *Mull*, 1924; Wells and Wells, *GM*, 1948; Holmes, Tomkeieff, Fermor, *GM*, 1949; Tilley, *QJ*, 1950.)

**Magmatic stoping.** See STOPING.

**Magmatism.** The origin, movement and action of magma, including the formation of igneous rocks (magmatites) by its cooling. Also the doctrine; particularly as to the question of the origin of granite (cf. 'transformism'). Hence 'magmatist' for one holding this doctrine. (Reynolds, *GM*, 1947; Tomkeieff, *QJ*, 1947.)

**Magmatite.** A rock formed on the solidification of a magma.

**Magnafacies.** A facies on a large scale, pre-eminently exemplified by the Old Red Sandstone. 'The Old Red Sandstone magnafacies (late Silurian-earliest Carboniferous) of southern Britain accumulated in a clastic wedge between a developing continent in the north and an expanding ocean in the south . . . a thickly developed complex of facies' (Allen in House and others, eds., *The Devonian System*, *PA*, 1979).

**Magnesian limestone.** A limestone containing a significant amount of the mineral dolomite. In the widest sense it includes dolomite-rock.

**Magnesian Limestone.** The magnesian limestone facies of the Permian of NE. England, containing distinctive species of fossils, mostly of genera occurring in the Carboniferous, e.g. the brachiopod *'Productus' horridus*, the lamellibranch *Schizodus obscurus*, and the bryozoan *Fenestella retiformis*. It rests with marked unconformity on the Carboniferous and the highest beds are hidden beneath the Trias (but known from borings) as a result of off-lap in the Permian succession. The Magnesian Limestone corresponds to the Zechstein Limestone of Germany; both are associated with shales and salt deposits and this association in each region represents the Thuringian or Upper Permian series. (Michell, *MS*, 1788, 'Roch Abbey and Brotherton Limes'; Farey, *Derbyshire*, 1811, 'Magnesian Limestone'.)

**Magnetic reversal.** When rocks are placed in chronological order against an isotopic time-scale a unique palaeomagnetic pattern may be revealed by observations of the remanent magnetism. This pattern is shown chiefly by repeated reversals of the polarity of the earth's magnetic field (geomagnetic field) as time goes on, and variations in the duration of each of the conditions (magnetozones). We thus have available a time scale based on magnetic-reversal pattern which can be applied to rocks which span a sufficiently long time but which cannot be dated directly by isotopic methods or correlated by other means with rocks already dated. This method is critically important in establishing the history of the crust in ocean basins and has provided proof of the symmetrical operation of sea-floor spreading on either side of a mid-ocean ridge. *Plate Tectonics and Geomagnetic Reversals* (Cox, 1973).

**Magnetic survey.** See GEOPHYSICS. Magnetic surveys may be made by instruments on land (e.g. 'Magnetic anomalies over the Askrigg block', Bott, *QJ*, 1961), at sea (e.g. 'A magnetic survey in the western English channel', Allan, *QJ*, 1961), or in the air (e.g. 'Aeromagnetic survey', Collar and Patrick in Moseley, ed., *Geology of the Lake District*, 1978).

**Magnetite.** A black mineral, iron oxide, $Fe_3O_4$, with a metallic lustre, often crystallizing as octahedra in the cubic system. Quite opaque, even in the thinnest section. A remarkable property is that it is magnetic. It occurs as a common primary accessory mineral, and also as a secondary alteration product, in many igneous rocks, particularly those rich in mafic minerals, and as a common heavy mineral in sands.

Occurs also as a workable ore in segregations of magmatic or metamorphic origin.

**Magnetozone.** 'It is possible in a long succession of continuous deposition to recognize patterns of geomagnetic reversals which may be used for correlation between sections or with the standard geochronometrical scale. A "magnetozone" is a unit of rock with specific magnetic character (not only polarity)' (*Stratigraphical Guide*, *GSSR* (11), 1978). Hence 'magnetostratigraphy', 'magnetostratigraphic': 'Magnetostratigraphic polarity units—a supplementary chapter of the ISSC *International Stratigraphic Guide* [Hedberg, ed., 1976]' (Hedberg, ed., *Geoly*, 1979).

**Major intrusion.** See MINOR INTRUSION.

**Malacostraca.** The subclass of the class Crustacea, phylum Arthropoda, that includes (among others) the important order Decapoda, the crabs, lobsters, and shrimps, which occur as fossils from the Trias onwards. [G. *malakos*, soft, *ostracon*, the shell.]

**Malvernian.** The Precambrian gneissic rocks that form the central axis of the Malvern Hills. 'Mineralogy of the Malvern Hills' (Horner, *TGS*, 1811). 'Malvernian' (Callaway, *PGS*, 1879).

**Malvernoid.** See CALEDONIAN.

**Mammalia.** The class of the mammals, of the subphylum Vertebrata, ranges from the Jurassic and becomes dominant in the Tertiary, the Age of Mammals. [L. *mamma*, the breast.]

**Man.** Modern man, though divisible into several races, is a very clearly defined species of animal which biologists have named *Homo sapiens*. His characteristics, however, are the result of mental power rather than lying in any very great skeletal peculiarity. In skeletal structure the modern apes are sufficiently like man for zoologists to place them together in the same superfamily, Hominoidea. Fossil skeletal remains of the Hominoidea are very meagre (fortunately, however, they are parts of the skull) but among them, in the Pleistocene, have been found a few that bridge the structural gap between modern man and the modern apes. At the same time we have the artifacts, chiefly flint implements, as a sign that work was being done which is not attempted by any living ape. Some of these skeletal remains of Hominoidea are so similar to modern man that they are placed in the same species. *Homo sapiens* seems to have become well established at the beginning of the Upper Palaeolithic period (perhaps about 20,000 years ago), and has probably ever since been the sole living species of the genus *Homo*. In the Lower Palaeolithic other remains have been placed in distinct species, the chief being *H. neanderthalensis*, of Mousterian age, and *H. heidelbergensis*, probably older. Also older than Mousterian are the remains of 'Java Ape-man' and 'Peking Ape-man' (and comparable remains elsewhere), all now placed in the genus *Pithecanthropus*. The genus *Homo* and the genus *Pithecanthropus* constitute the family Hominidae; and all members of the family Hominidae are called 'Man'. A link between the Hominidae and the apes (Pongidae), by some placed in the one family, by others in the other, is the fossil *Australopithecus* from S. Africa. This, however, occurs too late in the Pleistocene for the remains so far found of it to be considered those of the human ancestor. It is thought likely that the human line diverged from that of the apes in late Miocene or early Pliocene times, but there are no certain skeletal remains of man in the Pliocene. See PRIMATES. *Geological Evidences of the Antiquity of Man* (Lyell, 1863). *Descent of Man* (Darwin, 1871). *Early Man in Britain* (Dawkins, 1880). *Antecedents of Man* (Le Gros Clark, 1959). 'Dating the emergence of Man' (Oakley, *AdS*, 1962). *Man's Evolution* (Brace and Montagu, 1965). *Origins of Man* (Buettner-Janusch, 1966).

**Mantle.** Used specially in the following senses. **1.** The part of the earth's interior between the core and the crust, that is, from a depth of about 3,000 km. (nearly half-way to the centre) to the base of the crust. The uppermost part of the mantle thus overlaps the sub-crustal part of the lithosphere. It is solid, but probably flows, given time, like pitch (see CONVECTION THEORY). Its composition is thought to be that of the ferromagnesian silicates (ultrabasic), but to what extent crystalline or vitreous is still less certain. The mantle is hot. The term is not very appropriate here as this zone hardly 'cloaks' the earth-body. *The Earth's Mantle.* (Gaskell, ed., 1967). (Daly, *Strength and Structure of the Earth*, 1940.) **2.** = Regolith. Weathered rock-material is sometimes called 'waste-mantle'. **3.** As in the phrase 'mantled gneiss dome'. See BASEMENT GNEISS DOME. **4.** In a more general sense it is sometimes used for rocks unconformably overlying lower rocks, particularly as equivalent to the 'cover' in a compound structure. **5.** As 'sedimentary mantle': the sediments on the sea floor covering the solid crust.

**Mantle plumes.** 'A hypothetical system of fixed pipe-like channels of upwelling within the mantle which had their origin at the core-mantle boundary. The tops of these plumes were hotter than the surrounding material through which they had risen. Some plumes remaining fixed as plates passed over them, generated lavas which penetrated the plates' (Oxburgh, *PGA*, 1974, adapted). 'Convection plumes in the lower mantle' (Morgan, *N*, 1971). This hypothesis is used to explain the occurrence of linear groups of oceanic islands together with volcanic seamounts, the best example being that of which the Hawaiian archipelago forms a part. The age of the volcanic products in the chain increases from one end to the other. The supposition is that the whole volcanic chain results from the movement of the oceanic plate over an underlying static 'hot spot' at the top of a mantle plume.

**Manx Slates.** The formation of almost unfossiliferous flags and slaty mudstones that occupies the greater part of the Isle of Man. It appears to be a south-westerly extension of the Skiddaw Slates series and the exposed rocks probably range in age from the upper part of the Cambrian to the lower part of the Ordovician. (Lamplugh, *QJ*, 1895, *MGS*, *Isle of Man*, 1903; Simpson, *QJ*, 1963; Downie and Simpson, *PGS*, 1965.)

**Marble.** Although in use for over seven centuries for any decorative stone that will take a polish, it has become generally, and in technical geology entirely, restricted to limestones (e.g. 'crinoidal marble'). Still further restricted it applies to limestone that has become recrystallized (by thermal metamorphism or otherwise) to a hard, light-coloured mosaic of calcite crystals. 'A very hard black shining stone [dolerite] found at Powke-hill, which upon polishing proved a tolerably fair black marble' (Plot, *Staffordshire*, 1686). 'The limestone about Chudleigh and Ashburton takes so good a polish as to obtain the denomination of marble' (Maton, *Western Counties*, 1797). [G. *marmaros*, a sparkling crystalline rock.]

**Marcasite.** The form of iron sulphide ($FeS_2$) crystallizing in the orthorhombic system. Often occurs in nodules with a radiating fibrous structure. 'The name marcasite is from an Arabic word used for crystallized pyrites of any sort during the Middle Ages' (Miers, *Mineralogy*, 1929).

**Marginal modification.** The modification of the normal composition or texture (or both) of an igneous intrusion often occurring near its margin as a result of more rapid cooling there. Differentiation causes the composition to be more basic, sudden chilling causes the texture to be finer. 'Fractional crystallization seems to afford a complete explanation of differentiation in place as exemplified in some of the commonest phenomena of this order, and especially in the relatively "basic marginal modifications" of intrusive rockbodies' (Harker, *Igneous Rocks*, 1909). 'Very many of the basic dykes, and more especially those of small or moderate breadth, show some degree of "marginal modification" in texture, structure, etc., in consequence of the more rapid cooling of the edge as compared with the interior parts of the dyke' (Harker, *MGS*, *Tertiary Igneous Rocks of Skye*, 1904).

**Marine.** Of or pertaining to the sea; with aeolian, fluvial, and glacial, one of the four main geological agents and conditions.

**Marine band.** A thin bed with a marine fauna occurring among strata that, from their fauna or otherwise, are, for the greater part, taken to be of non-marine origin; as exemplified in the Coal Measures. 'Marine beds in the Coal Measures of North Staffordshire' (Stobbs and Hind, *QJ*, 1905, who give a full bibliography including references to works by Sowerby, Binney, and Phillips, 1839/45). 'Higher in the series [at Frastburg] . . . an interesting example occurs of a black shale full of marine shells, resting on a seam of coal about three feet thick' (Lyell, *Travels in N. America*, 1845). 'Marine bands and other faunal marker-horizons in relation to the sedimentary cycles of the Middle Coal Measures of Nottinghamshire and Derbyshire' (Edwards and Stubblefield, *QJ*, 1947).

**Marine geology.** See SUBMARINE GEOLOGY. *Marine Geology* (Kuenen, 1950). *Papers in Marine Geology* (Shepard commemorative volume; Miller, ed., 1964). *Introduction to Marine Geology* (Keen, 1968). 'A marine geological survey off the north-east coast of England' (Dingle, *JGS*, 1971).

**Marker horizon.** A band of rock, with a distinctive character, used throughout an area to mark what is presumed to be a stratigraphical horizon. Alternative words in the term are 'index', 'key', 'band', 'bed'. See quotation (Edwards and Stubblefield) under MARINE BAND (where each band is a separate faunal marker horizon). 'Electrical resistivity marker bands in the Lower and Middle Chalk of tbe London Basin' (Gray, *BGS*, 1958). 'Considerable attention has been given by geologists of continental Europe to a search for lithological marker bands in strata of Westphalian (Coal Measures) age. Amongst those which have been discovered, the most widespread and stratigraphically valuable are the thin layers of kaolinite-mudstone which have become known as tonstein bands' (Eden and others, *GM*, 1963). 'The *leptotheca* band [Silurian] is noteworthy: it is a dark shale in which the "green streak", an inch thick, forms a marker band of wide distribution not only in Wales but in the Lake District also' (George, *BRG, S. Wales*, 1970).

**Marker point.** See BOUNDARY STRATOTYPE and CHRONOSTRATIGRAPHY.

**Marl.** 'Marl is a mixed rock containing clay minerals and aragonite or calcite, usually together with accessory components, such as silt, in lesser quantity. It is friable when dry, and plastic when wet' (Sugden and McKerrow, *GM*, 1962). 'A fat chalke or kinde of marle' (Camden, *Britannia*, 1586, trans., 1610; of the Sussex downs). '"Marl" is a common term in North Staffordshire for any locally won clay' (Williamson, *GM*, 1946). The Keuper Marl is a red mudstone, only slightly calcareous; on the other hand, much of the Oxford Clay is really marl. [Old French, *marle* (in modern French replaced by *marne*).]

**Marmarosis.** (Marmorosis.) The conversion of a limestone into an even-grained marble. Normally this is done either (*a*) by the action of meteoric waters, causing solution and recrystallization, or (*b*) by recrystallization under thermal metamorphism. Also as 'marmorization'. (Geikie, *Textbook*, 1882; this early use of the term being for the metamorphic effect.) [G. *marmaros*, L. *marmor*, marble.]

**Mass.** A term long ago suggested for any occurrence of 'rock' in the widest sense. 'I employ the word masses, not venturing, as some writers have done, to apply the word rock, associated from time immemorial with the idea of hardness and solidity, to the contents of a sand or clay pit' (Greenough, *Geology*, 1819). All his material that the geologist deals with still lacks an appropriate comprehensive name. ('Rock-mass' is a more definite entity.)

**Mass movement.** Usually refers to the downward movement under gravity, on the land, of a mass of superficial or near-to-the-surface rock material, comprising landslide and creep in their widest senses. 'Recognition of the importance of mass-movement in the shaping of the lands has lagged far behind our knowledge of the action of running water, glaciers, winds, and waves' (Sharpe, *Landslides*, 1938).

**Massif.** 'Mountain heights forming a compact group' (*OED*). 'Pre-existing resistant masses, "stable blocks" or massifs, which are often the denuded relics of former mountain systems' (Wooldridge and Morgan, *Geomorphology*, 1937). 'The Carboniferous Limestone 'massif,' in the Pin Dale area of Derbyshire' (Eden and others, *BGS*, 1964). 'St. George's Land, the "massif" of palaeogeographical reconstruction. The origin of the massif as an upwarp above depositional level lay in the Caledonian movements' (George in Owen, ed., *Upper Palaeozoic and Post-Palaeozoic Rocks of Wales*, 1974). 'Structure, metamorphism and geochronology of the Arequipa Massif of coastal Peru' (Shackleton and others, *JGS*, 1979). [French.]

**Massive. 1.** Without stratification, cleavage, schistosity; particularly as applied to this usual character of igneous rock-bodies. **2.** Thickly or obscurely bedded; as applied to stratified rocks.

**Master joint.** See quotation (Jukes) under JOINT.

**Mathematical geology.** The application of mathematical methods to geological problems. Also called 'geomathematics'. *Studies in Mathematical Geology* (Vistelius, 1967). *Geomathematics* (Agterberg, 1974).

**Matrix. 1.** As a lithological and petrographical term; the relatively fine-grained interstitial matter between much larger particles, as in, for instance, a porphyritic igneous rock (though here 'groundmass' is more usual), a conglomerate, boulderclay. **2.** In a fossiliferous rock or rock fragment; the rock material as opposed to the actual fossil itself. See quotation under CAST (1).

**Meander.** Meanders are the sinuous curves of a river, especially those it makes as it flows over its flood plain. These develop and tend to become more pronounced by lateral corrasion undermining the river bank on the outer side of a bend. A meander may become a complete loop, thus short-circuiting the river and leaving the deserted loop as a 'cut-off' which may be partly occupied by a crescentic 'oxbow lake'. 'River meanders' (Leopold and Langbein, *ScA*, 1966). [L. *Maeander*, f. G. *Maiandros*, a river in Asia Minor.]

**Measures.** Strata in some districts are sometimes alluded to as 'measures', and the word occurs in such stratigraphical names as Coal Measures (most familiarly), Culm Measures, Red Measures, &c. The term may refer 'to the long-established practice of naming the different seams of a coalfield by their measure or thickness' (*OED*). See quotation (Murchison) under BALLSTONE.

**Mediosilicic.** See PERSILICIC.

**Megacryst.** See PHENOCRYST.

**Megacyclothem.** A cyclothem on a large scale, comprising minor cyclothems.

**Megascopic.** See MACROSCOPIC.

**Mélange.** A medley of materials of different kinds, characters, or origins. 'The Mélange. A peculiar rock-type occurring in parts of the Mona Complex [Precambrian] is composed of a tumbled mass of fragments, angular and ill-sorted, some many yards across, others microscopic, that lie in a green schistose matrix. The fragments are of phyllite, schist, greywacke, quartzite, limestone, pillow-lava, and other rocks, derived from a variety of sources. This "mélange", several hundred feet thick, crops out over many square miles in Anglesey and reappears in Lleyn; Bardsey Island is virtually composed of it. Greenly supposed the Mélange [so called by him] to be a gigantic fault breccia, [while] Shackleton has suggested that it is a primary rock-type and is better regarded as a subaerial landslip deposit' (Smith and George, *BRG*, *N. Wales*, 1961). 'The Pilton Beds pass by insensible gradations upwards into the Carboniferous shales and contain a mélange of fossils belonging to the two formations' (Dewey, *BRG*, *SW. England*, (1948). See PSEUDOPSEPHITE. [French.]

**Melanocratic.** Dark coloured; for an igneous rock rich in the mafic (dark coloured) minerals. 'Melano-', as a prefix to a rock-name, means that it is a variety darker than normal. 'It is as if there were a kind of repulsion between the ferromagnesian and alumino-alkaline constituents. Dark rocks rich in the former, and light rocks rich in the latter, represent the extreme forms of many intermediate types; and Prof. Brögger has recently proposed that this should receive expression by the application of the terms. melanocratic and leucocratic to these two strongly-contrasted varieties' (Teall, *QJ*, 1901). [G. *melas*, dark, *krateō*, predominate.]

**Melanosome.** The dark-coloured part of a rock-body, particularly of a metamorphic one. 'The migmatites [of the Barousse Massif] are composed of muscovite-biotite-sillimanite schists (the melanosome) with lenticular or lit-par-lit quartz-feldspar segregations (the leucosome)' (Harris, *GM*, 1974).

**Meltwater channel.** A channel cut by the meltwaters of a glacier, underneath or along the side of the glacier or by a stream issuing from the snout.

**Member.** See LITHOSTRATIGRAPHY. 'A member is part of a formation, recognised as such because of some individual peculiarity' (*Stratigraphical Guide*, *GSSR* (11), 1978).

**Menevian.** A stratigraphical division, in N. and S. Wales, containing fossils (including *Paradoxides davidis*) characteristic of the upper part of the Middle Cambrian series. They are mudstones, about 250 m. thick in Pembrokeshire and 90 m. round the Harlech Dome. (Geological Survey maps, 1855; Salter and Hicks, *BA*, 1865.) [Menevia =St. David's, Dyfed.]

**Mercalli scale.** See SEISMIC INTENSITY.

**Mercia Mudstone group.** A name formally introduced in 1980 for the middle of the three lithostratigraphic units of the Triassic system in Britain. 'The group comprises formations united by being predominantly argillaceous in character; these overlie the Sherwood Sandstone group and underlie the Penarth group and comprise essentially the units of the former "Keuper Marl"' (*A Correlation of Triassic Rocks in the British Isles*, *GSSR* (13), 1980). [Mercia, central kingdom of Anglo-Saxon England.]

**Merioneth series.** See CAMBRIAN SYSTEM. [Meirionydd district of Gwynedd.]

**Merostomata.** A class of the Chelicerata (Arachnida in the wide sense), phylum Arthropoda, comprising the aquatic subclasses (or orders) Xiphosura (of which *Limulus*, the King-crab, is the only living genus, but of which fossils occur rarely from the Cambrian) and the wholly extinct (Palaeozoic) Eurypterida. [G. *meros*, a part, *stoma*, the mouth.]

**Mesa.** See CIRCUMDENUDATION. [Spanish.]

**Mesocratic.** If this term can mean anything it must be that some (unspecified) quality—one present to a medium degree—is a predominant feature of the rock. It is, in fact, sometimes applied in the case of colour, the rock being intermediate between leucocratic and melanocratic. [G. *Krateoa*, predominate.]

**Mesolithic.** See NEOLITHIC. *The Mesolithic Age in Britain* (Clark, 1932).

**Mesostasis.** The last-formed, interstitial material of an igneous rock; glassy, cryptocrystalline, or microcrystalline. [G. *mesos*, middle, *stasis*, a standing.]

**Mesothem.** 'Mesothems are time-significant stratigraphical units, forming major cycles of deposition, which are normally bounded above and below by unconformities or shelf areas, but which have their limits defined by marker points at the bases of chronozones in continuously deposited sediments in basins. Each mesothem contains several smaller cycles, or cyclothems. Mesothemic deposition is typically found when eustatic sea level changes cause repeated flooding of, and withdrawal from, a shelf area' (Ramsbottom, *JGS*, 1978). 'Major cycles in the Namurian' (Ramsbottom, *PYGS*, 1977). 'The mesothem is regarded as having time significance comparable in duration to that of a stage or chronozone, but is outside the usual chronostratigraphical hierarchy. Much larger such units are called "synthems" and smaller ones "cyclothems" (in familiar usage)' (*Stratigraphical Guide*,

*GSSR* (11), 1978). (Hedberg, *International Commission on Stratigraphical Classification, Circular*, 1973.)

**Mesozoic.** Of the middle ages of life, between ancient and modern (Palaeozoic and Cainozoic) from about 225 to 70 million years ago. The main fossil groups are the ammonites, lamellibranchs, gastropods, brachiopods (terebratulids and rhynchonellids), echinoids, corals (hexacorals), reptiles, and gymnosperms; with fish, pteridophytes, sponges, crinoids, nautiloids, star-fish, and crustaceans subordinate. In a reconstruction of Mesozoic life the various groups of extinct reptiles, on land (dinosaurs), in the sea, and also in the air, are so important that the era has been called the Age of Reptiles. (Phillips, *Palaeozoic Fossils of Cornwall* [&c.], 1841.)

**Metabasite.** A metamorphosed basic igneous rock. 'On "metabasite"' (Elliott, *GM*, 1974). (Sederholm, Finland, 1907.)

**Metal.** This term has now no special usage in geology (that is, beyond its ordinary and chemical senses), but was formerly used for any hard rock (as in the modern usage 'road-metal'), particularly a hard stratum of sandstone or shale. See quotation (Sinclair) under SEAM.

**Metalliferous.** 'Bearing or producing metal' (*OED*). *A Treatise on Metalliferous Minerals and Mining* (Davies, 1880).

**Metallogeny.** The origin of metallic ore-deposits. 'Metallogenesis' is synonymous. 'The metallogeny of the British Isles. The recent development of the geographical method of studying ore-deposits is due to Professor L. de Launay, who first pointed out its applications in 1897. His object is to delineate the various regional types of ores, and to show their intimate genetic relation with tectonics and with petrographic provinces, the regional types being termed "metallogenetic provinces"' (Finlayson, *QJ*, 1910). 'Metallogenetic epochs' (Lindgren, *EG*, 1909). 'Metallogenesis in Mesozoic-Cenozoic orogenic belts' (symposium, *JGS*, 1978).

**Metamorphic assemblage.** See ASSEMBLAGE (5).

**Metamorphic aureole.** See AUREOLE.

**Metamorphic belt.** 'Regional metamorphism takes place in the deeper parts of orogenic belts, while regional metamorphic rocks are found in what are here called "metamorphic belts"' (Miyashiro, *JPt*, 1961).

**Metamorphic complex.** Has practically the same meaning as 'metamorphic assemblage' while emphasizing the manner of the assemblage. 'The Lewisian, Moinian and Dalradian metamorphic complexes of Scotland' (Giletti and others, *QJ*, 1961).

**Metamorphic facies.** See FACIES.

**Metamorphic grade.** See GRADE (2).

**Metamorphic rock.** See METAMORPHISM. *Petrology of the Metamorphic Rocks* (Mason, 1978).

**Metamorphic-plutonic cycle.** A sequence (rather than a 'cycle') which has as its terms: metamorphism, migmatization, magmatization, movement of magma (intrusion), solidification.

**Metamorphism.** Change in the character of a rock. There is considerable variation in the limitations attached to the use of the term, but it usually means mineralogical, textural, or very small-scale structural adjustments within rock-masses in response to changing chemical or (particularly) physical conditions. In the latter case, the metamorphism occurs normally as a result of an increase in either temperature or stress (or both,) the change remaining (probably in a metastable state) on their return to the previous value; any reverse change on cooling or release of stress is 'retrograde metamorphism' ('diaphthoresis'). Metasomatism due to chemically active gases from a cooling magma (pneumatolysis of country rock) is usually included, but metasomatism due to weathering or to the percolation of underground water is usually excluded. Largeand medium-scale structures of folding and faulting are excluded, but slaty cleavage (particularly when well developed) and schistosity are included.

(But Turner and Weiss, in their *Structural Analysis of Metamorphic Tectonites*, include all folded rocks other than those that are 'weakly deformed'.) The mere lithification of sediments is excluded, but the changes undergone by vegetable debris to become coal, and the change in rank of a coal, are sometimes included (e.g. Wellman, *GM*, 1950). The complete melting of a rock (magmatization, ultrametamorphism) obliterates the previous metamorphism. Hence 'metamorphic' and the verb 'metamorphose'; in fact, 'metamorphic' seems to be the form in which the term was first introduced: 'For altered stratified rocks the term "metamorphic" (from μετα [*meta*], trans, and μορφή [*morphē*], form) may be used' (Lyell, *Principles*, 1833). 'On an intrusion . . . Highlands of Scotland, and its accompanying metamorphism' (Barrow, *QJ*, 1893). *Treatise on Metamorphism* (Van Hise, 1904). *Metamorphism* (Harker, 1932/50). *Metamorphism and Metamorphic Belts* (Miyashiro, 1973). 'The end of the nineteenth and the beginning of the twentieth century saw the birth of the modern scientific study of metamorphism. Subsequent advances, important though they were, consisted largely in the elaboration of the principles established by Barrow and Van Hise in the field, by Becke and Harker with the aid of the microscope, by Goldschmidt and Eskola, who introduced physical chemical interpretations, and by Schmidt and Sander, who laid the foundations of petrofabrics' (Rast, *PGS*, 1962).

**Meta-(rock).** A metamorphosed rock which was originally of the kind or type included in the name. Thus 'metasediment' or 'metasedimentary rock', 'meta-igneous rock', 'metadolerite', 'meta greywacke', 'metamagmatite', &c.

**Metasomatism.** Change in the substance of a rock by the removal of chemical constituents or the introduction of new ones, resulting in chemical subtraction, addition, or replacement. This is most extensively and obviously brought about by weathering and by percolating meteoric waters containing matter in solution; but the term is most commonly in use for the production of new minerals by the introduction of new elements from igneous sources, by pneumatolysis or percolating solutions. [G. *sōma*, *sōmat-*, the body.]

**Meteor.** Meteors are small bodies of matter travelling through interplanetary space. Those entering the earth's atmosphere are heated to incandescence by friction and are either wholly or partially burnt up. In the latter case, the remaining piece of matter from the middle of the meteor reaches the earth as a meteorite.

**Meteoric water.** Water in, or (particularly) derived from the atmosphere, as distinct from juvenile water derived directly from igneous action. [G. *meteōros*, in the air.]

**Meteorite.** See METEOR. 'Meteorites are divided into three main groups which pass one into another: Siderites, the iron meteorites, consisting almost entirely of iron alloyed with nickel; Siderolites, mixtures of nickel-iron and heavy basic silicates, such as olivine and pyroxene; Aerolites, the stony meteorites, consisting almost entirely of heavy basic silicates, olivine and pyroxenes, and resembling some of the rare and most basic types of terrestrial igneous rocks' (Tyrell, *Petrology*, 1929). 'A description of some supposed meteorites found in seams of coal' (Binney, *MMLPS*, 1851). *Meteorites* (Mason, 1962). *Nature and Origin of Meteorites* (Sears, 1978). See also CHONDRITE [G. *meteōra*, things in the heavens.]

**Meteorite crater.** 'Meteorite craters are depressions filled by explosion breccias and impactites, and range in diameter from several metres to 100 km' (Khryanina and Ivanov, Russia, 1977, in translation. *D*). See CRATER and ASTROBLEME.

**Method.** Method in geological research, particularly as regards the taking of field notes and collecting of specimens, may be said to have been established by John Woodward in his *Natural History of the Earth* (1695) and in a pamphlet (1696) abbreviated in his *Fossils of all kinds digested into a Method* (1728). As North remarks about the latter (*TCNS*, 1931), this shows 'an astonishing appreciation of vital points, and is as worthy of the attention of field naturalists and museum workers as when it was written'. 'I made strict enquiry wherever I came and laid out for intelligence of all places where the entrails of the earth were laid open, either by nature or by art, and humane industry. And wheresoever I had notice of any considerable natural spelunca or grotto; and digging for wells of water, or for earths, clays, marle, sand, gravel, cole, stone, marble, ores of metals, or the like; I forthwith had recourse thereunto: and taking as just account of every observable circumstance, I entered it carefully into a journal which I carry'd along with me for that purpose' (1695). *Methods in Practical Petrology* (Milner and Part, 1916). *Methods for the Study of Sedimentary Structures* (Bouma, 1969). 'The method of multiple working hypotheses' (Chamberlin, *JG*, 1897). Detailed and exact methods are usually called 'techniques'.

**Miarolitic.** ' Some granites contain small irregular cavities lined with well-terminated crystals of the normal constituents of the rock, accompanied by some of the rarer accessories. Such granites are said to be "miarolitic". In Britain the granites of the Mourne Mountains and Lundy Island provide typical specimens' (Hatch and Wells, *Igneous Rocks*, 1965). (Rosenbusch, 1887; Judd, *QJ*, 1889.) [Italian, *miarolo*, a local name for a variety of granite.]

**Mica.** A family of rock-forming minerals, silicates of aluminium and potassium, with magnesium and iron in the dark varieties, all having hydroxyl (OH) in addition. They crystallize in the monoclinic system and have a perfect cleavage, splitting into thin elastic plates which have a splendent lustre on their surfaces. The chief micas are muscovite and biotite. [Probably L. *mico*, shine; possibly L. *mica*, a crumb, a grain.]

**Micrite.** The microcrystalline component of a sedimentary rock; applied chiefly to limestones.

**Microcline.** A form of potassium felspar, occurring in the highly potassic granites and pegmatites, crystallizing in the triclinic system but usually distinctive, under the microscope, by possessing a complex system of spindle-shaped twin lamellae in two sets nearly at right angles. In general physical characters microcline is indistinguishable from orthoclase, and the departure from monoclinic symmetry is only very slight (hence the name, given by Breithaupt, 1830).

**Microcrystalline.** See CRYSTALLINE.

**Microdiorite.** The medium-grained equivalent of a diorite, usually porphyritic (diorite porphyrite, or porphyrite). Usually occurring as minor intrusions.

**Microfabric.** See FABRIC.

**Microfacies.** 'The sum of the lithological, mineralogical, textural and fossil characters of a sedimentary rock as seen in thin section. The method of microfacies work was described by H. B. Milner in 1926, but the term was introduced by Professor J. Cuvillier in 1951' (Smout, *N*, 1958). This sense is, however, not very definitely established and the term might be held to imply, or might be used for, merely a facies on a small scale.

**Microfauna.** See FAUNA (1).

**Microfossil.** Microfossils comprise (*a*) the microscopic fossils of those groups of organisms of which the individuals are always, or usually, microscopic in size (e.g. the Foraminifera) and (*b*) the microscopic discrete fossils of particular skeletal elements which are always, or usually, microscopic (e.g. sponge spicules, conodonts). The following are the main groups of microfossils of these two kinds:

| | |
|---|---|
| Animal kingdom | |
| Protozoa | Foraminifera |
| | Radiolaria |
| Porifera | sponge spicules |
| Crustacea | Ostracoda |
| Incertae sedis | conodonts |
| Plant kingdom | |
| Algae | diatoms |
| (General) | coccoliths and rhabdoliths |
| | hystrichospheres (dinoflagellates) |
| | spores, pollen grains, and small seeds |

A third kind (*c*) comprises those that represent very young growth-stages of such normally macrofossil groups as brachiopods and molluscs. There is no definite size limit for the category of microfossil, but one recent worker (Craig, *QJ*, 1954) has taken 2 mm. 'The microfossils of the Upper Caradocian phosphate deposits of Montgomeryshire' (Lewis, *AMNH*, 1940). 'Microfossils from the Manx Slate series' (Downie and Ford, *PYGS*, 1966). *Microfossils* (Brasier, 1979). 'It was in the year 1837 that scientists were first made clearly aware of the immense abundance of microfossils. This came about through the presentation to the Berlin Academy of Sciences, on the 20th July, of a paper in which it was convincingly demonstrated that large masses of rock, whole strata, might be made up entirely of the fossilized remains of organisms too small to be distinguished by the naked eye. . . . The microscopist who had reported this discovery was already a renowned figure in the German scientific world, Christian Gottfried Ehrenberg (1792–1876)' (Sarjeant, *Pln*, 1978).

**Microgabbro.** See DOLERITE.

**Microgranite.** The medium-grained equivalent of a granite, sometimes restricted to the non-porphyritic variety (as distinct from granite-porphyry). Micro-adamellite and micro-granodiorite are the varieties equivalent to adamellite and granodiorite. Usually occurring as minor intrusions.

**Micrographic.** See GRAPHIC.

**Microlite.** 'Microlites are minute crystals of tabular or prismatic habit, occurring in microcrystalline or hemicrystalline rocks and groundmass. They are distinguished from crystallites by their capacity to give a reaction with polarised light' (Holmes, *Nomenclature of Petrology*, 1928).

**Micropalaeontology.** The study of microfossils. It does not now include the microscopic study of macrofossils, such as thin sections of plant-petrifactions or corals, or colonial forms, such as Bryozoa or Graptolithina, of which the component individuals are almost microscopic. Formerly this restriction did not apply, e.g. 'Contributions to micro-palaeontology: Notes on some species of Monticuliporoid corals' (Nicholson, *AMNH*, 1884). 'Ehrenberg (1795-1876) was one of the earliest investigators of the microorganisms of various geological formations. In his *Mikrogeologie*, 1854, the results of his principal investigations were published, but in 1838 he had described sundry "infusorial earths"' (Woodward, *History*, 1911). *Principles of Micropalaeontology* (Glaessner, 1945). *Marine micropalaeontology* (Haq and Boersma, eds., 1978).

**Microplankton.** See PLANKTON. 'Microplankton from the Kimeridge Clay' (Downie, *QJ*, 1956). 'Microplankton [and] stratigraphical problems in the Eocene of the Isle of Wight' (Eaton, *JGS*, 1971).

**Microsyenite.** The medium-grained equivalent of a syenite, usually porphyritic (syenite-porphyry or orthoclase-porphyry). Usually as minor intrusions.

**Mid-ocean ridge.** A ridge in the ocean floor, typically in the more central parts, along which sea-floor spreading is generated. One of the longest and most conspicuously defined is the Mid-Atlantic ridge which splits down the middle of the Atlantic Ocean. 'The World mid-Ocean Ridge System is a volcanically active area of the sea floor some tens of thousands of km in length, at which new ocean crust is being created' (Cronan, *JGS*, 1980).

**Migma.** An intimate mixture of magma and highly metamorphosed, but not melted, rock. Hence 'migmatization'. (Reinhard, 1934.) [G. *migma*, a mixture.]

**Migmatite.** A rock formed on the solidification of a migma. 'Sederholm (1907) coined the term migmatite to describe incompletely granitized rocks with a mixed appearance' (Reynolds, *GM*, 1958). 'Migmatites are mixed rocks in which two components are intimately mingled, a host material representing pre-existing rocks and a granitic component which is at least in part derived from an outside source. Migmatites pass insensibly into metamorphic rocks on the one hand and into more or less homogeneous granites on the other and thus link the two great divisions of the plutonic series into a connected whole' (Read and Watson, *Introduction to Geology*, 1962).

**Millet-seed sand.** Sand of the desert with grains very perfectly rounded by wind abrasion and having a surface like that of ground glass. 'The Penrith Sandstone is usually a coarse pinkish rock, many of the grains are of "millet-seed" type, suggestive of origin as desert sand' (Eastwood, *BRG, N. England*, 1963). [millet, an Indian grass.]

**Million years.** See TIME UNIT.

**Millstone Grit series.** The name long ago given to the thick beds of sandstone ('grit') and shale occurring in Britain between the Carboniferous Limestone and the Coal Measures. The sandstones were much used for millstones. Its thickest and most characteristic development is in the Pennines where it reaches a thickness of about 1,000 m. and the gritstone beds in the upper part of the whole series form conspicuous escarpments. Similar beds occur in the same stratigraphical position on the Continent. The grits and much of the shales are non-marine and contain fragments of fossil plants, but the marine shales have a characteristic fauna of goniatites. The whole appears to be of deltaic origin. Considered as a chronostratigraphic unit, it is equivalent to the Namurian series. (Whitehurst, *Strata in Derbyshire*, 1778.)

**Mine.** In addition to the usual meaning of an excavation or boring for digging out coal, salt, ores, &c., the term may also refer, particularly in earlier writings, to the material so mined, particularly coal and iron ore. 'The inland parts [of Cornwall] have rich and plenteous mines of tinne. . . . The Derwent [Cumberland] creepeth betweene the mountains wherin, at Newlands and elsewhere, copper mines were discovered' (Camden, *Britannia*, 1586, trans., 1610). See quotation (Robinson) under STRATUM.

**Mineral. 1.** Any substance obtained by mining. As an extension of this we have a use of 'mineral' as a comprehensive adjective applicable to every kind and state of matter entering into the composition of the earth-body (see EARTH). Thus, for example, coal and petroleum, even natural gas and ground-water are, in this sense, 'mineral substances'. **2.** A solid inorganic substance having a definite homogeneous chemical composition and occurring in the earth's crust. Although a shell (of a mollusc, &c.), composed of calcium carbonate or silica, is a 'mineral' shell (as distinct from a horny or bony substance), it is hardly 'a mineral'. Nearly all minerals are crystalline (in a wide sense). 'During the eighteenth century rocks as a rule had been included among minerals. Kirwan, for example, wrote: "All bodies found in or on the earth, and not of vegetable or animal nature, are called minerals" (*Elements of Mineralogy*, 1794)' (Tomkeieff, *GM*, 1939). *A Dictionary of the Names of Minerals* (Chester, 1896). *Mineral Names* (Mitchell, 1979). [medieval L. *minera*, a mine.]

**Mineral cleavage.** See CLEAVAGE (2).

**Mineral industry.** The economic exploitation of mineral resources. 'The mineral industry in Wales' (Thomas, *PGA*, 1972).

**Mineral kingdom.** May be said to be an appropriate phrase for the domain of geology. 'My principal intention was to get as compleat and satisfactory information of the whole Mineral Kingdom as I could possibly obtain' (Woodward, *Natural History of the Earth*, 1695). *The Natural History of the Mineral Kingdom* (Williams, 1789). Fossils belong to both the animal, or vegetable, kingdom and the mineral kingdom.

**Mineral spring.** A spring which, instead of being practically pure water, holds substance in solution. Such a spring may be saline (containing chlorides of sodium, potassium, magnesium), sulphurous, or chalybeate (rich in iron compounds), the term, particularly in the form 'mineral waters', being especially associated with supposedly healthgiving properties. 'Mineral springs—This term is applied in popular phraseology to springs which contain sufficient mineral matter in solution to give a distinct taste to the water' (Lake and Rastall, *Textbook*, 1910/41).

**Mineral vein.** See VEIN.

**Mineralize.** The *OED* gives, among other senses: to convert into a mineral substance;

to impregnate with mineral matters. These senses apply chiefly to fossilization, but in general geology the term is largely used for the production of aggregates and veins of minerals, particularly ores, by some secondary process, as distinct from their production by direct sedimentation or the solidification of a magma. Hence 'mineralization', 'mineralizing agents'.

**Mineralogy.** The science of minerals. Hence 'mineralogical', 'mineralogist'. 'Theophrastus (372-287 B.C.), Aristotle's pupil and successor, wrote a book *Of Stones*, which is the oldest treatise extant on minerals' (Crook, *History of the Theory of Ore Deposits*, 1933). 'The *De Natura Fossilium* (1546) is the first attempt at a systematic mineralogy, and in it Agricola describes many minerals for the first time' (Joan Eyles, *N*, 1955). 'This will not perhaps be unwelcome to some that love Mineralogy much better than they understand it' (Boyle, 1690; quoted in *OED*). 'Mineralogy' was used in former days also in the sense of the geology of a region or district: *Mineralogy of the Scottish Isles* (Jameson, 1800). 'On the mineralogy of the Malvern Hills' (Horner, *TGS*, 1811). *Elements of Mineralogy* (Kirwan, 1784/ 1810). *System of Mineralogy* (Dana, 1837; 18th edition by Hurlbut, 1971). *Elements of Mineralogy* (Rutley, 1874/1953). *Mineralogy* (Miers, 1902/29). *Mineralogy of Sedimentary Rocks* (Boswell, 1933). [A name long-established and euphonious, but etymologically a hybrid of L. and G. derivation.]

**Minerogenesis.** The origin and growth of minerals.

**Mining geology.** Geology applied to the mining of coal, salt, ores, &c. *Mining Geology* (Park 1906).

**Minor intrusion.** Igneous intrusions are sometimes classified on the basis of size into 'major' and 'minor', and, while intrusions may be of any size, there is some tendency towards a natural twofold grouping into large and small. 'Minor intrusion' is in commoner use than 'major intrusion'; such are the ordinary-size dykes and sills, and the intrusive (as distinct from the extrusive) parts of volcanic necks. The type of the major intrusion is the batholite (if, in fact, intrusive, and not a case of granitization *in situ*).

**Miocene.** The third general stratigraphical group of the Tertiary succession, usually ranking as a series; if a Neogene system is recognized, it is the lowest of the two, or the three, series within it. The Miocene is known chiefly from S. Europe and is generally considered to be absent from Britain; it was the time of the culmination of the Alpine movement. (Lyell, *Principles*, 1833; see PLIOCENE.) [G. *meiōn*, less, expressing 'a minority of recent species'.]

**Miogeosyncline.** See GEOSYNCLINE.

**Misfit.** 'A river too large or too small to have eroded the valley through which it flows (Davis)' (Cotton, *Landscape*, 1948). One too large is an 'overfit', one too small an 'underfit'; underfits are the commoner and the more obvious, and are the kind usually implied.

**Mississippian.** Lower Carboniferous. Used chiefly in America where it is held to constitute a separate system. (Winchell, 1870.) [Mississippi State.]

**Mobile belt.** 'There may be recognized in the earth's crust two types of regions–the "mobile belts" and the "resistant (stable, rigid) masses" or blocks. The mobile belts are long and relatively narrow zones, which in the earlier stages of their history have the form of troughs in which thick sedimentary formations are deposited throughout considerable periods of time, and which after compression become the sites of fold mountain chains. During the period of sedimentation they constitute geosynclines' (Hills, *Structural Geology*, 1940). 'We may term the circum-Pacific and Alpine-Himalayan zones "mobile belts" in contrast to the relatively stable tracts which make the remainder of the crust. It is evident that their topographical peculiarities are only the superficial expression of a state of instability affecting the crust and the underlying mantle' (Read and Watson, *Introduction to Geology*, 1962).

**Mode.** A quantitative statement of the actual mineral content of an igneous rock, as distinct from that of the chemical composition expressed as a norm or as proportions of elements or oxides. This statement is usually given as the percentage by weight of each of the component minerals.

**Mofette.** An exhalation of carbon dioxide, at air temperature, in regions of (usually otherwise extinct) volcanic activity. [Italian *mofeta*, a noxious exhalation.]

**Moffat series.** The beds in S. Scotland, some 200-250 ft. thick, of dark graptolitic shale which form a highly condensed succession from the Caradocian to near the top of the Valentian (Llandovery) series. The Moffat series includes the Glenkiln, Hartfell, and Birkhill Shales. (Lapworth, *GM*, 1872/6, *QJ*, 1878.) [Moffat, Dumfriesshire.]

**Mohorovičić discontinuity.** See EARTH'S CRUST.

**Moh's scale.** See HARDNESS (1). [F. Mohs, 1773-1839, a German-Austrian mineralogist.]

**Moinian.** The highly folded and highly metamorphosed series (metamorphic assemblage) of rocks, chiefly granulites and schists (metamorphosed sandstones and shales), occurring over about the northwest half of the Scottish Highlands, but southeast of the Moine thrust; now often called simply 'The Moines' (*MGS Ardnamurchan*, 1930; Green, *QJ*, 1935; Sutton and Watson, *QJ*, 1954). The deformation and metamorphism is that of the Caledonian orogeny (*sensu lato*); the age of the rocks themselves, considered as unmetamorphosed sedimentaries, is thought to be Precambrian (probably Torridonian). 'The Moine Series is the last great formation of sedimentary origin in the British Isles whose stratigraphical relations remain to be established' (Sutton, *SP*, 1961). 'New light on the Moine rocks of the Central Highlands of Scotland' (Piasecki, *JGS*, 1980). (Peach and Horne, *QJ*, 1888). [Moine, or A'Mhoine, district, Sutherland.]

**Molasse.** 'A provincial [French] name for a soft green sandstone, associated with marls and conglomerates, belonging to the Miocene period extensively developed in the lower country of Switzerland' (Lyell, *Principles*, 1833). The term has now taken on the expression of the connexion between this facies and the period immediately following the Alpine orogeny. The deposits appear to be largely continental or freshwater. See FLYSCH.

**Molecular palaeontology.** 'Molecular palaeontology is the study of the detailed molecular architecture of any organic molecules which may be found in rocks of all ages, but particularly in the rocks of Pre-Cambrian age' (Calvin, *PGS*, 1969).

**Molecular proportions.** The proportions of the numbers of molecules of various chemical compounds in a rock, such as the oxides or actual mineral substances. These proportions, usually expressed as percentages, are given by dividing the weights of the compounds by the molecular weights. (The chemical analysis of a rock is usually expressed as weight percentages, not molecular percentages.)

**Mollusca.** Now restricted to a particular phylum of invertebrate unsegmented animals most of which bear shells. There are a number of classes of which the Bivalvia, the Cephalopoda, and the Gastropoda are by far the most abundant in the fossil record. Other classes found fossil are the Amphineura, the Monoplacophora, the Rostroconchia, and the Scaphopoda. [L. *mollis*, soft.]

**Moment.** Ordinary time expressions are used in geology on the understanding that they are relative to the time-scale of the geological ages. Thus 'moment' means any short period during which some event or process occurs and which is not further analysable temporally; such as the fossilization of shells in a bed of rock an inch or so thick. See SECULE.

**Monadnock.** An upstanding residual hill on a plain, as yet not worn down to the general level; due either to the outcrop of a specially resistant rock mass, or to some more accidental cause (circumdenudation). 'I have fallen into the habit of calling a residual mound . . . a monadnock, taking the name from that of a fine conical mountain of south-western New Hampshire, which grandly overtops the dissected peneplain of New England' (Davis, *GJ*, 1895). 'The monadnock hills of Lleyn' (Smith and George, *BRG, N. Wales*, 1961). 'Monadnocks are by definition hills standing alone; the name of the original comprises two Gaelic words—*monadh*, mountain, and *cnoc*, a round hill. It is likely that the name was given by Scots settlers who came into southern New Hampshire in the early eighteenth century' (Linton, *GJ*, 1961).

**Monian.** The metamorphic (Precambrian) rocks of Anglesey (including an associated granite). (De la Beche, *Geological Observer*, 1851, 'Mona Series'; Blake, *BA*, 1886, *QJ*, 1888, 'Monian System'; Greenly, *MGS, Anglesey*, 1919, 'Mona Complex'.) [Môn, Mona = Anglesey.]

**Monocline.** A pronounced bend in a series of sedimentary rocks which, on either side of it, are horizontal or nearly so. 'Monoclines of the Colorado Plateau' (Kelly, *BGSA*, 1955). The term has also been used in the sense of 'homocline'.

**Monogenetic.** Derived from one source; as, for instance, in reference to pebbles in a conglomerate.

**Monomineralic.** For an igneous rock containing one essential mineral only; e.g. anorthosite, dunite.

**Montian.** See PALAEOGENE and EOCENE. [Mons, Belgium].

**Montmorillonite.** A clay mineral, probably originating in the devitrification of volcanic glass. The rock bentonite is formed almost entirely of this mineral, and fuller's earth is generally largely composed of it. (Mauduyt, 1847.) [Montmorillon, France.]

**Monzonite.** A coarse-grained igneous rock with both potash felspar and a plagioclase (excluding albite), the dominant mafic mineral being augite. Thus intermediate between syenite and gabbro, but often used in a rather wider sense to include rocks intermediate between syenite and diorite. (de Lapparent, 1864.) [Monzoni, Trentino, Italy. ]

**Moon.** See SELENOLOGY.

**Moraine.** Rock material being carried, or having been carried and deposited, by a glacier. Derived from the slopes above the glacier, as scree material, and also plucked and scraped from the rocks over which the glacier passes. Includes material of all sizes, from the largest chunks and boulders to the finest rockflour. Moraine in transport may be classified as lateral, medial (median), terminal (still in the ice), englacial, ground; that left behind by a retreating, shrinking, or completely vanished glacier may be lateral (ridges), terminal or end (mounds), or general ground moraine. Terms for the various kinds of moraine 'were introduced by Agassiz in 1838, though Hugi had already recognized the various types in 1830 from which time moraine studies may be said to have assumed a scientific aspect' (Charlesworth, *Quaternary Era*, 1957). *Moraines and Varves* (Schlüchter, ed., 1979). 'A moraine' is one distinct mound or ridge of moraine material. [French.]

**Morphogenesis.** The origin (perhaps including the early growth or development) of form, as a whole or of particular morphological features. Hence 'morphogenetic'. Can apply generally, e.g. to an organism (in either ontogeny or phylogeny) or a landscape.

**Morphological series.** In palaeontology, a graded series of fossils arranged to show variation either in the individuals as a whole or in some one particular variable character. Such a series may be closely or loosely graded and cover a narrow or a wide range of variation.

**Morphology.** The study of form. Often used for the more explicit 'geomorphology': *Morphology of the Earth* (King, 1962/67); 'The morphology of Charnwood Forest' (Ford in *Geology of the East Midlands*, 1968). It may be purely descriptive (cf. 'lithology'). It finds its commonest application in the geological sciences as the study (or description) of the form and structure of the different kinds of fossils.

**Morphotectonics.** 'Morphotectonics is the tectonic interpretation of the morphological features of the earth. It is particularly a study of the larger features of continents and oceans, the great elevations and depressions and the lineaments that transect them, or, in many instances, determine their outlines' (Hills, *PGS*, 1960, more fully discussed in *QJ*, 1961; term used by Hills, *JGSA*, 1956).

**Mortar structure.** A cataclastic structure wherein, at a certain stage of the breaking-down process, relatively large and rounded relics of the more resistant crystals remain embedded in a matrix of much smaller, more finely crushed, material.

**Mosaic.** 'A grain fabric with mosaic structure is realized when the crystals abut along relatively simple bounding planes' (Niggli, *Rocks and Mineral Deposits*, 1954).

**Mould.** See CAST (1).

**Mountain.** An individual mountain belongs to one of two quite distinct types as regards origin and development: (1), one of accumulation, a volcano, (2) one left standing as a result of the erosion of the surrounding rock. A mountain of type (2) may well be produced on the site of a volcano of a past geological age (see VOLCANO). Probably there is nowhere an individual mountain (as distinct from a mountain range) formed by local crustal upheaval. This simple classification of mountains has not been outmoded by the development of the science of geomorphology.

**Mountain building.** Literally and generally: the building of mountains, by accumulation (volcanoes) or by any kind of earth-movement. But 'mountain-building movements' is synonymous with 'orogenic movements'. *Mountain Building* (van Bemmelin, 1954). 'Mountain building' (Shackleton in Haslett and St. John, eds., *Science Survey* 2, 1961).

**Mountain system.** In the widest sense, all the folded mountain ranges produced during a particular span of earth-history, whether these ranges (1) are still mountainous, (2) have been worn down, (3) have been not only worn down but buried beneath later rocks, (4) have been re-exposed at the surface, either remaining as low-lying areas or being now upstanding again as a result of uplift and regional differential erosion. 'In what may be termed the Bertrand time-classification of folded mountain systems: "Caledonian" includes all folded mountains developed in early Palaeozoic times, not later than Devonian; the name is derived from Scotland. "Hercynian" includes all folded mountains developed in later Palaeozoic times, that is Carboniferous, extending into Permian; the name is derived from the Harz in Germany. "Alpine" includes all folded mountains developed in Mesozoic and Tertiary times; the name is derived from the Swiss Alps' (Bailey, *BA*, 1928).

**Mud.** Very wet clay or soil; clay material forming a sub-aqueous sediment. [Middle English *mode, mudde*].

**Mud belt.** The belt of marine deposition extending from the outer limit of the belt of variables to the outer limit of terrigeneous deposition. The deposits laid down in this belt are almost entirely muds.

**Mud cracks.** The system of cracks, usually polygonal, that develop in a layer of mud as it dries; sun-cracks, drying cracks, desiccation cracks. Hence the similar cracks, of like origin, in a mudstone, often represented by casts (moulds) in relief, on the underside of the immediately overlying bed (a sole-mark).

**Mud volcano.** A pile of mud ejected by eruptions of gas, water, and mud. The types may be said to be in the Caucasus; thus Scrope (*Volcanos*, 1862), referring to that region: 'Mud volcanos, as they are called, i.e. cones of a ductile unctuous clay . . . spurting up waves and liquid mud'. It may refer to a muddy hot spring or geyser in a region of volcanic activity and any resulting mound of mud. The term is more appropriate to the latter kind; but a mud volcano quite unconnected with any igneous activity may emit inflammable gases so that the term is then, incidentally, not inappropriate. (Oswald, *GM*, 1928; Wilson, *PGS*, 1965.) The term would apply to a body of mud produced under water in the manner of a 'sand volcano'. Also see quotation (Lees) under DIAPIR and quotation (Humphrey) under SEDIMENTARY VOLCANISM.

**Mudflow.** The more or less intermittent flow or creep of mud downhill.

**Mudstone.** A hardened mud (hardened clay), a hard argillaceous rock. A name used by Murchison (*Silurian System*, 1839), adopting local usage, for the argillaceous, largely shaly, strata of his Silurian system, the Lower Ludlow Rock in particular, implying its suitability for any similar kind of rock. The name is obviously appropriate for 'a mud which has become stone', though its original local use probably expressed 'stone which becomes mud' (on weathering and wetting). Such a rock is often more or less laminated and it thus seems that, on grounds of both sense and original definition, mudstone should include shale; but recent definitions often restrict it to unlaminated rock. Such a rock is also often cleaved and it seems that 'mudstone' might also include slate, at least where there is no appreciable mineral reconstitution. A mudstone, in the wide sense here adopted, thus comprises a natural group of rocks within which there are all degrees of development of certain structures, easily designated by qualifying adjectives; it takes its place with the names of the other two great classes of sedimentary rock, sandstone and limestone.

**Mullion structure.** Rods or columns (complete or incomplete in cross-section) of rock, suggesting the mullions of windows, occurring in rows, clusters, or larger aggregates. They are formed tectonically, but in various ways; by the crumpling and rupture of beds in folding, by the intersection of cleavage and bedding, and by regular boudinage. The mullions lie parallel in the fold direction (*MGS, Ireland, Donegal*, 1891.) The term has also been used for the casts of grooves on a fault plane, caused by the fault movement. (Wilson, *PGA*, 1953; Dearman, *GM*, 1966.)

**Multiple intrusion.** An intrusion (a sill or dyke in particular) which has formed as a result of successive injections of magma. These injections may be of the same kind (chemically) or of different kinds. Some authorities make a distinction between a 'multiple' and a 'composite' intrusion, the latter term being applied to one composed of different kinds of igneous material; of these authorities some would go further and reserve the term 'composite' intrusion for one (composed of different kinds of material) in which the parts were injected penecontemporaneously so that they 'do not chill against one another' (Thomas and Bailey, *MGS*, *Mull*, 1924). 'The occurrence of "composite sills and dykes", in which acid and basic rocks form parts of a single intrusive body, and seem to have been intruded almost contemporaneously. The existence of such composite intrusions in Arran seems to have been recognized more or less clearly by Bone (1820) and Delesse (1858), who describe a sill of quartz-porphyry resting upon, and apparently passing into, one of dolerite at Drumadoon Point' (Harker, *MGS*, *North Arran* [etc.], 1903). 'The Mesozoic-Cenozoic Coastal Batholith of Peru is a multiple intrusion of gabbro, tonalite and granite occupying the core of the Western Cordillera' (Pitcher, *JGS*, 1978). Such various and confusing usages result from the fact that 'multiple' and 'composite' are adjectives so nearly synonymous that they are unsuitable for denoting different concepts.

**Muschelkalk.** The middle of the three series of the German Trias; in Britain the time representative is shared between the Bunter and Keuper, but the exact correlation is unknown. It is a limestone with characteristic fossils, of which the ammonoid *Ceratites nodosus* and the crinoid *Encrinus liliiformis* are the best known. [German, shelly limestone.]

**Muscovite.** The commonest of the white micas, silicate of aluminium and potassium, with hydroxyl (OH). Characteristic of the most acid igneous rocks, such as certain granites and pegmatites; very common in metamorphic rocks and a common constituent of many sedimentary rocks, particularly sandstones. First obtained from Russia. (Dana, 1850.) [Muscovy.]

**Mussel-band.** A band or bed of shale in the Coal Measures, containing nonmarine lamellibranchs ('freshwater mussels'). 'We have a stratum of ironstone, plentifully abounding with the shells of fish; but it is easy to observe that these shells are not marine productions, but of freshwater lakes, rivers, etc. being actually the remains of horse-muscles' (Whitehurst, *Strata in Derbyshire*, 1778). 'The Tankersley or "Musselband" ironstone . . . so called from the great number of fossil shells (*Unio*) which characterize it' (Symth, *MGS*, *Iron Ores*, 1856).

**Mutation.** Change. Apart from its use in a general sense (as in 'the many mutations which these older rocks had undergone', Murchison, *Silurian System*, 1839) it has special senses in biology. **1.** 'The term mutation was introduced by W. Waagen in 1869, in his work *Die Formenreihe des Ammonites subradiatus*, for the express purpose of distinguishing variations in time, to which alone this term is applied, from contemporaneous variations in space, for which the term variety was and is in universal use. The time-series alone, that is to say, the original or root form with its successive mutations, constitutes the *Formenreihe*, a term subsequently translated by Waagen himself as developmental series. . . . A mutation, in the palaeontological and original sense, may be defined as a contemporaneous assemblage of individuals united by specific identity of structure *inter se*, and by common descent from a known pre-existing species, from which they differ in some minute but constant character or characters. Such mutations are successive steps of a genetic series, and each may be considered as of specific rank and may have its own contemporaneous varieties' (Bather, in Marr's address, *QJ*, 1905). The first part of this quotation makes 'mutation', in modern terminology, equivalent to 'transient', the second part equivalent chiefly to 'chronological subspecies'. The *Formenreihe* is our 'evolutionary series' or 'lineage', either of which would have been a better translation than 'developmental series', which suggests ontogeny at least as much as phylogeny. **2.** The term has since become firmly fixed in genetics for a discontinuous, inheritable, change in a genetic factor; and this usage (to the exclusion of the other) is the one now generally accepted by palaeontologists. (De Vries, 1901.) [L. *muto*, change.]

**Mutual solution.** See SOLUTION.

**Mylonite.** A metamorphic rock produced by grinding and rolling out. It is characterized by a very close texture with banding, platy fracture, and, perhaps, new minerals. Hence 'mylonitization'. (Described by Lapworth from the Durness-Eriboll district, *GM*, 1883/5, and named 'mylonite' by him in conversation in 1883 and in *N*, 1885.) [G. *mulē*, a mill.]

# N

**Nabarro-Herring creep.** See CREEP. [F. R. N. Nabarro, *PSL*, 1948; C. Herring, *JAP*, 1950.]

**Namurian.** The lowest of the three series into which the Silesian (Upper Carboniferous) subsystem is divided. See *A Correlation of the Silesian Rocks in the British Isles* (*GSSR* (10), 1978). (Purves, Belgium, 1883.) [Namur, Belgium.]

**Nannofossil.** Nannofossils = microfossils; for instance, and perhaps especially, minutes forms such as coccoliths. See quotation (Hay, Stover) under COCCOLITH. [G. *nannos*, a dwarf.]

**Nappe.** 'A *nappe* (Fr.) or *Decke* (Ger.) is a sheet of rocks, of large dimensions (of the order of miles), that has moved forward for a considerable distance (again of the order of miles) over the formations beneath and in front of it, finally covering them as a cloth covers a table. A nappe may be either the hanging wall of a great low-angle overthrust, or a recumbent fold of which the reversed middle limb has been completely sheared out as a result of the great horizontal translation. Classical regions of nappe structure are the Highlands of Scotland and the European Alps. (It should be noted that the terms nappe and Decke are used in French and German respectively for any covering sheet of rock, such as a layer of gravels or a basalt flow. In English, however, they are used only in a tectonic sense.)' (Hills, *Structural Geology*, 1953). A nappe is essentially a separate covering sheet of rock and thus it hardly seems permissible to include within the term an unbroken recumbent anticline, as has sometimes been done. 'The Iltay Nappe . . . the Moine Nappe . . .' (Bailey, *QJ*, 1922). [French, a cloth.]

**Natural arch. 1.** Formed in a protruding piece of sea-cliff by the cutting of a cave right through it, or by two caves, one from each side, meeting. Well-known British examples are 'London Bridge', Torquay (Devonian); 'The Green Bridge of Wales', Pembrokeshire (Carboniferous Limestone); Durdle Door, Dorset (Jurassic). **2.** An anticline in a well-bedded formation, conspicuously exposed in section.

**Natural gas.** 'Inflammable gas that occurs in porous rock of the earth's crust and is found with or near accumulations of crude oil [petroleum]. Being in gaseous form, it may occur alone in separate reservoirs. More commonly it forms a gas cap or mass of gas entrapped between liquid petroleum and impervious capping rock layer in a petroleum reservoir. Under conditions of greater pressure it is intimately mixed with, or dissolved in, crude oil. Typical natural gas consists of hydrocarbons having a very low boiling point' (*McGraw-Hill Encyclopedia of Science and Technology* 1960).

**Natural history.** The study of all natural objects, animal, vegetable, and mineral. Though primarily associated with living objects, especially animals, it is used in geology, not only in connexion with fossils (where, nowadays, it would probably imply, particularly, a consideration of the mode of life of the organisms) but also in connexion with rocks and minerals, where the more general and philosophical aspects of the matter, such as origin, formation, and manner of association, are being treated, as in Harker's well-known book, aptly entitled *The Natural History of Igneous Rocks* (1909), Arber's *The Natural History of Coal* (1911), and Teall's addresses on 'The natural history of cordierite' and 'The natural history of phosphatic deposits' (*PGA*, 1899 and 1900). Of early geological books embodying the term there are Woodward's *Essay towards a Natural History of the Earth* (1695) and *An attempt towards a Natural History of the Fossils of England [Catalogue]* (1728/9), Williams's *Natural History of the Mineral Kingdom* (1789), Plot's *Natural History of Oxfordshire* (1677), his *Natural History of Staffordshire* (1686), and Morton's *Natural History of Northamptonshire* (1712), these three last having important chapters and figures descriptive of the (organic) fossils. Hutton (1795) refers to 'the natural history of granite'. Miller's *Natural History of the Crinoidea* (1821) is a purely descriptive work. 'A sketch of the natural history of the Cheshire rocksalt district' (Holland, *TGS*, 1811). There is also the useful sense of signifying all that can be found out about the nature and history of a geological unit or body: 'The great labour of collecting materials for a complete natural history of each stratum' (Farey, *Derbyshire* 1811).

**Nautiloidea.** An order of the class Cephalopoda, phylum Mollusca, having a chambered, elongate-conical shell, either straight, curved, or, most commonly, coiled in a symmetrical (plane) spiral. The partitioning septa, where they meet the shell, form broadly undulating sutures, and there is a calcareous tube (siphuncle) which pierces the septa centrally. These two features distinguish the nautiloid shell from the ammonoid. They range from the Tremadocian to the present day, but now there is only the one genus *Nautilus* living. Their acme was in the Silurian. (Zittel, 1884.) [genus *Nautilus* (G. *nautilos*, a sailor).]

**Nebular hypothesis.** 'The hypothesis for the origin of the solar system presented by the great French mathematician Pierre Simon Laplace at the end of the eighteenth century. According to this, a rotating and therefore flattened nebula of diffuse material cooled slowly and contracted. In the plane of motion, successive rings of matter were supposed to have split off, to condense into the planets of our present solar system' (Whipple, *Earth, Moon, and Planets*, 1968).

**Negative.** **1.** See POSITIVE and EUSTATIC MOVEMENTS. **2.** A fossil in the form of an impression, either an external impression or the impression of internal features on the surface of an internal mould.

**Nekton.** The (animal) life swimming in the waters of the sea or in bodies of fresh water. See BENTHOS. [G. *nēktos*, swimming.]

**Nematoblastic.** See LEPIDOBLASTIC. (Becke, 1903.) [G. *nēma*, a thread.]

**Neocomian.** The four lower general stages of the Cretaceous are, from below the Berriasian (Ryazanian), Valanginian, Hauterivian, and Barremian. The first, named from Berrias, in S. France, was formerly included in the Valanginian. The Valanginian and the Hauterivian are named from places near Neuchatel (Neocomium) in Switzerland. The first three stages are (or at least were) thus grouped together as the Neocomian, and the fourth stage, from Barreme, in SE. France, is usually also included. In Britain the Neocomian, in this wider sense and simply as a time-division, with no implication of facies, is represented by all but the highest part of the Speeton Clay and equivalent strata in E. England and, in S. England, by the whole of the Wealden. See CRETACEOUS SYSTEM.

**Neogene.** The Miocene and Pliocene series combined thus constituting a stratigraphical system, the upper part of the Tertiary succession. The East Anglian Crag deposits, even when all are included in the Pliocene, are hardly representative of a system, so that the term is not much used in British geology. Sometimes, but not usually, the Pleistocene is included in the Neogene, with the Tertiary thus carried up to the present day. See *A Correlation of Tertiary Rocks in the British Isles* (*GSSR* (12), 1978). (Hoernes, 1864.)

**Neolithic.** 'The later or polished Stone age; a period characterized by beautiful weapons and instruments made of flint and other kinds of stone. This we may call the Neolithic period' (Lubbock, *Prehistoric Times*, 1865). This name was originally used for all that later part of the Stone-age which followed the Palaeolithic (at about 8,000 B.C. on modern computation) up to the time when Man took to using metal. It is now realized that the critical advance in the history of mankind, that from hunting to farming, which laid the foundations of civilization, occurred later—about 3,000 B.C. in Temperate Europe, but earlier than that (closely following the Palaeolithic) in parts of the Middle East. Similarly the use of metal began earlier in the Middle East (about 4,000 B.C.) than in Temperate Europe (about 2,000 B.C.). The name Neolithic now usually refers to this late Stone-age 'culture'; that between the Palaeolithic and Neolithic being the Mesolithic. The culture immediately succeeding the Neolithic is variously termed Bronze, Copper-Bronze, Chalcolithic (copper-stone, an overlapping). Thus these several cultures, following a constant order in any one region, do not correspond to general time-periods and should hardly be called 'ages' (as they usually are); the passage from Palaeolithic to Mesolithic is however still somewhat arbitrarily taken as occurring at about the same time (10,000 years ago) everywhere. See STONE AGE, PALAEOLITHIC. *The Neolithic Revolution* (Cole, *BM*, 1959/67).

**Neomorphism.** See DIAGENESIS.

**Neontology.** The study of present-day life, as opposed to palaeontology.

**Neotectonic.** Recent tectonic. Tectonic activity that has occurred during the 'recent' period including that still going on.

**Neotremata.** One of the two orders of the class Inarticulata, phylum Brachiopoda.

The pedicle opening is a slit in the pedicle valve. One valve, usually the pedicle, is flattened, the other conical. Cambrian to present day. (Beecher, 1891.) [G. *trēma, trēmat-*, an aperture.]

**Neozoic.** Of the ages of new forms of life. Not in general use, but sometimes used for Mesozoic and Cainozoic combined (Forbes, *QJ*, 1854; Lapworth, *GM*, 1883; Watts, *Geology for Beginners*, 1898/1957; Jukes-Browne, *Stratigraphical Geology*, 1912) or for Cainozoic alone (Neaverson, *Stratigraphical Palaeontology*, 1955).

**Nepheline.** See FELSPATHOID. (Haüy, 1800.) [G. *nephelē*, a cloud, by reason of its tendency of alteration from a clear state, particularly along cleavage planes.]

**Neptunian. 1.** Produced in a marine environment. For particular usages see NEPTUNIAN DYKE and quotation under PLUTONIAN. **2.** Pertaining to the doctrine of the Neptunists. 'The Neptunian system which ascribes the formation of all minerals to the action of water alone, and extends this hypothesis even to the unstratified rocks' (Playfair, *Illustrations*, 1802).

**Neptunian dyke.** So far, the only definitely distinguished form of Neptunian insertion; the filling in, by sedimentation, of a fissure so that the resulting body of sedimentary rock is in the form of a dyke. The fissure might be formed under subaerial conditions and the surface then sink beneath the sea to receive the sediment, or it might be formed under the sea and be immediately filled in. 'A fossiliferous limestone [Ordovician] associated with Ingletonian beds is of the nature of a "Neptunian dyke" similar to those in the Pre-Cambrian of the Church Stretton district or the sandstone "dyke" in the Jurassic of the Golspie area' (King, *QJ*, 1932). See SEDIMENTARY INSERTION. Professor Whittard writes: 'This term was used in conversation; probably another of those Lapworthian remarks which has found its way into print long after it was first used. I understood that the type occurrence for Neptunian dykes was that at Hazler Hill' (mentioned, but not so named, by Cobbold, *Church Stretton*, 1900, and by Lapworth in *Geology in the Field*, 1910; referred to by name by Strachan, Temple, and Williams in *GM*, 1948). [L. *Neptunus*, Neptune, god of the sea.]

**Neptunic rock.** Rock formed in the sea. If used at all, this seems the only appropriate meaning; but it has been proposed (Read, *PGA*, 1944) as a general name for all sedimentary rocks, wherever formed.

**Neptunist.** 'Within a few years [after the publication of Hutton's 'Theory of the Earth', *TRSE*, 1788] the geological world had split into two camps, the Neptunists, as Werner's followers were called, on account of the prominent part played by water in Werner's theories; and the Plutonists, as Hutton's supporters came to be known, after Pluto, God of the Underworld whence Hutton supposed his igneous rocks to have originated. With the Huttonians were associated the Vulcanists who believed in the igneous origin of the vast spreads of basalt and other types of lava, still thought by the Wernerians to be aqueous in origin' (Eyles, *HS*, 1964). Hence 'Neptunism', 'Plutonism', and 'Vulcanism'. 'Neptune and the flood', 'From Vulcanism to paleontology' (Gillispie, *Genesis and Geology*, 1951). 'Neptune versus Vulcan' (Fenton and Fenton, *Giants of Geology*, 1952).

**Neritic.** Pertaining to the shallow seas; for accumulations of shells, but sometimes for the whole environment of deposition on (approximately) the continental shelf. [G. *nēritēs*, the sea-mussel.]

**Neutral fold.** A sideways-closing fold. (Bailey and McCallien, *TRSE*, 1937.)

**Névé.** Snow which, by thawing at the surface and refreezing below, becomes a porous mass of ice. Pressure will convert this to solid ice. The névé fields are the gathering grounds of glaciers. [French.]

**New Red Sandstone.** The red sandstone (desert) facies of the Permian and Triassic systems, well exemplified in the British Permo-Trias. If the name is given a not too rigid time-significance (less rigid than Old Red Sandstone), then where the lowest beds of a local series of this facies are of Upper Carboniferous age it is allowable to include them. (Strachey, *PT*, 1725, 'Red Earth'; Fitton, in an anonymous review, *ER*, 1818 (Feb.), and Buckland in Phillips, *England and Wales*, 1818, 'New Red Sandstone'.)

**Newlands series.** The beds in the Girvan area of Scotland corresponding to the Birkhill Shales (Silurian) of the Moffat area. They comprise about 300 m. of mixed shelly and graptolitic facies, the former predominating. (Lapworth, *QJ*, 1882.) [Newlands, Ayrshire.]

**Nife.** See CORE.

**Nivation.** Snow-patch erosion. As the edge of the snow retreats by seasonal melting the underlying wet rock, with little or no vegetation, is easily attacked by weathering, particularly frost action. Snow patches thus tend to 'dig themselves in' and initiate the first stage in the formation of a cirque. [L. *nix, niv-*, snow.]

**Nodule.** A small lump of mineral or stony substance occurring in a bed of rock. 'Mineral nodules, such as . . . marcasite, flint; or metallick nodules, or lumps yielding copper, iron, tin, lead' (Woodward, *Method of Fossils*, 1728). 'The nodular character of limestones appears to be a matter deserving especial attention, since this condition may frequently denote the later stages of the process of dissolution of a limestone-bed. Such nodules, which I propose to term "residual", should be carefully distinguished from contemporary calcareous nodules. The conversion of a bed of jointed limestone into a band of residual nodules is a destructive process, while the formation of a band of concretionary calcareous nodules is a constructive one' (Rutley, *QJ*, 1893). 'In a nodule, composed for the most part of ferruginous matter, one or more of the shells may be calcareous. By the removal of a calcareous layer, by percolating water, from the interior of a nodule, a central ferruginous kernel becomes detached, and rattles when the concretion is shaken (Klapperstein or "rattle-stone")' (J. Geikie, *Structural and Field Geology*, 1953). See quotation (*BRG*) under FLOW-STRUCTURE.

**Nomenclature.** In geology; the system of names given to the particular kinds of minerals, rocks, fossils, &c., 'Nomenclature of rock-groups. . . . Nomenclature– . . .' (Jukes-Browne, *Stratigraphical Geology*, 1912). *Nomenclature of Petrology* (Holmes, 1920/8). 'The rules of nomenclature' (Davies and Stubblefield, *Introduction to Palaeontology*, 1961). *International Code of Zoological Nomenclature* [including fossils] (Stoll and others, eds., 1961). 'How should rocks be named ?' (Sabine, *GM*, 1974). 'It may be doubted whether any more fruitful advance has ever been made by the human species than that of proposing to distinguish between different objects by the use of words which serve as a sort of shorthand for things. These words are known as "names". Progress in science or art demands a constant increase in nomenclature' (Stobbs, N. Staffs. miners' glossary, *TNSFC*, 1916).

**Nonconformity.** See UNCONFORMABLE (table).

**Non-marine** Indicating an aqueous condition, product, association, or environment, not known precisely (especially as to whether the water was fresh or brackish), but evidently not typically marine. 'Revision of the non-marine lamellibranchs of the Coal Measures' (Davies and Trueman, *QJ*, 1927). 'The non-marine Mollusca of the Cambridgeshire gravels' (Kennard and Woodward, *QJ*, 1919). It is a useful non-committal term. 'The recognition of a marine fauna is not commonly disputed. But the [inference as to] more restricted bodies of water, of varying salinity, needs far more closely defined terms and criteria than are often given by palaeontologists—or even admitted to be lacking. The terms deltaic, estuarine, littoral, brackish, lagoonal, have their definitions (or at least their limits of connotation) and to use them indiscriminately—or worse, synonymously —is to abandon even an attempt at precision' (Rayner in Westoll, ed., *Fossil Vertebrates*, 1958).

**Non-sequence.** In a conformable series of beds, a gap in the record at a surface of contact between two beds, representing a time during which no permanent deposition took place. Non-sequences are presumably common but, owing to any one of them being normally of geologically short duration, can hardly be detected. 'It is quite impossible in any quarry to say of one bed that rests directly on another, that it was not only the next formed bed at that particular place, but that no other bed, or even no other set of beds, was formed anywhere else in the interval between them. If he place his finger on the plane of stratification between two beds, that little space may mark the lapse of years, centuries or millenniums' (Jukes, *Manual*, 1862). 'Some gaps in the Lias' (Walford, *QJ*, 1902). 'The evidence for a non-sequence between the Keuper and Rhaetic series in north-west Gloucestershire and Worcestershire' (Richardson, *QJ*, 1904). 'More gaps than record' (Ager, *The Nature of the Stratigraphical Record*, 1973).

**Norite.** A gabbro with an orthorhombic pyroxene as the dominant mafic mineral. (Esmark, 1823). [Nor (for the north). ]

**Norm.** In petrology, the expression of the chemical composition of an igneous rock in terms of the molecules of relatively simple minerals, the 'standard' or 'normative' minerals. It is an extension of the method of expressing the composition as oxides. The proportions of the normative minerals can be given by weight (weight-norm, the original form of expression) or by numbers of molecules (the molecular norm, the precise chemical formula for the molecule being agreed). The normative minerals are units for calculation; they may or may not be present in the rock as real minerals. Cf. 'mode'. 'In 1903 four American petrologists, Cross, Iddings, Pirsson, and Washington, advanced a classification which is in reality based on chemical composition (*Quantitative Classification of Igneous Rocks*, Chicago, 1903). Analyses are first calculated into a set of standard minerals (the "norm"). . . The norm is divided into a "salic" and a "femic" group' (Tyrrell, *Petrology*, 1929).

**Normal erosion.** 'The subaerial, as distinguished from marine, eroding agencies, fall into two groups, normal and special, and it is by the normal group, running water and the weathering processes, that the shaping of the land surface is mainly effected' (Cotton, *Geomorphology of New Zealand*, 1922, quoted in Stamp, ed., *Geographical Terms*, 1961). *Landscape as developed by the Processes of Normal Erosion* (Cotton, 1948). However, 'normal' and 'special' are hardly satisfactory here as (1) erosion always proceeds normally according to the conditions, and (2) glacial and aeolian erosion are not so rare as to be 'abnormal'.

**Normal fault.** A fault which dips to the downthrow side. (A vertical fault is usually included.) If the geometrical relations of bedding and fault, resulting from the fault movement, are alone considered, a fault may be said to be a normal fault if it dips to the side on which the vertical separation of beds is downwards. Called 'normal' because of the opinion as expressed, for instance, by Lapworth (*GM*, 1883): 'Faults and dislocations in gently inclined formations . . . the plane of fracture normally hades or inclines in that direction in which the rocks have been depressed'. The reverse relation defines the 'reverse fault' which is, in fact probably commoner than the 'normal fault'. For full discussion and history see Dennis, *Tectonic Dictionary*, 1967.

**Normal fold.** 'Associations of persistent, long and relatively narrow folds may be termed "normal fold structures", in contrast with the recumbent and highly complex folds of the *Decken* [nappes]. Seeing that the term "normal fold" is no longer in general use as a synonym for "symmetrical fold", it might be used with advantage in the sense here suggested, without leading to confusion' (Hills, *Structural Geology*, 1953).

**Normally consolidated deposit.** One 'that has never been under greater pressure than that existing at the present time' (Skempton, *PGS*, 1968).

**Nose.** The pitching (plunging) end of a fold, as seen in plan, on the ground.

**Nuée ardente.** 'A rather curious phenomenon has been observed to occur during the eruption of acid volcanoes. In addition to the effusion of lavas or the evolution of rising clouds of fragments, acid eruptions were observed to result in the emission of incandescent clouds moving at high speeds close to the ground surface. Such clouds have been called "nuées ardentes" on the island of Martinique where a whole town (St. Pierre) was devastated in 1902 by one of them. Once a nuée ardente comes to rest a chaotic mixture of incandescent particles and fragments settles in a thick sheet' (Rast, *NIW*, 1960). 'La production des grandes nuées ardentes destructrices' (Lacroix, 1908). 'The origin of the vulcanological concept of nuée ardente' (Hooker, *Is*, 1965). [French, glowing cloud.]

**Nummulitic formation.** See PALAEOGENE. In Egypt, the Sphinx is carved from an outcrop of Nummulitic Limestone, and the Pyramids are built from huge blocks of the same rock.

**Nunatak.** 'A nunatak (*nuna*, lonely; *tak*, a peak) is an Eskimo term for an island of rock or mountain peak in a sea of land-ice: A. E. Nordenskiold introduced the word into glacial literature' (Charlesworth, *Quaternary Era*, 1957). 'Camp was made . . . at the foot of a nunatak, the summit of which was 4,960 feet above sea level' (Whitney, Climatic Changes, 1882, quoted in *OED*). The Swedish plural nunatakker is sometimes used in English.

# O

**Obduction.** Term proposed by Coleman (*JGR*, 1971) for the emplacement of ophiolite complexes onto continental margins.

**Oblique fault.** A fault of which the strike is oblique to the strike of the beds it affects.

**Obsequent fault-line scarp.** A fault-line scarp where the higher ground is on the downthrow side of the fault.

**Obsequent stream.** A stream flowing in a direction 'against' that of the primary consequent streams; see CONSEQUENT DRAINAGE. It would normally be a counter-dip stream. 'These short streams have a direction opposite to that of the original consequents and may therefore be called obsequents' (Davis, *GJ*, 1895).

**Observe.** 'Like all other students, geologists know that their first business is to observe. Observing involves a great deal more than mere looking, and more even than just noticing: it demands accuracy in both of these coupled with an intelligent attitude towards the objects seen' (Hawkins in Evans, *The Observer's Book of Geology*, 1949). *The Geological Observer* (De la Beche, 1851 and later).

**Obsidian.** A glassy rhyolite. [L. *obsidianus*, natural glass.]

**Oceanic crust.** See EARTH'S CRUST.

**Oceanic ridge.** Primarily a purely descriptive term; but for its special significance see PLATE TECTONICS and SEA-FLOOR SPREADING.

**Oceanic trench.** For a special usage with a particular significance see PLATE TECTONICS.

**Ocellar.** A texture, in igneous rocks, in which mineral flakes such as biotite, are tangentially disposed round the borders of crystals, or groups of crystals, of later growth. It is as if the flakes had been pushed outwards against the sides of small liquid pockets in which the later crystals were growing. The lamprophyres of Wigtownshire 'occasionally show in perfection the ocellar structure' (Read, *GM*, 1926). 'The crystallization of the teschenite from the Lugar Sill, Ayrshire–The ocellar structures' (Phillips, *GM*, 1968). (Rosenbusch, 1887; Judd, *QJ*, 1889.) [L. *ocellus*, a little eye.]

**Ockham's razor.** (Occam's razor.) The logical principle of 'Ockham's razor' is 'that the fewest possible assumptions are to be made in explaining a thing' (*OED*). It is particularly pertinent in geology: when a condition or phenomenon is found to be in accordance with the operation of an already well-established principle, there is no need to seek any further explanation. See DEDUCTION, where examples will be found. [William of Ockham, Surrey, 1290–1349.]

**Offlap.** The successive contraction in the lateral extension of beds, in an upward sequence, due to their being successively deposited in a shrinking sea; the opposite of onlap.

**Offset.** Used in geology chiefly for the effect of a dip or oblique fault on the outcrops of beds. These are displaced or 'offset' along the line of the fault.

**Oilshale.** A shale containing complex organic matter (kerogen) which, on heating, decomposes to yield oil. The shale represents tbe slow accumulation of inorganic sediment together with the organic debris contributed by aquatic plants and animals. 'The origin of oilshale' (Cunningham-Craig, *PRSE*, 1916). *Oil-shales of the Lothians* (*MGS*, 1927, *IGS*, 1978).

**Old Red Sandstone.** A long-established name for the non-marine representative of the Devonian system, very well developed in Britain, but hardly elsewhere in Europe, though occurring in N. America. The rocks are chiefly red sandstones, conglomerates, and shales, and the fauna predominantly one of fishes, mainly the ostracoderms and the now nearly extinct ganoid-scaled fishes. These are chiefly found in Scotland. As regards the time-range assignment, the Old Red Sandstone overlaps at its base onto the latest part (post-Ludlow) of the Silurian and at its top onto the very earliest part (earliest Courceyan) of the Dinantian. The name was introduced by Jameson (1805) as a translation of Werner's *Aelter rother Sandstein*, with which he correlated it; but Werner's formation, known also in his days as *Rothe Tode Liegende* (Red Dead foundation), is, on the whole, Lower Permian. 'The name Old Red Sandstone flourished mightily in the land of its adoption, while it died out completely in its native country, where its place was taken by the alternative title, *Rothliegende*' (Bailey, *Tectonic Essays*, 1935). It was first tabulated (as 'Old Red Sandstone') in its proper stratigraphical position in 1818, but was for twenty years classed with the Carboniferous system. 'I venture for the first time in the annals of British geology to apply to it [the Old Red Sandstone] the term system, in order to convey a just conception of its importance in the natural succession of rocks' (Murchison, *Silurian System*, 1839). (Jameson, *Dumfries*, 1805, 'Old Red Sandstone', but confused; Smith, *Map*, 1815, 'Red and Dunstone', correctly placed; Fitton (anonymously), *ER*, 1818 (Feb.) and Buckland in Phillips, *England and Wales*, 1818, 'Old Red Sandstone', correctly placed; Hugh Miller, *The Old Red Sandstone*, 1841.) Often abbreviated to 'O.R.S.' See DEVONIAN SYSTEM and DOWNTONIAN.

There are four stratigraphical divisions of the Old Red Sandstone in the typical region of S.W. Britain: Downtonian, Dittonian, Breconian, Farlovian, named from places in the Welsh Borderland. The first three of these (from below upwards) correspond roughly to the Lower Devonian of the general classification, the fourth to the Upper Devonian. The Middle Devonian appears to be almost entirely unrepresented here.

**Old volcano.** See VOLCANO.

**Oldland.** An area of ancient rocks (Precambrian or Palaeozoic), particularly when this has been reduced by erosion to subdued relief.

**Olenellus series.** The Lower Cambrian series, characterized particularly by trilobites of the family Olenellidae (Mesonacidae). Other trilobites also occur, the families Redlichiidae and Protolenidae being confined to the series. In Britain the Lower Cambrian is well developed and well exposed in the Welsh Cambrian areas; as the Caerfai beds (Pembrokeshire), the greater part of the Harlech Grits (Merioneth), and the Llanberis Slates (Caernarvonshire). It is exemplified also in the lower part of the Comley Sandstone in Shropshire and in the quartzites of the NW. Highlands of Scotland. (Walcott, *IGC*, 1888, *USGS*, 1890; Hicks, *GM*, 1892/4.) [genus *Olenellus*.]

**Olenus series.** The Upper Cambrian series (excluding the Transition, or Tremadocian, series) in Europe and eastern N. America, characterized particularly by trilobites of the sub-family Oleninae of the family Olenidae. In the type area, Wales, the *Olenus* series is represented by the Lingula flags. (Walcott, *IGC*, 1888; Hicks, *GM*, 1892/4.) [genus *Olenus*.]

**Oligocene.** The second general stratigraphical group of the Tertiary succession usually ranking as a series; if a Palaeogene system is recognized, it is the higher of the two series within it. See *A Correlation of the Tertiary Rocks in the British Isles* (*GSSR* (12), 1978). In Britain the Oligocene rocks are freshwater (or brackish) clays, marls, and limestones, with a few marine beds; they occur only in the Hampshire tectonic basin. They were first described by Webster (*TGS*, 1814); the equivalent strata in the Paris basin had been described by Brongniart and Cuvier in 1808. The beds abound with beautifully preserved fossil shells, particularly in the Isle of Wight, for example: the lamellibranchs *Cordiopsis incrassata* and *Corbicula obovata*, and the gastropods *Batillaria concava*, *Limnaea longiscata*, *Planorbis discus*, *Potamides vagus*, and *Viviparus lentus*. (Beyrich, 1854; 'The Oligocene strata of the Hampshire Basin', Judd, *QJ*, 1880.) [G. *oligos*, few, expressing 'few recent species'.]

**Oligoclase.** See PLAGIOCLASE and (for etymology) FELSPAR.

**Olistostrome.** A sedimentary deposit consisting of a chaotic mass of allochthonous exotic components (olistoliths) of different kinds which have slipped down amongst themselves, the whole being the result of gravity sliding on a massive scale over a wide area. 'Olistostromes and olistoliths' (Abbate and others, *SG*, 1977). 'The Moni Mélange, Cyprus: an olistrostrome at a destructive plate margin' (Robertson, *JGS*, 1977). [G. *olisthos*, slipperiness, *strōma*, something spread out.]

**Olivine.** A family of rock-forming minerals, silicates of magnesium and iron, being an isomorphous series from a magnesium end-member, forsterite ($Mg_2SiO_4$), to an iron end-member, fayalite ($Fe_2SiO_4$). They are olive-green and glassy-looking crystals, in the orthorhombic system; usually somewhat ellipsoidal (olive-shaped), with cleavage in the form of irregular cracks. They are essential constituents of many basic and ultrabasic igneous rocks. (Werner, 1790.) [L. *oliva*, an olive.]

**Onion weathering.** See SPHEROIDAL.

**Onlap.** The spreading of higher beds, in a conformable sequence, over wider and wider areas than those below them due (normally) to the progressive submergence of the land. This process results in the condition of overlap within the onlapping series.

**Ontogeny.** 'The course of development of an individual' (*SA*, 1956). 'The ontogeny of trilobites' (Whittington, *BR*, 1957). Though almost exclusively a biological term, it has been used in other connexions, e.g. *Ontogeny of Minerals* (Grigor'ev, 1965).

**Ooid.** Synonymous with 'oolith' and preferable as being less likely to be confused with 'oolite'.

**Oolite.** A sedimentary rock usually a limestone, sometimes an ironstone, resembling in texture the roe of a fish, composed, entirely or largely, of small rounded concentrically layered grains (ooliths). 'This stone which is brought from Kettering in Northamptonshire, and digg'd out of a quarry, has a grain altogether admirable. It is made up of an innumerable company of small bodies, not much differing from a globular form They appear to the eye like the cobb or ovary of a herring or some smaller fishes' (Hooke, *Micrographia*, 1665). 'The oolite or roe-stone, a species of limestone which derives its name from small globules embedded in it, supposed to resemble the roes of fishes' (Bakewell,

*Geology*, 1813). 'Oolites formed at the present time are calcareous—aragonite or calcite. "Fossil" oolites, however, may be siliceous, dolomitic, haematitic, pyritic and so forth' (Pettijohn, *Sedimentary Rocks*, 1957). Hence 'oolitic'. Oolite, as a formation name, refers to one or other (or all) of the oolitic formations of the Jurassic. [G. *ōon*, an egg.]

**Oolith.** A particle of an oolite. The term was first used for the oolite rock itself, by Bruckmann (1721), who thought that the rock was actually formed of petrified eggs. See OOID.

**Ooze.** In geology, a pelagic deposit composed entirely or predominantly of the hard parts of small planktonic organisms. There are four main types, named after the predominant constituent. Calcareous: Globigerina ooze, formed of foraminifera (particularly of the genus *Globigerina*) is the commonest, by far the most widespread ooze, covering about 50 million square miles of the ocean floor; Pteropod ooze, forming particularly in parts of the Atlantic. Siliceous: Radiolarian ooze, beneath the tropical belts of the Pacific and Indian oceans; Diatom ooze, beneath cold waters. The oozes form as distinctive deposits only where they are not masked by the more rapidly deposited land-derived material. The calcareous and siliceous skeletons tend to settle to the bottom together, but the calcareous are more abundant than the siliceous so that the resulting deposit, where both accumulate together, is dominantly calcareous. However, in water more than about 3.5 km. deep, calcareous skeletons become dissolved, leaving the siliceous organisms to form a deposit on their own. (Murray and Renard, *Voyage of H.M.S. Challenger, 1872-76*, 1891 and earlier.) [Old English, *wáse*.]

**Open fold.** See CLOSE.

**Opencast.** Open to, and worked from, the surface (with any merely superficial overburden removed); as distinct from underground. For mines and mining.

**Ophiolite.** 'Ophiolite sequences include ultrabasic rocks (often serpentinised), basic dyke and pillow lava complexes and cherts, commonly thought to represent slices of oceanic crust and mantle caught up within or between continental rocks during subduction or continental collision. Such sequences can mark the suture between two continental plates that have collided. However, oceanic crust can also be thrust, or obducted, on to a continental plate and emplaced some distance away from the suture, or be formed in small rifts that close without ever developing into true ocean basins' (Anderton and others, *A Dynamic Stratigraphy of the British Isles*, 1979). 'Ophiolite: it's definition [etc.]' (Church in Irving, *MRC*, 1972). *Ophiolites* (Coleman, 1977). *North American Ophiolites* (Coleman and Irwin, eds., 1977). 'Ophiolites and related rocks' (symposium, *AJS*, 1980). 'Ophiolite' was first introduced by Brongniart (France, 1813) for a rock composed mainly of serpentine with the addition of other minerals such as garnet, diallage, etc. The meaning was expanded by Steinmann. [G. *ophis*, a serpent.]

**Ophitic.** A texture of igneous rocks (particularly dolerites) where, typically, small unorientated lath-shaped crystals of plagioclase felspar lie enclosed in a body of larger crystals of a mafic mineral which is typically augite but may be another mineral, commonly hornblende, olivine, or ilmenite. Each individual crystal of the mafic mineral is thus deeply penetrated by felspar laths and may completely enclose some of them. When the individual mafic crystals are very large compared with the felspars and thus completely enclose many of them, the texture merges into the poikilitic. The term 'ophitic' is sometimes applied, with doubtful propriety, to the mineral itself when it occurs as rather large enclosing crystals. Less typically, the body of enclosing mafic minerals is composed of crystals which are not much larger than the idiomorphic felspars. In hand specimens the rocks are characteristically a dark greenish colour speckled with the light felspars, an appearance which has suggested a snake's skin. (Michel-Lévy, 1877.) [G. *ophis*, a serpent.]

**Ophiuroidea.** The brittle stars, a subclass of the Asterozoa, phylum Echinodermata.

The rays are long and flexible and are sharply marked off from the central disk. [genus *Ophiura* (G. *ophis*, a serpent, *oura*, the tail).]

**Orbicular.** 'Orbicules consist of a central core surrounded by one or more concentric shells of contrasted texture and mineralogy. They are (when unbroken and undeformed) spheroidal or ellipsoidal. They range from less than one inch to more than one foot in diameter and are closely crowded together or sparsely scattered. "Orbicular rocks" are igneous, metamorphic, or migmatitic rocks that consist of orbicules, matrix (the rock between and immediately surrounding orbicules), and country rock (the formation containing orbicules and matrix). Orbicular rocks are world-wide in occurrence' (Leveson, *BGSA*, 1966; a composite quotation). (Delesse, 1849.)

**Order.** In the classification of animals and plants, recent or fossil, an order is a division of a class, itself comprising one or more families.

**Ordovician system.** (Here considered as excluding the Tremadocian.) It overlies the Cambrian and is succeeded by the Silurian. The base has been defined in two parts of N. Wales; in the Arenig district, at the base of the Basal Grit and in the Dolgellau district, at the base of the Basement series. If there is any unconformity with the Tremadocian in these districts it is very slight. By far the commonest fossils are trilobites and brachiopods, in the shelly facies, and graptolites in the graptolitic facies. Many of the characteristic families have their earliest representatives in the Tremadocian. The chief characteristic trilobite families are the Trinucleidae and Asaphidae and, for the brachiopods, particular subgroups and genera of the superfamilies Orthacea, Clitambonacea, Syntrophiacea, Strophomenacea, and Dalmanellacea. The chief graptolite families confined to the Ordovician are the Dichograptidae, Leptograptidae, and Dicranograptidae; but, in addition, the Diplograptidae are abundant.

The following are the series of the Ordovician system:

Bala { Ashgill
Caradoc
Llandeilo (s.s.)
Llanvirn
Arenig (s.s.)

The classic ground of the Ordovician is N. Wales where the outcrop extends through the Snowdon mountains, round the outskirts of the Harlech Dome, through and to the south of the Cader Idris range; also over much of the Lower Palaeozoic part of S. Wales. Much igneous rock occurs with the sedimentaries (hence the rugged mountains). The system also outcrops over other extensive areas of Britain: the Lake District and Scotland.

The system was proposed and named by Lapworth in 1879 (*GM*) to comprise those stratigraphical groups in the middle part of the Lower Palaeozoic succession which were found to form, appropriately, from purely geological considerations, a third system between a Cambrian below and a Silurian above. It also had the advantage of bringing to an end the long dispute as to the boundary between Sedgwick's Cambrian and Murchison's Silurian. Lapworth's Ordovician, now a name of almost world-wide currency, comprised Sedgwick's Arenig and Bala and Murchison's Llandeilo and his (redefined, restricted) Caradoc. See CAMBRIAN SYSTEM and SILURIAN SYSTEM. *A Correlation of Ordovician Rocks in the British Isles* (*GSSR* (3), 1973). [Ordovices, an ancient British tribe in N. Wales.]

**Ore.** ' "Solid, naturally occurring mineral aggregate of economic interest from which one or more valuable constituents may be recovered by treatment". After a long discussion this definition was agreed by the Council of the Institute of Mining and Metallurgy (1955)' (Stamp, ed., *Geographical Terms*, 1961). *History of the Theory of Ore Deposits* (Crook, 1933). 'Reflections on ore genesis and exploration' (Pereira, *MgM*, 1963). [Old English *ora*.]

**Ore-shoot.** 'The individual mineral deposit, which may be regarded as a coarsely crystalline rock made up of one or more introduced minerals. A mineral vein may carry several ore-shoots, separated by barren stretches' (Dunham and Stubblefield, *QJ*, 1944).

**Organic.** In geology, always in the sense 'of animals or plants', to distinguish organic remains or traces (fossils in the modern sense) from inorganic material (minerals and rocks) or markings. But although all fossils are 'organic' in this wide sense, the material of the hard parts may be 'organic' in the narrower sense of being composed of some carbon compound, such as chitin or woody tissue, as distinct from being composed of mineral material, such as calcium carbonate or silica.

**Organic deposit.** A deposit formed of the remains of organisms. Includes (1) shelly accumulations and coral reefs (2) the oozes, and (3) peat. The term is naturally extended to the corresponding hard compacted rocks: many limestones, some siliceous rocks other than quartzite or quartz-sandstone, and coal and lignite. For such a deposit the term 'biodepositional' has been suggested.

**Original crust.** 'The oldest known rocks on the surface of the earth are normal metamorphic rocks about 1,000 million years younger than the estimated age of the earth [see PRECAMBRIAN]. We are still far from getting down to the original crust, if indeed it is anywhere still preserved' (Ager, *Introducing Geology*, 1975).

**Ornithischia.** One of the two orders of dinosaurs (Mesozoic land reptiles). Among the characteristic features are the four-rayed structure of the pelvis and the absence of teeth from the front of the jaws. It comprises the suborders Ornithopoda, Stegosauria, Ankylosauria, and Ceratopsia. [G. *ornis, ornith-*, a bird, *ischion*, a hip-joint.]

**Ornithopoda.** A suborder of the order Ornithischia. Herbivorous, with a prevailingly bipedal mode of locomotion. Jurassic and Cretaceous.

**Orocline.** 'Structural or mountain arc owing its form to differential horizontal displacement after the main features of the structural zone originated' (*AGI*, *Dictionary*, 1976). 'The orocline concept in geotectonics' (Carey, *PRST*, 1958).

**Orogeny.** Mountain-making, the formation of mountain ranges. It is implied that this is done by folding and thrusting, as distinct from epeirogeny which, by its broad crustal movement, might in fact produce mountainous country, either directly or (more likely) by subsequent erosion of the uplifted region. It would include, particularly in the form 'orogenic movements' ('mountain-building movements'), movements of the earth's crust of this kind which might not actually have produced mountains (because erosion kept pace with the rise of folds). 'Orogenesis' is synonymous but might be held to refer more particularly to the early development of the movements. Hence also 'orogen' and 'orogenic belt'. 'The great mountain systems, or orogens, are zones of extreme compression of the earth's crust weaving a complex pattern of majestic sweeps around the world. The great Alpine—Himalayan—Rockies–Andean ranges and the island arcs of the East and West Indies are physically the most spectacular examples, because the most recent; whereas older mountain systems, such as the Hercynian or the Caledonian, have been largely eroded almost out of existence as physical features, leaving exposed only their deep roots as testimony to their earlier presence' (Lees, *QJ*, 1952). *Time and Place in Orogeny* (Kent and others, eds., 1969). 'Concept of orogeny' (Cebull, *Geoly*, 1973). 'The cause of such mountain-building movements or "orogenics" is now much more clearly understood in the light of plate tectonics' (Ager, *Introducing Geology*, 1975). Further discussion and history in Dennis, *Tectonic Dictionary*, 1967. See DIASTROPHISM and EPEIROGENY. (G. *oros*, a mountain.]

**Orthochemical.** Pertaining to the material of a chemically-characterised sedimentary rock, this material having been formed by direct chemical or mechanical precipitation. The rock is commonly a limestone, composed of lime-mud and/or calcareous micrite. Cf. ALLOCHEMICAL.

**Orthoclase.** The chief of the potassium felspars, potassium aluminium silicate. Very common in granite, and other acid igneous rocks, where it may occur as large porphyritic crystals as well as a component of the ground mass; sometimes pink in colour. Monoclinic system. See FELSPAR. [G. *klaō*, break.]

**Orthoclinal mountains.** See PLAGIOCLINICAL MOUNTAINS.

**Orthogenesis.** 'A term proposed by Eimer in 1895 to describe evolution in a definite direction under the influence of some internal factor' (Trueman, *TGSG*, 1940). The rather vague meaning implied in the etymology and various ideas as to what exactly might be meant by an 'internal factor' or similar phrase have led to various usages. For instance, by some it has been taken to mean that evolution goes straight ahead, not wandering in a random manner; by others, to mean a tendency for all lines of descent in a group to evolve in the same direction; by others, again, to imply a goal predetermined by some inherent vital impulse.

**Orthogneiss.** The ordinary gneiss, presumably derived by metamorphism from a plutonic igneous rock. A gneiss supposedly derived from a sedimentary rock is sometimes called a 'paragneiss' (which hardly seems a suitable term).

**Oryctology.** 'The term "oryctology" was in use in the middle of the eighteenth century for the study of fossils, which then included almost every substance dug out of the earth' (Woodward, *History*, 1911). When the term 'fossil' became restricted, so did the term 'oryctology'; but it very soon became superseded by 'palaeontology'. 'Outlines of Oryctology' (subtitle of Parkinson's *Fossil Organic Remains*, 1822). [G. *oruktos*, dug up.]

**Os** (pl. **osar**). One of the two types of esker (in the wide sense; the other being the kame). 'Osar are tortuous or gently sinuous ridges, their appearance is often remarkably artificial, resembling railway embankments. Their crest is typically so narrow that it can just accommodate a path or road. The cross-section, like that of a goat's back, is roughly an isosceles triangle' (Charlesworth, *Quaternary Era*, 1957). Their origin is obscure and may be different in different cases; they seem to be connected in some way with streams which flowed on the surface of, within, or underneath the ice. See ESKER. [Swedish *As, Asar*.]

**Osseous.** 'Abounding in fossil bones' (*OED*). 'In the [Kirkdale] cave most of the bones are broken into small angular fragments and chips, the greater part of which lay separately in the mud, whilst other were wholly or partially invested with stalagmite; and others again mixed with masses of still smaller fragments and cemented by stalagmite, so as to form an osseous breccia' (Buckland, *Reliquiae Diluvianae*, 1823). Bone-beds of any age are 'osseous'. [L. *osseus*, bony.]

**Ossiferous.** 'Containing or yielding bones' (*OED*). 'Ossiferous caves and fissures' (Buckland, *Reliquiae Diluvianae*, 1823).

**Osteichthyes.** The bony fishes, a class of the superclass Pisces. They range from the Devonian and become the dominant fishes from the beginning of the Mesozoic. The enamelled 'ganoid' scales of certain kinds are fairly common fossils. [G. *osteon*, a bone, *ichthus*, a fish.]

**Osteolith.** A fossil bone.

**Ostracoda.** An aquatic subclass of the class Crustacea, phylum Arthropoda, in which a small oval bivalved carapace completely encloses the body, the size varying from about 0·25 mm. to 12 mm. in length. Ranges from the Cambrian to the present day. [G. *ostracon*, the shell.]

**Ostracoderm.** See AGNATHA.

**Otolith.** An ear-stone of a fish; primarily a biological, not a geological term, but otoliths do occur as fossils. 'Otoliths of fishes from the Upper Kimmeridgian of Buckinghamshire and Wiltshire' (Frost, *AMNH*, 1924). [G. *ous, ōto-*, the ear.]

**Outcrop.** The area, strip of ground, line, or place where a rock-body (stratum, formation, intrusion, vein, metamorphic belt, or any other) or a rock-surface (bedding surface, fault, or any other) emerges at the earth's surface. We may imagine any mere thin layer of soil to be removed; if we know the body or surface to be buried beneath a superficial deposit such as beach sand, boulder-clay, or peat, we may say that the outcrop is concealed, though this perhaps sounds contradictory. Where such a concealment is extensive and substantial, but not quite continuous,'incrop' has been used. 'Most of our coal has been discovered . . . by exploring their outcrops' (*ER*, 1805, *OED*). 'Outcrop' is also used as a verb, derived from the substantive, but some prefer to use 'crop out'. In one form or another it is one of the most frequently used verbs in geology. 'To outcrop' seems etymologically satisfactory and has been in use for over a hundred years, whereas 'to crop out', though perhaps more elegant, involves an awkward transposition of word elements. 'Below the chalk formation the upper green-sand and sometimes also the gault "crop out". We use this term, borrowed from our miners, to express the coming up to the surface of one stratum from benenth another' (Lyell, *Principles*, 1833). 'Crop' is an earlier form. Jameson (*Geognosy*, 1808) uses the term 'outgoing'.

In a geological map of a piece of country two factors govern the form of the outcrop of a plane rock-surface: (1) the steepness of the dip, (2) the relief of the ground. If the rock-surface is vertical its outcrop will be entirely controlled by the first factor; it will be a straight line in the direction of strike across the country regardless of the relief. If the rock-surface is horizontal its outcrop will follow the contour lines of the ground. The more marked the relief the stronger will be its power of control over the form of outcrops; the steeper the dip the greater will be the power of its control. The actual form of an outcrop of a plane surface in any particular case is thus the resultant effect of these two independent controls.

**Outlier.** An outcrop of an upper (normally a newer) rock surrounded by lower (normally older) rocks, the upper rock being an outlying patch separated from its main outcrop. The outlying rock may be conformable or unconformable to the rock on which it rests. This separation is commonly due simply to erosion. 'These insulated portions of strata of the plastic clay [Lower London Tertiaries] that have been noticed at Seaford and Newhaven appear to be outlying fragments . . .' (Buckland, *TGS*, 1817). ' "Outliers" of those strata placed at considerable distances from their continuous range, with which they have every appearance of having been once connected' (Conybeare and Phillips, *England and Wales*, 1822). 'The discovery of a great outlier of Lias in the north of Shropshire and adjacent part of Cheshire' (Murchison, *Silurian System*, 1839). 'New Red Sandstone outlier of Leek [unconformable on Carboniferous]' (Hull and Green, *MGS*, *Stockport* [&c.], 1866). Outliers of this ordinary kind are normal features of the evolution of topography and of the resulting geological map. An outcrop of a rock let down between faults, or strongly bent down in a syncline, and thus surrounded by older rocks, is also an outlier (of a structural kind) if at the same time it forms a patch outlying from a main outcrop. 'The Silurian outlier west of Caer Caradoc' (Cobbold, *MN*, 1892). 'Careg Cennen castle is built on an outlier of Carboniferous Limestone let down in the midst of Old Red Sandstone country by an eastward continuation of the Llandyfaelog belt of disturbance' (Pringle and George, *BRG*, *S. Wales*, 1948). (In topographical, but not geological, description, an 'outlier' may refer purely to a feature of the relief, regardless of the geology.)

**Outwash deposit.** A glaciofluvial deposit resulting from the escape of drainage from the front of an ice sheet or large glacier. It may be in the form of an extensive 'outwash plain' or a more limited 'outwash fan' ('outwash apron').

**Overburden. 1.** The rock material, loose or consolidated, overlying a seam, vein, or bed of rock, which has to be removed before that seam, &c., can be exposed for opencast working. **2.** The upper part of a mass of sediment, compressing and consolidating the sediment below.

**Overflow channel.** A feature of drainage under glacial conditions: a channel or notch cut by the overflow waters of a glacier lake. Also called 'spillway'.

**Overfold.** A fold in which the steeper limb is overturned beyond the vertical. An anticline is normally implied, a syncline in this condition (and it should be just as common as an anticline) seems to be nameless, but would presumably also be an 'overfold' ('underfold' seems not to be used). 'Overfolds', for a succession of anticlines and synclines, that is, a succession of folds with alternate middle limbs overturned, is quite usual. 'In consequence of the flexures by which they are often bent backwards, a bed really superior in its general position, may appear to be inferior in partial observations: thus [giving a figure of a typical overfold]' (Conybeare and Phillips, *England and Wales*, 1822).

**Overlap.** 'The extension of one bed, or set of beds, beyond the original termination of the bed or set of beds below it . . . occurs among large groups in a conformable series, which overlap each other successively, in consequence of the newer groups spreading over wider and wider areas than those below them' (Jukes, *Manual*, 1862). The spreading is due (normally) to the progressive submergence of the land. See ONLAP. This is the accepted usage, and to be distinguished from overstep. However, in early geological descriptions, 'overlap' was used for what is now known as 'overstep'. (De la Beche in his *Sections and Views*, 1830, so uses it; but in his *Geological Observer*, 1851, he fully explains 'overlap' in its modern exclusive sense. Murchison uses 'overlap' for unconformity in all editions of *Siluria*, 1854/72. It is used for the effect of the Moine Thrust in *QJ*, 1888 and *MGS*, 1907.) 'As the beds of the old mountain mass of Lower Palaeozoic rocks (St. George's Land) were followed with gross unconformity and overstep by the Carboniferous Limestone, so within the Limestone, as the area subsided and the Carboniferous sea gradually crept up the ancient hillsides flanking the depressions, successively higher beds progressively overlapped those below to cover greater and greater areas' (Smith and George, *BRG, N. Wales*, 1948).

**Overprint.** See RELICT. 'Thermally overprinted Dalradian rocks near Cleggan, Connemara' (Ferguson and Harvey, *PGA*, 1979).

**Override.** In connexion with the movement of rock-masses. 'Thrusts and reversed faults, showing overriding from the north' (McKerrow and Campbell, *PGS*, 1959). 'Overriding southerly rock transport' (Simpson, *GM*, 1962).

**Oversaturated rock.** In petrology; an igneous rock is 'oversaturated' when there is more silica content in the chemical composition of the original magma than is required to make silicate minerals on solidification. There is thus excess silica which appears as free quartz (or some other form of free silica).

Further application of 'saturation' terms is not so simple as the above and may refer to a mineral or to the rock as a whole. Certain minerals (olivine and the felspathoids in particular) are poor in silica content and normally form only when there is not enough silica in the composition of the magma to form minerals of a higher silica content; the crystallization together of such 'unsaturated' minerals and free quartz is therefore incompatible. Rocks containing unsaturated minerals (and no free quartz) are said to be 'undersaturated' or 'unsaturated'. 'On saturated and unsaturated igneous rocks' (Shand, *GM*, 1913).

**Overstep. 1.** An effect of unconformity on outcrop. Where there is unconformity the base of the upper series, along its outcrop, 'oversteps' from one to another of the members of the lower, truncated, series (except when the two series have the same strike-direction and the outcrop is parallel to it). The term is chiefly used when an unconformity is not obvious but is made evident by detailed mapping. Where there is overlap within the upper series, the overstep of this series as a whole is carried on from point to point, along the line of unconformity, by the successive beds in the upper series. The overstep of the base of the Cretaceous in Yorkshire, across the underlying Jurassic formations, was recognized by Henry Cavendish in a letter to

John Michell written in 1788. 'Detailed mapping of the outcrop has shown that in the extreme east of Sussex the Sandgate Beds overstep the Hythe Beds to rest on the Weald Clay' (Kirkaldy, *QJ*, 1937). **2.** Any section showing an angular discordant unconformity may be said to exhibit 'overstep' of the base of the upper series from one member of the lower series onto another.

**Overthrust.** See THRUST.

**Ox-bow lake.** See MEANDER. Originally an American term.

**Oxford Clay.** The Jurassic clay formation (180 m. thick in S. England), only the upper part of which is placed in the general Oxfordian stage, the lower and middle parts being placed in the Callovian. Characteristic fossils are the ammonites *Kosmoceras*, *Quenstedtoceras*, and *Cardioceras* and, among other invertebrates, well-known species such as the lamellibranch *Gryphaea dilatata* and the belemnite *Hibolites hastatus*. The vertebrate remains are chiefly those of the marine reptiles *Ichthyosaurus*, *Plesiosaurus*, and *Pliosaurus*. (Smith in Farey, *Derbyshire*, 1811, 'Clunch Clay'; Buckland in Phillips, *England and Wales*, 1818, 'Oxford Clay'.) [Oxford.]

**Oxfordian.** See JURASSIC SYSTEM.

**Oxygen.** The element which, in chemical combination, is the most abundant of all in the mineral substance of the earth's crust.

# P

**Pacific suite.** 'A general term for the whole assemblage of calc-alkali rocks, notably represented by andesites, granodiorites, and associated rocks; directing attention to their distribution around the Pacific, to their association with Pacific types of coastline, and more generally to their association with tectonic structures of the mountain-building type due to compression building, and overthrusting' (Holmes, *Nomenclature of Petrology*, 1928). See ATLANTIC SUITE.

**Packing.** The mode of association of the particles in a sedimentary rock. See TEXTURE. 'Even in a rock composed wholly of uniformly sized spherical elements, there are several ways in which these spheres can be arranged or packed. When the sizes and shapes are more varied the manner of packing becomes more complex' (Pettijohn, *Sedimentary Rocks*, 1957). 'The packing texture of clastic sediments' (Smalley, *N*, 1964).

**Pahoehoe.** A Hawaiian name for one of the two chief forms of lava emitted from volcanoes of the Hawaiian type; the other being aa. 'The first form is called by the Hawaiians pa-hó-e-hó-e. Imagine an army of giants bringing to a common dumping-ground enormous cauldrons of pitch and turning them upside down, allowing the pitch to run out, some running together, some being poured over preceding discharges, and the whole being finally left to solidify. The surface of the entire accumulation would be embossed and rolling, but each mass by itself would be slightly wrinkled, yet, on the whole, smooth, involving no further impediment to progress over it than the labor of going up and down the smooth-surfaced hummocks' (Dutton, *USGS*, 1883). Bulbous protrusions in front of pahoehoe flows are 'lava toes' (Wentworth and Macdonald, *USGS*, 1953). 'The term signifies "having a satin-like aspect" ' (Dana, *Characteristics of Volcanoes*, 1890). See AA.

**Palaeoanthropology.** This, the science of 'old' mankind, would usually refer to the mankind of those times before he began leaving deliberate monuments or inscriptions, that is, to 'pre-historic' or 'early' man. It has also been defined as 'the study of human evolution through analysis of the fossil record leading to modern man' (*N*, 1968).

**Palaeobiogeography.** The distribution of the various forms of life, over the earth, in past geological times. *Atlas of Palaeobiogeography* (Hallam, ed., 1973). *Paleobiogeography* (Ross, ed., 1976).

**Palaeobiology.** Generally, has the same meaning as 'palaeontology', but definitely excluding biostratigraphy; that, is, it comprises palaeobotany and palaeozoology. It is sometimes used specially for the study of the mode and conditions of life of organisms in the past. *Paleobiology of the Invertebrates* (Tasch, 1973). More specifically still it may denote the study of those fossils that are the remains of the 'soft' parts of organisms, together with the consideration of the biological significance of the presence of organic substances in rocks. Such material is supplied chiefly by micro-organisms (principally bacteria, actinomycetes, fungi, and algae) and so we have the science of palaeo-microbiology, or geo-microbiology.

**Palaeobotany.** The science of the plant life of past geological ages, as revealed by fossils; a major branch of palaeontology. 'The science of palaeobotany or palaeophytology' (Nicholson, *Palaeontology*, 1872). '*Antediluvian Phytology*, illustrated by a collection of the fossil remains of plants peculiar to the Coal Formation of Great Britain' (Artis, 1825). 'The internal structure of *Fossil Vegetables* found in the Carboniferous and Oolitic deposits of Great Britain' (Witham, 1833). The history of the study of fossil plants is given by Gordon (*BA*, 1934). 'Palaeobotany, the study of fossil plants, provides a fragmentary and tenuous evolutionary picture of plant life from as far back as about 3,000 million years ago' (Bassett, *Amg*, 1973). Palaeobotany is appropriately studied more by those who are primarily botanists rather than by those who are primarily geologists (see VERTEBRATE PALAEONTOLOGY for an analagous placing).

**Palaeocartography.** See PALAEOGEOGRAPHY. 'Palaeocartographical reconstructions' (Arkell, *Jurassic System*, 1933).

**Palaeocene.** (Paleocene.) See EOCENE. (Schimper, 1874, 'Paléocéne'.)

**Palaeoclimatology.** The study of the climates of past geological ages. Palaeogeography, in a wide sense, would include it. An early essay is that by Hopkins (*QJ*, 1852) and a well-known book that by Brooks (*Climate through the Ages*, 1926/49). *Fossil Plants as Tests of Climate* (Seward, 1892). 'Palaeoclimates' (Durham in Ahrens and others, eds., *Physics and Chemistry of the Earth*, 1959). *Problems in Palaeoclimatology* (Nairn, 1964). *Climates throughout Geologic Time* (Frakes, 1979).

**Palaeocurrent.** A current which existed in a former geological age (inferred from the sedimentation features of rocks of that age). 'Palaeocurrents of Dalradian metasediments' (Knill, *PGA*, 1959). 'Sedimentary structures and palaeocurrent directions from the Silurian rocks of Kirkcudbrightshire' (Craig and Walton, *TEGS*, 1962). *Paleocurrents and Basin Analysis* (Potter and Pettijohn, 1963/77).

**Palaeodelta.** See DELTAIC.

**Palaeoecology.** 'Palaeoecology is a branch of geology devoted to the understanding of relationships between ancient organisms and their environments' (Imbrie and Newell, *Approaches to Paleoecology*, 1964). 'The fossil and its environment' (Bather, *QJ*, 1928, who used the term 'palethology'). *Paleoecology* (Ladd, ed., *MGSA*, 1957; there are 1077 pp.). *Principles of Paleoecology* (Ager, 1963). *Introduction to Quantitative Paleocology* (Reyment, 1971). *The Ecology of Fossils* (McKerrow, ed., 1978). [G. *oikos*, an abode.]

**Palaeoenvironment.** Environment in the geological past, particularly the environment in which organisms lived and/or sediments were laid down somewhere at some time. 'The palaeoenvironment of the Bembridge Marls (Oligocene) of the Isle of Wight' (Daley, *PGA*, 1973).

**Palaeogene.** The Eocene and Oligocene series combined, thus constituting a stratigraphical system, the lower part of the

Tertiary succession. Although they form a natural unit in Britain, it is more commonly recognized as a system on the Continent. In S. Europe, N. Africa, and other parts of the Tethys region it is characterized by the large foraminifer *Nummulites* and is thus known also as the Nummulitic formation.

The system can be classified thus:

| | | *General stages* | *British Eocene formations* |
|---|---|---|---|
| Oligocene | | Aquitanian | |
| | | Chattian | |
| | | Stampian | |
| | | Sannoisian | |
| | | Ludian | |
| Eocene | | (=Bartonian) | Barton beds |
| | | Ledian<br>Lutetian | Bracklesham beds |
| | | Ypresian | Bagshot Sands<br>London Clay |
| | Landenian | Sparnacian | Woolwich and Reading beds |
| | | Thanetian | Thanet Sands |
| | | Montian | |

See *A Correlation of Tertiary Rocks in the British Isles* (*GSSR* (12), 1978). 'Radiometric dates from N.W. European glauconites and the Palaeogene time-scale' (Odin and others, *JGS*, 1978).

**Palaeogeography.** The historical reconstruction of the physical geography of past geological ages. 'Palaeophysiography' is synonymous. Early essays in palaeogeography are incorporated in the celebrated works (both dated 1836) of Phillips ('Circumstances attending the deposition of the Mountain Limestone formation', *Yorkshire*) and of Fitton ('Theory of the Wealden', in 'Strata between the Chalk and the Oolites', *TGS*). 'While we owe to Ramsay the most striking of the early attempts to picture for us certain of our phases of palaeogeography, it was Clifton Ward and Prof. E. Hull, who endeavoured to express the facts with regard to British deposits upon maps. They have been followed by Mr. A. J. Jukes-Browne, whose work on *The Building of the British Isles* [1911] has been of great service' (Watts, *QJ*, 1911). 'Paleogeography of North America' (Schuchert, *BGSA*, 1910). *Palaeogeography of the Midlands* (Wills, 1948). *Palaeogeographical Atlas* (Wills, 1951). 'Lower Carboniferous palaeogeography' (George, *PYGS*, 1958). Hence 'palaeocartography', the making of palaeogeographical maps.

The one major part of earth history, throughout geological time, that goes largely unrecorded in the rocks is that concerning the physiography of the land during the working of the geological cycles in operation at various stages in the several regions of the earth. We know much, and in detail, about the marine-deposition phase and the deformation phase of the cycle (also about igneous activity and metamorphism), but comparatively little about the land phase. 'Buried landscapes' provide some rare glimpses.

**Palaeogeology.** The historical reconstruction of the geological, as distinct from the geographical, features of a region in a past geological age, particularly the reconstruction of a palaeogeological map, which would show the areal geology of an ancient surface. See SUBCROP MAP. 'Palaeo-geological and geographical maps of the British Isles' (Hull, *TRDS*, 1882). *Paleogeologic Maps* (Levorsen, 1961). *A Palaeogeological Map of the Palaeozoic Floor below the Upper Permian and Mesozoic Formations in England and Wales* (Wills, *MGSL*, 1974). Hence 'palaeostructure', the structure of a region or rock-series at some past period (arrived at by substracting the later deformation).

**Palaeogeophysics.** See GEOPHYSICS. *Palaeogeophysics* (Runcorn, ed., 1970).

**Palaeogravity.** The gravity force at the earth's surface in past geological times. 'Quantitative limits to palaeogravity' (Stewart, *JGS*, 1977).

**Palaeolatitude.** An example from among the many terms that might be coined for particular aspects of palaeogeography. 'Palaeolatitude of evaporite deposits' (Irving and Briden, *N*, 1962). 'Inferred palaeolatitudinal positions [during the] growth of the E. Antarctic ice sheet' (Drewry, *JGS*, 1975).

**Palaeolithic.** The Old Stone age of man's activities, roughly equivalent to the Pleistocene (excluding Recent). The implements ('palaeoliths') are more or less rudely fashioned but show a general succession of 'cultures' of increasing degrees of workmanship. The names of these cultures (applying to Europe and, less surely, to other regions of the Old World) are taken from French places and are: Chellean (more precisely, Abbevillian), Acheulian, Mousterian, Aurignacian, Solutrian, Magdalenian. Of these the first three, occupying by far the greater length of time, are usually grouped as Lower Palaeolithic, the second three as Upper Palaeolithic. In the Upper Palaeolithic (beginning, perhaps, about 20,000 years ago) bone was extensively used in making implements, and man's artistic talent at this time is shown in the mural paintings of the French rock shelters. The several cultures, though successive in a general way, were probably not strictly contemporaneous in different regions. The Palaeolithic passes into the Mesolithic about 10,000 years ago (10,000 B.P., 8,000 B.C.). See STONE AGE and NEOLITHIC. 'Firstly, the period of the Drift, when men shared the possession of Europe with the Mammoth and other extinct animals. This we may call the Palaeolithic period' (Lubbock, *Prehistoric Times*, 1865).

**Palaeomagnetism.** The magnetism of a rock acquired in a past geological age. Many rocks, both igneous and sedimentary, of all ages, are found to be slightly magnetized due to the magnetism of their iron oxide mineral particles ; and the direction of this magnetization is found to be independent of the present direction of the earth's magnetic field. The following hypothetical principle is invoked in explanation. In the case of an igneous rock, these particles, at the time of their crystallization, or very soon after, acquired a magnetization in the direction of the earth's magnetic field existing at that time, the rock as a whole thus acquiring such a magnetization. Further, this magnetization, once acquired, has remained permanently fixed in direction relative to the particles themselves and thus to the rock-body as a whole. In the case of a sedimentary rock, detrital iron oxide minerals, with their magnetism acquired when first formed in the ultimate parent igneous rock, tend to settle each with its magnetization directed according to the direction of the earth's magnetic field existing at the time of the deposition of the sediment, so that the resulting sedimentary rock as a whole tends to acquire its own permanent direction of magnetization, dating from the time of its formation. The direction of magnetization of a rock (igneous or sedimentary) being thus supposed to be a permanent primary 'fabric' of the rock, the present lack of correspondence between this direction and that of the earth's magnetic field must be due to a relative displacement of these two directions since the rock was formed. It seems that this displacement must be due either to a change in the earth's magnetic field relative to the crust as a whole, or to a relative displacement of the continents themselves (continental drift), or to both. The data for considering such questions obviously lie in the measurements that may be made of the present directions of magnetization of rock masses of all ages in all parts of the world, correcting the values for observed tilting, folding, or large tear-faulting. 'The ancient magnetic field may be found "fossilized" in the form of a permanent magnetization in certain rocks. This weak but nevertheless detectable magnetization is termed natural remanent magnetization (N.R.M.)' (Coulomb and Joubert, trans., Nairn, *Physical Constitution of the Earth*, 1963; quotation slightly recast). The term 'remanent magnetism' is often used. *History of the Earth's Magnetic Field* (Strangeway, 1970). *Palaeomagnetism and Plate Tectonics* (McElhinny, 1973/79). *History of the Earth's Magnetic Field* (Strangway, 1970). *Palaeomagnetism and Plate Tectonics* (McElhinny, 1973/79). 'Palaeomagnetism: state of the art 1976' (symposium papers, *JGS*, 1977). 'In principle, identification of the direction of original remanent magnetism in a suite of rocks can be translated into an inference of its palaeolatitude and palaeomeridional orientation at the time of its formation' (Fuller and Briden in Moseley, ed., *Geology of the Lake District*, 1978). 'It is perhaps worth recalling that it is over thirty years since Dr. McLintock and Dr. Pheminster noted that the Lornty dyke in Perthshire was magnetized in a direction opposed to that of the earth's present field' (Taylor, *PGS*, 1960).

**Palaeomicrobiology.** See PALAEOBIOLOGY.

**Palaeontographical.** Descriptive of fossils, concerned with the description of fossils; e.g. the illustrated monographs of the Palaeontographical Society. Lhwyd's *Lithophylacii Britannici Ichnographia* (1699) was the first work devoted entirely to the description of British fossils; Brander and Solander's *Fossilia Hantoniensia* (1766) may perhaps claim to be the first scientific palaeontographical monograph; Miller's *Crinoidea* (1821) must be one of the first to deal with one restricted group. The substantive 'palaeontography', for the description of fossils (somewhat analogous to 'petrography'), is not common: 'Palaeontology, the *discussion* of past organisms, must always be preceded by Palaeontography, their *description*' (Hawkins, *Invertebrate Palaeontology*, 1920).

**Palaeontological record.** See FOSSIL RECORD.

**Palaeontology.** The science of the life of past geological ages, as revealed by fossils. The word was introduced, in the form *palaeontologie*, by de Blainville (*Manuel de Malacologie*, 1825) and was later used in a Latinized adjectival form, in the title *Bibliographia Palaeontologica*, by Fischer de Waldheim, in Poland in 1834. (Cox, *PGA*, 1958.) In 1834 it was also used in England, by Lyell (*Principles*), as follows: 'Paleontology. The science which treats of fossil remains, both animal and vegetable [giving the etymology]'. In 1838 (*Elements*) he uses the spelling 'palaeontology'. 'The name Palaeontology gradually displaced the older terms Oryctology, *Petrefaktenkunde* and *Petrefaktologie*. Palaeontology is the study, in the field and in the laboratory, of the remains of organic life preserved in the rocks' (Stubblefield, *BA*, 1954). The first works to be published in Britain which may be called textbooks of palaeontology were Parkinson's *Organic Remains of a Former World* (3 vols., 1804/11, with, in vol. i, a scholarly review of the nature of fossils, probably the first historical résumé, in Britain at least, of any branch of geology), the same author's *Fossil Organic Remains* (1822), and Martin's *Attempt to establish a Knowledge of Extraneous Fossils on Scientific Principles* (1809). In France, at about the same time, the great names are Lamarck (invertebrates) and Cuvier (vertebrates). *Palaeontology* (Owen, 1860). *The Early History of Palaeontology* (Edwards, *BM*, 1931/67). 'Palaeontology in literature' (Lamont, *QMJ*, 1947). 'British Palaeontology: a retrospect and survey' (Cox, *PGA*, 1956). 'Discoveries in British palaeontology' (Challinor, *WGQ*, 1967). *Principles of Paleontology* (Raup and Stanley, 1971/78). The three main aspects of palaeontology are comparative morphology (systematic), stratigraphical palaeontology (including palaeoecology), and evolutional palaeontology. The two branches are palaeobotany and palaeozoology. Biostratigraphy, an application of the study of fossils, is hardly 'palaeontology'. Palaeontology is a department of biology but is often considered to be, also, a department of geology. [the G. *onto-* is here taken to imply living, not merely being. It is thus curious that neither 'fossil' nor 'palaeontology' are precisely appropriate terms etymologically.]

**Palaeontology of the present.** The study of the life and organic remains of the present day from the point of view of the palaeontologist; to see how this would provide material for the palaeontologist of a future age and so, at the same time, how it throws light on the significance and imperfections of the fossil record. 'An important part of such work is to observe carefully how the hard parts of animals and plants accumulate as potential fossils, especially on the sea floor, and how they are buried in sediment' (Ager, *Paleoecology*, 1963). (Richter, Germany, 1928, as *Aktuopalaontologie*.)

**Palaeopedology.** The study of palaeosols. *Paleopedology: Origin, Nature and Dating of Paleosols* (Yaalon, ed., 1971).

**Palaeophysiography.** See PALAEOGEOGRAPHY. 'The subject of Palaeo-physiography, or the geography of past geological times' (Hull, *Physical History of the British Isles*, 1882).

**Palaeophytology.** The same as palaeobotany, but not often used. See quotations under PALAEOBOTANY. 'Palaeophytology from 1860 to 1900' (Green, *History of Botany*, 1914).

**Palaeoplain.** This term, though hardly to be found in British writings, seems a useful one for, in particular, a near-planal surface of unconformity considered from the palaeogeographical point of view. It was suggested in America: 'Ancient buried destructional plains veneered by constructional formations might be appropriately termed paleoplains' (Hill, *Atlas of the U.S.*, 1900, quoted in *AGI Glossary*, 1957).

**Palaeosalinity.** The salinity of a body of water in the geological past, used especially for that of such a body at the time of deposition of a particular sediment in it. 'How measuring paleosalinity aids exploration' (Frederickson and Reynolds, *OGJ*, 1960). 'Paleosalinity in the Yoredale formation' (Walker, *BAAPG*, 1964).

**Palaeoslope.** Any slope in palaeogeography; especially a submarine slope on which sedimentation took place, considered particularly in connexion with the influence this slope may have had in determining the final primary form and internal structure of the sedimentary body. See quotation (Davies and Cave) under CLEAVAGE. 'Directions of slumping are consistent indicators of regional palaeoslopes' (Rupke, *JGS*, 1976).

**Palaeosol.** An old soil, particularly one formed during an interglacial period and covered by later deposits. 'Quaternary buried palaeosols' (Valentine and Dalrymple, *QtnR*, 1976). 'Paleosols capping regressive carbonate cycles in the Pennsylvanian Black Prince Limestone, Arizona' (Goldhammer and Elmore, *JSP*, 1984). See PALAEOPEDOLOGY.

**Palaeostructure.** See PALAEOGEOLOGY.

**Palaeotemperature.** See THERMOMETRY (2, 3). *Paleotemperature Analysis* (Bowen, 1966).

**Palaeothermometry.** See THERMOMETRY (2, 3).

**Palaeovolcanalogy.** The study of the volcanoes and volcanic action of past geological times. 'Paleovolcanology' (Cook, *ESR*, 1966).

**Palaeozoic.** Of the ages of ancient life; applied collectively to the stratigraphical systems from the Cambrian to the Permian and to the era of time they represent. The Palaeozoic strata and ages are divided into two parts, the Lower and Upper Palaeozoic. ('Older' and 'Newer' would be etymologically more consistent, but are seldom used; the combination of stratigraphical and time terms shows how these conceptions, though quite distinct, are intimately interchangeable.) The main fossil groups of the Lower Palaeozoic are the trilobites, brachiopods, corals (rugose and tabulate), and graptolites, with crinoids, cystids, and nautiloids subordinate; of the Upper Palaeozoic, brachiopods, rugose corals, crinoids, goniatites, fishes, pteridophytes, and pteridosperms, with trilobites, lamellibranchs, gastropods, and nautiloids subordinate. The name appears to have first been used by Sedgwick in 1838 (*PGS*) for the 'Cambrian' and 'Silurian' systems combined, that is, for the Lower Palaeozoic. Sedgwick and Murchison, at the end of their Devonian paper (read 1839, *TGS*, 1840) used 'Palaeozoic' for the Cambrian to Carboniferous inclusive. Phillips (*Palaeozoic Fossils of Cornwall* [&c.], 1841) included the Permian Magnesian Limestone at the top using 'Lower Palaeozoic' in the sense it has kept ever since and tentatively suggesting 'Middle Palaeozoic' for the Devonian and 'Upper Palaeozoic' for the Carboniferous and Magnesian Limestone. Murchison (*Siluria*, 1854) used 'Upper Palaeozoic', in its modern sense, for the Devonian, Carboniferous, and Permian combined, implying that Edward Forbes had already taken this view of the nomenclature.

**Palaeozoology.** The science of the animal life of past geological ages, as revealed by fossils. 'Palaeontology' is used more frequently than 'palaeozoology' for the study of animal fossils.

**Palagonite-tuff.** 'Pyroclasts consisting of basaltic glass. . . . Palagonite-tuff falls in this category. The basaltic glass fragments of which it was originally composed have suffered devitrification and conversion into a dull dark green substance of doubtful composition named palagonite (Penck, 1879)' (Hatch and Wells, *Igneous Rocks*, 1965).

**Paleo-.** See PALAEO-.

**Palichthyology.** The science of fossil fishes. An example from among the terms that may be coined for the palaeontology of particular organic groups. 'Palichthyologic notes' (Egerton, *QJ*, 1850).

**Palimpsest structure.** See RELICT.

**Palingenesis. 1.** In biology generally: =recapitulation. (Haeckel, 1879). **2.** In petrology: 'Widespread re-fusion of rocks with or without intimate interpenetration by granite' (Tyrell, *Petrology*, 1929, where the meanings of 'palingenesis' and 'anatexis' are discussed). (Sederholm, 1907.) [G. *palin*, over again.]

**Palinspastic.** 'The original position of early Paleozoic sediments in eastern North America has been altered by later folding and thrusting so that the present map gives a foreshortened base relative to the ancient geography. On the paleogeographic base map . . . each slice has been moved back from the foreland to the position from which it is believed to have been displaced. . . . The resulting map displays the several slices in their conceived relative original positions. The name"palinspastic"(G. παλι(ν)σπαστιϰ [*sic*], stretched back) is proposed for base maps of this character' (Kay, *BGSA*, 1937). 'Paleogeographic and palinspastic maps' (Kay, *BAAPG*, 1945).

**Paludal.** Of or pertaining to a marsh; applied particularly to deposits. [L. *palus*, *palud-*, a marsh.]

**Palynology.** The study of fossil spores, pollen, and certain microscopic bodies of uncertain affinity (e.g. acritarchs); also the analysis of the different kinds in a rock or deposit. This is used in the geochronology and correlation of rocks and deposits, in the reconstruction of environmental conditions of deposition, and in the reading of the history of successive floras. Among strata, it is applied to the content of rocks of any age, particularly to coal-seams; but the term is especially used for the examination of the pollen content ('pollen analysis') of peats, &c., in the study of Pleistocene chronology, palaeogeography, and plant-history. *A Palynological Investigation of the Dalradian Rocks of Scotland* (Downie and others, *IGSc*, 1971). 'The microspores in the coal-seams of North Staffordshire' (Millott, *TIME*, 1939/46). *Aspects of Palynology* (Tschudy and Scott, eds., 1969). *Textbook of Pollen Analysis* (Faegri and Iversen, 1975). (Hyde and Williams, *PAC*, 1945.) [G. *paluno*, sprinkle (as dust, pollen).]

**Pan.** A hard cement-like layer ('hard pan'), usually of iron hydroxide ('iron pan'), deposited in the soil some inches or feet below the surface as a result of solution and deposition (with some chemical reaction) by water percolating through the uppermost layers of the soil. A pan may be impervious alike to root growth and to further downward drainage of the water; it is especially common in podsols.

**Pangaea.** Wegener's name for the (hypothetical) more or less coherent landmass of Upper Palaeozoic times, when, as supposed, the present continents were bunched together. See CONTINENTAL DRIFT. [G. *gaia*, earth, land.]

**Panidiomorphic.** An igneous rock-texture in which all the conspicuous minerals are idiomorphic or nearly so (e.g. some lamprophyres).

**Paper-shale.** Very finely laminated shale; the laminae easily separable. 'The Wealden strata may be subdivided into two groups, the upper consisting, for the most part, of dark grey shaly clay, thinly laminated, and splitting on exposure to the weather into paper-shales' (Bristow, *MGS*, *Isle of Wight*, 1862). 'Above the Watchet Beds come the "Paper-Shales" of the Lower Lias' (Richardson, *QJ*, 1911). Carbonaceous shales often have this quality.

**Paraconformity.** See UNCONFORMABLE (table). (Dunbar and Rodgers, *Principles of Stratigraphy*, 1957.)

**Paradoxides series.** The Middle Cambrian series, in Europe and eastern N. America, characterized particularly by trilobites of the family Paradoxididae and abundant particular species of *Agnostus* and *Eodiscus;* other genera of fossils are also characteristic, particularly the sponge *Protospongia*. In the type area, Wales, the *Paradoxides* series comprises the Solva and Menevian beds, and the upper part of the Harlech Grits. In England it is exemplified in the Upper Comley Sandstone and the shales at Nuneaton, with many species of *Agnostus*. (Lapworth, *GM*, 1881; Walcott, *IGC*, 1888; Hicks, *GM*, 1894.) [genus *Paradoxides*.]

**Paragenesis.** The genetically determined association, or (less commonly) the growing-together, of particular rocks, minerals, or (in crystals) chemical elements. Usually, the community of origin of a set of minerals. 'Note on paragenetic formations of carbonate of lime and oxide of iron, and of quartz and oxide of iron, at the Mwyndy Iron Mines, Glamorganshire' (Vivian, *MM*, 1876). 'Groups of [certain] minerals are found associated in pegmatites in a manner indicating a significant "paragenesis" ' (Harker, *Igneous Rocks*, 1909). 'The paragenesis of kyanite-eclogites' (Tilley, *MM*, 1936). 'The paragenesis of sylvine, carnallite, polyhalite and kieserite in Eskdale borings, Yorkshire' (Armstrong and others, *MM*, 1951). (Breithaupt, 1849.)

**Paragneiss.** See ORTHOGNEISS.

**Paralic.** By the sea. for deposits laid down on the landward side of a coast, in shallow water subject to marine invasions. Thus marine and non-marine sediments are interdigitated; typically, as exemplified in the lower part of the Coal Measures, the non-marine (paralic) predominate, with relatively thin marine bands. Krumbein and Sloss (*Stratigraphy and Sedimentation*, 1963) adopt a usage according to which 'interfingered marine and continental' sediments are all 'paralic'. [G. *paralos*, near the sea.]

**Parallel folding.** See CONCENTRIC FOLDING.

**Parallel roads.** Old beaches at several levels on the sides of a valley. It is now accepted that these were formed by a lake which was dammed up in the valley by ice (a glacier lake). Each beach ('road') corresponds with a temporary level of overflow, such overflows being themselves marked, perhaps, by definite overflow channels. The classic example of this phenomenon is in Glen Roy in the Scottish Highlands. 'Parallel Roads in Glen-Roy', appendix in Pennant's *Tour in Scotland*, 1776. 'George Greenough was the first to suggest that the "Parallel Roads" of Glen Roy were the successive beach-levels of a former lake (in the MS journal of his geological tour of Scotland in 1805)' (Rudwick, *BJHS*, 1962). Louis Agassiz was the first to recognize that they were 'the successive strand-lines of a former ice-damned lake . . . in a letter written on 3 October [1840] at Fort Augustus . . . published in *The Scotsman* on 7 October' (Davies, *AS*, 1968). 'On the parallel roads of Glen Roy' (Jamieson, *QJ*, 1863).

**Parametamorphism.** A direct transitional change from one mineral or kind of rock into another chemically similar mineral or kind of rock. 'Sometimes the augites are found undergoing direct paramorphic change and transition into horneblende' (Judd, *QJ*, 1886, two quotations combined).

**Parameter.** Primarily a mathematical term. It has a definite usage in crystallography (involving other technical terms). Geologists have recently appropriated the term in a large way to refer to the factors, data-categories, or quantity-indicators in a problem.

**Paramoudra.** Paramoudras are 'large flints, barrel-shaped, cylindrical, or pear-shaped, standing erect in the chalk, NE. Ireland, Yorkshire, & Norfolk' (Arkell and Tomkeieff, *Rock Terms*, 1953). They were first described, and the term introduced, by Buckland from Ireland (*TGS*, 1817) who thought they were true animal-fossils and treated the name as a generic name in the usual Latin form (with plural -ae). Those from Norfolk (where they are also called potstones) were described by Taylor

(*TGS*, 1824) and a general view of their occurrence in a chalk-pit near Norwich is given in the later editions of Lyell's *Student's Elements of Geology* (e.g. 1885). Arkell and Tomkeieff give several possible derivations of the term. 'It is said that when Dean Buckland came across these objects in Antrim he asked his guide what name they went by. The Irishman, who had heard the Dean calling stones by strange names, was equal to the occasion, and invented "paramoudra" without a moment's hesitation' (Sollas, *Age of the Earth* [etc.], 1905). A more circumstantial anecdote is related in the *Life* of Buckland (Gordon, 1894). 'Near Norwich flint takes the form of the familiar paramoudras (the name was probably derived from Erse words meaning "sea-pears"), large cylindrical masses from about 1 to 2 ft. in diameter and as much as 4 ft. in height and hollow like a pipe. In their natural position in the Chalk, the tubular cavity is vertical' (Chatwin, *BRG, East Anglia*, 1964). See also Woodward, *England and Wales*, 1887.

**Parataxitic.** See EUTAXITIC.

**Parataxon** (pl. **parataxa**). A very recently coined word for a taxonomic category (taxon) in palaeozoology to accommodate discrete skeletal parts that are always fossilized separately; these parts, while among themselves showing likenesses and differences that invite classification on normal biological lines, being inadequate for the identification of the true genus or species (&c.) of the whole animal. The chief animal fossils for which parataxa are called for are conodonts and aptychi. It does not appear to be proposed that the term should be used in palaeobotany, where it is equivalent to the 'form-genus', 'form-species' &c. (terms that were also used in palaeozoology for what it is now proposed to call parataxa). The question of the nomenclatute of discrete parts was considered by Martin in his *Extraneous Fossils*, 1809: 'Parts in a detached state are not to be considered as the foundation of permanent species; when the parts in question thus occur, they must be arranged as "temporary" species, except they be ascertained to belong to species already established, to which they must be referred as "imperfect specimens".'

**Parautochthonous.** See AUTOCHTHONOUS (5).

**Paravolcanic.** Associated with volcanic activity but not a part of that activity itself. 'The North Elgon Depression [Uganda] is a paravolcanic downwarp between the early Miocene volcanoes of Kadam and Elgon. The downwarp is filled with pyroclastic rocks from the two volcanoes' (Clark, *QJ*, 1969).

**Parent magma.** In one sense: the magma from which a particular igneous rock-body has solidified. In another, and wider sense: the hypothetical common magma from which the several rock members of a petrographical province are supposed to have been derived.

**Parent rock.** The rock from which detached pieces or ingredients have been derived. 'At a stream work near Roach, in Cornwall, are found crystals of tin so large, that it became an object to follow them to their source, and the miners have discovered the parent vein', and 'bowlder-stones are evidently not in situ; they are, for the most part, traceable to the parent rock' (Greenough, *Geology*, 1819). The following gives an allied use of 'parent': 'At Aberystwyth on the beach are vast varieties of transported pebbles, the parent localities of which are somewhat problematical' (Ramsay, *MGS*, vol. 1, 1846).

**Pari passu.** Together with and at the same time. To illustrate by a question that might arise: was the contorted bedding of a particular rock produced *pari passu* with deposition (from a turbidity current), penecontemporaneously (by slumping), or by a quite separate later process (tectonic deformation) ? [L., with equal step.]

**Particle.** The smallest separable or distinct unit in the composition of a rock. All rocks, except those purely of glass, are made up of particles, and the vast majority, of mineral particles. 'The shape of rock particles' (Barrett, *Sd*, 1980).

**Particle concentration.** In a porous rock the particle concentration, or 'fractional concentration of solids '(C) is given by the equation: C= space occupied by solids ÷ total space. 'The porosity is simply equal to (I-C), but is a less fundamental quantity' (Allen, *GM*, 1969).

**Particulate texture.** Of a sedimentary rock; 'results from the accumulation of discrete particles; it is characterized in its unaltered state by point contacts between particles and by the associated intergranular spaces' (Knox, *Petrology for Students*, 1978).

**Parting.** 'A plane by which a hard rock is naturally divided into layers; a thin stratum, generally soft, following such a plane' (Mason, *BGS*, 1957).

**Passage beds.** Beds which in their lithological and faunal characters are transitional between those below and those above, thus implying transitional conditions of deposition. (Cf. transitional series.) 'On the passage-beds from the Upper Silurian rocks into the Lower Old Red Sandstone at Ledbury, Herefordshire' (Symonds, *QJ*, 1860). ' "Passage-beds" between Trias and Lias in Raasay and Skye' (Woodward, in *QJ*, 1901). There is also the lateral 'passage' of one kind of rock into another: 'The passage of a seam of coal into a seam of dolomite' (Strahan, *QJ*, 1901).

**Past and present.** Geology deals with the past and the present and looks to tbe future. *Principles of Geology, being an Attempt to explain the former Changes of the Earth's Surface by reference to Causes now in Operation* (Lyell, 1st edition, 1830/3). The book opens, however, with a statement of the complementary proposition: 'Geology is the science which investigates the successive changes that have taken place in the organic and inorganic kingdoms of nature. . . . By these researches into the state of the earth and its inhabitants at former periods we acquire a more perfect knowledge of its present condition, and more comprehensive views concerning the laws now governing its animate and inanimate productions'. The subtitle of the last three editions (10th-12th, 1867-75) reads : *or the Modern Changes of the Earth and its Inhabitants considered as illustrative of Geology*. 'It has often been insisted upon that the Present is the key to the Past' (Geikie, *Textbook*, 1882). 'The well-tried Lyellian principle of studying the present as a key to the past' (Hollingworth, *QJ*, 1962).

What can be learnt from our observations of the world around us at the present day? The answer is: very little. We can see the inexorable processes of erosion on land and deposition in the sea actually going on, but for the most part and on anything but a small scale this conclusion is based on inference, not on direct observation. If we assert that it is nevertheless obvious it must be remembered that scientists did not realize it (chiefly because they could not allow the time it would take) before Hutton (*Theory of the Earth*, 1795, and an earlier exposition) and Lyell proclaimed that it must be true. Volcanoes in action have always been visible and earthquakes felt. The principles of geology must be derived not so much from what we see to-day as from our interpretation of the geological record; the past is the key to the present even more than 'the present is the key to the past'.

**Pattern of earth history.** A phrase that usefully emphasizes the orderly operation and inter-relation of the geological processes and the orderly recurrence of geological events. 'The evolution of sedimentary, volcanic, metamorphic and plutonic rocks is linked to produce the pattern of earth history' and 'All geological processes react on one another and thereby provide a pattern of earth history' (Read and Watson, *Introduction to Geology*, 1962).

**Patterned ground.** 'A group term for the more or less symmetrical forms such as circles, polygons, nets, steps, and stripes that are characteristic of, but not necessarily confined to, mantle subject to intensive frost action [permafrost]' (*AGI Dictionary*, 1976). 'Patterned ground by the Worcestershire Avon' (Shotton, *GM*, 1960). 'Subsurface large scale patterned ground structures observed as cropmarks in cereal fields' (Christensen, *Bo*, 1974).

**Paulopost.** Acting or formed just a little after, penecontemporaneously. Has been specially used in the case of the autopneumatolysis of an igneous rock, producing the final mineralogical modifications from its own residual fluids, after the general solidification of the mass. 'It is concluded that the greisen [Grainsgill, Skiddaw] is of secondary origin, having been derived from the normal granite through the action of paulopost processes' (Hitchen, *QJ*, 1934). [L., a little after.]

**Peat.** A superficial accumulation consisting of vegetable matter which has become decomposed to a certain limited extent. Forms chiefly in cold and temperate regions. The composition and character of the peat varies slightly according to the conditions of its formation; 'hill peat' and 'fen peat' are two types distinguishable in Britain. [Middle English.]

**Pebble.** A small stone (smaller than a boulder or a cobble) worn and more or less rounded by the action of water (usually and typically), ice, or blown sand.

**Pebble bed.** Any conglomerate with pebbles may be called a 'pebble bed', but the term is used chiefly for those conglomerates in which the pebbles weather conspicuously and fall loose, as perfectly exemplified in the Bunter Pebble Beds.

**Pebidian.** A group of volcanic rocks of Precambrian age visible in several outcrops in Pembrokeshire. Associated Precambrian granitic and dioritic intrusions form the Dimetian group. 'I propose now to divide the Pre-Cambrian rocks into two distinct series under the local names of Dimetian (Dimetia being the ancient name for a kingdom which included this part of Wales) and Pebidian (Pebidiauc being the name of the division or hundred in which these rocks are chiefly exposed)' (Hicks, *QJ*, 1877). 'In the neighbourhood of St. David's, the Pebidian tuffs . . . are well displayed in the cliffs to the south and west of the city, and the sections in the area became classical on account of the controversy respecting the age of the tuffs and associated intrusions.' (Pringle and George, *BRG, S. Wales*, 1948).

**Pedestal block.** See PERCHED BLOCK.

**Pedestal rock.** An upstanding rock in the form of a pedestal supporting a cap. Produced as the undercutting effect of localized erosion or of differential erosion or of both. If the rock is all of the same kind and structureless, curious shapes may be formed (such as the mushroom shape); if the cap is harder than the pedestal, or if there is strong horizontal stratification (or jointing), a more regular form, with abrupt distinction between cap and pedestal is likely to result. Pedestal rocks may be produced on any expanse of bare rock, but particularly in deserts ('zeugen') or on the sea-shore.

**Pediment.** The geomorphological conception of a surface, veneered with detritus, spreading away from the foot of a scarp and cut nearly horizontally in the rock as the scarp retreats through erosion. Though nearly flat it would be slightly concave upwards, meeting the foot of the scarp rather abruptly; and it would be relatively stable, being adapted to, and moulded by, sheet-flow of water which has very little erosive effect. Land forms ascribed to the operation of this principle are particularly characteristic of a climate of spasmodic rainfall, such as S. Africa. Where the sheet-flow is changed by obstruction or channelling into turbulent flow, gorges ('dongas') tend to be cut in the pediment. A pediment may be cut into and destroyed by the retreat of another scarp below it, just as it is itself probably cutting into a higher pediment. Hence 'pedimentation'. (Bryan, 1922, 1925; King, *GM*, 1949, *Morphology of the Earth*, 1967.)

**Pediplain.** A widely extended plain of low relief but multi-concave form, produced by the coalescing of ever-increasing pediments, with scarps meeting from opposite sides and the consequent near-obliteration of the upland between. Hence 'pediplanation', 'the development of landscapes by the twin processes of slope retreat and pedimentation' (King, *GM*, 1949). A pediplain is a kind of peneplain; if it is to be contrasted with a 'peneplain' the writer should (it would seem) state in what restricted sense he is using the term 'peneplain'.

**Pedogenic.** Concerning the formation of the ground or soil and the production of structures therein.

**Pedology.** 'The writer ventures to hope that the convenient term [pedology] (G. πέδον [pedon] = soil or earth) will be more generally used to describe the scientific study of soils. There seems to be no satisfactory alternative' (Robinson, 'Pedology as a branch of geology', *GM*, 1924). 'In recent years the outlook has been considerably changed by the outstanding work of the Russian School which has put a backbone into the study of soils and turned that study into Pedology. The word Pedology has been almost unanimously accepted on the Continent and to a large extent in America' (Comber, *Scientific Study of Soil*, 1927). It has since become similarly accepted in Britain.

**Peel technique.** A method of reproducing certain details of a flat surface, etched with acid (or otherwise abraded), of a fossil, rock, mineral, or metal by covering it with a liquid substance (such as a dispersion of cellulose acetate) which, when allowed to dry, can be peeled off (as a plastic 'peel'). The peel will either (1) bear a simple impression of the relief of the etched surface, (2) contain within it parts of the etched surface which have been peeled off, or (3) retain pigment with which minute crevices and pits in the surface have been filled. The technique has been most extensively used in the study of the plant petrifactions contained in coal-balls; here the peel contains plant material separated from the matrix (case 2, above). Details and references in Kummel and Raup (eds.), *Handbook of Paleontological Techniques*, 1965.

**Pegmatite.** A rock showing a conspicuous graphic texture, the commonest variety being granite-pegmatite, chiefly of quartz and felspar (Haüy). Also for any coarse-grained quartz-felspar igneous rock (Haidinger). [G. *pēgma, pēgmat-*, a framework.]

**Pelagic.** Of or pertaining to the open ocean. Applied particularly to the deposits which have fallen from the upper waters and settled to the floor of the sea. (Murray and Renard, *Voyage of H.M.S. Challenger 1872-76*, 1891 and earlier.) The term 'pelagic' is sometimes used for the whole mass of water of the sea as distinct from the sea floor and in this sense is the whole of the marine environment of the plankton and nekton. [G. *pelagos*, the open sea.]

**Peléan type.** See VOLCANICITY. Extremely violent explosions with emission of nuées ardentes and extrusion of very viscous lava tending to consolidate as a protruding plug. 'On the probable peléan origin of the felsitic slates of Snowdon, and their metamorphism' (Dakyns and Greenly, *GM*, 1905). (Lacroix, *Montagne Pelée*, 1904 and 1908.) [Mt. Pelée, Martinique, eruption of 1902.]

**Pelecypoda.** See LAMELLIBRANCHIATA. [G. *pelekus*, a double-axe.]

**Pelite.** An argillaceous rock, a mudstone. Hence 'pelitic'. 'Clay rocks (pelites)' (Geikie, *Textbook*, 1882). See quotation (Read) under DIFFERENTIATION (2). There is a tendency for the Greek-derived names 'pelite', 'psammite', and 'psephite', and their corresponding adjectives, to be used particularly when referring to the metamorphic equivalents of the normal argillaceous, arenaceous, and rudaceous rocks—terms derived from the Latin. (Tyrell, *GM*, 1921.) [G. *pēlos*, clay.]

**Pelmatozoa.** One of the two subphyla of the phylum Echinodermata. The animal is fixed. It includes the classes Crinoidea, Cystidea, Blastoidea, and Edrioasteroidea; the last three being extinct. [G. *pelma, pelmat-*, stalk (alluding to the fixed habit).]

**Penarth group.** A name formally introduced in 1980 for the uppermost of the three lithostratigraphical units of the Triassic system in Britain. 'The argillaceous, calcareous and locally arenaceous formations of predominantly marine origin which occur between the Mercia Mudstone group and the base of the Lias are here constituted the Penarth group' (*A Correlation of Triassic Rocks in the British Isles, GSSR* (13) 1980). This group is equivalent to the Rhaetic, but is proposed so as to obviate confusion with the chronostratigraphic name Rhaetian which includes (as part of the Triassic) the basal few metres of the (lithostratigraphic) Lias. [Penarth, South Glamorgan.]

**Penecontemporaneous.** Almost contemporaneous; usually applied to a process which cannot be quite contemporaneous with another process, but which follows very soon. Thus penecontemporaneous erosion of a bed recently deposited (Buckman, *QJ*, 1901), slump-structures as following on deposition on a slope, &c. See PARI PASSU.

**Peneplain.** Almost a plain, a nearly flat surface of country. This near-flatness can be caused in many ways, by erosion or deposition. The term is most often used to describe a low-lying country of faint relief which has been worn down by long-continued subaerial erosion (Davis, *AJS*, 1889). It is sometimes implied that a marine abrasion surface is excluded by the term. Peneplains well above sea level may be formed by erosion of a uniformly resistant rock, by the uplift of a peneplain formed at a lower level, or in some other way. See HILL-TOP SURFACE. The term is sometimes spelt 'peneplane', and 'peneplanation' is always the derivative term.

**Penetrative.** 'Anything uninterrupted and uniformly distributed in space (or within an arbitrarily defined part of it) is penetrative' (Oertel, *BGSA*, 1962). The word implies something produced dynamically, excluding such facts as static mineralogical composition or bedding. In the above, the widest, sense it might comprise any such structural or textural feature pervading a rock-mass as distinct from one localized or on a surface only. However, it is used particularly for some features where this is constant in attitude or orientation, thus for 'penetrative axial-surface cleavage' (Dewey, *QJ*, 1967) or 'penetrative fabric' (Sanderson and others, *JGS*, 1980).

**Pennant series.** A series chiefly of sandstone (Pennant Sandstone or Pennant Grit) between the Lower and Upper Coal Measures in the S. Wales and Bristol coalfields. The stratigraphical range is that of the freshwater lamellibranch zones of *Anthraconauta phillipsi* and *A. tenuis* (Smith, *MS*, 1797/9, 'Pennant Stone'.) [For question of derivation see Arkell and Tomkeieff, *Rock Terms*, 1953.]

**Pennsylvanian.** Upper Carboniferous. Used chiefly in America where it is held to constitute a separate system. (Williams, 1891.) [Pennsylvania State, U.S.A.]

**Peperite.** Peperites are a class of volcanic fragmental rocks of varied constitution mixed with calcareous-argillaceous cement exemplified at the type-locality of the hill of Gergovia in Central France. This is an exhumed volcano of Oligocene age, and the peperites composing it appear to be the product of the explosive projection of basaltic ejectamenta into the steadily accumulating lime-mud of a lake. The term 'peperite' (French *pépérite*) is derived from the Italian *peperino* which was given to the rocks by Scrope (*Extinct Volcanoes of Central France*, 1858), adopting this name from the Italian geologists who had used it as a technical term for, in particular, a somewhat similar rock at Montecchio Maggiore in northern Italy. Scrope envisaged the now generally-accepted circumstances of the formation of the Gergovia rocks. 'A lacustrine volcano of Central France and the nature of peperites' (Jones, *PGA*, 1969).

**Perched block.** In a special sense, a large detached piece of rock which is perched precariously on a hillside and which has presumably been brought to, and let down gently on, that spot by the shrinking of a glacier on which it was being carried. Also used in a wider sense to include any glacial erratic, which may be said to be perched in any way, particularly a 'pedestal block' (Hughes, *QJ*, 1886), one perched on a pedestal which it has protected from erosion (e.g. those at Norber Brow, Ingleborough). Rocks forming ice-tables on a glacier are sometimes called 'perched blocks' or 'pedestal blocks'. A connexion with glaciers is implied in all these cases, but as a purely descriptive term it might be used for any 'block' that had become 'perched', and has been so used for one capping an earth pillar. The term appears to have arisen in the French form *blocs perchés* for occurrences in the Alps.

**Percolation.** Used specially for rain water percolating down through surface soil and subsoil.

**Pericline.** A fold in which the dips are away from or towards, a centre; that is, a dome structure or a basin structure (more commonly used for the former; see QUAQUAVERSAL). Hence 'periclinal'. 'The geological structure of the Kingsclere pericline' (Hawkins, *QJ*, 1939). The term is sometimes applied to a very short anticline (e.g. the Poxwell Pericline, Dorset).

**Peridotite.** A coarse-grained ultramafic igneous rock very largely or entirely composed of olivine. Felspar is typically absent, but other mafic minerals may be present. (Rosenbusch, 1877.) [peridot, f. French *péridot*, jewellers' name for olivine.]

**Periglacial.** Surrounding or bordering the glacial; for the zone, and the physical conditions within it, immediately surrounding or bordering a region that is under glacial conditions. 'The term "periglacial" has generally been confined to the effects of heavy frost. The "periglacial zone", or "area", therefore, should be regarded as that zone surrounding an ice-sheet, in which the cooling effect of the ice produced a frost climate' (Zeuner, *Pleistocene Period*, 1959). 'Periglacial structures in the Aberystwyth region of Central Wales' (Watson, *PGA*, 1965). *Periglacial Processes and Environments* (Washburn, 1973 and, as *Geocryology*, 1980). *Periglacial Geomorphology* (Embleton and King, 1975). *The Periglacial Environment* (French, 1976).

**Period.** Common time-terms—period, age, and epoch among them—have been used in special senses in geology; thus 'period' has been used for the time corresponding to a stratigraphical system (see CHRONOSTRATIGRAPHY). But it hardly seems legitimate, or convenient, to take a perfectly ordinary word and restrict it to a very special technical sense not, perhaps, very often required. 'Period' should be available for any use in its ordinary sense of 'a course or length of time' (*OED*).

**Peritidal.** Used for deposits formed 'around' the tidal zone. 'Peritidal carbonate facies models: A review' (Wright, *GeolJ*, 1984).

**Perlitic.** Like a pearl; for the small globular shapes within a glassy or devitrified rock and for the convolute and spheroidal cracks themselves (due to contraction) which define these shapes. 'Perlite, spherulitic perlite, perlitic pitchstone, and perlitic obsidian . . .' (Allport, *QJ*, 1877). 'Perlitic and spherulitic structures in the lavas of the Glyder Fawr' (Rutley, *QJ*, 1879). 'Perlitic cracks in quartz' (Watts, *QJ*, 1894).

**Permafrost.** Perennially frozen ground. (Muller, *USGS*, 1943.) 'Permafrost structures on Sully Island, Glamorgan' (Bradshaw and Smith, *GM*, 1963). *Bibliography on Snow, Ice and Permafrost* (*USLG*, 1965). *Permafrost Terminology* (Brown and Kupsch, 1974).

**Permeability. 1.** For rocks in general but particularly for sedimentary rocks. 'Permeability is the property of a rock which allows the passage of fluids without impairment of its structure or displacement of its parts. A rock is said to be permeable if it permits an appreciable quantity of fluid to pass through in a given time, and impermeable if the rate of passage is negligible. Obviously, the rate of discharge through a given cross-section depends not only on the rock, but also upon the nature of the fluid and the hydraulic head or pressure' (Pettijohn, *Sedimentary Rocks*, 1957). 'Except in the case of solids which are themselves permeable, permeability cannot exist without porosity, for without porosity there would be no openings through which fluids might flow. However, porosity can exist with no permeability if the pores are disconnected. The permeability of connected pores is determined primarily by their size' (Russell, *Petroleum Geology*, 1960). The term is sometimes used in a wider sense, synonymous with 'perviousness'. **2.** A special sense in igneous action as in 'a remarkable permeability of the gabbro by the granitic magma' (Harker, *MGS*, *Tertiary Igneous Rocks of Skye*, 1904).

**Permeation.** Used specially for the intimate penetration of country rock by metamorphic agents, particularly of an already metamorphosed rock by, for instance, granitizing agents so that the rock becomes completely recrystallized. Hence 'permeation gneiss' (Read, *MGS*, *Central Sutherland*, 1931).

**Permian system.** The lower part of that great group of rocks that intervenes between the Carboniferous and the Jurassic. It is found that this lower part has a fauna of Palaeozoic affinities so that it is placed as the topmost Palaeozoic system. Named from the province of Perm in Russia where its marine facies is well developed. The marine facies is also exemplified in the Magnesian Limestone of NE. England, but otherwise the system is in Britain inseparable from the Trias and the whole is called the Permo-Trias. Apart from particular general of brachiopods, lamellibranchs, and other groups, the most distinctive element in the Permian invertebrate fauna is the presence of ammonoids with septal sutures intermediate in complexity between those of the Carboniferous goniatites and the Triassic ammonites. In Britain the Coal Measures pass up into the Permo-Trias with least sign of any break in the Midland coalfields (but the question of this break is still outstanding). Here the base of the Permo-Trias is usually taken at the top of the Enville beds. In France, e.g. at Autun, Coal Measure conditions persist into strata that are the equivalents of Permian red sandstone strata elsewhere, so that we can select here, within a constant type of lithology and fauna, a general level, one not dependent on a local facies or local structural change. This level to define the top of the Carboniferous and base of the Permian is taken between the Stephanian (below) and Autunian (above), distinguished by their floras. In the type region of Perm the base of the lowest formation, the Artinskian, appears to be diachronous but the Artinskian is in general equivalent to the Autunian. The history of the founding of the Permian system is given by the Permian Subcommittee in America (*BGSA*, 1960). *A Correlation of Permian Rocks in the British Isles* (*GSSR* (5), 1974). (Murchison, *PM*, 1841, *Russia in Europe*, 1845.)

**Permitted intrusion.** See FORCEFUL INTRUSION.

**Permo-Carboniferous. 1.** The whole of the Permian and Carboniferous systems combined to form one system. (E.g. Gignoux, in translation, *Stratigraphic Geology*, 1955.) **2.** Belonging to the Permian and Upper Carboniferous combined; particularly for that part of the Gondwana series which can be correlated only in this general way. **3.** Belonging to either the lowest Permian or the uppermost Carboniferous or both; for formations of which the age-assignment is doubtful (such as the Enville beds), or which are known to span the boundary between the two systems. In America a formation in a similar case is 'Permo-Pennsylvanian'.

**Permo-Trias.** (Permo-Triassic.) See PERMIAN SYSTEM and NEW RED SANDSTONE. 'Permo-Trias in Britain' (Sherlock, *PGA*, 1926). *The Permo-Triassic Formations* (Sherlock, 1947). *Triassic and Permian Rocks of the Midland Counties* (Hull, *MGS*, 1857). Also applicable to a formation of which the exact age-assignment is doubtful or which is known to span the boundary.

**Perpendicular.** This is perhaps a more appropriate term than 'vertical' for indicating a direction at right angles to some previously defined, or understood, direction, such as the direction (attitude) of stratification. In an early account of such directions, the term was used for the direction of joints (though as 'horizontal' was used for the one direction, 'vertical' might have been the natural choice for the other): 'I call those fissures, which distinguish the stone into strata, horizontal ones; and those which intersect these perpendicular; not so much with respect to the present site of the strata, which is alter'd, in many places, and now much different from the original situation' (Woodward, *Natural History of the Earth*, 1695).

**Persilicic.** 'A term suggested to replace the term "acid" as applied to igneous rocks; for "intermediate" and "basic" the corresponding terms are "mediosilicic" and "subsilicic" respectively. Clarke, 1911' (Holmes, *Nomenclature of Petrology*, 1928).

**Perthite.** An intergrowth of potassium and sodium felspar, probably crystallized as a product of exsolution. Hence 'perthitic'. See FELSPAR and ALKALI-FELSPAR. (Thomson, 1832.) [Perth, Ontario, Canada.]

**Perviousness.** The ease of flow of fluids through a rock-mass, depending on the permeability of the rock and the degree to which the water (or other fluid, such as oil) can circulate along joints, bedding, fissures, faults, or any other structural or erosional channels. Hence rock-formations are pervious to a certain degree, or absolutely or relatively impervious.

**Petralogy.** See PETROLOGY.

**Petrifaction.** Turning or turned into stone. An old name for fossilization or for a fossil. Now chiefly used for a particular kind of fossilization (especially plant fossilization), where there is permeation of the substance by mineralizing fluids (usually calcareous or siliceous) so that often even the most delicate structures may be preserved (encased or replaced). The best-known plant petrifactions make up the coal-balls where 'they represent, as it were, parts of the raw material of coal which have been saved by petrifaction from carbonization, and have consequently retained their structure' (Scott, *Studies in Fossil Botany*, 1920). 'The last known mode of inducing stony hardness is substitution; that is, the introduction of stony and, sometimes, of metallic substances, into organic bodies whether of the vegetable, or of the animal kingdoms, in proportion as the particles of these organic substances are destroyed by putrefaction, so as to assume the place, and, consequently, the form and figure of these, as if cast in the same mould. The mineral substances, thus moulded, are called "petrifications" in the most proper sense of the word; but, by many, particularly by the Germans, this word is used in a looser sense, to denote any organic substance found buried at great depths in the earth, or embodied in stone, whether converted into stony matter or not' (Kirwan, *Geological Essays*, 1799). 'Those petrifactions of wood, where, though the vegetable structure is perfectly preserved, the whole mass is siliceous' (Playfair, *Illustrations*, 1802).

**Petrochemistry.** A department or aspect of geochemistry; the chemistry of rock composition and rock formation. 'Petrochemistry of the Scottish Carboniferous-Permian igneous rocks' (Tomkeieff, *BV*, 1937). 'The petrochemistry of the Ardara aureole' (Pitcher and Sinha, *QJ*, 1957). 'A petrochemical study of the Pen-y-Gader dolerite' (Davies, *GM*, 1956). 'Petrochemistry of Upper Proterozoic glacigene rocks' (Bowes, *PGA*, 1970). 'Petrochemistry . . . of the Dalradian metabasaltic rocks of the S.W. Scottish Highlands' (Graham *JGS*, 1976).

**Petrofabric.** The full term for the 'fabric' of a rock; but it usually implies a microscopic fabric. It is specially used in connexion with metamorphic rocks. 'The foundation of the petrography of metamorphic rocks was laid by Becke in 1904, at the time when the study of igneous rocks was passing from purely descriptive petrography to the study of chemistry and petrogenesis. It was at a much later date that new methods were applied to the study of the texture of metamorphic rocks. These methods were the statistical method of Schmidt (1917, 1932) and the geometrical method of Sander (1930), which he calls *Gefugekunde* and which is usually rendered in English as "petrofabric analysis" or "structural petrology" (Loewinson-Lessing, trans. Tomkeieff, *Historical Survey of Petrology*, 1954). 'New work on petrofabrics' (Voll, *LMGJ*, 1960). 'The interpretation of petrofabric diagrams' (Phillips, *SP*, 1960).

**Petrogenesis.** The origin and mode of formation of rocks, particularly of igneous rocks. (Iddings, 1872.)

**Petrographical province.** A region within which the igneous rocks of one general period of activity have certain peculiarities in common so that a degree of relationship is implied. The conception and the term were first put forward by Judd (*QJ*, 1886) who showed that the British and Icelandic Tertiary igneous rocks belong to one great petrographical province. 'The rock members making up a province are believed to be comagmatic or consanguineous, implying that they are all derived from a hypothetical common magma, often called a parent magma' (Barth, *Theoretical Petrology*, 1962).

**Petrography.** Descriptive petrology, the study of rock-specimens and rock-types. This can be done in detail and with precision only by the use of the microscope. Hence 'petrographical' or 'petrographic'. The chief landmarks in the development of microscopic petrography are Sorby's paper (*QJ*, 1858), Zirkel's *Lehrbuch der Petrographie* (1866), and Rosenbusch's *Mikroskopische Physiographie* (1873/7). *British Petrography* (Teall, 1888). *Descriptive Petrography of the Igneous Rocks* (Johannsen, 1931/50). *Petrographic Methods and Calculations* (Holmes, 1921). *Sedimentary Petrography* (Milner, 1922/62). 'The petrography of the Millstone Grit series of Yorkshire' (Gilligan, *QJ*, 1919). 'The petrography of the Portland Sand of Dorset' (Latter, *PGA*, 1926). *The Petrography of Freshwater Ice as a Method of Glaciological Investigation* (Shumskii, trans., Kraus, 1964). 'A brief history of the petrographic microscope' (Sigsby, *Cmp*, 1966).

**Petroleum.** 'A naturally occurring, oily, flammable liquid composed principally of hydrocarbons and occasionally found in springs and pools but usually obtained from beneath the earth's surface by drilling wells. Formerly called rock oil, unrefined petroleum is now usually called crude oil' (*McGraw-Hill Encyclopedia of Science and Technology*, 1960). *Petroleum Formation and Occurrence* (Tissot and Welte, 1978). [medieval L., rock-oil.]

**Petroleum geology.** The geology of petroleum (and of natural gas); that is, the detection and exploration of its occurrence and extent within the rocks of the earth's crust, the examination of the conditions under which it is found, the consideration of its origin and mode of accumulation, and the study of all aspects of the geological sciences which apply to these ends. 'Petroleum or rock oil of western Pennsylvania [&c.]' (Rogers, *PPSG*, 1860). *Geological Results of Petroleum Exploration in Great Britain* (Falcon and Kent, *MGSL*, 1960). *Geology of Petroleum* (Levorsen, 1961). *Petroleum Geology* (Chapman, 1973). *Introduction to the Petroleum Geology of the North Sea* (Glennie, ed., 1984).

**Petrology.** The science of rocks; the study of their mineral composition, texture, and internal structure (petrography), and of their origin and mode of formation, particularly of the igneous rocks (petrogenesis). The term was first introduced, in the form 'petralogy', by Pinkerton (*Petralogy, a Treatise on Rocks*, 1811), with its modern meaning, but in the middle of last century it was sometimes used with a very wide connotation, thus: 'The study of rock-masses . . . characters to be observed in the field . . . modes of stratification . . . denudation and its results . . . origin and growth of mountain chains' (Jukes, *Manual*, 1862), departments of geology now definitely excluded from petrology. 'Petrology traces its origin back both to mineralogy and to geology. In this double parentage is reflected the very essence of rocks, which are at the same time both geological bodies and mineral associations' (Loewinson-Lessing, trans., Tomkeieff, *Historical Survey of Petrology*, 1954). 'Petrology has become physico-chemistry applied to the crust of the earth' (Barth, *Theoretical Petrology*, 1952). 'The evolution of petrological ideas' (Teall, *QJ*, 1901/2). *Petrology for Students* (Harker, 1895/1960). *Principles of Petrology* (Tyrrell, 1926/9). *Nomenclature of Petrology* (Holmes, 1920/8).

**Petrophysics.** The physics of rock behaviour. 'The petrophysical characteristics of sedimentary and igneous rocks are important . . . in deformation' (Harland, *GM*, 1956).

**pH value.** A quantitative expression of soil acidity and alkalinity. 'pH is defined as the negative index of the logarithm of the hydrogen-iron concentration . . . pH7 means neutrality, pH values below 7 signify increasing acidity, and pH values above 7 increasing alkalinity' (Jacks, *Soil*, 1954).

**Phacolite.** (Phacolith.) 'In the ideal case of a system of undulatory folds there is increased pressure and compression in the middle limbs of the folds, but in the crests and troughs a relief of pressure and a certain tendency to opening of the bedding surfaces. A concurrent influx of molten magma will therefore find its way along the crests and troughs of the wave-like folds. Intrusive bodies corresponding more or less closely with this ideal case are common in folded districts. Since some distinctive name seems to be needed, we may call them phacolites' (Harker, *Igneous Rocks*, 1909). The form 'phacolith' has been largely, but unwarrantably, substituted for the original 'phacolite' (*see* LACCOLITE). [G. *phakos*, a lentil.]

**Phanerocrystalline.** See CRYSTALLINE. [G. *phaneros*, plainly visible.]

**Phanerozoic.** On the basis of the character of the present material evidence of the passage of geological time, particularly that supplied by fossils, geological history falls into two major periods, Precambrian and Cambrian onwards. The former is probably some seven or eight times as long as the latter. For this latter period, which often has to be referred to as a whole, there has until recently been no generally agreed name in use, so that such awkward names as 'Post-Precambrian' or 'Cambrian-to-Recent' have been employed. However, in 1930, Chadwick (*BGSA*) had proposed 'Phanerozoic' (clear evidence of life) for this period (with 'Cryptozoic', hidden evidence of life, for Precambrian), and this name has recently been revived by, e.g. Gilluly, *QJ*, 1963, and in *The Phanerozoic Time-scale* (Harland and others, eds., 1964).

**Phase.** Apart from its ordinary meaning, and its technical meaning in other sciences, there are two chief special uses in geology. **1.** For 'facies', particularly when this is on a small scale, being either a variety within a main facies or one of short duration or local occurrence. **2.** In the history of one period of igneous activity in one region, a chapter in which one kind of activity is dominant. See IGNEOUS CYCLE.

**Phenoclast.** Phenoclasts are conspicuous pebbles or fragments in a rudaceous rock, a conglomerate or breccia, embedded in a fine-textured matrix. The Hertfordshire Pudding-stone provides a good example. [G. *phainō*, bring, or come, to light.]

**Phenocryst.** 'Lavas that reach the surface of the earth in a fluid condition consolidate upon cooling rapidly to a more or less perfect glass. The glass in most cases contains crystals of several minerals scattered uniformly through it. Usually both large and small crystals occur in the same glass. The large ones that stand out prominently from the mixture of glass and small crystals are said to be "phenocrysts", the remainder of the rock is called the "groundmass". (The writer, after consultation with Prof. J. D. Dana and others, suggests the term "phenocryst", from φαίνω [*phainō*] =[be] conspicuous' or eminent, and κρύσταλλος [*krustallos*] =crystal.) Lavas may also cool in such a manner that the whole mass becomes crystalline and no glass left. When there are porphyritical minerals ("phenocrysts"), they are scattered through a groundmass wholly made up of crystals; but surface lavas seldom attain so high a degree of crystallization' (Iddings, *BPSW*, 1892, quoted by Mather and Mason, *Source Book*, 1939). The term is hardly used for the large crystals in a porphyritic granite because all crystals there are more or less conspicuous; here the appropriate term is 'megacryst'.

**Philosophy of geology.** The knowledge or study, in the widest senses, of the science of geology; the study of principles of geology. 'James Hutton and the philosophy of geology' (Tomkeieff, *TEGS*, 1948).

**Phonolite.** A trachyte especially rich in alkali, containing nepheline. The fine-grained equivalent of a nepheline-syenite. 'The name *Klingstein* [clinkstone] was used by Werner for certain rocks on account of the resonance produced by striking slabs of it with a hammer. Later, Klaproth (1801) changed the name to phonolite. At that time the mineralogical composition was unknown and the name referred simply to a rock of certain physical characteristics' (Johannsen, *Petrography*, 1938). [G. *phōnē*, a sound.]

**Phosphorite.** 'The term "phosphorite" is applied to rocks composed essentially of calcium phosphate minerals. The terms "rock phosphate" and "phosphate rock" have a similar meaning. Rocks in which phosphate is a distinct, but minor, component are described as phosphatic' (Knox, *Petrology for Students*, 1978).

**Photogeology.** In the widest sense: photography applied to the study of any branch or aspect of geology. It would, however, hardly be used for such specialized techniques as photomicrography in petrology, and it would then be: the recording and interpretation of geological features by means of photography. *British Geological Photographs* (*BA*, 1902/4). *Classified Geological Photographs* (*GS*, 1963). In present usage it refers particularly, if not exclusively, to the procedures of aerial photography for geological purposes and the geological interpretation of aerial photographs. 'A photogeological analysis ... Ecuador' (Marchant, *QJ*, 1961). *Photogeology* (Miller and Miller, 1961). 'Uses of aerial photographs by geologists' (Rich, *PE*, 1947). *Photogeology and Regional Mapping* (Allum, 1966).

**Phreatic.** Pertaining to underground water; as in 'phreatic surface' (water table), 'phreatic volcanic eruption' (akin to geyser activity).

**Phyletic.** Pertaining to a line of descent in the organic world. [G. *phulon*, a tribe.]

**Phyllite.** 'A compact lustrous schistose rock with its minerals less well defined than in a mica-schist, the characteristic mineral by which the foliation is controlled being sericite [a mica]' (Holmes, *Nomenclature of Petrology*, 1928). (Naumann, *c.* 1850.) [G. *phullon*, a leaf.]

**Phylogeny.** The evolution, the course of evolution, of a group of animals or plants. The group may be of any taxonomic size from species to phylum. Hence 'phylogenetic series' (see EVOLUTIONARY SERIES). [G. *phulon*, a tribe.]

**Phylum** (pl. **-a**). A 'tribe or phylum' (Haeckel, 1876), now used chiefly in the special sense of a prime division (taxonomic category) of the animal kingdom. (In the plant kingdom 'division' takes its place.) [L., f. G. *phulon*, a tribe.]

**Physical geography.** See GEOGRAPHY. 'The branch of Geography which deals with the physical features of the earth, including land, water and air' (Moore, *Dictionary of Geography: Terms used in Physical Geography*, 1967). *Class-book of Physical Geography* (Geikie, 1877). *Modern Physical Geography* (Strahler and Strahler, 1978). Discussion in Stamp, ed., *Geographical Terms*, 1961. The facts of the physical geography of the present day obviously form the essential and preliminary part of the more philosophical and historical science of Geomorphology.

**Physical geology.** Comprises those aspects of geology that are concerned with the physical geological processes; their action and results. The study of these processes is fundamental to the study of most branches of geology; structural geology and geomorphology may be said to lie wholly within physical geology. 'Researches in Physical Geology' (Hopkins, *TCPS*, 1838). *Physical Geology*, (Green, 1876/98). *Principles of Physical Geology* (Holmes, 1944/65/78). *Physical Geology* (Mallory and Cargo, 1979). 'Physical geology' can also mean the application of the science of physics to the problems and practice of geology.

**Physico-chemical geology.** The application to geology of the principles of physical chemistry. *Physico-chemical Geology* (Rastall, 1927). See PETROLOGY.

**Physics of deformation.** The application of the laws of physics to the interpretation of deformed rocks in terms of the forces involved. 'Physics of deformation' (Hills, *Structural Geology*, 1963).

**Physil.** See PHYSILITE.

**Physilite.** 'The term physil, an abbreviated form of phyllosilicate, is proposed to describe all sheet minerals regardless of grain size. Physilites are rocks with a high content of physils' (Weaver, *SG*, 1980).

**Physiography.** **1.** A description of the physical nature of objects. 'Microscopical Physiography of the rock-making minerals' (translation by Iddings, 1888, of Rosenbusch's *Mikroskopische Physiographie* . . . ). **2.** More or less synonymous (according to varying usage) with 'physical geography'. Huxley (*Physiography*, 1877) introduced the term in this sense. Hence 'physiographical', as, for instance, in Wills's *Physiographical Evolution of Britain* (1929), a work in palaeophysiography (=palaeogeography). Full discussion of all meanings of 'physiography' in Stamp, ed., *Geographical Terms*, 1961.

**Phytolith.** See BIOLITH.

**Picrite.** 'The picrites were so named by Tschermak (1866) on account of their high content of magnesia (bitter-earth). The original picrites occur as intrusions in the Cretaceous and Eocene in the highlands of Moravia and Silesia. These were augite-olivine rocks. Since they were described the definition has been widened to allow of the occurrence of other coloured silicates in the place of augite, and according to the current English usage, the essential feature is the presence of small amounts of plagioclase' (Hatch and Wells, *Igneous Rocks*, 1952). [G. *pikros*, bitter.]

**Pictograph.** A diagram combining picture and graph. 'Variation diagrams, or pictographs, of non-marine lamellibranch communities' (Eager, *LMGJ*, 1953).

**Piedmont.** Neither an ordinary English, nor (apparently) an ordinary French, word, but widely used as an adjective in physical geography with the sense of lying or formed at the foot of mountains. 'Piedmont profiles in the arid cycle' (Balchin and Pye, *PGA*, 1955).'Rivers which discharge from mountainous country on to flat plains usually deposit the bulk of their coarser debris in a more or less well-defined "piedmont belt"' (Hatch and Rastall, *Petrology of the Sedimentary Rocks*, 1972). '"Piedmont glaciers", consisting of sheets of ice formed by the coalescence of several valley glaciers which have spread out below the snow line—like lakes of ice' (Holmes, *Physical Geology*, 1965/78; the best known example is the Malaspina glacier in Alaska).

**Piercement fold.** A diapiric anticlinal fold. See DIAPIR.

**Pillar.** Used particularly in 'earth pillar' and (e.g. Lhwyd *in* Camden, 1695, and Banks *in* Pennant, 1774) for a column of columnar jointing.

**Pillow-lava.** Lava showing 'pillow structure'. 'The structure is commonly irregular, the masses resembling pillows or soft cushions pressed upon and against one another' (Cole and Gregory, *QJ*, 1890). 'Variolitic pillow-lava from Newfoundland' (Daly, *AG*, 1903). 'I was anxious to observe the process of formation of pillow-lava [at the present day, in the Samoa Islands, subaerial lava flowing into the sea]' (Tempest Anderson, *QJ*, 1910). 'The pillow-lavas are a group of basic igneous rocks that occur, in our experience, only as submarine flows' (Dewey and Flett, *GM*, 1911). The marginal part of a 'pillow' tends to differ, both texturally and chemically, from the inner part.

**Pingo.** 'In Arctic Canada, Greenland and northern Siberia there are sporadic mounds and cones, covered with deeply fissured layers of mantle deposits which have been upheaved by intrusions of ground-ice. They have [large] dimensions, their elevation above the surrounding country being anything up to 43 m. Many of them have a crater at the summit, sometimes containing a shallow lake during the summer months. These remarkable structures are called "pingos" (their Eskimo name), and they have also been referred to as "hydro-laccoliths" and as "cryo-laccoliths"' (Holmes, *Physical Geology*, 1978, with photographs).

**Pipe.** A more or less tubular hollow in a rock in which rock material has later accumulated; the 'solution pipe' is the commonest of these. Another kind is the following: a rock surface with subaerially produced pot-holes may subsequently be covered with a deposit which fills them, the resulting 'pipes' thus being preserved. 'Sandstone pipes infilling potholes. Lower Carboniferous, Trwyn Dulban, Anglesey' (Bates and Kirkaldy, *Field Geology*, 1976).

**Pipe clay.** See BALL CLAY.

**Pisces.** The fishes, a superclass of the subphylum Vertebrata. They are (cold-blooded) aquatic vertebrates with the limbs in the form of fins and the skin bearing scales (including teeth) or bony plates. They are the first to appear of the vertebrates, some fossils having been found in the Ordovician (N. America). In Britain they become definitely established in the fossil record at the base of the Downtonian (but see Squirrel, *GM*, 1958). This series is now generally placed in the Old Red Sandstone, but some authorities place it in the Silurian; so that records of 'Silurian' for the occurrence of groups of fishes are apt to be ambiguous. Teeth and bony plates are readily fossilizable; but the internal bones are fossilizable only in rather exceptionally favourable circumstances. [L. pl., fishes.]

**Pisolite.** A coarse-grained oolite. 'A stone possessing a structure like an agglutination of pease' (Lyell, *Principles*, 1833). [G. *pisos*, a pea.]

**Pisolith.** A particle of a pisolite.

**Pit-and-mound structure.** Small pits or mounds (about 1/10 in. to 1/2 in. across), or mounds with a central pit, on bedding surfaces. The various forms are probably produced in various ways. Generally, they appear to be due to processes of sedimentation; some have been taken to be the impressions of raindrops. 'On recent and fossil semi-circular cavities caused by air-bubbles on the surface of soft clay, and resembling impressions of raindrops' (Buckland, *BA*, 1842). 'Small pit and mound structures developed during sedimentation' (Kindle, *GM*, 1916).

**Pitch.** As a structural term. **1.***a*. The dip along the line of the crest of an anticline or the trough of a syncline. Where this dip is appreciable, the fold is a 'pitching fold'; where the crest or trough is practically horizontal, it is 'non-pitching'. Anticlines and synclines are usually somewhat boat-shaped, virtually non-pitching over the greater part, with opposed pitch at each end. *b*. This dip may be the expression of a general tilt of a fold or a folded series, being the vertical angle between the (spatial) fold direction and the horizontal. **2.** Used in mining, particularly in America, in connexion with a linear orebody lying within a vein-surface (considered as a plane), and adopted by geologists in a more general sense during the last two or three decades. It refers to a straight line, or a series of parallel straight lines (a lineation), on a plane surface, and is the angle between the line or lines and the strike of the plane. measured in the plane itself. See PLUNGE. (Clark and McIntyre, *AJS*, 1951.) Jukes (*Manual*, 1862) defines 'pitch', among mining terms, as 'a portion of a vein prepared and set apart for working'.

Unfortunately, the more specialized and, in general geology, seldom required, sense (2) has lately become so much insisted on as the exclusive use that geologists in general have, it seems, too readily allowed themselves to give up the firmly established and universally understood sense (1), which is one of the most fundamental and familiar concepts in structural geology. But it is curious that this sense, particularly that of a general tilt (1*b*), seems to have had no term attached to it till about the early years of this century. O. T. Jones's use of 'pitch' for this, in connexion with the folds of the Plynlimon district (*QJ*, 1909), is the earliest the present writer has found in British writings; but it appears that Willis (*USGS*, 1893) had so used the term (Dennis, ed., *Tectonic Dictionary*, 1967) and see PUMPELLY'S RULE. Not only has 'pitch' been taken away, but 'plunge' has been used instead, which (in the writer's view) makes matters worse, because of the ordinary use of 'plunge'. So long as 'pitch' was used in both cases, its meaning in any particular instance could be unambiguously interpreted according to the context; but if a term is to be used instead of 'pitch' in one of its meanings (1), another term should be used for it in its other meaning (2).

Thus the term 'rake', always synonymous, in one of its meanings, with 'pitch' in the mining sense (2), has been recommended by, e.g. a *USGS* committee (undated, 195-) and Hills (*Structural Geology*, 1963).

An early account (perhaps the first) of a pitching fold is given by John Williams (*Mineral Kingdom*, 1789): 'The coal and other strata in this place fall over in a kind of trough, so that the middle of the tongue is by far the deepest part of it; and it rises and crops out upon both sides, that is, towards the south-west and towards the north-east, so that there is a dip of declivity from both these sides towards the middle of the trough, and the middle of the trough dips away towards the north-west. Perhaps I may not be perfectly understood in pointing out these different dips; but as Gilmerton is near Edinburgh, young gentlemen can go out and investigate circumstances upon the spot, which is very practicable, as a fine lime-quarry lies below the coal, and the quarry is wrought upon both sides of the tongue'. **3.** In the early days of geology, 'pitch' was sometimes used simply as an alternative term for 'dip' (e.g. Maton, 1797; Conybeare and Phillips, 1822).

**Pitchstone.** A predominantly glassy acid igneous rock, usually with crystallites and sometimes porphyritic. It has a dark pitch-like lustre. Most pitchstones are intrusive, but some may be extrusive lavas. 'In ascending the hill towards Corygills [Arran] a very considerable vein of dark leek-green pitchstone makes its appearance' (Jameson, *Mineralogy of the Scottish Isles*, 1800). (Schulz and Poetsch. 1759.)

**Pivotal fault.** A fault in which the blocks are pivoted about an axis at right angles to the fault line. The sides of upthrow and downthrow are reversed in the two directions away from the axis.

**Placer.** A superficial deposit, particularly an alluvial gravel or sand, containing precious minerals. Examples are gold placers, the gold being derived from rocks containing auriferous quartz-veins; and tin placers (e.g. those of Cornwall), the tin being derived from tin-bearing granites. Placers may be recent or may occur 'fossil' in the geological formations; the 'banket' gold-bearing conglomerates of the Rand, S. Africa, occur in rocks possibly of Precambrian age.

**Placodermi.** Primitive jawed fishes, an extinct class of the superclass Pisces. They are almost entirely confined to the Old Red Sandstone (including Downtonian) but one group extends through the Carboniferous. The most abundant and striking group is that of the Arthrodira, armoured, with a jointed neck. [G. *plax, plak-*, a flat plate, *derma*, skin.]

**Plagioclase.** The isomorphous series of soda-lime felspars, including the end-members albite and anorthite. Stages between albite and anorthite are represented by oligoclase, andesine, labradorite, and bytownite. Very common in the diorites and andesites and in the basic igneous rocks. Triclinic system. [G. *plagios*, oblique, *klaō*, break.] See FELSPAR.

**Plagioclinal mountains.** A mountain chain or line of hills in which the strike of the rocks is markedly oblique to the trend of the chain or line. 'Plagioclinal mountains' (Callaway, *GM*, 1879, citing British hills composed of Precambrian rocks, e.g. the Caer Caradoc chain in Shropshire). Callaway uses the adjective 'orthoclinal' for the usual case where rock-strike and relief-trend more or less coincide.

**Plain of marine denudation.** An extensive wave-cut platform, including one that has been uplifted. See WAVE-CUT PLATFORM and HILL-TOP SURFACE. The term is also applicable to a palaeoplain so formed. 'The figure is a diagram representing no particular section, but simply the general form of the country across the Lower Silurian strata of Cardiganshire, as shown by myself [*MGS*, vol. 1, 1846]. . . . If we take a straight-edge, and place it on the topmost part of the highest hill, and incline it gently seaward, it touches the top of each hill in succession. . . . It occurred to me, when I first observed this circumstance, that at a period of geological history of unknown date, this inclined line that touches the hill-tops must have represented a great "plain of marine denudation"' (Ramsay, *Physical Geology and Geography of Great Britain*, 1894). 'On the coast at Watchet, in Somerset, a plain of rock has been formed that extends seaward for some distance at low water. Along the coast formed of Old Red Sandstone near Arbroath in Forfarshire, the broad inland plain ends abruptly in vertical precipices that rise from 150 to 250 feet above the waves at their base, and while the tide is retreating to its lowest ebb, long reefs and skerries of hard-edged strata tell of the progressive cutting back of a great modern "plain of marine denudation", similar to that old one which stretches inland from the high edge of the existing cliff' (Ramsay, 1894). 'Such a surface was called by Sir Andrew Ramsay a "plain of marine denudation"; the title is a long one, but is far more expressive than the attempts made in some sciences to cover up each conception in a single word of Greek' (Cole, *Open-air Studies*, 1895). 'Concept first recognized by Ramsay, A. C., 1846, but term first applied by Geikie, A., *TGSG*, 1871' (*AGI Glossary*, 1957). This is, of course, a genetic term, and therefore its applicability to particular cases of hilltop surface is debatable; it is certainly debatable in the case of Ramsay's 'type area' of Cardiganshire (Challinor, *WGQ*, 1969); also in the case of Snowdonia (Challinor and Bates, *Geology Explained in North Wales*, 1973) and of the Weald (Worssam, *IGSc*, 1973).

**Planar.** Arranged as a plane or in planes, spread out ('plane' in the loose geological sense of a surface, whether truly plane or not); usually, more or less parallel planes are implied, such as bedding or cleavage. This two-dimensional arrangement is to be distinguished from the one-dimensional, the linear, arrangement; for instance, are the lines or bands seen on a surface of an igneous rock, chance linear structures (such as strings of crystals), or are they the edge views of planar structures (such as platy bands of flow) ?

**Planation.** In geomorphology, the production of a plane (or nearly plane) surface. This is most commonly and extensively done by erosion ('deplanation'), but is also done by deposition ('applanation'). See PENEPLAIN.

**Plane.** Geological terminology is at a loss for a general word to denote what is usually the most important of all the features of rock-masses, their boundaries in contact one with another, and for other, more imaginary, conceptions of extension without thickness. The correct word must surely be 'surface'; but unfortunately this word usually, in common parlance, implies the outside of some body. Therefore to denote surfaces that are essentially internal (exposed as lines in plan and section) the geologist is apt to use the term 'plane'. These surfaces are in fact quite commonly near plane, plane enough for practical purposes, and in many theoretical considerations they are taken as true planes, so that the terms, 'bedding-plane', 'fault-plane', 'axial plane' are then appropriate; but can geologists be allowed such licence as to speak of, for instance, a 'curved bedding-plane' ? 'Plane' has one not inconsiderable advantage over 'surface'; it is shorter and more euphonious. See SURFACE. Further discussion in Dennis, ed., *Tectonic Dictionary*, 1967.

**Planetesimal hypothesis.** The hypothesis of the formation and growth of the earth formulated by Chamberlain and Moulton (both of Chicago) during the first few years of this century. The distinctive feature of it is that the earth, the matter of which was torn from the sun by the pull of a passing star, has grown by the gravitational attraction and accretion of solid particles, 'planetesimals'. Other versions of an 'encounter' hypothesis have since been put forward. The Planetismal hypothesis is also one of the 'accretion' hypotheses.

**Planetology.** The scientific study of the planets is now not only of astronomical interest. 'For the first time in the history of geology we are facing the possibility of comparing the Earth with other neighbouring planets' (Katterfel'd and Beneš, *IGC*, 1968). 'An introduction to the geology of Mars' (Ronca, *PGA*, 1970).

**Plankton.** A collective name for all the organisms (mostly microscropic, 'micro-plankton') floating or drifting in the waters of the sea or in bodies of fresh water. Fossils of planktonic organisms are the best potential indicators of stratigraphical horizon over a wide area. [G. *plagktos*, drifting.]

**Plantae.** The plant kingdom comprises the following main divisions: Thallophyta, Bryophyta, Pteridophyta, Pteridospermae, Gymnospermae, Angiospermae. The first three are the seedless plants, the last three the seed-bearing plants. The Pteridospermae are an extinct group. [L., plants.]

**Plastic.** Readily mouldable. Used particularly in 'plastic flow' (see FLOW) and 'plastic deformation' ('plastesis') of rocks.

**Plate tectonics.** A group of concepts that postulates in essence, that the earth's crust, together with the upper part of the underlying mantle, is entirely made up of 'plates' continuously in motion with respect to each other. The size of the plates is variable but there are about half a dozen particularly large plates. The base of the plates is the base of the lithosphere, at its junction with the asthenosphere. Plates are generated by the process of sea-floor spreading at mid-ocean ridges and are absorbed at their opposite margins, typically by the process of subduction at oceanic trenches. 'This great unifying concept draws sea-floor spreading, continental drift, crustal structures and world patterns of seismic and volcanic activity together as aspects of one coherent picture' (Oxburgh in Gass and others, eds., *Understanding the Earth*, 1972). 'In 1965 the theory of plate tectonics was announced in essentially complete form by J. Tuzo Wilson (*N*), thus beginning one of the most dramatic revolutions of geological thought ever' (Dott, *ESR*, 1978). Outline discussions, and history of the theory are many, for instance: Oxburgh (quoted above), 'The plain man's guide to plate tectonics' (Oxburgh, *PGA*, 1974), 'The emergence of plate tectonics' (Bullard, *AREPS*, 1975), *Plate Tectonics and Crustal Evolution* (Condie, 1976). See BENIOFF ZONE, COLLISION ZONE, EARTH'S CRUST, ISLAND ARC, LITHOSPHERE, MAGNETIC REVERSAL, MID-OCEAN RIDGE, OCEANIC TRENCH, SEA-FLOOR SPREADING, SUBDUCTION ZONE, TRANSFORM, TRANSFORM FAULT.

**Plateau basalt.** 'The term "plateau basalt", long familiar in geological literature, and Tyrrell's more appropriate expression "flood basalt" are applied synonymously to basaltic lavas occurring as vast composite accumulations of subhorizontal flows which, erupted in rapid succession over great areas have at times flooded sectors of the earth's surface on a regional scale' (Turner and Verhoogen, *Igneous and Metamorphic Petrology*, 1960). These flows (later covering rocks having been removed by erosion) now tend to form plateaux. In some cases the basalt, instead of reaching the surface, may have been intruded as extensive sills.

**Plateau gravel.** Spreads or patches of superficial gravel occupying flat areas, or the higher parts of a district, at heights differing in different places, usually from about 30 to 100 m. above sea level. The gravels are of doubtful origin. (Woodward, *England and Wales*, 1887.) Playfair (*Illustrations*, 1802) describes the flint gravels capping higher ground in E. Devon as being relics of the Chalk.

**Platform.** A level, or nearly level, rock-surface, on any scale; particularly one receiving, capable of receiving, or having received, an unconformable cover of sediments. See COMPOUND STRUCTURE. 'The Palaeozoic platform upon which the Secondary rocks of England rest' (Strahan, *QJ*, 1913). The term 'geoplatform' is sometimes used for a concept exemplified in the follwing: 'The Mediterranean depression is part of a global belt of deformation. This depression is regarded as the result of immense en echelon displacements between the Eurasian and Afro-Arabian geoplatforms. Opposing north-south directed movements of the two platforms contributed towards a complex fracture pattern' (Kostov, *PGA*, 1978). In some contexts the term 'platform' is synonymous with 'shelf', particularly in connexion with facies. See also WAVE-CUT PLATFORM.

**Platy fracture.** The property of fracturing into thin plates, usually applied to the kind of fracture sometimes shown by mylonites and schists (metamorphic rocks).

**Pleistocene.** The latest period of geological time, and the corresponding deposits and events, extending back about 1.5 to 2.0 million years and taken either to include the Recent or Holocene (thus extending forward to the present day) or as excluding that period. On the whole, it covers the time since the ordinary marine stratigraphical record ceases to be largely available, because of the establishment of the present general distribution of land and sea, so that the material evidence of the period is of a special kind and will be chiefly in the form of superficial land deposits. It so happens, however, that the Pleistocene period largely coincides with The Ice Age so that the physical records of the time are largely records of glaciation. But the Pleistocene includes pre-glacial formations, particularly in East Anglia. The period also largely coincides with the advent of man, so is characterized by a unique and highly peculiar faunal element. In Britain the boundary between Pliocene and Pleistocene has been customarily taken at the top of the Cromer Forest Bed series; but everywhere else in the stratigraphical record we seek a general boundary in a continuous marine succession. This is provided in Italy, where the Calabrian, with what is taken to be a Pleistocene marine fauna, overlies the Astian with a Pliocene fauna. (The terrestrial deposits, in France and Italy, corresponding to the Calabrian are the Upper Villafranchian.) The corresponding horizon in Britain is considered to be in the middle of the East Anglian Crags, between the Coralline Crag below and the Red Crag above, all of which, with a distinctive individuality in common, have always hitherto been called 'Pliocene'; though Prestwich (*Geology*, 1888) had questioned whether the base of the Pleistocene might not with greater propriety be drawn at the base of the Red Crag. If the horizon in Italy is taken as the plane of reference, the Red Crag (with the deposits above it) becomes 'Pleistocene'. (*IGC*, 1948; Baden-Powell, *N*, 1953, *PGA*, 1955; Flint, *SPGSA*, 1965; Grichuk and others, *RICQ*, 1965; Zagwijn, *BO*, 1974.) 'The term Pleistocene was introduced by Lyell in 1839 in a French translation of his *Elementary Geology*, but it was not until 1873 (in *Antiquity of Man*, 4th ed.) that it was adopted in the sense in which we now use it' (North, *PGA*, 1943). 'In 1823 Buckland published his *Reliquiae Diluvianae* and the Pleistocene chapter of geology was begun' (Kennard, *PGA*, 1945). *The Pleistocene Period* (Zeuner, 1945/59). *Pleistocene Geology and Biology* (West, 1968/77). [G. *pleistos*, most, expressing 'mostly recent species .]

**Pleochroic.** Showing more than one complexion of colour; applied, in mineralogy, to this quality ('pleochroism') as a crystal is rotated in polarized light. Pleochroic haloes are the dark strongly pleochroic zones round certain radio-active inclusions (e.g. zircon) in certain minerals (e.g. mica). [G. *pleiōn*, more, *chros*, complexion.]

**Plexus.** An interwoven network of lines; applied particularly in palaeontology to such a conception of an evolving group, a 'plexus of descent' (Swinnerton, *Palaeontology*, 1923).

**Plication.** 'Small-scale fold, or process of folding into small folds' (Dennis, ed., *Tectonic Dictionary*, 1967).

**Pliensbachian.** See JURASSIC SYSTEM. [Pliensbach, Württemberg.]

**Pliocene.** In Britain, long applied to all the uppermost beds of the regular geological column, chiefly the Crags of E. Anglia. Among the abundant shell fauna of the Crags, the following are some representative species: the lamellibranchs *Astarte omalii*, *A. borealis*, *Cardita senilis*, *Cardium edule*, *C. parkinsoni*, *Glycimeris glycimeris*, *Spisula arcuata*, *Nucula cobboldiae*, *Macoma obliqua*, *Venus casina*, and the gastropods *Liomesus dalei*, *Nassa reticosa*, *Neptunea contraria*, *Scaphella lamberti*, *Turritella incrassata*, *Littorina littorea*, *Nucella tetragona*. But the upper of these Crags, the Red and Norwich Crags and associated beds, are now on a wider correlation considered to be Pleistocene. See PLEISTOCENE. (Lyell, *Principles*, 1833; 'I have been much indebted to my friend, the Rev. W. Whewell, for assisting me in inventing and anglicizing the terms Pliocene, Miocene, and Eocene'.) [G. *pleiōn*, more, expressing 'a majority of recent species'.]

**Plug.** See SALT DOME, SEDIMENTARY INTRUSION, VOLCANIC PLUG, and quotation (Lees) under DIAPIR.

**Plumes.** See MANTLE PLUMES.

**Plunge. 1.** The diving effect of an anticline whose axial surface has become bent over and which is, so to speak, heading (closing) downwards. (e.g. Bailey and Weir, *Introduction to Geology*, 1939, fig. 51.) **2.** Proposed in a specialized sense by Smyth in 1908 (*TAIME*) for an angle where it is necessary to distinguish from 'pitch' (sense 2.). It refers to a straight line, or a series of parallel straight lines (a lineation), on a plane surface, and is the vertical angle between the line or lines and the horizontal plane. See PITCH. (For a figure illustrating these senses of 'pitch' and 'plunge' see Phillips, *Stereographic Projection*, 1954.)

Unfortunately the term 'plunge' has been extended beyond this specialized sense (2) and used for other vertical angles, particularly (**3**) for what has long been known as 'pitch' in folding, thereby also becoming confused with 'plunge', sense (1). This appears to have been first proposed by Clark and McIntyre (*AJS*, 1951), and geologists have largely but uncritically (or so it seems) adopted it. McIntyre uses the phrase 'pitch of the fold-axes' in the first summary of his paper read on 29 Nov. 1950 (*PGS*, 22 Dec. 1950), but changes it to 'plunge of the fold-axes' (*QJ*, 27 Dec. 1951). He uses 'plunge' throughout the printed paper but all the contributors to the printed discussion, except the author himself, use 'pitch'. McIntyre used 'plunge' in *GM*, Dec. 1950. Violence and impetuosity (to some degree at least) are inherent in the word 'plunge', so it is hardly suitable for a gentle slope; but it is not inappropriate for a particular behaviour of 'pitch' (sense 1*a*): 'Anticlinal folds have a beginning and an end; some may plunge steeply at the ends' (Park, *Geology*, 1914). In this sense (3) it is inclination that is being expressed; now 'inclination' is one of the ordinary, non-geological meanings of 'pitch', but hardly one of the ordinary meanings of 'plunge'. See ATTITUDE. **4.** The same meaning as 'dip', particularly in the sense 'dip beneath'. 'Strata are said to dip when they form an angle with the horizon; the point towards which they plunge being considered the dip' (De la Beche, *Manual*, 1831). 'The easterly regional dip causes the Carboniferous strata to plunge beneath the younger New Red Sandstone of the Cheshire Plain' (George, *BRG, N. Wales*, 1961).

**Pluton.** A plutonic rock-body. Thus the definition of 'pluton' depends on the definition of 'plutonic'. There are 'magma plutons' ('orthoplutons') and 'migma plutons' ('metaplutons') (Rittmann and Cloos, 1936; Rittmann, *Volcanoes*, 1962). A very wide definition is: 'A term which embraces all intrusive bodies of igneous rock. It is convenient when the intrusion conforms to none of the above definitions [sills, etc.] or its configuration has not been determined' (Turner and Verhoogen, *Igneous and Metamorphic Petrology*, 1960). 'Plutonite' is synonymous.

**Plutonian.** Pertaining to igneous action beneath the earth's surface. Used particularly in considering questions of the genesis of ore bodies. 'In its mineralogical application, the Plutonian concept invokes a process of ore deposition from hydrothermal solutions originating at depth from an igneous magmatic source. . . . The Neptunian concept implies ore genesis by the syngenetic or diagenetic concentration of sulphides in sediments' (King, *MG*, 1966).

**Plutonic. 1.** Pertaining to the action of heat deep within the earth's crust, particularly as applied to the deep-seated 'plutonic rocks'. These include igneous and migmatitic rocks (granitic complexes and gneisses), and even such metamorphic rocks as schists, if there is a conception of one great plutonic group in classification. How 'deep' is a matter of varying usage but the igneous rocks formed at relatively small depths, the hypabyssal rocks, are usually excluded. When the term first came into use it was practically the same as 'igneous' and included volcanic rocks. It seems to have been first used by Kirwan (*Mineralogy*, 1796) : 'There is another system which attributes not only to basalts, but to all stony substances, an igneous origin. This may be called the Plutonic system'.

But the term soon became restricted so as to exclude the volcanic rocks and at least those minor intrusions that seemed to be directly connected with volcanic outpourings. Thus the general term 'igneous' included the 'plutonic' and the 'volcanic'. 'The unstratified crystalline rocks have been very commonly called Plutonic, from the opinion that they were formed by igneous action at great depths, whereas the volcanic, although they have risen up from below, have cooled from a melted state upon or near to the surface' (Lyell, *Principles*, 1833). Today it is usual (at least in Britain) to consider three modes of occurrence of igneous rocks; plutonic, hypabyssal, volcanic. **2.** Pertaining to the doctrine of the Plutonists. 'Dr. Hutton's theory very well deserves to be distinguished by a particular name; and . . . we may join Mr. Kirwan in calling this the Plutonic System. For my own part, I would rather have it characterized by a less splendid, but juster name, that of the Huttonian Theory' (Playfair, *Illustrations*, 1802). [G. *Ploutōn*, Pluto, god of the infernal regions.]

**Plutonic association.** See VOLCANIC ASSOCIATION.

**Plutonism. 1.** All those operations that give rise to the plutonic rocks. 'A commentary on place in plutonism' (Read, *QJ*, 1948). 'A contemplation of time in plutonism' (Read, *QJ*, 1949). **2.** See NEPTUNIST.

**Plutonist.** See NEPTUNIST.

**Plutonite.** See PLUTON. 'The coarse-grained deep-seated rocks or plutonites' (Rittmann, *Volcanoes*, 1962).

**Plutonometamorphism.** Metamorphism under extremely deep-seated conditions. See ECLOGITE.

**Pluvial.** Pertaining to rain, its action and effects; such as a landslide, or the cutting of a gully in a hillside and the consequent spreading out of the eroded material below. 'Pluvial denudation; Langtoft, Yorkshire' (*BA* photograph, *c*. 1903). Also as 'pluvial periods', the specially rainy periods in the more recent geological history of a region in low latitudes; supposedly corresponding in some way with the general glacial and interglacial periods. [L. *pluvia*, rain.]

**Pneumatolysis.** Chemical action between the minerals in a rock and the volatiles given off by a magma in the later stages of cooling and solidification. The process may act within the igneous rock itself, affecting the minerals already formed ('autopneumatolysis') though in this case it is a part of the actual production of the completed igneous rock. It may also refer to the metasomatic change in the country rock through which the volatiles have percolated. The autopneumatolysis of granite can nowhere be better studied than in Devon and Cornwall, where the chief effects are tourmalinization, greisening, and kaolinization. See METASOMATISM and CONTACT METAMORPHISM. [G. *pneuma, pneumat-*, air (&c.), *lusis*, a setting free.]

**Pocket deposit.** A cavity in a rock filled up with subsequently-deposited material. Examples are 'the pocket deposits of Derbyshire' (Ford and King in Sylvester-Bradley and Ford, eds., *Geology of the East Midlands*, 1968), where the pockets are solution-cavities in the Carboniferous Limestone, up to several hundred yards across, filled with gravels, sands, and clays washed in during (probably) either Triassic or Tertiary times (or which have later collapsed into the pockets).

**Podsol.** 'The term is derived from a Russian folk name for a grey or ash-coloured soil. Podsols occur typically in cool, humid climates under coniferous forest or heath vegetation. . . . Podsols may be seen on the Bagshot Sands of SE. England; they are widespread on the sandy soils of northern and middle Europe' (Robinson, *Soils*, 1949). The upper, eluviated, layer (below the humus) is intensely leached, while the lower, illuvial, layer is greatly enriched with iron compounds sometimes forming a pan. Also spelt 'podzol'.

**Poikilitic.** (Poecilitic.) **1.** A texture of igneous rocks where numerous small crystals of one of the essential minerals are enclosed, 'spotted about', with no common orientation, in much larger ones of another essential mineral. When the smaller crystals are felspar and the larger ones a mafic mineral not so much larger as to enclose many of them, the texture merges into the ophitic. In a way, the poikilitic texture is the converse of the porphyritic. **2.** Has been applied (in the sense of many-coloured) for the variegated marls of the Trias. [G. *poikilos*, spotted.]

**Poikiloblastic.** See CRYSTALLOBLASTIC.

**Polar wandering.** 'Until about 1956 most geophysicists seem to have favoured 'polar wandering' (without continental drift) as a sufficient explanation of the changes of latitude disclosed by the palaeomagnetic data then available. The polar wandering hypothesis is generally understood to mean that an outer shell of the earth, involving the crust and probably part of the mantle, has shifted as a whole relative to the axis of rotation, which remains almost fixed, relative to the stars. If this were an adequate explanation by itself, then the geographical positions of the poles at any given time would be the same for all the continents. Assuming the continents to have remained fixed relative to one another, there would be a single 'polar-wandering curve'. But every continent has its own polar wandering curve and these are so different that it is obvious that the continents have changed their relative positions through time' (Holmes, *Physical Geology*, 1978).

**Pollen analysis.** See PALYNOLOGY.

**Polycyclic.** Many-cyclic; applied particularly to an entity such as a geomorphological feature, a structure, or a metamorphic rock, of which the present character has resulted from the repeated operation of a process that is 'cyclic' in one sense or another.

**Polygenetic.** Having many origins, derived from many sources; thus may be applied, for instance, to pebbles in a conglomerate. (Sometimes as 'polygenic'.)

**Polymetamorphism.** Polyphase (multiple) metamorphism. 'Rocks which have suffered two or more separate acts of metamorphism are said to be "polymetamorphic" ' (Read and Watson, *Introduction to Geology*, 1962). 'The importance of polymetamorphism in the Highlands' (Sutton, *SP*, 1961).

**Polymorphism.** In mineralogy, the property of a chemical element or compound whereby it occurs in more than one crystal form (and with correspondingly different physical properties). The differences are due to the different arrangements of the atoms. Includes trimorphism (e.g. titanium dioxide, $TiO_2$, as anatase, brookite, and rutile) and the commoner dimorphism (e.g. calcium carbonate, $CaCO_3$, as calcite and aragonite, and carbon, C, as graphite and diamond). See quotation (Chinner) under ZONE (2).

**Polyphase.** Proceeding or accomplished in several distinct phases, episodes, stages. 'The Caledonian orogeny involves several phases of deformation and metamorphism. . . . Clough [1897, *MGS*, Cowal] was perhaps the first to recognize the polyphase deformation of the Dalradian rocks of the Highlands' (Rast, *PGS*, 1962).

**Polyphyletic.** Derived along several distinct lines of descent. Chiefly used in a case where some similarity of organic form (e.g. the genus *Monograptus*) is supposed to have been achieved by evolutionary convergence.

**Polyzoa.** See BRYOZOA. 'Polyzoa versus Bryozoa' (Moore, ed., *Treatise on Invertebrate Paleontology*, 1953). 'An account of the work on Polyzoa' (Pitt, *PGA*, 1961).

**Porcelainous.** (Porcellaneous, Porcellanous.) Having the appearance of porcelain. Applied to invertebrate shells, particularly foraminifera (Williamson, 1858), and to rocks ('porcellanite', e.g. the Lias clay metamorphosed by a Tertiary dolerite sill at Portrush, N. Ireland).

**Porifera.** The phylum of the sponges, the simplest animals that have the body differentiated into tissues. They are aquatic (nearly all marine) and sessile, the body being essentially of a sac-like form. Most of them have skeletons, siliceous, calcareous, or of the organic material, spongin; the latter (to which the ordinary bath sponge belongs) not being fossilizable. The mineral skeletons may be coherent, in the form of the animal, or occur as loose spicules. Spicules occur in many rocks, and the coherent sponges are common in certain rather restricted formations, particularly in the Cretaceous. [L. *porus*, a pore, *fero*, bear.]

**Porosity.** The condition of being porous, of containing pores; quantitatively, the proportion of pore space in the space occupied by the rock as a whole ('void ratio'). See PARTICLE CONCENTRATION and PERMEABILITY.

**Porphyrite.** 'A term which has been variously used for pre-Tertiary andesitic rocks, altered andesite rocks, and hypabyssal rocks of marked porphyritic texture and andesitic composition. The last usage referred to is that now customary' (Holmes, *Nomenclature of Petrology*, 1928).

**Porphyritic.** See PORPHYRY. In spite of the derivation of the word, it invariably refers to the texture, not the colour. 'Porphyritic structure. When one of the constituent parts of the mountain-rock is dispersed through a basis in the form of grains or crystals, the rock presenting this appearance is said to be "porphyritic" ' (Jameson, *Geognosy*, 1808). 'Porphyritic, consisting of a compact ground in which distinct crystals are imbedded, or of a granitic ground in which some of the crystals are much larger than the rest' (Bakewell, *Geology*, 1813). The term 'porphyritic' is also applied to the 'distinct' or 'much larger' crystals themselves.

**Porphyroblastic.** See CRYSTALLOBLASTIC.

**Porphyroid.** The etymology does not provide a definite meaning and some local groups of porphyritic igneous rocks have been called 'porphyroids' (e.g. Charnwood Forest, Precambrian), but the term has tended to become restricted to a metamorphosed porphyritic igneous rock in which the original phenocrysts have largely escaped granulation. This latter meaning is essentially different from that of the porphyroblastic structure, although Holmes (*Nomenclature of Petrology*, 1928) defines 'porphyroid' as referring to certain porphyroblastic metamorphic rocks.

**Porphyry.** Originally, the purplish rock quarried anciently in Egypt, which was also characterized by felspar phenocrysts. The name gradually came to be applied to all igneous rocks (plutonic, hypabyssal, or volcanic) with comparatively large crystals of any mineral in a distinctly finer-grained ground mass, that is, having the porphyritic texture; but a porphyritic granite (such as the Shap or Cornish granite) would hardly be called a 'porphyry'. See quotation (Playfair) under WHINSTONE. [G. *porphura*, purple.]

**Portland beds.** The British, and the typical, representatives of the Portlandian stage (Jurassic) of NW. Europe (or of the lower part of that stage). In the British area, they appear to have been deposited only within the triangle whose corners are now in Dorset, Oxfordshire, and Sussex, and they are there mostly concealed beneath the newer rocks. However, they are well exposed in the Isles of Portland and Purbeck where they are about 70 m. thick. There are two divisions, the Portland Sand below and the Portland Stone above, the latter being the famous limestone used for building. The limestone has huge ammonites (e.g. *Titanites giganteus*) and other characteristic fossils such as the lamellibranch *Trigonia gibbosa* and the gastropod *Aptyxiella portlandica*. (Michell, *MS*, 1788, 'Portland Limes'; Webster, *TGS*, 1814, 'Portland Oolite'; Smith, *Stratigraphical System*, 1817, 'Portland Rock'.)

**Portlandian.** See JURASSIC SYSTEM. In English usage the name is restricted to tbe Portland beds, thus excluding the Purbeck beds.

**Positive.** As noted under EUSTATIC MOVEMENTS, Suess used 'positive' and 'negative' for rise and fall, respectively, in sea-level. But 'positive' and 'negative' are also used for the rising and sinking, respectively, of crustal elements relative to the neighbouring crustal elements, and where this rise or fall occurs over coastal regions the terms have the opposite meaning to Suess's.'The Lewisian basement of the Outer Hebrides is today almost devoid of overlying cover-rocks because these islands have functioned over hundreds of millions of years as parts of a positive block standing above sea-level or at any rate receiving only limited thicknesses of sediment' (Watson, *PGA*, 1977). 'This basement complex [Midland Valley of Scotland] may have been an area of positive relief in late Precambrian and early Palaeozoic times' (Graham and Upton, *JGS*, 1978).

**Post-glacial.** See ICE AGE. 'Problems of post-glacial denudation' (Strahan, *QJ*, 1914). 'The post-glacial deposits of the Lincolnshire coast' (Swinnerton, *QJ*, 1931). 'Studies of the post-glacial history of British vegetation' (Godwin and others, *PT*, 1938 onwards).

**Posthumous.** In tectonics. Chiefly appears to be used in the case of a recurrence of forces and movement along lines or over areas affected by similar forces in a previous period. It seems, however, from the ordinary meaning of the word, that it would be more appropriately reserved for faulting (or folding) caused by the delayed action of a stress the setting up of which had long ago ceased. A new stress might well cause movement to occur over an old fault-plane, this being a plane of weakness (the movement might be said to be 'prelocated'); and if the new stress were of the same kind, and in the same sense, as the old one, the stress and the resulting structure might then be said to be 'revived'; the fault would be 'reactivated'. Incidentally, the dictionary tells us that the ultimate derivation is from the L. *postumus* (superlative of *post*,) latest, so that the customary use of tbe term may not be inappropriate after all.

**Post-Precambrian.** Later than Precambrian. See PHANEROZOIC. 'The post-Pre-Cambrian rocks laid down during the last 600,000,000 years' (Kirkaldy, *Study of Fossils*, 1963).

**Potassium-argon dating.** See RADIOACTIVITY AGE-METHOD. *Potassium Argon Dating* (Schaeffer and Zähringer, 1966). 'Potassium-argon ages for minerals from the Ross of Mull' (Beckinsale and Obradovich, *SJG*, 1973).

**Pot-hole. 1.** A bowl-shaped or cylindrical hollow scooped out or bored by pebbles swirled round by eddies. 'The geologist must examine with care these rock-basins, or "pot-holes", of which examples in every stage of development may be found. The beginning is very slight. In some ledge there happens to be a notch so placed that when flood waters pour down, the stones they bring are bumped over a particular spot more frequently than elsewhere. At this spot they pound until they have hollowed out a hole. While the hollow is shallow the stone bounce out again and others come in, just as one may see a strong trout swept down the current into the "pot" and out again. Presently the hole is abraded deep enough to retain the stones, which spin and grind in the current until the excavation is deep. The gorge at the Strid owes its existence to a series of such pot-holes all enlarged until the party-walls between them have broken down' (Kendall and Wroot, *Geology of Yorkshire*, 1924). **2.** See SWALLOW.

**Practical geology.** This means (1) work in the field and in the laboratory as distinct from that in the study, or (2) geology in the service of man (see ECONOMIC GEOLOGY). If the finding and utilizing by man of various kinds of rocks and minerals could be called 'geology'—he has at least appreciated their qualities and has discovered where they occur—'practical geology' began in the earliest stages of man's existence. Are palaeolithic flint implements 'fossils' or are they evidences of 'practical geology' ? Both, no doubt it might be said. All stone buildings depend for their construction on 'practical geology' in that sense. However, we can hardly properly speak of 'practical geology' as being pursued before man's knowledge and thought had brought the science of geology into existence. *Practical Geology* (Harrison, 1878). *Aids in Practical Geology* (Cole, 1890). 'Practical geology and the natural environment of man' (Dunham, *QJ*, 1967). The Museum of Practical Geology, the institution created by Sir Henry De la Beche, Director of the Geological Survey, and opened by the Prince Consort in Jermyn Street, London, in 1851.

**Poured-in deposit.** 'The Ordovician [and other Lower Palaeozoic] sediments fall into three main facies: a graptolitic shale facies, usually pelagic and off-shore; a mixed shelly, sandy, muddy, and calcareous facies, relatively near-shore; and a facies of 'poured-in' detritus, often carried by turbidity currents in accumulations of great thickness' (George, *BRG*, *S. Wales*, 1970).

**Precambrian.** (Pre-Cambrian.) 'I met Sir H. on Saturday at Bangor. We had a short rap at Anglesey at very old rocks–older than the Cambrian' (Ramsay, in a letter dated 1849, quoted in Geikie's *Life*, 1895). Sedgwick placed the metamorphic, and associated igneous, rocks of Anglesey and south-western Lleyn (Caernarvonshire) below his Cambrian system (*PGS*, 1838, 1843, *QJ*, 1847, 1852; successively more decisively). 'Mr. Hughes pointed out that the conglomerate [Cambrian of Pembrokeshire] contained fragments of the hornstone and quartz of this older series, which he considered was probably part of an old ridge or shoal, possibly of Laurentian, but certainly of Pre-Cambrian age' (in Harkness and Hicks, *QJ*, 1871). 'The Precambrian rocks of Shropshire' (Callaway, *QJ*, 1879). 'The term "pre-Cambrian" has no doubt been, and may still be, a convenient designation for that undefined and imperfectly known portion of the geological record which underlies the Cambrian system. But it is obviously too indefinite for purposes of precise stratigraphical classification. It unites a vast succession of rocks differing widely from each other in structure and age, and possessing, indeed, only the one common character of being older than the Cambrian age' (Geikie, *QJ*, 1891). In this general sense (a sense which the term itself seems to demand) it became synonymous with Archaean in its wide sense (as in Geikie's *Textbook*, 1903); but with the restriction of Archaean, and in countries where that term is in use, Precambrian is often used only for the overlying rocks (older than the Cambrian), i.e. as equivalent to Algonkian (=Proterozoic). In British usage the term means all the time and all the rocks before and below the Cambrian (but see INFRA-CAMBRIAN). Probably the first geological description of any British area of Precambrian rocks is that by Farey of Charnwood forest in his *Derbyshire* (1811). This was very soon followed by Bakewell's description of the same region (*Geology*, 1813).

The chief outcrop of the Precambrian is over the region of the Canadian Shield. 'In 1863 Sir William Logan described two Precambrian systems in eastern Canada. The younger one consisting of schist and phyllite was named Huronian and mistakenly identified as Cambrian. The older, or Laurentian system, was composed of gneissic granite. Later studies showed that the Precambrian is much more complex. It is now known to consist of several great sequences of sediments, each of which in turn was more or less metamorphosed and granitized' (Weller, *Stratigraphic Principles and Practice*, 1960). *A Correlation of Precambrian Rocks in the British Isles* (*GSSR* (6), 1975). Precambrian time extends from the formation of the earth about 4,600 million years ago up to the beginning of Cambrian time about 600 million years ago.

**Preferred orientation.** A tendency towards a particular orientation in any of the elements in the structure, make-up, or fabric of a rock or deposit. 'The microfabric . . . a strong preferred orientation of quartzes' (Sutton and Watson, *QJ*, 1956). 'The axial planes of the folds show a poor preferred orientation' (Johnson, *QJ*, 1957). 'The preferred orientation of glacial till pebbles has been used for many years as an indicator of the direction of ice-movement (e.g. Holmes, *BGSA*, 1941; Lundgvist, Sweden, 1948; West and Donner, *QJ*, 1956; Norris, *GM*, 1962)' (Banham, *PGA*, 1966). ('Preferred' is sometimes similarly used in reference to other geological states and values.)

**Pre-glacial.** Immediately before the Glacial Period and as distinct from that, or the Post-glacial, period. 'Preglacial floras from Castle Eden' (Eleanor Reid, *QJ*,

1920). Could also refer locally to the time preceding the onset of glacial conditions there. See PLEISTOCENE.

**Prehistory.** Used in archaeology to mean the account of events or conditions prior to history written or deliberately recorded by man. Such accounts usually concern 'early man', man himself being thus already on the scene. But 'history' in geology has a much wider meaning, and 'prehistory', if it were used, would refer to times prior to the events or conditions recorded in the rocks ('Pre-Precambrian' ?).

**Pre-located.** Suggested use in tectonics: see POSTHUMOUS.

**Pressure solution.** 'Sorby, in a series of classic papers published in the late nineteenth century, described observations he had made in the field and with the newly invented petrological microscope which led him to the conclusion that in many rocks deformed by tectonic processes "mechanical force had been resolved into chemical action". These conclusions have been fully supported by recent work and the process of solution and precipitation of carbonates and silicates in deformed rocks is generally termed pressure solution' (Ramsay, in conference report, *JGS*, 1977).

**Primary.** **1.** In stratigraphy, used in the early years of the 19th century as equivalent to Primitive, later including the Lower Palaeozoic and, later still, becoming equivalent to the whole Palaeozoic, which latter term has now almost completely superseded it. **2.** For a rock of which all the constituents are newly formed rock-particles, that is, particles that have never been constituents of a previously formed rock and that are not the products of alteration or replacement. Thus rocks solidified from a magma, chemical precipitates, and even, perhaps, purely organic deposits are primary (as distinct from secondary or derived), though the term is usually reserved for igneous rocks, solidified from a magma. **3.** For structures and textures formed, in the case of sedimentary rocks, during the initial deposition of the sediment (depositional structures and textures) and, in the case of igneous rocks, during the solidification of the magma. *Atlas and Glossary of Primary Sedimentary Structures* (Pettijohn and Potter, 1964). *Primary Sedimentary Structures and their Hydrodynamic Interpretation* (Middleton, ed., *SEPM*, 1965). **4.** For minerals which have been produced directly from a melt or a solution as distinct from secondary minerals which have replaced those already formed. **5.** For the elements of a fold or fault structure that constitute its geometrical form, as distinct from the features and values that appear, as it were, secondarily by virtue of the setting of the structure in relation to the earth's horizon (Challinor, *PGA*, 1945/6).

**Primates.** The order of the class Mammalia which includes man himself. Fossil remains are thus of special interest but unfortunately are rarer than for any other large group of mammals. This rarity is due to the animals being mostly tree-dwellers, geological deposits not normally being formed in forested regions, and also to their being mostly tropical creatures, whereas the known Tertiary fossil beds are in what are today temperate zones. However, in the Eocene, these regions, as shown by their vegetation, were tropical, and fossil primates (though necessarily the more primitive forms at the outset of their evolution) are less rare. *History of the Primates* (Le Gros Clark, 1954).

The following is a classification of the order (groups not containing *Homo* left undivided):

(Order) Primates.
- (Suborder) Lemuroidea, lemurs. Eocene.
- Tarsioidea, tarsiers. Eocene.
- Anthropoidea, monkeys, apes, and man.
  - (Superfamily) Ceboidea, New-world monkeys. Miocene (?Eocene).
  - Cercopithecidea, Old-world monkeys, Oligocene.
  - Hominoidea, apes, and man.
    - (Family) Pongidae, apes. Oligocene.
    - Hominidae, man. Pleistocene (?Pliocene).
      - (Genus) Homo, 'true man'.
        - (Species) H. sapiens, modern man. Ur. Palaeolithic.
        - H. neanderthalensis;*
        - H. heidelburgensis*
      - Pithecanthropus, 'ape-man'*

(The dates are those of first appearance.)
*Lr. Palaeolithic (extinct).

[L. *primas*, *primat-*, a thing that is first.]

**Primitive.** One of Werner's groups of rocks (towards the end of the 18th century), including granite and other crystalline rocks which he supposed to be chemical precipitates thrown down with extremely irregular surface from an uninterrupted 'primitive' ocean. Apart from the question of the particular mode of formation, they were supposed to have been formed in some way peculiar to the earliest times (to which they were all referred). The rocks which were called primitive roughly include (*a*) what later came to be known as Archaean, (*b*) all highly metamorphosed rocks, and (*c*) the larger plutonic masses.

**Principles of geology.** Those broad conclusions as to process and history arrived at by reasoning from the occurrence and explanations of generalized facts. For instance: the principle of igneous intrusion, of the superposition and consequent relative age of strata, of the correlation of strata by means of fossils, of regional metamorphism, of differential erosion. 'Plate tectonics', a concept recently formulated, appears to be already established as a new valid principle. The 'geological cycle' is a very widely embracing principle, while 'uniformitarianism' may be said to be the most comprehensive geological principle of all. *A critical examination of the First Principles of Geology* (Greenough, 1819). *Principles of Geology* (Lyell, 1830/33-1875). *Principles of Geology* (Gilluly and others, 1951/75).

**Prismatic jointing.** See JOINT, COLUMNAR JOINTING, and BASALTIFORM.

**Problematicum** (pl. **-ica**). A marking, object, &c., in the rocks, of which the nature presents a problem. It may be doubtful whether the object is organic or inorganic (a quasi-fossil), but the term is used chiefly for undoubted organic remains which are nevertheless *incertae sedis*. 'A characteristic fossil of the Arenig base in most of the outcrops is the problematical organism *Bolopora undosa*' (Smith and George, *BRG, N. Wales*, 1948). 'A species of the problematical *Atikokania* . . . described from the Pre-Cambrian of Ontario . . . first ascribed to the Archaeocyathinae and later referred to a spongoid form, is probably inorganic' (Whittard, *PBNS*, 1953). 'A new genus of problematicum, *Hensonella* . . . whatever its true nature . . . a conspicuous and easily recognized micro-fossil' (Elliott, *QJ*, 1959). 'Trace fossils and problematica' (Häntzschel in Moore, ed., *Treatise on Invertebrate Paleontology*, 1962/75). [modern L.]

**Process and pattern.** There is a principle involving constancy of process and variety of 'pattern' or effect. This is exemplified in geomorphology, where the physical and chemical processes (working according to the immutable laws of nature) produce an endless variety of land-form according to the rock-materials and rock-structure on which they act. In a very different field, palaeontology, we have another example: 'This does not imply that the Pattern of mantle secretion (in the Brachiopoda] has never changed. On the contrary there is as much conclusive proof to show that it has, as there is circumstantial evidence indicating that the Processes of secretion have not' (Williams, *Let*, 1968).

**Profile. 1.** The surface of the ground or a structural surface may be considered to be cut through along some plane so that the edge appears as a profile. In geology this plane is either some vertical plane or a plane perpendicular to a significant direction. It may usually be assumed to be a vertical plane; but, for instance, the profile of a folded surface may sometimes be taken to be the trace of that surface on a plane at right angles to its fold-direction (in space), that is, on the 'normal cross-section' of the fold (Challinor, *PGA*, 1945). This latter tectonic usage is adopted by McIntyre (*GM*, 1950, and *QJ*, 1951). **2.** Whereas the cross-profile of a valley is an ordinary profile, the longitudinal profile, long profile, or *thalweg* is a special kind. Here the imaginary vertical section is not a plane, but follows the winding of the valley and the profile is the upper edge considered as being straightened out. If the section follows the winding of the river itself within the valley floor another, and flatter, profile results. **3.** A 'projected profile', in geomorphology, usually means the profile of the highest ground over the length of a selected strip of country projected on to a vertical plane parallel to the strip (giving a kind of skyline). **4.** A vertical column to scale of beds exposed on the face of a cliff or quarry, made semi-diagrammatic by adjusting the dip to the horizontal and showing the lithology by conventional marking. One side of the column is an irregular line (a profile) drawn in accordance with the resistance to weathering of the several beds. See, for instance, the examples figured by Arkell, *MGS, Weymouth*, 1947 (Jurassic) and by Dineley, *QJ*, 1966 (Devonian). **5.** See SOIL PROFILE.

**Pro-glacial.** Applied to features immediately in front of, or round the margin of, a mass of ice, particularly to glacier lakes ('pro-glacial lakes') and their associated spillways. It is implied that the features are caused by the presence of the ice. Used by Wooldridge and Morgan (*Geomorphology*, 1937). 'Pro-glacial features in Middlesex' (Lawrence, *PGA*, 1964).

**Projection.** The representation, usually on a plane surface, of a figure, by drawing rays through every point of it so as to fall on the surface. A map is a projection. Various methods of projection give various representations. A form of projection used particularly in crystallography and structural geology is the stereographic projection. (Phillips, *Stereographic Projection*, 1954.)

**Proterogenesis.** The biological hypothesis according to which an individual, in the early stages of its growth, tends to look forward to the future of its evolutionary history instead of looking to the past as in the theory of recapitulation (palingenesis); in other words, new evolutionary features appear first in the early growth stages, lasting longer in successive generations until they become established as features of the adult. 'The tendency for conditions seen in "youth" sometimes to anticipate the later course of evolution occurs not uncommonly in fossil sequences and is an example of an important principle, that of Proterogenesis' (Swinnerton, *Fossils*, 1960). (Schindewolf, 1925.) [G. *proteros*, forward.]

**Proterozoic.** Usually, a part of Precambrian time, and the corresponding rocks. Sometimes applied to the whole, definitely or vaguely, but usually more specifically, to the later part of that time in the rocks of which there are undoubted traces of life. In this latter sense Proterozoic is equivalent to Algonkian. ' "Proterozoic" rock groups include those Precambrian rocks that are least deformed and metamorphosed' (Gill, *Proterozoic in Canada*, 1957). 'In its restricted sense, the Proterozoic [in N. America] consists of two parts commonly considered to be systems. The lower is the Huronian and the upper was named Keweenawan by Brooks in 1876' (Waller, *Stratigraphic Principles and Practice*, 1960). However, the etymology (the comparative 'former', 'earlier') has invited the more general usage of the equivalent of 'Lower Palaeozoic', as in Jane Donald's paper on 'Proterozoic Gasteropoda' (*QJ*, 1902); indeed 'Protozoic' would be the more appropriate for these 'first life' Precambrian rocks, as suggested by Sedgwick (*PGS*, 1838). [G. *proteros*, former.]

**Proto-Atlantic ocean.** The concept of an early ocean which widened to form the present Atlantic. If the concept of a quite separate pre-Devonian ocean (Iapetus) is accepted, the name Proto-Atlantic would refer specifically to a later ocean which, after a new rift, enlarged continuously during the Mesozoic and Tertiary up to the present time. See IAPETUS OCEAN.

**Protoclastic.** 'Protoclastic textures are produced by the continued flow of an almost solid igneous body; the original minerals may be realigned and often partly crushed during this process' (Read and Watson, *Introduction to Geology*, 1962).

**Proto-Earth.** The substance, and any form it may have had, which eventually became the one coherent body, Planet Earth. 'When the proto-Earth was growing by the accretion of planetismals . . .' (Hayashi and others, *EPSL*, 1979).

**Protozoa.** The phylum of the simplest animals, most of them minute. They are marked off from all the rest of the animal kingdom by not being differentiated into specialized parts of the body. Those possessing hard parts are the Foraminifera and Radiolaria. [G. *prōtos*, first.]

**Protozoic.** Proposed independently, but at about the same time, by both Sedgwick and Murchison; by the former (*PGS*, 1838) for what is now known as Precambrian, 'should organic remains appear unequivocally in any parts of this class', and by the latter (*Silurian System*, 1839) for what is now known as Lower Palaeozoic.

**Provenance.** Source of origin, particularly that of the components and material of a sedimentary rock, whether the main substance or, for instance, the pebbles in a conglomerate or the heavy minerals in a sandstone. 'The history of any given bed involves determination of the source rocks and the source area from which the sediment came, in a word, its provenance' (Pettijohn, *Sedimentary Rocks*, 1957). 'Dartmoor detritals: a study in provenance' (Brammall, *PGA*, 1928). 'Provenance of far-travelled pebbles in the pre-Anglian Pleistocene of East Anglia' (Hey, *PGA*, 1976).

**Province.** In animal and plant geography, and thus in palaeontology, usually refers to a region of land or sea that, within one major climatic or environmental belt, is characterized by elements of the fauna or flora (more especially, particular species or genera) that are there chiefly because they have evolved there and have not mingled with elements outside. The special character of the fauna is due to isolation rather than environment.

**Psammite.** An arenaceous rock, a sandstone. Hence 'psammitic'. 'Gravel and sand rocks (psammites)' (Geikie, *Textbook*, 1882). See quotation (Read) under DIFFERENTIATION (2), and PELITE. [G. *psammos*, sand.]

**Psephite.** A rudaceous rock, a conglomerate or breccia (etc.). 'A rock composed of large rock-fragments, pebbles or blocks set in a groundmass varying in kind and amount. The psephites include the conglomerates, breccias, and tillites or boulder-clays' (Read, *PGA*, 1958). Hence 'psephitic'. See quotation (Read) under DIFFERENTIATION (2), and PELITE. [G. *psēphos*, a pebble.]

**Pseudo-anticline. 1.** An upward buckling of the superficial layers of the ground due either to changes in volume brought about by pedogenic processes or to some other cause. (Price, *BAAPG*, 1925; Jennings and Sweeting, *AJS*, 1961.) **2.** See quotation (*BRG*) under FAN STRUCTURE.

**Pseudo-bedding.** This term, and 'pseudo-stratification', are sometimes used for structural features that have the appearance of bedding or stratification but are of quite a different nature, particularly for horizontal jointing in igneous rocks or, more rarely, for certain occurrences of shearing or cleavage. 'Pseudo-bedding in the Dartmoor granite' (*TRGSC*, 1952).

**Pseudobreccia.** See BRECCIA (2).

**Pseudo-conglomerate.** A sedimentary rock, looking like a conglomerate, containing more or less rounded inclusions which have grown within the rock by some sort of concretionary process after, or perhaps penecontemporaneously with, its deposition as a sediment. 'The pseudo-conglomerate in the Lower Calcareous Grit' (Arkell, *QJ*, 1936).

**Pseudo-fossil.** See QUASI-FOSSIL. 'Pseudo-fossils from the Silurian of Tipperary' (Geikie, *QJ*, 1899). 'Pseudofossils: a plea for caution' (Cloud, *Geoly*, 1973).

**Pseudomorphism.** The assumption by one mineral substance of a form proper to another; particularly for a crystal which by infiltration, replacement, or alteration has come to occupy the space previously occupied by another and thus to take on its shape. Hence 'pseudomorph', for the assuming crystal.

**Pseudoplankton.** A collective name for animal organisms attached to floating and drifting vegetation.

**Pseudo-psephite.** 'Pseudo-psephites are produced in a variety of ways and comprise such rocks as fault-breccias, crush-conglomerates and tectonic mélanges' (Read, *PGA*, 1958).

**Pseudo-syncline.** See quotation (*BRG*) under FAN STRUCTURE.

**Pseudo-tectonic.** See TECTONIC and quotation (Sylvester-Bradley) under SUPERFICIAL STRUCTURE.

**Psilophytales.** An extinct group of the division Pteridophyta; the earliest and most primitive group of vascular plants, occurring in Lower and Middle Devonian rocks. [genus *Psilophyton* (G. *psilos*, bare, smooth, *phuton*, a plant).]

**Pteridophyta.** One of the main divisions of the plant kingdom including, as the chief groups, the Filicales (ferns), Equisetales (horsetails), and Lycopodiales (clubmosses), and the extinct groups, Psilophytales (Devonian) and Sphenophyllales (Carboniferous and Permian). [G. *pteris, pterid-*, a fern.]

**Pteridospermae.** The seed-ferns, an extinct group of plants which had fernlike fronds but bore seeds ; usually placed as one of the main divisions of the plant kingdom. All the typical adundant forms are Carboniferous and Permian; some Mesozoic plants are provisionally placed in the group.

**Pterodactyl.** (Pterodactyle.) An old, familiar name for any member of the order of Mesozoic flying reptiles, the Pterosauria. 'A great sensation was produced when, in the Jurassic shales of Solenhofen [Bavaria], a complete skeleton of a perfectly preserved small saurian was found with winglike appendages. Cuvier (1812) [1809?] recognized the skeleton as essentially reptilian in structure, called it *Pterodactylus*, and described it as a flying reptile' (Zittel, *History*, 1901). 'In the same blue lias formation at Lyme Regis, in which so many specimens of Ichthyosaurus and Plesiosaurus have been discovered by Miss Mary Anning, she has recently found the skeleton of an unknown species of that most rare and curious of all reptiles, the Pterodactyle, an extinct genus, which has yet been recognised only in the upper Jura limestone beds of Aichstedt and Solenhofen, in the lithographic stone' (Buckland, *TGS*, 1829). [G. *pteron*, a wing, *daktulos*, a finger.]

**Pteropod ooze.** See OOZE.

**Pteropoda.** Pelagic gastropods with conical, urn-like, or spirally coiled shells. Undoubted pteropods range from the Cretaceous, but certain Palaeozoic fossils, of which *Hyolithes*, of the Cambrian, is one of the commonest, have been considered by some to be pteropods.

**Pterosauria.** An order of Mesozoic reptiles (Jurassic-Cretaceous) that became adapted for flight by modification in the body structure, particularly the forelimbs, and the possession of membranous wings. Possibly this group may be placed more properly as a separate class of the subphylum Vertebrata. [G. *pteron*, a wing, *saura*, a lizard.]

**Ptygmatic vein.** An extremely contorted quartzo-felspathic vein in, typically, a migmatite. It seems that the vein material, when in a highly viscous state among less viscous surrounding material, buckled plastically on being urged forward by injection from behind. A ptygmatic vein may occur as the result of flow-folding in an acid lava or minor intrusion (e.g. in the pitchstone of Arran). 'Ptygmatic structures and their formation' (Wilson, *GM*, 1952). (Sederholm, Finland, 1907.) [G. *ptugma*, folded matter.]

**Pudding-stone.** See CONGLOMERATE and quotations (Hutton) under BASAL CONGLOMERATE and CONSOLIDATION. An old popular term for any conglomerate (*OED* gives it as current in 1753) but now chiefly restricted to particular occurrences, such as the Hertfordshire Pudding-stone (Eocene), which have long been so named and in which the pebbles suggest plums or raisins. Also called 'plum-cake' stone.

**Pulsatory sequence.** See CYCLIC SEQUENCE.

**Pumice.** 'A rock known to the ancients. The word occurs in Pliny. It is a rock froth which forms crusts on more compact lava or occurs in the form of volcanic ejectamenta. It is glass so filled with air bubbles that the pore space may be much greater than the glassy material' (Johannsen, *Petrography*, 1932). [L. *pumex*, *pumic-*.]

**Pumpelly's rule.** 'The degree and direction of the pitch [plunge] of a fold are often indicated by those of the axes of the minor plications on its sides' (Pumpelly and others, *USGS*, 1894). The *AGI*, *Glossary* (1977) expands this into: 'The generalization that the axial surfaces of minor folds of an area are in accord with those of the major fold structures'.

**Purbeck beds.** The British, and the typical, representatives of the Purbeckian stage (Jurassic) of NW. Europe (or, if a Purbeckian stage is to be 'lost'—see Lloyd, *GM*, 1964—of the upper part of the Portlandian stage). Limestones, marls, and clays, mostly of lacustrine or lagoonal origin, with old soils ('dirt beds'), and with intercalations of marine strata, the chief of which, the 'cinder bed', in the middle of the formation, is largely made up of *Ostrea distorta*, The fossils of the non-marine beds are extraordinarily varied. Ostracods are abundant and are used for zoning. Freshwater gastropods are distinctive, particularly in the famous Purbeck Marble, 'a greenish or reddish limestone crowded with shells of the freshwater pond snail *Viviparus*' (Arkell, *MGS*, *Weymouth*, 1947). The lamellibranch *Neomiodon medius* is so abundant as to be a rock-former in places. There are beds with the alga *Chara*, beds with various insects, and beds with the isopod crustacean *Archaeoniscus brodiei*, and in the 'dirt beds' occur cycad-like plants, conifers, and jawbones of small mammals. The Purbeck beds occur in S. England only, and the whole sequence (about 120 m.) is fully exposed in the Isle of Purbeck. (Webster, *TGS*, 1814; Smith, *Map*, 1815.)

**Puy.** 'The third type [of volcano] is distinguished by the formation of groups of cinder-cones or lava-domes, which from their admirable development in Central France have received the name of "puys"' and 'The Carboniferous puys of Scotland . . . we have now to study the records of another phase of volcanism, where scattered groups and rows of puys, or small volcanic cones . . .' (Geikie, *Ancient Volcanoes*, 1897). [French, a peak.]

**Pyrite.** (Iron pyrites.) The form of iron sulphide ($FeS_2$) crystallizing in the cubic system. A brass-yellow mineral with a bright metallic lustre. Occurs commonly in veins. [G. *puritēs*, a thing connected with fire, from its making sparks when struck.]

**Pyrites.** This name, by itself, in modern usage, is synonymous with pyrite (=iron pyrites). But, qualified, it is used for other sulphides of iron and copper, as 'copper pyrites', 'white iron pyrites' (= marcasite).

**Pyroclastic.** Applied to fragmentary materials produced by explosive volcanic action. The materials may have been either liquid or solid when ejected. A recent pyroclastic deposit may be loose or have become consolidated. Pyroclastic rocks occur among the rocks of past geological ages. 'The word "ash" is not a very good one to include all the mechanical accompaniments of a subaerial or subaqueous eruption, since ash seems to be restricted to a fine powder, the residuum of combustion. A word is wanting to express all such accompaniments, no matter what their size and condition may be, when they are accumulated in such mass as to form beds of "rock". We might call them perhaps "pyroclastic materials", but I have endeavoured in vain to think of an English word which should express this meaning' (Jukes, *Manual*, 1862). The completely appropriate word is still wanting; meanwhile, Juke's suggestion is generally adopted. There is a distinction between a rock composed of subaerially erupted pyroclastic material deposited as true sediment in the neighbouring sea ('tuffite') and one composed of the same kind of material deposited on land; the latter being a purely igneous rock. Though 'pyroclastic rocks' may be said to apply exclusively to extrusive products, a similar mechanism ('pyroclastic action') may produce bodies of fragmental material that are intrusive (see Patterson, *QJ*, 1963; and TUFF). The individual fragments and particles of a pyroclastic rock are 'pyroclasts'. 'Volcaniclastic' (or 'volcanoclastic') is sometimes used. (For discussions of the nomenclature of pyroclastic rocks see Williams, *PLGS*, 1926; Wentworth and Williams, *BNRC*, 1932; Anderson, *BGSUC*, 1933; Blyth, *BV*, 1940; Ross and Smith, *USGS*, 1961, Le Bas and Sabine, *GM*, 1980.) (G. *pur*, *puro-*, fire.]

**Pyrogenetic mineral.** 'The minerals formed from igneous magmas are termed "pyrogenetic" ' (Tyrrell, *Petrology*, 1929). Minerals of pneumatolytic, hydrothermal, or dynamothermal origin are excluded. This is the obvious meaning but the term is sometimes restricted to the anhydrous minerals of igneous rocks, usually developed at high temperatures, e.g. felspars, pyroxenes, and olivines. The hydroxyl-bearing minerals, such as the amphiboles and micas, are then termed 'hydatogenetic minerals'.

**Pyromeride.** A conspicuous nodular spherulite occurring in certain rhyolitic rocks. 'Among the various secondary changes to which the Caernarvonshire rhyolites have been subject, none is more striking than that evinced by the nodular rocks which we proceed to discuss. Under such names as "*roches globuleuses*", "*pyromerides*", "globular porphyry", "ball rock", "concretionary felstone", "porphyry with agate nodules","nodular felsite", and "coarsely spherulitic lava" these rocks have been described by numerous geologists, and more than one theory has been advanced to account for their peculiar structures [citing the literature]' (Harker, *Bala Volcanic Series*, 1889). 'The pyromerides of Boulay Bay, Jersey' (Parkinson, *QJ*, 1898). Holmes (1928) gives 'Monteiro 1814'. [French *pyromeride*, f. G. *pur*, *puro-*, fire, *meros*, part.]

**Pyrometamorphism.** Metamorphism produced by extreme heat. The term is applied particularly to xenoliths, so altered, in basaltic rocks. 'The most abundant xenoliths [Surtsey, Iceland] consist of . . . fragments of pyrometamorphic rock' (Sigurdsson, *GM*, 1968).

**Pyroturbidite.** See PYROCLASTIC. 'Submarine pyroclast flows (described from Japan and U.S.A. by Fiske and Matsuda, *AJS*, 1964, and Fiske, *BGSA*, 1963) are genetically related to ignimbrites, but as rocks they are petrographically and structurally distinct, and have radically different palaeogeographical significance. They might be called "pyroturbidites", to distinguish them from ignimbrites, for logically the term "ignimbrite" (literally "showers of fire") can only be applied to pyroclast flow deposits that accumulated on a land surface under subaerial conditions' (Fitch, *BV*, 1967).

**Pyroxene.** A family of rock-forming minerals, mainly silicates of magnesium, calcium, and iron. Augite (magnesium, iron, aluminium, and calcium) is the commonest member. Among others are enstatite (magnesium) and hypersthene (magnesium and iron), crystallizing in the orthorhombic system, and diopside (magnesium and calcium) and aegerine (iron and sodium), in the monoclinic system. ['strange to fire', apparently so named from the opinion that those minerals were only accidentally caught up in the lavas that contain them.]

**Pyroxenite.** A coarse-grained ultramafic igneous rock consisting essentially of one or more pyroxene minerals. 'The term was used originally simultaneously by Senft [Germany] and Coquand [France]' (Johannsen, *Petrography*, 1938) Plutonic.

# Q

**Quantum evolution.** 'The initial evolution of Paleocene mammals . . . is an example of quantum evolution, of the sudden appearance within a short space of time of large taxonomic units—units in this instance at the level of orders. This phenomenon is encountered frequently in the stratigraphic record, and has been accorded much attention by paleontologists interested in evolutionary problems. One characteristic of quantum evolution, as preserved in the geologic record, is the lack of annectent forms. Thus new groups of high taxonomic rank suddenly appear, commonly without any fossils to show the transition between them and their putative ancestors. In such instances, we can only suppose that there were exceedingly rapid evolutionary trends in very definite directions to fill ecological vacancies, developments taking place at such speed that we are unable to find traces of them in the record of the rocks. In effect, there was an evolutionary explosion of mammals as a result of quantum evolution with the advent of the Paleocene Epoch' (Kay and Colbert, *Stratigraphy and Life History*, 1965). The term 'quantum' is here being used in a sense somewhat analogous to its use in physics. See BURST.

**Quaquaversal.** Pointing in every direction; chiefly as the 'quaquaversal dip' of a periclinal structure; also for the structure itself. See quotation (Phillips) under CROSS-FOLDING. The term is sometimes restricted (e.g. J. Geikie, *Structural and Field Geology*, 1905/53) to a dome-structure, a basin-structure being then called a 'centroclinal fold'. [L. *quāquā*, whithersoever, *versus*, towards.]

**Quarry.** In addition to the usual meaning of an excavation for stone, made by cutting or blasting, the term may also refer to the unquarried stone itself; this is the usual sense in 17th-century writings.

**Quarry-water.** The moisture contained in newly quarried stone. 'It is desirable to shape the stones which are to be used in architecture while they are yet soft and wet, and while they contain their "quarry-water", as it is called' (Lyell, *Elements*, 1838). See quotation under GREENSTONE (2).

**Quartz.** A very hard glassy-looking mineral, the common crystalline silica, $SiO_2$. Often in hexagonal forms. A constituent of all over-saturated igneous rocks, that is, all acid rocks and some intermediate and basic types. Common in metamorphic rocks and as a veinstone, and the dominant constituent of sandstones (the sand grains being quartz). 'Apparently the first direct reference to a substance called "quertz" is to be found in a little booklet entitled *Eyn Nützlich Bergbüchlein* published anonymously in Augsburg probably in 1505 . . . The Saxon miners called large veins, *Gange*, and the small cross veins or stringers, *Querklüfte*. The name ore (*Erz*, *Ertz*) was applied to the metallic minerals, to the gangue or to the vein material as a whole. In the Erzgebirge silver ore is frequently found in small cross veins composed of silica. It may be that this ore was called by the Saxon miners *Querklüftertz* the cross-vein-ore. Such a clumsy word could easily be condensed to *Querertz* and then to *Quertz* and eventually become *Quartz* in German, *quarzum* in Latin, and "quartz" in English' (Tomkeieff, *MM*, 1942). 'The name quartz (*quraz*) is a German word of unknown origin, and has come into universal use in all languages since the 16th century' (Miers, *Mineralogy*, 1929). See CRYSTAL.

**Quartzite.** A rock almost completely siliceous, being a nearly pure quartz-sandstone cemented by interstitial quartz into a hard mass, or recrystallized (by metamorphism or other means) into a mosaic of quartz crystals. 'The regularly stratified quartzy mountain rock . . . there are large and high mountains of this stone in the shires of Ross and Inverness, which on a clear day appear at a distance as white as snow' (Williams, *Mineral Kingdom*, 1789; evidently alluding to the Cambrian Quartzite).

**Quartzose.** Mainly or entirely composed of quartz. In a rather special sense, for 'quartzose conglomerates', those in which pebbles of quartz or quartzite predominate (e.g. Murchison, *Silurian System*, for those in the Old Red Sandstone and the New Red Sandstone).

**Quartz-porphyry.** An (acid) igneous rock with conspicuous phenocrysts of quartz; includes the granite-porphyries and the quartz-porphyritic rhyolites.

**Quasi-fossil.** A marking, object, &c., in the rocks, which looks as if it might be a fossil but which is very doubtfully of an organic nature or origin. 'Quasi fossils of the Longmynd [Precambrian]' (Cobbold, *Church Stretton*, 1900). Also called 'pseudo-fossil'.

**Quaternary.** Of or pertaining to post-Pliocene time; that is, the Pleistocene, including the Recent (up to the present day). It is a sub-era of the Cainozoic era (see TERTIARY). (Desnoyers, 1829.) 'I have two of yours unanswered: the first is as to whether "Quaternary" would not be a better word than "Post-Pliocene". Most decidedly so' (letter from Godwin-Austen to Prestwich, dated 1859, quoted in *Life and Letters of Sir Joseph Prestwich*, 1899). *Quaternary Ice Age* (Wright, 1914/37). *Quaternary Era* (Charlesworth, 1957). *A Correlation of Quaternary Deposits in the British Isles* (*GSSR* (4), 1973). *Quaternary Geology* (Bowen, 1978). 'Quaternary time scale' (Berggren and others, *QtnR*, 1980).

# R

**Radioactivity age-method.** A method for determining the age of certain substances by finding the proportions in them of a certain kind (isotope) of an element to a certain kind (isotope) of another element from which it is known that it must have been derived by radioactive decay at a known 'half-life' rate. Used chiefly to determine the age of formation of certain rock-forming minerals and thus of the rocks first containing them. 'In 1902, only seven years after the discovery of radioactivity, Rutherford and Soddy had provided evidence for the atomic disintegration theory of radioactive phenomena, and three years later Rutherford, in the 1905 Silliman Memorial Lecture, made the first clear suggestion of using radioactivity to measure geological time (*Radioactive Transformations*, 1906). An earlier mention of this possibility was made in a lecture Rutherford gave in 1904 to the International Congress of Arts and Sciences in St. Louis. Shortly after this, Boltwood (*AJS*, 1907) began using the amount of lead in uranium minerals as a means of dating them'. (Wager in Harland and others, eds., *The Phanerozoic Time-scale*, 1964). 'Since 1907 other parent and daughter pairs of elements have been discovered and pressed into service, with the discovery of isotopes and the invention of the mass spectrograph and the mass spectrometer' (Hawkes, *QJ*, 1957). The outstanding name in connexion with the application of radioactivity age-methods during the last fifty years is Arthur Holmes. His first paper was 'The association of lead with uranium in rock-minerals, and its application to the measurement of geological time' (*PRS*, 1911) and among his last, 'A revised geological time-scale' (*TEGS*, 1959). This kind of measurement is 'radiometric' or 'isotopic', hence e.g., 'isotopic age-determination', 'radiometric date', 'radio-isotopic dating'. 'Measuring geological time', Moorbath in Gass and others, eds., *Understanding the Earth*, 1972. *Geochronology:*

*Radiometric Dating* (Harper, ed., 1973). See GEOCHRONOLOGY.

**Radiocarbon dating.** A special application of the radioactivity age-method. This activity is due to the presence of carbon-14 which has a half-life very much shorter than that of the other isotopes used in this method. It is thus applicable only to carbon-bearing material of the late Pleistocene (less than about 100,000 years old.) *Radiocarbon dating* (Libby and associates, 1955/65). 'Radiocarbon dating and Quaternary history in Britain' (Godwin, *PRS*, 1960). 'The radiocarbon method' (Shotton, *QJ*, 1966).

**Radiogeology.** 'The term radiogeology was introduced into geological literature in 1924 by the Russian geochemist Vernadsky, to denote the branch of knowledge concerned with the distribution pattern of the radioactive elements throughout the earth's crust and the role of radioactive processes in geological phenomena. . . . One cannot of course write on radiogeology without reference at some length to the part it has played in establishing the age of the earth and the geochronology of the stratigraphical succession; but in addition radioactivity studies have revolutionized some of our conceptions on the mode of formation of ore bodies, they have shed new light on the origin of some igneous and metamorphic rock complexes, and they have demonstrated that hitherto unsuspected geochemical transformations are commonplace in the history of sedimentary strata. . . . No other field of geological study has grown quite so vigorously in the post-war years' (Davidson, 'Some aspects of radiogeology', *LMGJ*, 1960). 'A radiogeologic study of the granites of SW. England' (Tammemagi and Smith, *JGS*, 1975).

**Radiolaria.** Protozoa possessing a siliceous skeleton of elements which are united to form a lattice-like structure or some other arrangment of rods and spines. As recognizable individuals they are not common as fossils. [L. *radiolus*, a little spoke of a wheel.]

**Radiolarian earth.** See SILICEOUS ORGANIC DEPOSIT.

**Radiolarite.** See SILICEOUS ORGANIC DEPOSIT.

**Radiolarian ooze.** See OOZE.

**Radiometric.** See RADIOACTIVITY AGE-METHOD. *Radiometric Dating for Geologists* (Hamilton and Farquhar, 1968).

**Raft.** Generally: a flat-floating structure made by the binding together, or aggregation, of separate pieces (primarily, logs). Recently, the term has been used in geology for any, not too small, relatively thin piece of rock caught up in a moving liquid, particularly a magma, and drifting freely, often more or less vertically, within that liquid. Thus the qualilies of thinness and drifting have been seized on, but those of flat-floating and binding-together ignored. 'The most remarkable phenomenon displayed by the [Donegal] Granite is that provided by the trains of definitely free-swimming rafts of country rocks' (Pitcher and Read, *QJ*, 1958).

**Rag.** 'Rag' and 'ragstone' are old and rather vague terms, signifying some hard or rubbly rock. They give well-established names to certain stratigraphical formations, such as Kentish Rag (Cretaceous) and Coral Rag (Jurassic).

**Rain and rivers.** As a combined operation, practically the sole agent of subaerial removal, transport, corrasion, and deposition in temperate climates. *Rain and Rivers* (Greenwood, 1857/76). *Work of Rain and Rivers* (Bonney, 1912).

**Raindrop impression.** See PIT-AND-MOUND STRUCTURE. 'An account of impressions and casts of drops of rain, discovered in the quarries at Storeton Hill, Cheshire' (Cunningham, *PGS*, 1839). 'On fossil rain-marks of the Recent, Triassic and Carboniferous periods' (Lyell, *QJ*, 1851). Also called 'rain-print', 'rain-pit'. The 'raindrop' interpretation of such impressions should be received cautiously in view of the fact that the coincidence of circumstances which would allow a few (only) scattered raindrop impressions to be immediately covered by a protective permanent layer of sediment must be exceedingly rare; yet so-called raindrop impressions are not very rare !

**Raised beach.** A beach, typically including the wave-cut platform itself (in which case 'raised platform' is more appropriate), occurring above, and separated from, the present beach, out of reach of present wave action; thus an old beach representing a former level of the sea. 'There are many other marks [in Scotland] of a sea beach upon a higher level than the present' (Hutton, *Theory of the Earth*, 1795; after describing some examples). 'Description of a raised beach on the north-west coast of Devonshire' (Sedgwick and Murchison, *TGS*, 1840).

**Rake.** See PITCH (2).

**Ramp valley.** 'For the Dead Sea trough, Willis postulates boundary faults that are high angle thrusts caused by lateral compression, a view which is essentially similar to Wayland's interpretation of the Lake Albert rift. For valleys of this type Willis has coined the term "ramp valleys" (*BGSA*, 1928)' (Hills, *Structural Geology*, 1953).

**Ranges of fossils.** Ranges given from system to system are apt to be ambiguous because certain stratigraphical series are placed by some authors in one system and by others in another. This particularly applies to the Cambrian, Ordovician, Silurian, Devonian, Triassic, and Jurassic systems because it is not universally agreed in which of two systems (in each case) the Tremadocian, Downtonian, and Rhaetic series should be placed.

**Rank in coal.** Coal assessed on the percentage of carbon it contains. 'The stage reached by a coal in the course of its carbonification' (*Nomenclature of Coal Petrology*, 1957). The chief ranks are represented by lignite, sub-bituminous coal, bituminous coal, and anthracite, in that order of increasing carbon content. 'Origin and development of coal rank' (Caldwell, *CG*, 1966).

**Rate of process.** Questions as to the rates at which geological processes may work under various conditions are obviously important. Relative rates of processes going on at the same time in the same region are particularly so. For instance, what is the relation, in any particular case, between the rate of uplift in orogeny and that of the simultaneous degradation by erosion? It seems that the evolution of land-form must depend primarily on this relation. 'Rate of deposition' and 'rate of crust movements' (Sollas, *Age of the Earth*, 1905). 'Rates of change within orogenic belts' (Sutton in Kent and others, eds., *Time and Place in Orogeny*, 1969). 'Rates of deformation', 'The rate of emplacement of batholiths' (Price, Pitcher, *JGS*, 1975). 'How long does it take for a sheet intrusion to crystallize?' (Phillips, *GT*, 1976). 'Rates of marine transgressions and regressions' (thematic papers, *JGS*, 1979).

**Rattle-stone.** See quotation (J. Geikie) under NODULE.

**Reaction rim.** An outer zone (rim in section) of, or bordering, a crystal where chemical reaction has occurred between it and its surroundings to produce mineral material of a different composition. This would tend to result from incomplete chemical change as a magma crystallized on the reaction principle, or from the reaction between a xenocryst and its (molten) surroundings.

**Reaction series.** A series of minerals successively formed (and successively lost) by chemical reaction during the cooling and crystallization of a magma. 'On account of the continual reaction relation between crystals and liquid in a solid solution series such as the plagioclases, it is proposed to call such solid solutions a "continuous reaction series" . . . by a "reaction pair" is meant that crystals of the first compound react with the liquid to produce the second during the normal course of crystallization. A reaction relation of this latter type may exist between three or more compounds and the compounds, arranged in proper order, may then be said to constitute a "discontinuous reaction series" . . . the series olivine-pyroxene-amphibole-mica is a prominent example among the rock-forming minerals' (Bowen, 'The reaction principle in petrogenesis', *JG*, 1922; also *Evolution of the Igneous Rocks*, 1928).

**Reactivate.** Bring to a state of renewed activity; hence reactivation. 'The movements . . . on this main fault were pre-Visean in age, though there may have been post-Visean reactivation' (Pitcher and others, *QJ*, 1964). 'A regional synthesis suggests that certain structural trends in the concealed Palaeozoic rocks of southeastern England and adjacent parts of north-western Europe have been reactivated several times in the subsequent history of the area, and in some cases control the alignment of present-day surface features' (Shephard-Thorn and others, *PT*, 1972). See BASEMENT CONTROL and POSTHUMOUS.

**Recapitulation.** In biology, the repetition during the development of the individual (ontogeny) of stages passed through during the ancestral evolution of the race (phylogeny). The phylogenetic evidence is supplied by palaeontology; it may be shown in one structural feature or several features concurrently. Recapitulation may happen to occur in particular cases for various reasons (e.g. see DEDUCTION), but the term usually refers to the theory that there is some innate tendency towards it. Translated into Greek, 'recapitulation' becomes 'palingenesis'; and this, as a theory, is the direct opposite of the theory of 'proterogenesis'. It does not seem likely that both general theories can be valid at the same time, and the conflicting evidence throws doubt on each.

**Recency.** This term is a useful one when 'the state or quality of being recent' (*OED*) is being emphasized. 'In the last few decades proofs have been discovered of the recency (previously unsuspected) of mountain uplift in various parts of the world. In New Zealand the dating within the last few years by microfloral evidence, largely by the study of fossil pollen grains (palynology), of deposits laid down in early Pleistocene times has made it evident that at least some of our mountains have arisen in that period —and quite probably within the last few hundred thousand years' (Cotton, *T*, 1962).

**Recent.** The 'Recent' period comprises the last 10,000 years or so, since the establishment in the temperate zones of the conditions with which we are familiar and since, in particular, the latest departure of the ice sheets from these regions. When the Pleistocene is taken to extend to the present day, the Recent is the latest, minute, fraction (of the order of one-hundreth) of the Pleistocene. 'Recent' was introduced by Lyell in 1833, but at first included the whole of the Pleistocene; when the latter was established it came to be restricted to its present meaning. Recent, Holocene, the Flandrian stage of the British Quaternary, and (roughly in most contexts) post-glacial are synonymous as regards the period of time referred to. In giving the range in time of a species, genus, &c., 'recent' means, in effect, 'living today'.

**Reconstruction.** Geology is very largely concerned with the reconstruction, in imagination, of the physical conditions and the life of past geological ages. Thus reconstructions (representations), by sketch, picture, diorama, or model, are used to illustrate the geological indications and inferences. The paintings and dioramas in the Geological Museum (London), the 'Reconstructions of ancient landscapes' by Vulliamy in Seward's *Plant Life through the Ages* (1931), and the restorations of extinct animals by Alice Woodward in Knipe's *Evolution in the Past* (1912) are beautiful and authoritative examples. Richey's 'Reconstruction of the Mull Volcano, in section' (*BRG*, *Tertiary Volcanic Districts*, 1961) and Smith and George's 'Map showing lines of ice-flow' (*BRG, N. Wales*, 1961) are two among innumerable examples of other kinds.

**Recrystallization.** The formation, in a rock, of new crystals from old. The new crystals may be of kinds of minerals new to the rock, or of kinds present before.

**Recumbent fold.** Typically, a fold (overfold) having an axial surface more or less nearly horizontal ('subhorizontal'; how far removed from the truly horizontal this may be allowed to be is a matter of varying usage). But a fold which pitches (plunges) more or less nearly down the dip of its axial suface, even when this surface is quite steeply inclined, may, in a certain condition, be considered as a pitching (plunging) recumbent fold. To include such a case a recumbent fold may be defined as one whose axial surface gives a subhorizontal trace in the plane of the 'vertical cross-section' (Challinor, *PGA*, 1945) of the fold. 'Recumbent folds in the schists of the Scottish Highlands' (Bailey, *QJ*, 1910). 'Steeply plunging recumbent folds' (Naha, *GM*, 1959).

**Red bed.** 'In simplest terms, a "red bed" is a sandstone, siltstone or mudstone made of detrital grains set in reddish-brown mud matrix or cemented by precipitated reddish-brown ferric oxide' (van Houten, 'Origin of red beds', in Nairn, ed., *Problems in Palaeoclimatology*, 1964). 'It is the diversity of origins of the colouring of red beds that explains in part the failure to produce a simple generally applicable answer to the question of their significance' (Friend, 'Some Devonian red beds in the Catskill Mountains, U.S.A.', *QJ*, 1966). The distinctive colour makes red beds conspicuous, so that, to the eye, they form a 'class' of rocks. The sub-New Red Sandstone surface in Britain (commonly a surface in Carboniferous rocks) is usually stained red, due largely to percolation from the red beds above, but perhaps also to desert weathering occurring before the deposition of those beds.

**Red clay.** This descriptive term is applicable wherever appropriate, for instance for a weathering product (see TERRA ROSSA and Prestwich, *QJ*, 1891). But it is, in particular, the name for an abyssal deposit. 'By far the most widely spread and most characteristic of all the the abyssal deposits is that to which the name of brown clay is applied. The older name, red clay, is now considered obsolete (Shepard, *Submarine Geology*, 1963) because the deposit typically has a chocolate brown colour. It occurs in all the deepest parts of the oceans . . . where the depth of water exceeds 2,600 fathoms' (Hatch and Rastall, *Sedimentary Rocks*, 1965). This abyssal deposit is of very variable composition and accumulates exceedingly slowly. 'Red' in descriptions of rocks and deposits is allowed a convenient latitude of interpretation, and the familiar name 'red clay' is likely to persist. See ABYSSAL DEPOSITS. (Murray and Renard, *Voyage of H. M. S. Challenger, 1872-1876*, 1891 and earlier.)

**Reef. 1.** A ridge of rock, shingle, sand, coral, or shelly material of any kind lying at or just above or just below the surface of the water. The term also applies to a mass of rock which evidently formed such a ridge at the time it was made. A 'reef', in this wide sense, may thus be formed of organic or inorganic material and the material of either kind may be accumulated either in a more or less loose state or in a firm state resistant to wave action. In a restricted sense, reef is confined to the latter state, a state resulting chiefly from the growth of an organically constructed framework, and of this the coral-reef is the best-known kind. (See Stubblefield, *QJ*, 1960.) The following table shows the relation between various senses of 'reef' and some associated terms:

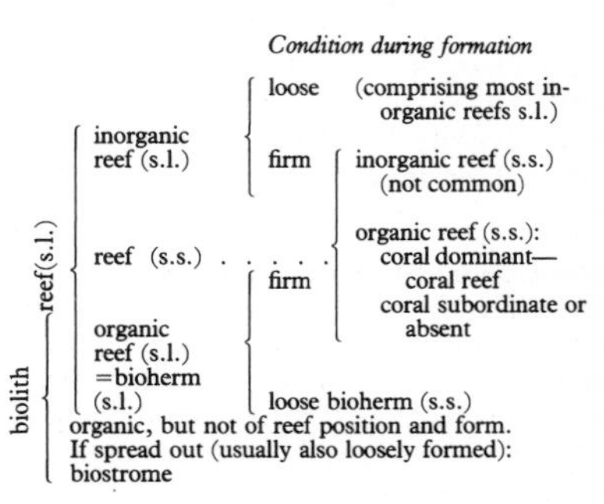


Among several special geological usages the chief is: **2.** A vein of gold-bearing quartz (originally, Australia).

**Reef facies.** See REEF LIMESTONE.

**Reef limestone.** Limestone that originally accumulated as a reef (s.l.). The Carboniferous reef limestones of England and Wales, for instance, show distinctive lithological and faunal facies, the reef facies or phases, in which calcite mudstones, crinoidal limestones, and brachiopod beds occur (all usually very light coloured). 'The nomenclature of Lower Carboniferous "reef" limestones in the north of England' (Bond, *GM*, 1950). 'Permian reef-limestones, which are regarded as marking the line of a local barrier reef near the edge of the Permian sea' (Smith, *BGS*, 1958).

**Reef-knoll.** A present-day knoll (small hill) composed of material that originally accumulated as a reef (s.l.; usually the material is reef limestone). The knoll may be the exhumation of an original feature of the reef (that is, the reef is a knoll-reef) or be a feature produced by erosion at a later time. Tiddeman (1889/1900) used the term 'reef-knolls' with their origin as original knolls in mind—he used 'reef-knoll' and 'knoll-reef' interchangeably—but this need not prevent 'reef-knoll' from being adopted as the term required for the wider, purely descriptive, meaning. The type areas are along the western Pennines. 'The origin and structure of the Lower Viséan reef-knolls of the Clitheroe district, Lancashire' (Parkinson, *QJ*, 1943). If the term were applied to a present-day reef it would denote one which, wholly or in part, was in the form of a knoll. See KNOLL-REEF. As the limestone composing a knoll might not necessarily be a 'reef', a more general term is 'limestone knoll'.

**Refold.** 'No structural correlations are yet possible between the Dalradian and the Moine series, but there is a general similarity in that in both areas large-scale recumbent folds are "refolded" about axes trending approximately at right angles to those of the first episode' (Giletti and others, *QJ*, 1961). 'Refolded folds in the Silurian rocks of Eyemouth, Berwickshire' (Dearman and others, *PYGS*, 1962). 'In the northern part of Pembrokeshire we may have the complicated forms resulting from a twist of rocks in a new direction over an older one' (De la Beche, *MGS*, vol. 1, 1846). A later folding is 'superimposed' on an earlier. Hills (*Structural Geology*, 1963) does not recommend the use of 'refold', 'as it implies some previous loss of fold structures'.

**Refractory.** Resistant; particularly to fusion by heat, and usually also to chemical reaction. *Refractory Materials* (several vols., *MGS, Mineral Resources*, 1920).

**Regeneration.** A profound change, a renewal, in the character of a large rock-complex. 'The effects of deep-seated regeneration as displayed by rocks of the Lewisian gneiss complex which underlies most of northwest Scotland. Now that the antiquity of much of the continental crust has been established it is clear that the metamorphic and tectonic processes of reworking at depth have played a major role in the geological evolution of the crust. These processes have provided the means by which old rock complexes have been reshaped and readjusted, without loss of coherence, in response to new crustal regimes; they have enabled the crust to adapt itself repeatedly to global changes in the distribution of stable and mobile regions and in the siting of zones of high temperatures' (Watson, *JGS*, 1973).

**Régime.** (Regime). A condition, system, or style having widespread influence or prevalence. Thus 'desert régime', 'periglacial régime', 'tectonic régime', 'volcanic régime'. 'The structural history of the earth's crust, as revealed by abundant evidence, shows that compression has been the dominant regime' (Lees, *QJ*, 1952). [French.]

**Regional geology.** All aspects of the geology of a region. Among the first systematic accounts of the geology of countries are those by Ami Boué of Scotland (1820) and by Conybeare and Phillips of England and Wales (1822). *British Regional Geology*, 1935/7 and later editions; nineteen handbooks, one to a region, written by officers of the Geological Survey.

**Regional metamorphism.** Metamorphism affecting collectively the rocks of a large region. From the kind of metamorphism we can see and by deduction from geological principles we also infer that regional metamorphism is a combination of dynamic and thermal ('dynamothermal') processes. The crystalline schists are the characteristic products of intense regional metamorphism. 'The distinction between contact and regional metamorphism was established in the middle of the nineteenth century through the work of three French petrologists, Durocher (1845-1846), Delesse (1857-1858) and Daubrée (1860)' (Loewinson-Lessing, trans., Tomkeieff, *Historical Survey of Petrology*, 1954). Over some extensive areas (e.g. the Grampian Highlands) zones of metamorphism may be recognized, possibly depending on the depth of origin of the metamorphism, or the action of separate metamorphic episodes combined with subsequent folding (see ZONE (2), end).

**Regmatic.** Pertaining to fracture. [G. *rhēgma*, a fracture.]

**Regolith.** 'Everywhere on the land areas, with the exception of the comparatively limited portions laid bare by ice or stream erosion, or on the steepest mountain slopes, the underlying rocks are covered by an incoherent mass of varying thickness. In places this covering is made up of material originating through rock-weathering or plant growth *in situ*. In other instances it is of fragmental and more or less decomposed matter drifted by wind, water, or ice from other sources. This entire mantle of unconsolidated rock material, whatever its nature or origin, it is proposed to call the "regolith" from the Greek words ῥῆγος [rēgos], meaning a blanket, and λίθος [lithos], a stone' (Merrill, *Rocks, Rock-weathering and Soils*, 1897).

**Regression.** In palaeogeography, this means a marine regression, that is, a retreat from the land by the sea (the opposite of transgression). Offlap in stratigraphy implies regression. 'Rates of marine transgressions and regressions' (thematic papers, *JGS*, 1979).

**Rejuvenation.** Used chiefly in geomorphology; the renewal of the power of downcutting by the subaerial agents (particularly running water) to that of a former, more 'youthful', stage. This would be brought about by relative uplift which would restore the general height above sea level, or by tilting so that the general slope in one direction was restored to something of its former steepness. It is doubtful whether a mere change in physical conditions, such as climate, or an accident, such as river-capture, would justify the term 'rejuvenation'. The most general and obvious effect of rejuvenation would be the cutting of steep-sided valleys within more open valleys.

**Relative age.** See CHRONOLOGY.

**Relict.** Chiefly used in metamorphic geology for primary mineralogical, textural, and structural characters of rocks that remain to some extent recognizable after these rocks have become metamorphosed. Such relict features are also called 'relicts', 'relics', 'residual features'. 'The interpretation of relics was compared by Sederholm to the reading of palimpsests, parchments used for the second time after original writing was nearly erased' (Barth, *Petrology*, 1952). 'Overprint' is also used as an analogy. 'The older dikes [Precambrian, Bighorn Mountains, Wyoming] are metadolerites which possess granoblastic margins that grade into interiors characterized by relict subophitic and locally, porphyritic texture' (Heimlich and others, *GM*, 1974).

**Relict hill.** Specially appropriate for a residual hill composed of a formation which has been denuded from the lower ground surrounding it. 'Relict hills of Torridon Sandstone resting on Lewisian Gneiss' (Phemister, *BRG, N. Highlands*, 1965).

**Relief.** The configuration of a part of the surface of the earth considered with regard to differences in height and steepness of slopes.

**Remanent magnetism.** See PALAEOMAGNETISM.

**Remanié.** Literally 'available for rehandling' or 'done over again'. Applied particularly to fossils which have been washed out of an older bed and re-buried in a new one ('derived fossils'); also applicable to pieces or masses of rock derived from an older formation and which have become incorporated in a new one, such as the pebbles in a conglomerate, boulders in boulder-clay, pieces of country rock caught up in a lava, large masses of country rock that have foundered during the formation of a volcanic caldera, &c. [French.]

**Remobilize.** A rock becomes 'remobilized' when, having at some previous time been mobile (such as when represented by a fluid magma), it becomes mobile again (such as by remelting). Hence a 'state of remobilization', a 'remobilization process'.

**Replacement.** This term finds many usages in geology. **1.** In connexion with the method of emplacement of igneous bodies, the country rocks may be replaced by being (*a*) incorporated in invading magma, (*b*) converted into magma on the spot (magmatized), or (*c*) granitized. **2.** The original material of the hard parts of an organism may be replaced by fresh mineral matter during fossilization. **3.** One mineral may be chemically replaced by another.

**Reptilia.** This class of the subphylum Vertebrata ranges from the Upper Carboniferous; and in the Permian and Mesozoic, the Age of Reptiles, it was represented by many groups, now extinct, which were then the dominant forms of life. Among these groups were the following:

| *Order* | *Suborder* |
|---|---|
| Saurischia | Theropoda<br>Sauropoda |
| Ornithischia | Ornithopoda<br>Stegosauria<br>Anklyosauria<br>Ceratopsia |
| Ichthyosauria | |
| Sauropterygia | |
| Pterosauria | |

The main orders of modern reptiles, all of which are fairly well represented in the fossil record, are: Chelonia (tortoises and turtles), Trias onwards; Squamata (lizards and snakes), Cretaceous onwards; and Crocodilia, Jurassic onwards. [L. *repo*, creep.]

**Resedimentation.** Generally, the deposition of sediment derived from a pre-existing sediment or sedimentary rock. However, the term has been used in a specialized sense: 'Mechanical deposition in cavities of post-depositional age. . . . The aspects of this process here treated refer to deposition of carbonate muds and silts, by bending of laminae or by internal mechanical erosion or solution (Sander, 1936, and *Depositional Fabrics*, 1951)' (Bathurst, *LMGJ*, 1958).

**Reservoir rock.** A rock-body capable of containing and storing a fluid—liquid (water, oil) or gas. 'Petrophysics of reservoir rocks' (*BAAPG*, 1950).

**Residual deposit.** The residue of loose (mineral) material left behind, in place, as a result of the weathering of a solid rock and the removal of some of the weathered, particularly the soluble, material. Sometimes the soil which may develop in the superficial layer of such a deposit is included, sometimes excluded. Also called 'saprolite'.

**Residual feature.** See RELICT.

**Residual hill.** See INSELBERG, MONADNOCK, RELICT HILL, CIRCUMDENUDATION.

**Residual nodule.** See NODULE. Residual nodules correspond among limestones, though usually on a smaller scale, to the core-stones produced by the weathering of granites.

**Resistant mass.** See MOBILE BELT.

**Resurrected.** See EXHUME.

**Resistate.** An accumulation, sediment, or solid rock composed of a residue of particles resistant to chemical weathering. Thus, for example, sand or gravel in contrast to clay.

**Retrograde metamorphism.** The mineralogical readjustments of a metamorphic rock, formed at a high degree of temperature or pressure (or both combined), to the conditions of a lower degree. This may occur on the reversal of the conditions which caused the original metamorphism, that is, on falling temperature and relief of stress; but such changes are usually very slight, if any occur at all, because the metamorphic minerals nearly always remain in a metastable state. Reversal may, however, be brought about by the metastability of minerals formed in thermal metamorphism being disturbed by a superimposed dynamic metamorphism, so that they revert to minerals stable at a lower metamorphic grade. Also called 'diaphthoresis'.

**Reverse fault.** A fault which dips to the upthrow side. If the geometrical relations of bedding and fault, resulting from the fault movement, are alone considered, a fault may be said to be a reverse fault if it dips to the side on which the vertical separation of beds is upwards. Also called 'reversed fault' and 'overfault'. See NORMAL FAULT.

**Revived.** Suggested use in tectonics: see POSTHUMOUS.

**Revolution. 1.** The working of a geological cycle or rhythm, or one such cycle or rhythm. **2.** A profound change, as in 'mountain-building revolution'. 'The earth has been the theatre of many great revolutions' and 'Those archives where nature has recorded the revolutions of the globe' (Playfair, *Illustrations*, 1802).

**Rhaetic series.** A series most clearly marked in NW. Europe as a set of passage beds between the continental Keuper (Triassic) beneath and the marine Lias (Jurassic) above. In Britain the series is now placed in the Triassic. The typical British Rhaetic beds are lagoonal deposits, shales, marls, and limestones with, as fossils, thin-shelled lamellibranchs (e.g. *Pteria contorta*) and bone-beds with fragments of fishes, amphibians, and reptiles. The Rhaetic outcrops across England (and into S. Wales) as a narrow band from the Yorkshire coast to the Devon coast, bordering the outcrop of the lowest Jurassic beds. [Rhaetian Alps.]

**Rheology.** The science 'of flow-phenomena in general' (Boswell, *SP*, 1951). Its importance in geology is obvious, in connexion with the action of running water, moving ice, magma, and rock-material in a plastic state. [G. *rheō*, flow.]

**Rheomorphism.** The liquefaction of a pre-existing rock, during the migmatization-granitization process, so that it may, at least potentially, flow and intrude into the surrounding rocks. (Backlund, 1937.) (Hills, *Structural Geology*, 1963, uses 'rheomorphic' for 'folds in highly incompetent beds that have undergone very considerable plastic deformation and translation . . . effects that reflect the great mobility of the beds'. Carey, *JGSA*, 1954, called these 'rheid' folds.)

**Rhizolite.** 'A rock showing structural, textural and fabric details determined largely by the activity, or former activity, of plant roots. The term rhizolite would appear to be equivalent to 'root rock', an informal term used by Perkins (1977)' (Klappa, *Sd*, 1980).

**Rhizoliths.** 'Organosedimentary structures resulting in the preservation of roots of higher plants, or remains thereof, in mineral matter' (Klappa, *Sd*, 1980).

**Rhyolite.** Acid lava; the fine grained, including vitreous, equivalent of a granite. Usually porphyritic, often showing flowbanding (hence the name). (Richthofen, 1860.) [G. *rhyas*, fluid.]

**Rhythm.** Although 'rhythm' and 'cycle' are used interchangeably by Barrell in his, important paper (*BGSA*, 1917), 'rhythm' suggest the recurrence, at more or less frequent and regular intervals, of one thing in particular, or an alternation, or a repetition of a sequence, on a rather small scale. 'The two lower groups [Maentwrog and Ffestiniog, of the Lingula flags series] consist of rhythmically alternating layers of dark shale and compact fine-grained light-grey micaceous sandstone. The individual beds are usually not more than a few inches thick' (Smith and George, *BRG, N. Wales*, 1948). 'One of the most striking features of the greater part of the Carboniferous succession in the Pennines is the frequent alternation of a sequence of certain different types of sediment accompanied by an equally marked variation in the kind of fossils they contain. There is in places a continual repetition of a more or less complete suite comprising successive beds of limestone, shale, sandstone, fireclay and coal; and such a repetition has been appropriately termed a "rhythmic succession" [Hudson, *PYGS*,

1924]. It bespeaks a series of rapidly alternating physiographic changes' (Wray, *BRG*, *Pennines*, 1948). 'Rhythm in sedimentation' (Allen, ed., *IGC*, 1948). On the other hand, 'rhythm' is sometimes used for large-scale repetitions: *Rhythm of the Ages* (Grabau, 1940); 'On rhythms in the history of the Earth' (Umbgrove, *GM*, 1939).

**Rhythmic sequence.** A cyclic sequence, but usually one on a comparatively small scale is implied; for instance, a repeated sequence of a few lithological or petrographical types, forming stratal bands or beds, or igneous layers, each sequence constituting a 'rhythmic unit' or 'rhythmite'. 'In the Helman Head Beds [Old Red Sandstone] we find a constant recurrence of certain phases marking very different conditions of sedimentation. The sequence consists of dark limestones, pale sandstones, and greenish-white mudstones, repeated again and again . . . the rhythmic sequence . . . an example of the sedimentary rhythm . . .' (Crampton, *MGS*, *Caithness*, 1914). 'The rhythmic succession of the Yoredale Series in Wensleydale' (Hudson, *PYGS*, 1924) 'Rhythmic layering in the ultrabasic rocks of Rhum' (Wager and Brown, *GM*, 1951). 'The rhythmites [in varved deposits] of proglacial lakes record a succession of annual cycles of sedimentation' (Shotton, *QJ*, 1966). See RHYTHM.

**Ria.** In the wide sense, any arm of the sea, stretching into a land area, resulting from the drowning of the lower part of a river-valley; fjords, however, being excluded. (It is presumed that the drowning is due to submergence.) In the narrow sense, restricted to coasts with structure transverse to the trend of the coast (Richthofen, 1886). (Cotton, *GJ*, 1956.) [Spanish.]

**Ribbon banding.** An appearance at outcrop suggesting bands of ribbon; seen chiefly in the bedding of sedimentary rocks. 'The lower part of the Maentwrog Beds includes many beds of tough fine-grained light grey micaceous sandstones ("ringers") that contrast in colour with the dark bluish grey shales. The individual beds are not more than a few inches thick and produce in any extended section of the group an appearance of ribbon banding' (Smith and George, *BRG, N. Wales*, 1961).

**Ribbon diagram.** A geological section through the country, not along a straight line, but along a curved, bent, or sinuous line. The section may be shown either as straightened out or in perspective; the latter is more particularly implied. 'Vertical and lateral variations in the Permian rocks of Yorkshire and the East Midlands, illustrated by a ribbon-diagram' (Edwards and Trotter, *BRG, Pennines*, 1975).

**Riebeckite.** See AMPHIBOLE. This deep-blue amphibole mineral is characteristic of strongly sodic acid igneous rocks, particularly well-known in the distinctive riebeckite-microgranite of Mynydd Mawr (N. Wales) and Ailsa Craig (Firth of Clyde). 'A. Sauer, 1888, after Dr. E. Riebeck who collected it' (Chester, *Names of Minerals*, 1896).

**Rift structure.** Two faults, or two sets of faults, roughly parallel, with the region between them structurally depressed. Also called 'fault trough'. Hence 'rift faulting', 'trough faulting'. A purely structural feature, irrespective of any resulting landform features.

**Rift valley.** A valley formed as the direct result of the depression due to rift faulting (trough faulting). *The Great Rift Valley* (Gregory, 1896). See FAULT-TROUGH VALLEY OF EROSION.

**Ring-complex.** An association of igneous intrusions which are ring-shaped in plan: ring-dykes and cone-sheets. 'Tertiary ring structures in Britain–Fig. 2. Ideal ring-complex composed of cone-sheets and ring-dykes, in plan and section' (Richey, *TGSG*, 1932). *Ring-complexes in the Younger Granite Province of Northern Nigeria* (Jacobson and others, *MGSL*, 1958). See quotation under IGNEOUS COMPLEX.

**Ring-dyke.** 'The conception of a ring-dike was first published by Clough, Maufe, and Bailey in their classic Survey memoir on the Glencoe district, Scotland [1916]. However, the name originated in the memoir on the island of Mull [1924]. The authors defined it thus: "A ring-dyke is a dyke of arcuate outcrop, where there is good reason to believe that the arcuate form is significant rather than accidental. Only in rare instances are ring-dykes so completely developed as to show an entire ring-outcrop"' (Daly, *Igneous Rocks*, 1933). Ring-dykes usually occur, several together and steeply dipping, in a ring-complex, associated with cone-sheets. 'The first carefully mapped and described pre-Quaternary ring-dyke seems to be one surrounding Slieve Gullion in Ireland; and for scenic distinction it remains unsurpassed. Not unnaturally in the 1870's, when it was discovered by the Geological Survey, it was regarded as a freak rather than a type-specimen' (Bailey, *TGSG*, 1958). See RING-COMPLEX and CAULDRON SUBSIDENCE.

**Ringers.** 'The Maentwrog or Rusty Flag series. . . . Interbedded with the slates [shales] are numerous coarser felspathic or gritty "ringers", which are massive in their bedding and well defined from their surrounding. These ringers range from something less than an inch up to a foot in thickness' (Fearnsides, *QJ*, 1910); and see quotation under RIBBON BANDING. Other formations have flagbeds which may 'ring' under the hammer and so merit the term (but the derivation is doubtful, see Arkell and Tomkeieff, *Rock Terms*, 1953).

**Ring-fault.** A steep ring-shaped fault, complete or incomplete. 'A great dislocation is associated with the Tertiary igneous complex of Rhum. It is now claimed as a Tertiary ring-fault with central uplift' (Bailey, *QJ*, 1944).

**Ripple-mark.** Ancient ripples, produced by the wind or by water currents and waves, as can be seen today on surfaces of sand or mud, preserved, as structural features of original deposition, on a rock surface. 'We have everywhere among the rocks many surfaces of the erected strata laid bare, in being separated. Here [Berwickshire coast] we found the most distinct marks of strata of sand modified by moving water. It is no other than that which we every day observe upon the sand of our own shore, when the sea has ebbed and left them, in a waved figure, which cannot be mistaken' (Hutton, *Theory of the Earth*, 1795). 'The ripple mark, so common on the surface of sandstones of all ages' (Lyell, *Elements*, 1838). 'Ripple-mark in the Rhinns of Galloway' (Kelling, *TEGS*, 1958). 'Ripple marks in Carboniferous Limestone' (Shiells, *GM*, 1963).

**Rise. 1.** The opposite of dip. As the rise (of strata, &c.) is immediately and precisely implied in a statement of dip, it is unnecessary to have both terms in use and 'rise' has dropped out. 'The three things most remarkable [about the Coal Measures of Lothian] are their "dipp", and "rise", and their "streek"' (Sinclair, *Short History of Coal*, 1672). **2.** 'An elongated, gently sloping elevation rising from the sea floor' (Monkhouse, *Dictionary of Geography*, 1965). This elevation may be due either to an upward movement of the (submarine) earth's crust, or to the accumulation of material on it's surface (e.g. the 'Continental rise').

**Rising.** 'Dry valleys are extremely common on this rock [Carboniferous Limestone]. Sinks are numbered in hundreds. Some of them swallow quite large streams. Drainage passes underground to ramifying systems of caves and galleries, which direct it to numerous springs. These, the complement of sinks, are called "risings"' (Dury, *Face of the Earth*, 1959).

**River capture.** The diversion of drainage from one river to another as the result of the competition between the rivers of a region, arising from differences in their natural advantages as regards power of erosion. Two rivers flowing in opposite directions from a general divide will seldom be quite equal in their power to grow headwards so that one will tend to tap, divert, and incorporate to itself the headwaters of the other as it pushes back its source beyond the general divide. Two rivers flowing parallel will similarly usually be unequal so that the valley of the stronger will grow laterally at the expense of the weaker and may even capture the weaker river itself. The most striking case of river capture is that in which the capturing river is visualized as flowing more or less at right angles to the direction of flow of the river to be captured and working backwards, and cutting downwards, so as to intercept that river at a lower level and divert it into its own course at an 'elbow of capture'.

**River terrace.** 'River terraces may be defined (following Gilbert, *Henry Mountains*, 1877) as terraces that border river valleys and mark former levels of flat valley floors' (Cotton, *Landscape*, 1948). The upper surface of the terrace represents the flood plain or valley floor when the river was at a higher level and the bottom of the slope the level at which the river cut into this plain. There may be several terraces above one another. Terraces are normal features of valley development (Challinor, *G*, 1932), but there may be special peculiarities of a particular set of terraces that (perhaps) indicate a fall in base level or a change in climate. 'The great mass of gravel which forms the successive terraces on each side of the river' (Playfair, *Illustrations*, 1802).

**Roche moutonnée.** 'De Saussure (*Voyages dans les Alpes*, 1779-96) gave the name "roche moutonnée" to the distinctive, rounded forms which abound in glaciated terrain (he himself failed to associate them with ice) and give the effect of a thick fleece or the wavy wigs styled "moutonnées" in his day. It became general after Agassiz adopted it in 1840. They are erosional forms in the glacial landscape and have occasionally been seen associated with modern ice. Their gracefully moulded contours, often oval in plan, are ever varying as to dimensions and endlessly repeated as to their features. They range from low shields to steep-sided eminences and considerable hills. Their iceward sides are well rounded and severely scoured. Striae curve round a boss if it is high and narrow, but arch over low and broad summits' (Charlesworth, *Quaternary Era*, 1957).

**Rock.** In the ordinary sense the word refers to those consolidated masses of mineral (including carbonized) material which enter into the composition of the earth's crust. But the geologist needs a term that will be more comprehensive and he has accordingly extended (with doubtful propriety) the meaning of 'rock' to include all those natural incoherent materials such as gravels, sands, muds, shellbanks, soil, peat, and boulder-clay. 'The term is used by geologists not only for the hard substances usually thus termed, but also for sands, clays, etc.' (De la Beche, *Manual*, 1831). 'Our older writers endeavoured to avoid offering such violence to our language, by speaking of the component materials of the earth as consisting of rocks and "soils". But there is often so insensible a passage from a soft and incoherent state to that of stone, that geologists of all countries have found it indispensable to have one technical term to include both, and in this sense we find *roche* applied in French, *rocca* in Italian, and *felsart* in German' (Lyell, *Elements*, 1838). Nevertheless, although according to the adopted technical definition it would be admissable to refer to 'alluvial rock', 'glacial rock', or 'blown rock', the geologist would usually prefer 'alluvial deposit', 'glacial drift', or 'blown sand' for Pleistocene and Recent loose accumulations. Magma is rock-material but not 'rock'. Engineers tend to make the old distinction in their terminology; see SOIL MECHANICS.

When we look at or think of a piece of ordinary country there are three chief elements present: (1) the works of man (buildings, roads, etc.), (2) vegetation (woods, pastures, gardens, etc.), and (3) the underlying substance of the land–that is, 'rocks' (with our geologically wide meaning). This last will be largely hidden by the other two, but everyone knows that it must be there, forming the body of the view. The most familiar exposures of rocks are the natural ones along the coast, in stream beds, and on mountain sides; and the artificial ones in quarries and cuttings.

'The fundamental procedure in geology is to observe and describe rock' ('Recommendations on stratigraphical usage', *PGS*, 1969). Rocks are studied in the field, as hand specimens, and as thin-section preparations under the microscope. A rock is certainly one of those objects which, when closely examined, are 'full of beauty, meaning, and curious details' (Hazlitt).

**Rock cleavage.** See CLEAVAGE (1).

**Rock flour.** See CLAY.

**Rock surface.** This usually means an exposed surface of rock, whether or not this surface is structurally significant. See SURFACE.

**Rock terms.** For history see Arkell and Tomkeieff, *English Rock Terms chiefly as used by Miners and Quarrymen*, 1953.

**Rock unit.** A rock-body defined exclusively on the basis of distinctive objective character. It may be stratigraphical, structural, or of any other nature; and it may be composed of any kind of rock material.

**Rock-basin.** 'Rock-basins, as we now know, may be hollowed out in many ways; by wind as shallow deflation basins, by inequal weathering under peat, by solution in limestones, by subsidence, by swirling waters, by anchor-ice, or by ice expanding in crevices and lifting out the loosened blocks. Nevertheless the vast majority were not made in any of these ways . . . glacial erosion of preglacially weathered material and of some solid rock is in the vast majority of cases the most satisfactory theory' (Charlesworth, *Quaternary Era*, 1957). 'Below Twll-du lie the sombre waters of Llyn Idwal, partly dammed up by a terminal moraine. The water is also, I incline to think, partly retained because it lies in a rock-basin, ground out by the old glacier' (Ramsay, *Old Glaciers of Switzerland and North Wales*, 1860).

**Rock-body.** A distinct body of rock of any kind. 'The granite of these veins and of the granite body itself' (Hall, *TRSE*, 1790).

**Rock-fall.** A fall of pieces of rock or individual boulders down a steep slope such as a mountain side or sea cliff.

**Rock-forming mineral.** About 2,000 different minerals are known but of these only a small proportion enter into the composition of rocks. Of these again, only a few, especially if we group together those with special affinities, make up the vast majority of rock-masses. The more important of these rock-forming minerals are: quartz, felspars, micas, amphiboles (hornblende, &c.), pyroxenes (augite, &c.), olivine, calcite and dolomite, iron compounds, and clay minerals. 'Rock-forming minerals' (Sullivan in Jukes, *Manual*, 1872). *Rock-forming Minerals* (Deer and others, 1962/3). 'The nature of rock-forming minerals' (Smith and Wells, *Minerals and the Microscope*, 1964).

The study of the rock-forming minerals (and to some extent the ores) is of course the concern of both the mineralogist and the geologist, but other minerals lie almost entirely in the domain of the mineralogist.

**Rock-mass.** A much-used term with an obvious meaning. It tends to imply something larger, less individualized, less distinctly marked off from its surroundings, and perhaps more shapeless, than 'rock-body'.

**Rock-salt.** The mineral, halite, common salt, sodium chloride, NaCl. It crystallizes in cubes and is colourless (when pure), soft, very soluble, and salty to the taste. It occurs chiefly as an evaporite. 'Rocks of salt of a vast thickness are frequently found in Cheshire' (Leigh, *Lancashire* [&c.], 1700). 'Two pieces of transparent rock-salt' (Woodward, *Catalogue*, 1729).

**Rock-type.** See TYPE (2).

**Rodding structure.** Mullion structure in general or (whether regarded as a variety of, or as something distinct from, that) a structure 'consisting of quartz-rods derived from silica secreted from the rocks during movement and metamorphism' (Wilson, *PGA*, 1953).

**Roof.** Particularly, in geology, for that of (1) a coal-seam and (2) an igneous intrusion. 'Typical coal seams are found embedded between rocks of a more or less definite character known as the "roof" and the "floor"' (Stopes, *Ancient Plants*, 1910). See quotation under UNDERCLAY. 'Floored eruptive bodies . . . each was hot and charged with gas, and was therefore capable of gas-fluxing here and there in its own roof' (Daly, *Architecture of the Earth*, 1938). 'The unroofing of the Dartmoor granite' (Groves, *QJ*, 1931).

**Roof pendant.** The inverted pyramids or downwardly directed wedges of country rock into the upper part of an igneous intrusion such as a batholite or stock. (Daly, *BGSA*, 1906.)

**Root.** Used in a tectonic sense. **1.** 'Certain isolated mountains, and even continuous ranges, which occur [in the Alps] are "mountains without roots"; they have no genetic connection with the pavement rocks on which they rest, for the latter are much younger than the masses above them' (J. Geikie, *Mountains*, 1913). 'The root of a recumbent anticline, or of a nappe, is the core of the anticline in the region where it is more or less vertical and gives the impression of rooting to the depths' (Collet, *Alps*, 1928). 'The roots of mountains' (Hills, *GM*, 1944). **2.** Orogenic root. See TECTOGENE. 'The deep structure of orogenic belts–the root problem' (Oxburgh in Kent and others, eds., *Time and Place in Orogeny*, 1969).

**Rose diagram.** A diagram indicating graphically values or quantities in the several directions of bearing. Used (but not so called) by Phillips, *Yorkshire*, 1836. 'Rose diagrams showing distribution of flute cast directions' (Cummins, *GM*, 1957). Also called 'rosette diagram' (Firman, *QJ*, 1960), 'direction-rose', or, simply, 'rose'.

**Rossi-Forel scale.** See SEISMIC INTENSITY.

**Rottenstone.** A much weathered but still coherent rock resulting from the removal of one or more of its constituents, and most commonly the result of the weathering of an impure, particularly a siliceous, limestone, the calcium carbonate having been dissolved. The siliceous powder may be used for polishing metal. Plot (*Oxfordshire*, 1677) refers to 'a sort of stuff they call "rotten stone", used for brightening copper'. A highly shelly sandstone usually becomes a rottenstone where weathered. 'The working of Rottenstone [a special variety] in Derbyshire' (Ford, *BPDMHS*, 1967).

**Roundness.** Applied to rock particles; the degree to which their outlines are devoid of angularity. Roundness and sphericity are independent properties. For a chart and history see Hatch and Rastall, *Petrology of the Sedimentary Rocks*, 1978. See PARTICLE (reference).

**Rudaceous.** Rubbly. For sediments and sedimentary rocks. [L. *rudus*, a rubbly mass of stones.]

**Rudite.** A rudaceous rock, a conglomerate or breccia (etc.), a psephite.

**Rugosa.** An extinct Palaeozoic (Ordovician-Permian) order of corals (class Anthozoa, phylum Coelenterata). The radial plates (septa) are typically of two orders, major and minor, alternating, and there is some degree of bilateral symmetry, particularly in the early stages of growth. Transverse plates are of two distinct kinds, tabulae and dissepiments. Very common fossils in the Silurian, Devonian, and Carboniferous limestones. (Milne-Edwards and Haime, 1850.) [L. *rugosus*, wrinkled.]

**Run.** A branch-like body of ore or igneous rock. 'The magma, after ascending from lower levels, may have spread laterally along numerous separate channels or "runs", as they may be called, approximately or strictly parallel to the bedding-planes, assumed to have been then horizontal, and thus given rise to finger-like, subparallel bodies on several different stratigraphical levels' (Jones and Pugh, *QJ*, 1948).

**Rupture.** 'Deformation characterized by loss of cohesion. Frequently flow grades into rupture, with a progressive loss of cohesion, until complete separation occurs' (Barth, *Petrology*, 1952).

**Rutile.** The commonest and most stable of the several crystalline forms of titanium oxide ($TiO_2$), occurring as an accessory mineral in the more acid igneous rocks, and in the crystalline schists; also as an accessory 'heavy mineral' in sedimentary rocks. It forms slender prisms and needles in the tetragonal system and has the highest refractive index of any of the normal rock-forming minerals. (Werner, 1803.) [L. *rutilus*, yellowish-red, which is its colour.]

# S

**'S' numbers.** See 'F' NUMBERS.

**Saccharoidal.** Having the appearance of sugar; for certain white or nearly white rocks with an equigranular texture, such as marble and aplite. [G. *sakcharon*, sugar.]

**Saddle.** In structural geology, a structural depression along the course of an anticline. 'Fig. Saddle in Lower Silurian rocks between Clarach Bay and Aberystwyth, formed by the junction of anticlinal and synclinal curves' (Reade, *Origin of Mountain Ranges*, 1886). See quotation under CULMINATION.

**St. David's series.** See CAMBRIAN SYSTEM. [St. David's, Preseli district of Dyfed.]

**St. George's Land.** The palaeogeographical land region that lay in Lower Carboniferous (particularly Tournaisian) times over most of the area that is now Wales and the southern part of the Irish Sea. 'At the commencement of Carboniferous times we find Leinster and Wales united across the Channel as an island, which we may name St. George's Island' (Sollas, *PGA*, 1894). The name 'St. George's Land' was given by Jukes-Browne (*Building of the British Isles*, 1911). 'St. George's Land' (George in Owen, ed., *The Upper Palaeozoic and Post-Palaeozoic Rocks of Wales*, 1974). See quotation under OVERLAP.

**Salic.** One of the two main groups into which normative minerals are divided (the other being the femic group). It includes chiefly quartz (silica), felspar, and felspathoids, the term being a portmanteau term to express this. It is used only when these minerals are being considered as components of the norm of the igneous rock, not for the minerals themselves (the 'felsic' minerals). The adjective is also used for the corresponding rock (see quotation under HYPERSOLVUS), and even for the corresponding magma.

**Saliferous.** Salt-bearing; for strata containing salt-deposits. 'Saliferous marls and sandstones' (Murchison, *Silurian System*, 1839; of the Keuper series).

**Salopian.** A stratigraphical series-name proposed by Lapworth (*AMNH*, 1879) for Murchison's Wenlock and Lower Ludlow (including Aymestry Limestone) series combined, based on the graptolites which constitute a succession of distinctive faunas. The graptolites (Graptoloidea) do not occur in the British rocks above the Lower Ludlow. The term is not now often used. 'Middle Silurian' has been applied to this range of strata: Boswell, *Middle Silurian rocks of North Wales*, 1949. [Salop = Shropshire.]

**Salt dome.** The whole structure resulting from a mass of salt (usually rocksalt), a 'salt plug' mass, forcing its way upwards (by some process of 'salt tectonics') through a series of sedimentary strata, partly breaking through them and partly pushing and bending them upwards. 'Diagrammatic cross-section through a salt dome' (Illing, *PGA*, 1942). See DIAPIR.

**Saltation.** Proceeding by jumps. Applied to such diverse happenings and conceptions as a movement of a rock particle along a river bed and the sudden appearance of a new feature in the course of evolution of an organic stock. [L. *salto*, leap.]

**Saltfield.** An area beneath which salt occurs, with the implication (usually) that the salt is workable. 'A sketch of the natural history of the Cheshire rock-salt district' (Holland, *TGS*, 1811). 'The saltfield of north Cheshire' (Wray and others, *BRG, Pennines*, 1975).

**Sand.** Material having a grain-size between that of small pebbles (a gravel) and that of clay (but see SILT). All ordinary sands are composed mainly of quartz grains. Looseness is implied, but 'sands' is incorporated in the names of such well-known stratigraphical formations as the Hastings Sands, Bagshot Sands. 'A common Teutonic word unchanged from Old English and recorded back to A.D. 825' (Arkell and Tomkeieff, *Rock Terms*, 1953).

**Sand volcano.** A body of sand produced by the extrusion of sand-laden water. From the occurrence of certain sandstone bodies among strata, it seems that such extrusions do occur under water. The word 'volcano' is hardly an appropriate one in this connexion. 'Sand volcanoes on slumps in the Carboniferous of County Clare, Ireland' (Gill and Kuenen, *QJ*, 1957).

**Sandstone.** A consolidated sand. In use in the 17th century. Sands grade (through silts) into muds, sandstones (through siltstones) into mudstones. 'While all or any considerable portion of the rock remains in the form of distinct grains, we might call it an argillaceous sandstone; the passage from that to a sandy clay, and then to a pure clay or shale, being often an insensible one' (Jukes, *Manual*, 1862). 'Sandstone classifications' (Klein, *BGSA*, 1963). 'A classification of common sandstones' (McBride, *JSP*, 1963). *Sand and Sandstone* (Pettijohn and others, 1973).

**Sandstone dyke.** See DYKE, NEPTUNIAN DYKE, and SEDIMENTARY INSERTION. 'Sandstone dikes' (Diller, *BGSA*, 1890). 'On some quarzite-dykes in Mountain Limestone near Snelston, Derbyshire' (Arnold-Bemrose, *QJ*, 1904).

**Sandstone injection.** See INJECTION and SEDIMENTARY INSERTION. 'On some remarkable dikes of calcareous grit, at Ethie in Ross-shire' (Strickland, *TGS*, 1840; some of the 'dikes' are sill-like injections). 'The sandstone injections of Eathie Haven' (Waterston, *GM*, 1950).

**Sannoisian.** See PALAEOGENE. [Sannois, France.]

**Saprolite.** See RESIDUAL DEPOSIT. [G. *sapros*, decayed.]

**Sapropel.** 'For the unconsolidated product of the decomposition of aquatic plants and associated organisms under neutral or mildly alkaline conditions, the term "sapropel", though not entirely satisfactory, appears to be the best to use. This is the organic sludge which accumulates, where conditions are favourable, on the bottom of lakes and of the sea. The algae, which are the main source of this material in many aquatic environments, differ significantly in composition from land plants in that they are rich in proteins and poor in carbohydrates, whereas the reverse is the case with land plants. Thus sapropel normally contains more nitrogen than humus' (Dunham, *AS*, 1961). Sapropel appears to be largely the modern equivalent of kerogen in the rocks. [G. *sapros*, putrid, *pelos*, mud.]

**Sarsen.** Sarsens or sarsen stones, also called 'greywethers' (from their likeness to sheep), are boulders of hard silicified sandstone scattered over the Chalk of SE. England, particularly Wiltshire. They are probably the residue of the once overlying Eocene Reading beds. The name may be derived either from the nomad Saracens, meaning 'outlandish', or from Old English *sarstan*, 'a grievous stone', from the trouble caused to agriculture ('Saracens' stones', Symonds, 1644; 'sarsdens or sarsdon stones', Aubrey, 1656/91; 'sarsens', Stukeley, 1743; these references given in *OED* and by Arkell and Tomkeieff). 'In the south of England the sandstone boulders known as sarsens, greywethers, druid stones, and bridestones, are familiar objects of the countryside. Solitary for the most part, they not infrequently occur in small groups of natural or artificial origin, and in a few restricted areas they are congregated in greater numbers: nowhere, however, are they so plentiful as in the Chalk country near Marlborough. Here, after centuries of exploitation, these stones still lie thick on some of the downland ridges, and thicker still in the adjacent winding bottoms, where their disposition suggests the idea of rivers of stones. There is something in their grey recumbent forms, half hidden in long grass and scrub, that awakens a lively interest in the beholder, and even when their nature is known they continue to stir the imagination, their bulk, their legendary associations, and a touch of melancholy in their wild surroundings investing them with a kind of glamour' (White, *MGS, Marlborough*, 1925).

**Saturation.** See OVERSATURATED ROCK.

**Saurischia.** One of the two orders of dinosaurs (Mesozoic land reptiles). Among the characteristic features is the three-rayed structure of the pelvis. It comprises the suborders Theropoda and Sauropoda. [G. *saura*, a lizard, *ischion*, a hip joint.]

**Sauropoda.** A suborder of the order Saurischia. The animals are very large herbivorous quadrupeds. Jurassic and Cretaceous.

**Sauropterygia.** One of the two orders of Mesozoic swimming reptiles. There is often a long neck, the animal evidently being slow-swimming, rowing through the water with the large paddle-like limbs. [G. *saura*, a lizard, *pterux*, a flapper.]

**Saussuritization.** A metamorphic alteration of an intermediate plagioclase felspar, in an igneous rock such as a dolerite, whereby the sodic and calcic components become divorced. The former separates as albite, while the latter gives rise to new aluminosilicates of lime. The whole compact product was originally throught to be a specific mineral, saussurite. The alteration may be due to autometasomatism, contact metamorphism, or low-grade dynamic metamorphism. 'N. T. Saussure, 1806, after Prof H. B. de Saussure who had described the mineral earlier' (Chester, *Names of Minerals*, 1896). The process was described by Williams (*BUSGS*, 1890).

**Scalar.** This usually means having degree, magnitude, 'scale', as distinct from having directional (vectorial) properties. Thus size, weight, &c., are scalar properties; and properties of shape are sometimes included.

**Scaphopoda.** A small class of the phylum Mollusca, having a slightly curved tubular shell, ranging from the Ordovician to the present day. [G. *skaphē*, a boat.]

**Scapolite.** An isomorphous series of minerals parallel to the plagioclase felspars from which they are derived by pneumatolysis. In addition to the plagioclase atomic constituents they contain, particularly, chlorine (Cl), or the carbonate or sulphate radicle ($CO_3$ or $SO_4$). They crystallize in the tetragonal system. Hence 'scapolitization'. (d'Andrada, 1800.) [G. *skapos*, a shaft.]

**Scar.** 'A bluff precipice of rock; hence the term "Scar Limestone" applied to the mountain limestone [Carboniferous] as it occurs in the hills of Yorkshire and Westmorland' (Page, *Geological Terms*, 1859).

**Scarp.** See ESCARPMENT.

**Scarplet.** A miniature scarp. 'The coarser sediments of the Denbighshire Grits tend to be impersistent, the harder beds eroding to surface scarplets running for a few hundred yards and then being replaced by scarplets at higher or lower horizons' (Smith and George, *BRG, N. Wales*, 1961). The term is also appropriate for features on a wave-cut platform formed by the outcropping of hard beds.

**Scenery.** The study of scenery, the form of the physical features of a region, is the most obvious aspect of the study of geology. 'To trace back, if that might be, the origin of the present surface of the country, and by working out the structure of the rocks, to contrast the aspect of the land to-day, with its condition in former geological periods, has been to the author of these pages the delightful occupation of years' (Geikie, *Scenery of Scotland*, 1865). 'No subject is more fascinating to students of Geology than Scenery' (Woodward, *England and Wales*, 1887). 'In England and Wales we are singularly placed to appreciate the relationship of scenery and structure, for few other parts of the earth's surface show in a similar small area so great a diversity of rock types and of landscape features' (Trueman, *Scenery of England and Wales*, 1938). 'The scenery we so much enjoy is no chance happening but an expression of the present stage in an interplay of processes that have built mountains only to reduce them, in course of time, to level plains, and have created seas, only to fill them with the debris of worn-down continents, in preparation for a new cycle of uplift and scenery-forming denudation' (North, *MJ*, 1953). 'The origin of the scenery of the British Islands' (Geikie *Landscape in History*, [&c.], 1905). *Scientific Study of Scenery* (Marr, 1900/26). *The Scenery of England* (Avebury, 1902). *Coast Scenery of N. Devon* (Arber, 1911). *South African Scenery* (King, 1942/51). *Britain's Structure and Scenery* (Stamp, 1946). Welsh *Scenery* (Evans, 1972). See STRUCTURE AND SURFACE.

**Schillerization.** 'Alike in the felspars, the pyroxenes, the olivines, and the biotites of plutonic rocks, there is evidence of progressive change taking place at gradually increasing depths. This change consists in the development along certain planes within the crystals of tabular, bacillar, or stellar enclosures, which, reflecting the light falling upon them at certain angles, give rise to the peculiar phenomenon expressed by the term "schiller" [f. German, play of colours]. It will be convenient to have a general name for this kind of change, and I propose to employ the term "schillerization" to express it' (Judd, *QJ*, 1885). The term is now used for the effect without necessarily implying any particular mode of origin of the enclosures. See quotation under LARVIKITE.

**Schist.** A rock largely or completely recrystallized (at a moderately high degree of regional metamorphism), structurally characterized by fine-scale foliation resulting from the parallel disposition of lamellar minerals, most commonly the micas. Hence 'schistosity'. Schistosity is, on one (perhaps the most generally held) view, closely allied to slaty cleavage, and its direction is essentially independent of original bedding in the rocks affected (tending to be parallel to axial planes), though where recumbent folding and schistosity were both produced from the same cause, the directions may nearly coincide in the limbs. On another view, schistosity tends to follow original bedding. Particular rock-types are named mica-schist, hornblende-schist, &c. In the older works, 'schist', 'schistus', and 'slate' were all rather vague terms and meant much the same thing (and included greywacke). See quotation (Hutton) under UNCONFORMITY. 'Fossil fish of the Caithness Schists [Caithness Flags]' (Sedgwick and Murchison, *TGS*, 1829). Later, 'schist' became more clearly delimited: 'This term should be restricted to such rocks as mica-schist, and the like, which have a foliated structure and split up in thin irregular plates, not by regular cleavage as in slate-rocks' (Page, *Text-book*, 1859). [G. *schizō*, split.]

**Schorl-rock.** A rock consisting of tourmaline and quartz, resulting from the complete tourmalinization of granite. The occurrences in Devon and Cornwall are the best known and are typical. [Cornish *schorl*, tourmaline.]

**Schuppen.** 'In regions of compression it is common to find some stratigraphical group or groups broken up into a multitude of minor thrust-slices, which are packed in characteristic imbricate fashion. The slices, so arranged, are called *schuppen*, a German word meaning scales. Individual schuppen incline steeply towards the direction from which the overthrust has occurred. Their relative movement is trivial, as concerns adjacent members, but may in the aggregate lead to impressive telescoping of the affected zone. Schuppen structure is generally attributable to the passage overhead of some major thrust-mass. It is abundantly represented in the North-West Highlands of Scotland' (Bailey, *QJ*, 1938).

**Scleractinia.** An order of corals (class Anthozoa, phylum Coelenterata) common today and as fossils, in certain formations, from early Mesozoic times onwards. The radial plates are inserted, as the coral grows, in cycles of multiples of six and there is radial symmetry throughout. Transverse partitions are usually not clearly differentiated into tabulae and dissepiments. Also (and, until recently, usually) called Hexacoralla. [G. *sklēros*, hard, *aktis*, a ray.]

**Scolecodont.** A jaw, with denticles, of an annelid worm. Scolecodonts are found fossil from the Ordovician onwards. They are composed of silica and chitin, the latter becoming carbonized to a jet black appearance in the process of fossilization. 'On Annelid jaws' (Hinde, *QJ*, 1879/80). (Croneis and Scott, *BGSA*, 1933.) [G. *skōlēx*, a worm.]

**Scoria.** 'Rough clinker-like masses formed by the cooling of the surface of molten lava on exposure to the air, and distended by the expansion of imprisoned gases' (*OED*). Also the similar material erupted and falling as pyroclastic material. A collective term; but the plural ('scoriae or, occasionally, scorias', *OED*) refers to fragments. Hence 'scoriaceous'. 'When the upper surface of a lava is rugged and full of steam-vesicles of all sizes up to large cavernous spaces, it is said to be "scoriaceous", and fragments of such a rock ejected from a volcanic vent are spoken of as "scoriae"' (Geikie, *Ancient Volcanoes*, 1897).

**Scour and fill.** See WASH-OUT.

**Scourian.** The Lewisian metamorphic complex (Precambrian, NW. Scotland) shows the effects of metamorphism at two distinct periods. Over parts of the outcrop the effects of the first metamorphism remain undisturbed, over other parts these effects have been obliterated by the second metamorphism. The effects of these two periods of metamorphism are discriminated by the state, in the several areas, of a widely distributed group of dykes (presumed to have been intruded all at one period). Where these dykes are unaltered, the metamorphism of the surrounding complex is evidently pre-dyke in age; where they themselves are metamorphosed, the metamorphism is evidently post-dyke in age. Having been thus discriminated as to age, each metamorphism is found to have its own peculiar mineralogical and structural features. This sequence of events was first established by the Geological Survey (*MGS*, *North-west Highlands*, 1907) and has been investigated in detail by Sutton and Watson (*QJ*, 1950) who introduced the names Scourian and Laxfordian for the first and second metamorphisms respectively. The names refer to particular metamorphisms (and associated orogenies), to stages of structural growth, not to rock formations (like 'Lewisian') or chronological divisions. Recent radiometric age-determinations have confirmed the reality of these two distinct metamorphisms and revealed the immense interval of time separating them, the Scourian occurring about 2,600 million years ago and the Laxfordian about 1,600 million years ago (Giletti and others, *QJ*, 1961; but see Bikerman and others, *JGS*, 1975). The dykes have also been dated as having been intruded about 2,000 million years ago. See also Dearnley and Dunning, *QJ*, 1967, with discussion on nomenclature, and Watson, *PGA*, 1977. [Scourie, Laxford; two localities on the W. coast of Sutherland.]

**Scree.** An accumulation, at the foot of a cliff, of weathered rock fragments, of all sizes, and mostly angular, fallen from above, piled up often at a very even angle of rest. In form, an individual pile of scree material tends to be part of a cone tapering up into a gully; at the foot of a line of cliff there is often a line of these coalescing apron-like screes. 'Talus' is another term. 'As the road proceeds along the margin of the lake [Wastwater], the screes on the opposite side form a striking object' (Otley, *English Lakes*, 1823). Screes of past geological ages are known, the most familiar British examples being those of the New Red Sandstone. [Old Norse *skritha*, a landslip.]

**Sea-floor spreading.** 'The hypothesis which is now known as "sea-floor spreading" was first formulated by the late Professor Harry Hess of Princeton University in 1960. Hess postulated that the mid-ocean ridges are situated over the rising limbs of convection currents in the earth's mantle, and that the thin oceanic crust is nothing more than a surface expression of the mantle, derived from it by simple chemical modification and continuously created by a process of lateral accretion or "spreading" away from the ridge crests.

In one giant leap forward, this single, revolutionary and elegant hypothesis provided us with a much greater understanding of the earth. It simultaneously revived the older, but largely rejected, concept of continental drift and paved the way for the more modern and highly successful concept of plate tectonics' (Vine, 'Sea-floor spreading', in Gass and others, eds., *Understanding the Earth*, 1972). See PLATE TECTONICS.

**Seam.** 'Strictly speaking, the line of separation between two strata, but loosely applied to subordinate strata occurring in any series, as "seams of coal" in the coal-measures' (Page, *Textbook*, 1856). 'All metals, as stone and tilles (which are seems of black stone, and participat much of the nature of coal) ly one above another, and keep a regular course' (Sinclair, *Short History of Coal*, 1672). 'The strata of coal are usually called "seams"' (Kirwan, *Geological Essays*, 1799). 'An upper portion [of the Hartfell Shales] is mainly composed of mudstones with seams of black shale' (Pringle, *BRG, S. Scotland*, 1948). 'Seam of gypsum, Keuper Marl, Aust Cliff' (Bates and Kirkaldy, *Field Geology*, 1976).

**Seamount.** 'The ocean floors are studded with steep mounds. Some may be so high that they emerge above sea level as islands. Those that do not emerge are called "seamounts". Some seamounts known as"guyots" are truncated and have flat tops. In almost every case, these islands and seamounts are built of volcanic rocks and debris' (Harris in Gass and others, eds., *Understanding the Earth*, 1972). 'Seamounts', including guyots, in the Pacific were described by Shepard (*Submarine Geology*, 1948).

**Seaquake.** An earthquake originating in a part of the crust beneath the sea.

**Seat-earth.** A bed of rock representing an old soil that supported vegetation. Examples are the underclays in the Coal Measures and the dirt-beds of the Purbeck beds. Includes soil-beds. 'Each bed of coal is supported by a layer known as underclay or seat-earth' (Huxley, *Physiography*, 1877).

**Secondary. 1.** In stratigraphy, used at first (particularly at the beginning of the 19th century, but also earlier) for all the strata above the Transition Rocks, that is from the 'Old Red Sandstone' to the 'Upper Freshwater Beds', Devonian to Oligocene in modern terminology; the Pliocene, if recognized then, would have been included. Later (c. 1820 in Britain) the Tertiary rocks (strata above the Chalk) were separated and, later still, the Upper Palaeozoic; for when the Mesozoic became established as a group (1841), 'Secondary' became equivalent to 'Mesozoic' and was gradually superseded by the latter name. **2.** For the ordinary clastic rocks in which the particles have been derived from pre-existing rocks. Some authorities (e.g. Tyrrell, *Petrology*, 1929) use 'secondary' for 'sedimentary' (in the wide sense). **3.** For structures and textures imposed on rocks after they have become consolidated. **4.** For minerals which have replaced others of earlier formation. **5.** For certain features and values in a fold or fault structure; see PRIMARY (5).

**Secretion.** In addition to its main use, which is in biology (for an example see ACCRETION), 'a secretion' in general geology is a mineral that has been deposited from solution in a cavity; e.g. a mineral vein, an amygdale, a geode. 'Calcareous spar and siliceous crystals are often found in stratified rocks, forming veins of secretion, or lining close cavities' (Playfair, *Illustrations*, 1802). [L. *secretio*, a separation.]

**Section.** In geology nearly always used in the sense of 'something cut across', not in the sense of 'a portion cut from' (except in schemes of classification). **1.** An exposure (exposed section) of rock in one place, such as a sea-cliff, stream-bank, or quarry-face. 'As there is nothing so effectual as local examples to enable us to form proper ideas in enquiries, let us go to the rocky shores of the ocean, or to some river which has cut deep into the rock, and chuse out a fair and lofty section of the strata' (Williams, *Mineral Kingdom*, 1789). **2.** A series of exposures, or one continuous exposure, of a rock succession, such as along a strip of coast or along a stream-bed.

'Without sections of mountains their internal structure cannot be perceived' (Hutton, *Theory of the Earth*, 1795). **3.** The reconstruction from the information on a geological map, or from a special traverse across the country, or from other information, of the vertical section of the structure along a particular line. The Geological Survey used to call this a 'horizontal section', to distinguish it from our no. 4; and geologists still sometimes use this expression. Strachey (*PT*, 1719/25) gives what is probably the first geological section of any kind across any part of Britain, and calls it 'A section of a coal country in Somersetshire'. Farey, during 1806 to 1808, drew (in *MS*) what are probably the first detailed and extended sections of British regions, across Derbyshire to Lincolnshire and across the Weald (redrawn and discussed by Ford, *MG*, 1967). White Watson's sections in 1811 (*Delineation of the Strata of Derbyshire*) and 1813 (*A Section of the Strata . . . in Derbyshire*) were perhaps the first reasonably accurate geological sections to be published in Britain (see Challinor, *TNSFC*, 1947, Ford, *PGA*, 1960, and Ford, *BPDMHS*, 1962). The drafting of geological sections is the subject of a work by De la Beche, *Sections and Views illustrative of Geological Phenomena* (1830). 'Geological sections should be drawn to the maximum depth permitted by the surface control and where there are interpretational difficulties several alternatives should be submitted; and it is important that sections should be drawn to true scale in order that the essential structural character should not be distorted. Sections should also be drawn to show early stages in the structural growth to ensure that the end picture is acceptable' (Lees, *QJ*, 1953). **4.** The reconstruction from the information on a geological map, or from information derived directly from borings, of the vertical section of the structure as it would appear if pierced downwards below a particular point. (This becomes a 'geological column' at the point, if the dips are adjusted to the horizontal to show the succession and thicknesses the more clearly, though 'section' is still often used for such a representation.) **5.** A geological map is, in fact, a geological section of a solid structure cut on an uneven, but more or less horizontal, surface and projected on to the horizontal plane ('the map represents a random section', Jones and Pugh, *QJ*, 1948); but, in practice, it is usually implied that a map is being expressly excluded when a 'section' of the structure of a region is specified. **6.** Of a specimen: (*a*) one cut across a piece of rock or fossil so as to make a smooth or polished surface, or (*b*) a thin slice of a rock, mineral, or fossil, being usually one prepared for examination under the microscope.

**Secular.** Lasting or going on persistently for an indefinitely long time; progressive, in contrast to cyclic or periodic. 'On the secular cooling of the earth' (Thomson, *TRSE*, 1864). [L. *seculum*, an age.]

**Secule.** The time equivalent to a zone (Jukes-Browne, *GM*, 1903). This conception had already been named 'moment' (*IGC*, 1881) but 'it may with justice be objected that the word "moment" was already long ago preoccupied in ordinary parlance and to misapply it to a period of many thousands of years is absurd' (Arkell, *Jurassic System*, 1933).

**Sediment.** Material deposited on the floor of the sea, a lake, or other mass of water by sinking and settling under the influence of gravity. Although the Latin derivation would seem to allow the inclusion of material, such as volcanic ash, settling from the atmosphere onto a dry land-surface, and even aeolian accumulations and boulder-clay, the term does not usually embrace these deposits. But see SEDIMENTARY ROCK. Hence 'sedimentation', the deposition of sediment. [L. *sedeo*, sink, settle.]

**Sedimentary dyke.** See DYKE and SEDIMENTARY INSERTION. 'Sedimentary dykes in the Dalradian of Scotland' (Smith and Rast, *GM*, 1958).

**Sedimentary environment.** The geographical, physical, chemical, and biological conditions under which a sediment accumulates. *Sedimentary Environments and Facies* (Reading, ed., 1978).

**Sedimentary injection.** See INJECTION, SEDIMENTARY INSERTION, and AUTO-INTRUSION.

**Sedimentary insertion.** There appears to be no term for the general phenomenon of a body of sedimentary rock-material which has become emplaced, inserted in one way or another, among deposits or rocks already formed. 'Sedimentary insertion' is here suggested, it being understood that 'sedimentary' refers to the character of the rock-material, not necessarily to the process involved. The process of insertion would include infilling of a fissure or hollow by sediment under water (Neptunian insertion), probably penecontemporaneously with, or not long after, the deposition of the sediment affected, and injection from below or above (injection insertion), a dyke being the commonest form in either case; and also an insertion due to subsequent, perhaps long subsequent, localized solution subsidence, of which the solution pipe is the commonest form. Sand, becoming sandstone, is the commonest material. The terminology may be set out thus:

| | sedimentary insertion |
|---|---|
| infilling (Neptunian) | Neptunian dyke |
| injection (= intrusion) | sedimentary injection (dyke, sill, &c.) |
| subsequent solution | solution pipe (or other form) |

Each of these kinds may be further specified as to its lithology, e.g. Neptunian limestone-dyke, sedimentary sandstone-injection (or, simply, sandstone injection), sandstone solution-pipe. (Discussion in Geikie, *Textbook*, 1903.) The following are additional commonly used terms referring to Neptunian or sedimentary injection dykes (the terms do not specify which), but excluding subsequent-solution structures: 'sedimentary dyke', 'sandstone dyke', 'clastic dyke', composed respectively of any sedimentary material, sandstone, any clastic material. The gash breccia may perhaps be considered a form of subsequent-solution structure.

**Sedimentary intrusion.** See INTRUSION. In a general sense, has the same meaning as 'sedimentary injection', but it is more particularly applicable to a sedimentary diapiric plug, where the action and resulting rock-body are on a relatively large scale.

**Sedimentary petrography.** See SEDIMENTOLOGY and PETROGRAPHY. The more recent history of research throughout the world is given in the latest edition (1962) of Milner's *Sedimentary Petrography*.

**Sedimentary rock.** In the true, restricted sense, a rock resulting from the consolidation of mechanically formed sediment; but it is usually extended to comprise all rocks other than purely igneous and completely metamorphic rocks, thus constituting one of the three major kinds of rock. In this wide sense it includes, among rocks of aqueous origin, organic accumulations and chemical precipitates; among volcanic rocks all pyroclastic ejectamenta; also the non-igneous land-formed rocks such as aeolian rocks and tillite. Another term for the 'sedimentary' class in this wide sense is 'derivative'. 'Sedimentary rock' is usually distinguished from loose 'sediment', notwithstanding the fact that geologists have agreed to call loose unconsolidated material 'rock'. *Petrology of the Sedimentary Rocks* (Hatch and Rastall, 1913/Greensmith, 1978). *Deposition of the Sedimentary Rocks* (Marr, 1929). *Sedimentary Rocks* (Pettijohn, 1949/75). *Origins of Sedimentary Rocks* (Blatt and others, 1972). See INTERPRETATION (Allen).

The most familiar rocks in Britain, as in many parts of the world, are varieties of sandstones, limestones, and mudstones. These names in themselves state the obvious; that they are hardened sands, lime deposits, and muds.

**Sedimentary structure.** A structure produced during sedimentation; that is, a primary structure in a sedimentary rock.

**Sedimentary volcanism.** Though there is an obvious contradiction in the component words of the term, this is sometimes applied to the production or phenomena of sand 'volcanoes' or any similarly formed mud 'volcanoes'. 'Contribution to the knowledge of sedimentary volcanism in Trinidad' (Kugler, *JIP*, 1933). 'Sedimentary volcanism, as a surficial manifestation of diapirism, may be either queiscent or explosive or a combination of the two. It may give rise to mud volcanoes or mud flows either alone or in association' (Humphrey, *BGSA*, 1963).

**Sedimentation.** See SEDIMENT. 'The strata of the globe bear all the appearance of a subsidence of a fluid; from a fluid we know that solid bodies naturally subside, stratum superstratum; thus we may see every day in the bed of the ocean and rivers, strata of sand, clay, gravel, etc. disposed exactly in the same manner and on the same principles with the strata over the whole globe' (Walker, *MS*, before 1780, printed in Scott, *Lectures on Geology by John Walker*, 1966).'In an internal zone of the Canadian Rockies, continuous and untroubled sedimentation extended from the later Precambrian Beltian Series, which has a thickness of over 30,000 feet, through the whole of the Palaeozoic and Mesozoic until interrupted by the intensive Laramide movements in the uppermost Cretaceous-Lower Eocene interval' (Lees, *QJ*, 1953). 'The sedimentation and sedimentary history of the Aberystwyth Grits' (Wood and Smith, *QJ*, 1958). 'Marine and continental processes combine to concentrate sediment in the vicinity of the coast. The site of maximum deposition is related to "sedimentation factors", including sediment supply, slope of land and sea floor, and hydrodynamic conditions' (Hoyt, *IGC*, 1968). *Principles of Sedimentation* (Twenhofel, 1939/50). *Stratigraphy and Sedimentation* (Krumbein and Sloss, 1951/63). *Physical Processes of Sedimentation* (Allen, 1970).

**Sedimentation unit.** A bed or lamina resulting from one distinct act of deposition. 'The beds are not, in themselves, individual sedimentation units as they contain numerous laminations' (Knill and Knill, *GM*, 1961).

**Sedimentology.** That branch of the science of geology which treats of sediments; their character and mode of formation. (The description of sedimentary rock types is 'sedimentary petrography'.) *An Introduction to Sedimentology* (Selley, 1976). *Fluvial Sedimentology* (Miall, ed., 1978). *Principles of Sedimentology* (Friedman and Sanders, 1978).

**Segregation.** A segregation (segregation-body, segregation-patch, segregation-vein) is a portion of a rock-mass separated from the rest either as the result of the processes of formation (such as crystallization during the formation of an igneous body), or as the result of subsequent processes (such as the formation of a mineral vein). [L. *segrego*, set apart from the main body.]

**Seismic.** Pertaining to earthquakes; primarily to natural earthquakes, but also to artificial quakes produced by explosions in seismic sounding and prospecting. (Mallet, *BA*, 1858.) [G. *seismos*, a shaking.]

**Seismic focus.** The centre of origin, within the earth's crust (or mantle), of an earthquake. Also called the 'hypocentre'. (Terminology discussed in Leet, *Practical Seismology*, 1938.) (Mallet, *BA*, 1858.)

**Seismic intensity.** See EARTHQUAKE. The intensity of an earthquake at any particular point can only be measured quantitatively by the intrumental (seismographic) records made at that point. But seismographs are sparsely distributed. Scales have therefore been devised based on the effects on people, objects, buildings etc.; of these the best-known are the Rossi-Forel scale (1883, a combination of scales used independently by Rossi in 1874 and Forel in 1878) and the Mercalli scale (1902). 'Scales of seismic intensity' (Davison, *BSSA*, 1921, and *Manual of Seismology*, 1921).

**Seismic survey.** See GEOPHYSICS. 'A seismic investigation of the history of the River Rheidol in Cardiganshire' (Coster and Gerrard, *GM*, 1947). 'Seismic prospecting in the English Channel' (Hill and King, *QJ*, 1953). 'Seismic refraction studies of geological structure in the inner part of the Bristol Channel' (Brooks and Al-Saadi, *JGS*, 1977).

**Seismic velocity.** See SEISMIC WAVES.

**Seismic waves.** The waves, originating at the seismic focus, produced in the earth-body by an earthquake. Also called 'earthquake waves'. They are separable into three kinds, and speeds depend on the kind of wave and certain physical properties of the material being passed through. P ('push-and-pull') waves move fastest, their speed depending on density and compressibility. S ('shake') waves are slower, their speed depending on density and rigidity. L waves, 'surface waves', are of a complex kind and travel through the earth's crust near the surface along complex paths ; they are the slowest waves but their energy is not so quickly dissipated with distance. When a 'seismic velocity' is given without qualification, it refers to the velocity of the P waves.

**Seismograph.** An instrument for recording earthquake-vibrations. Hence 'seismogram', the record so made. 'The name of the instrument is of some interest. Palmieri's use of the word "seismograph" (1859) is the earliest with which I am acquainted. Since his time, the meaning of the word has changed. [His] instrument belongs more nearly to the class now called seismoscopes, the name seismograph being reserved for one that registers the movements of the ground with exactness and in detail' (Davison, *Founders of Seismology*, 1927).

**Seismology.** The science of earthquakes. (Mallet, *BA*, 1858.) *An Introduction to the Theory of Seismology* (Bullen, 1963/79).

**Selenite.** See GYPSUM. 'J. G. Wellerius (*selenites*), f. G. σεληνη [*selēnē*], "the moon", probably alluding to its pale, bluish reflections' (Chester, *Names of Minerals*, 1896).

**Selenology.** The science of the moon, particularly 'lunar geology'. *Geology applied to Selenology* (Spurr, 1945). *Lunar Geology* (Fielder, 1965). *Geology of the Moon* (Guest and Greeley, 1977). The observation and recording of these features is 'selenography': 'Observational selenography began three and a half centuries ago, when Galileo turned his first tiny telescope toward the Moon' (Moore, *AdS*, 1960).

**Selvedge.** (Selvage.) A marginal zone, of distinctive character, of a rock-body such as an igneous intrusion (particularly a chilled contact) or of a mineral vein (particularly a thin layer of clayey material separating ore from country rock).

**Senonian.** See CRETACEOUS SYSTEM. [ Sens, France.]

**Sensu lato, Sensu stricto.** 'In the wide sense', 'in the restricted sense'; such qualifying phrases, whether in Latin or English, are often required to give precision in using names. They are particularly useful in palaeontology where, in the names of genera and species, it may not only be impossible to avoid, but may be convenient to keep, a dual usage. The abbreviations 's.l.' and 's.s.' are then added. Also used in other branches of geology, for instance 'granite (s.l.)'; and used in some of the tables of names in this Dictionary. [L.]

**Separate creation**. The biological doctrine that in the course of life throughout geological times new forms have been separately created, not continuously evolved. In spite of the somewhat ambiguous evidence provided by the fossil record (see QUANTUM EVOLUTION) this doctrine is no longer credited. 'It is possible to put into words the proposition that all the animals and plants of each geological epoch were annihilated and that a new set of very similar forms was created for the next epoch; but it may be doubted if any one who ever tried to form a distinct mental image of this process of spontaneous generation on the grandest scale, ever really succeeded in realizing it' (Huxley, *Darwiniana*, 1893).

**Septarium** (pl. **-ia**). A septal arrangement; extended to denote a concretion which shows this, in that the part near the centre has radiating cracks filled with calcite or some other mineral. These concretions are sometimes called 'beetlestones' or 'turtlestones'. 'Ironstones of an oblate or much compressed sphere, and the size from two or three inches diameter to more than a foot. In the circular or horizontal section, they present the most elegant septarium. There are only two ways in which the septa must have received the spar with which they are filled, first, by insinuation into the cavity of the septa after these were formed, or, secondly, by separation from the substance of the stone, at the same time that the septa were forming' (Hutton, *TRSE*, 1788). [L. *septum*, a partition.]

**Sericite.** The form of white mica produced by the alteration of the chemically similar potassic felspars. Hence 'sericitization'

(List, 1852). [G. *sērikos*, silken, alluding to the lustre.]

**Series.** Often used, as required, with its ordinary general meanings. Apart from this, it is used in several rather special senses. **1.** A set of igneous rock-types of the same general form of occurrence (plutonic, hypabyssal, or volcanic) and having a like general community of petrographic character, but varying along certain lines so that we may suppose they have resulted from various degrees of differentiation (Harker, *JG*, 1900.) **2.** A stratigraphical unit, particularly one throughout which the strata are conformable. **3.** In strict stratigraphical nomenclature it is the name for a prime division of a system, the series being not merely local but recognizable over a wide area and defined principally on palaeontological grounds; that is, it is used in an attempt to realize a time-rock unit. See CHRONOSTRATIGRAPHY.

**Serpentine.** A greenish or reddish mineral, magnesium silicate with hydroxyl (OH). It has a greasy lustre and a slightly soapy feel; the fracture is conchoidal and tough. It results from the alteration of minerals rich in magnesium, such as olivine, pyroxene, or amphibole, and sometimes forms rock-masses. The name, in various forms, seems to have come in gradually during the 16th and 17th centuries. [L. *serpens, serpent-*, a serpent, from the markings and colour suggesting a serpent's skin.]

**Serpentinization.** The metasomatic conversion of certain mafic minerals, especially olivine, to the mineral serpentine. Hence 'serpentine-rock', 'serpentinite', when an ultramafic igneous rock as a whole is so converted. 'Serpentinization: a review' (Moody, *L*, 1976).

**Serpulite.** A general name for a fossil that is a tortuous tube apparently made by an annelid worm. Typical serpulites, found from the Silurian onwards, are so similar to those made by the modern *Serpula* that they are placed in the same genus.

**Set.** For use in descriptions of faulting and jointing see FAULT, JOINT.

**Shale.** Laminated argillaceous rock, laminated mudstone; it being understood that the lamination, which is typically fragile and uneven, represents original bedding. See LAMINA. (The *OED* gives Hooson, *Miners' Dictionary*, 1747.) 'Slaty clay, shale' (Kirwan, *Mineralogy*, 1796). 'Origin and use of the word "shale," (Tourtelot, *AJS*, 1960). 'Towards a classification of shales' (Spears, *JGS*, 1980). [(probably) Old English *scealu*, shell, husk.]

**Shard.** A curved spindle-like fragment of volcanic glass formed by the disintegration of pumice.

**Shatter belt.** A belt of shattered, but not intensely crushed, rock. It may usually be considered as taking the place of a clean-cut fault.

**Shear strain.** The usual conception is of strain, within a rock-body, resulting from sliding along an almost infinitely large number of parallel planes almost infinitely close together. However, the term would apply to sliding along a smaller number of planes more distinctly separated, and even to sliding along one plane only. 'Progressive simple shear deformation on the Laxford Shear Zone, Sutherland . . . shear belt . . . shear strain . . . shear displacements . . . minor shear zones. (Davies, *PGA*, 1978).

**Sheath and core structure.** In certain predominantly glassy igneous rocks, particularly pitchstones: the glassy parts occur as rounded, discontinuous masses or 'cores' (some inches or feet across) separated by sinuous 'sheaths' of crystalline rock. 'Sheath and core structure in the Mull pitchstones' (Drysdale, *GM*, 1979). (Geikie, *PRPSE*, 1980.)

**Sheet.** A word appears to be needed for a broad and relatively thin rock-body, on any scale, without any special implication of horizontality. 'Sheet' (with any such implication removed) would perhaps serve. A sheet of rock is then one (of any size) that has a relatively great extent in two dimensions but only a small thickness; it may be planar, bent, or irregularly disposed. A stratum or bed is a sheet of sedimentary rock, an extensive lava-flow forms a sheet, a dyke or sill is a sheet of intrusive igneous rock, and a piece of slate part of a thin sheet produced by cleavage. In connexion with the forms of igneous bodies, various authorities have tentatively suggested 'sheet' in various special senses, the 'cone-sheet' is a definite form of one of these, or 'sheet' is made synonymous with 'sill' (e.g. : 'The basaltic rocks of the north of England occur in two forms—as "sheets" lying amongst the sedimentary strata and as "dykes" cutting through', Topley and Lebour, *QJ*, 1877; also Rastall, *Textbook*, 1941). There is also 'thrust-sheet', and there are 'intrusive sheet swarms' (Walker, *JGS*, 1975).

**Shelf edge.** See CONTINENTAL SHELF.

**Shelf facies.** The sedimentary facies corresponding to the environment of the shelf seas. See BASIN FACIES.

**Shelf seas.** The seas overlying the continental shelf, corresponding to the belt of variables. *Geology of Shelf Seas* (Donovan, ed., 1969).

**Shell.** An outer covering of an animal, fruit, &c.; particularly, and in palaeontology almost exclusively, that of an invertebrate animal, and usually calcareous. One form of skeletal structure. 'Stones clerly fascioned lyke cokills, and mighty shells of oysters turned in to stones' (Leland, *Itinerary, c.* 1538; of fossils in the Oolites). *Recent and Fossil Shells* (Woodward, 1851/6).

**Shell-marl.** Of the shelly deposits which form a calcareous gravel (shell-gravel), sand (shell-sand), or marl (shell-marl), the last is the most distinctive. Characteristically it is a freshwater deposit of mud containing abundant remains of calcareous algae, and shells of microscopic animals and molluscs (lake-marl). The molluscan shells are often perfectly preserved owing to quiet deposition. Examples are well-known among the Oligocene strata of the Isle of Wight, and are found here and there on the present land-surface as deposits from lakes now dried up. 'White shell-marl [Oligocene]' (Webster, *TGS*, 1814). 'A shell-marl deposit in Montgomeryshire' (Pugh, *MC*, 1928).

**Shelly facies.** A facies of sedimentary rocks characterized by 'shelly fossils'. The term is used chiefly in connexion with the Lower Palaeozoic rocks, in contrast to the 'graptolitic facies', the shelly fauna (in which brachiopods and trilobites are the most conspicuous elements) being associated with a mainly calcareous or calcareous-arenaceous lithology. See POURED-IN DEPOSIT and quotation (Smith and George) under FACIES.

**Shelly fossils.** In a conveniently very wide sense: all fossils of invertebrate animals with skeletons of mineral substance.

**Sherwood Sandstone group.** A name formally introduced in 1980 for the lowest of the three lithostratigraphic units of the Triassic system in Britain (it's base may lie locally in the Permian). It is arenaceous and 'broadly encompasses formations previously assigned to the "Bunter" and the arenaceous (lower) part of the "Keuper" in Britain' (*A Correlation of Triassic Rocks in the British Isles*, *GSSR* (13), 1980). [Sherwood Forest, Nottinghamshire.]

**Shield.** 'A term used for geologically very old and stable parts of the crust' (Krishnan, *India and Burma*, 1956). Usually a large exposed mass of Precambrian (Archaean) rock, e.g. Canadian Shield, Baltic Shield, Peninsular Shield (India). 'It is to the exposed Archean surface that we give the name of the Canadian shield' (Suess, *Das Antlitz der Erde*, 1888, '. . . *canadische Schild*', English trans., 1906). See quotation under CRATON.

**Shield volcano.** See HAWAIIAN TYPE.

**Shift.** Where a fault is accompanied by flexing of the strata, the actual slip along the fault will be only a part of the total dislocation between the rock masses on each side of the zone of displacement. 'The term "shift" is used to denote the relative displacement of points far enough removed from the fault to be unaffected by local disturbance in the fault zone, and in broad discussions of faulting it is more important than the "slip" (Hills, *Structural Geology*, 1953). As in the case of slip, there is the total net shift, which would in the general case be oblique, and the components, the (net) 'strike shift' and the (net) 'dip shift'.

**Shineton Shales.** The formation outcropping between Shineton and The Wrekin, Shropshire, and representing here the Tremadocian series of the Cambrian. It is faulted at the base and is overlain unconformably by considerably newer beds but, curiously perhaps, the whole of the Tremadocian, neither more nor less (as nearly as can be judged), is encompassed. Fossils are very well preserved, though to be found only in rather awkward rock-exposures in the beds and banks of streams: the graptolites *Dictyonema flabelliforme* and *Clonograptus tenellus;* the trilobites *Asaphellus homfrayi* and *Shumardia pusilla;* the brachiopods *Lingulella nicholsoni* and *Acrotreta sabrinae* and the cystid *Macrocystella mariae* are among the Tremadocian species specially common in the Shineton Shales. (Salter, *Cambrian and Silurian Fossils*, 1873; Callaway, *QJ*, 1874/7; Stubblefield and Bulman, *QJ*, 1927.)

**Shingle.** 'Small, roundish stones; loose, waterworn pebbles such as are found collected upon the sea-shore' (*OED*). 'The foreland of Dungeness is remarkable for its great expanses of shingle, which extend inland for distances of 2 miles or more from the present shore-line. Their surfaces are marked by systems of parallel ridges which have long been recognized as ancient beach-ridges, each one abandoned by the sea as a new one developed in front of it' (Hey, *GM*, 1967). See STORM BEACH.

**Shore.** In ordinary usage (for the seashore), it is the zone between low and high water-mark of extreme tides, extending upwards to the foot of cliffs or sandhills and including any stormbeach.

**Shore platform.** See WAVE-CUT PLATFORM. 'Shore platforms' (Hills, *GM*, 1949).

**Shoreline.** 'The line where shore and water meet' (*OED*). 'We will find it profitable to divide shorelines into four main classes: I, "Shorelines of submergence", or those shorelines produced when the water surface comes to rest against a partially submerged land area; II, "Shorelines of emergence", or those resulting when the water surface comes to rest against a partially emerged sea or lake floor; III, "Neutral shorelines", or those whose essential features do not depend on either the submergence of a former land surface or the emergence of a former subaqueous surface; IV, "Compound shorelines", or those whose essential features combine elements of at least two of the preceding classes' (Johnson, *Shore Processes and Shoreline Development*, 1919). 'An early Ordovician shoreline in Radnorshire' (Jones and Pugh, *QJ*, 1949). *Annotated Bibliography of Quaternary Shorelines* (Richards and Rhodes, *ANSP*, 1965). 'Vertical displacement of shorelines in Highland Britain' (Walton and others, *TIBG*, 1966).

**Shrinking earth.** The hypothesis that the earth is shrinking. 'The existence of folded mountains implies a contraction of the earth's interior' (Jeffreys, *The Earth*, 1929). 'The evolution of a shrinking earth' (Lees, *QJ*, 1953). There is thus the 'Contraction hypothesis' as opposed to the 'Expansion hypothesis'.

**Sial.** The masses of comparatively light rock—granitic rock, and sedimentary rocks in all degrees of deformation and metamorphism—which form the continental masses and neighbouring parts, projecting downwards into the sima. The sial together with the sima (in a restricted sense) constitute the earth's crust (the 'M crust', above the Mohorovičić discontinuity). The term is sometimes restricted to the deep-seated igneous rocks of these great masses. 'The "light" rocks of the continents are chiefly felsic. They belong to the interrupted earth shell called "the sal" by Suess. Wegener, following Pfeffer, objected to this portmanteau word on the ground of possible confusion with the Latin for "salt" and introduced "sial" as a preferable contraction for "silica-alumina rocks"' (Daly, *Igneous Rocks*, 1933).

**Siderite. 1.** = Chalybite. **2.** See METEORITE. [G. *sidēros*, iron.]

**Siderolite.** See METEORITE.

**Siderose.** A term, meaning iron-containing, proposed by Sollas (*PGA*, 1924) for use instead of 'ferruginous' in certain cases (e.g. for a cement of iron carbonate in a sandstone), to avoid any implication that the iron was necessarily in the form of the oxide.

**Siegenian.** See DEVONIAN SYSTEM. (Kayser, 1881). [Siegen, Germany.]

**Silcrete.** See CALCRETE. *Silcrete in Australia* (Langford-Smith, 1978).

**Silesian.** A now preferred name for the Upper Carboniferous, being a subsystem of the Carboniferous system. *A Correlation of Silesian Rocks in the British Isles* (*GSSR* (10), 1978). (van Leckwijck, Holland, 1960.) [Silesia, Poland.]

**Silica.** Silicon dioxide, $SiO_2$; in the natural crystalline form, occurs chiefly as the mineral quartz.

**Silicate.** A chemical compound of silica with one or more metallic oxides. Hence 'silicate minerals'.

**Siliceous.** Containing, or consisting of, silica. In geology, it is implied that a siliceous rock has a large, or at least a significant, amount of free silica, not merely silica in the form of silicates; that is, the rock is sandy (quartzitic) or an igneous rock with quartz crystals (or, possibly, glassy silica). 'This siliceous substance [quartz]' and 'siliceous strata [sandstone]' (Hutton, *TRSE*, 1788).

**Siliceous organic deposit.** As regards, particularly, the Recent and Tertiary organic siliceous accumulations, Greensmith (*Petrology of the Sedimentary Rocks*, 1978, and in his earlier revisions of 'Hatch and Rastall') has the following: 'The soft, entirely unconsolidated organic deposits are called "oozes". Deposits which have remained unconsolidated although no longer in process of accumulation are termed "earths", such as "radiolarian earths" and "diatomaceous earths". The consolidated equivalents of these purely organic accumulations are termed "radiolarites" and "diatomites".

**Silicify.** To make siliceous, or to make more siliceous; in the form 'silicification', particularly the replacement of some mineral (usually calcium carbonate) by silica.

**Silicon.** The element (Si); next to oxygen, the most abundant element in the crust of the earth.

**Sill.** In its modern technical usage, a concordant sheet-form igneous intrusion often intruded in a more or less horizontal position. The property of concordance with stratification is essential. A sill that does not follow the stratification exactly but changes its position among the strata gradually or abruptly is a 'transgressive sill'. 'The term "sill" [as now used] is derived from the remarkable example in the north of England, which has long been known as the Great Whin Sill. The word "sill" was probably applied to this flat cake of dark stone at the base of the hills, from its fancied resemblance to the sill or threshold of a house' (Geikie, *Textbook*, 1903). 'The great whin sill is a basalt, coarse-grained in texture . . . [its] thickness is very variable' (Winch, *TGS*, 1817). 'The "Whin sill", as it is called, of Yorkshire, Durham, and Northumberland, is a great eruption of melted rock interpolated in the limestone series' (Phillips, *Yorkshire*, 1836). 'The great Whin Sill of north-east England' (Cullen, *AMG*, 1968). In this Carboniferous series of the north of England, certain sedimentary beds are still named as 'coal sills', 'slate sills', 'grit sills', 'firestone sills' (Forster, *Section of the Strata*, 1821; Johnson and Dunham, *Geology of Moor House*, 1963).

**Sill complex.** 'The meta-igneous suite [of the S.W. Scottish Highlands] comprises a thick succession of frequently pillowed lavas and intercalated tuffs, and a broadly contemporaneous and comagmatic complex of basaltic, doleritic and gabbroic sills intruded into the thick underlying (largely middle Dalradian) sedimentary pile. This paper describes the petrochemistry and petrogenesis of the sill complex' (Graham, *JGS*, 1976).

**Sillar.** See IGNIMBRITE. (Fenner, *BGSA*, 1948.) [A local Peruvian name.]

**Sillimanite.** See ANDALUSITE. 'G. T. Bowen, 1824, in honor of Prof. B. Silliman (Chester, *Names of Minerals*, 1896).

**Silt.** This usually refers, in the ordinary way, to a deposit that is choking a river, harbour, &c., but geologists sometimes use the term for a deposit of which the average grain-size is somewhere between that of sand and clay. [probably Anglo-Saxon *sealt*, salt; if this is correct it apparently denoted originally a salty deposit of mud, &c.]

**Siltstone.** A consolidated silt.

**Silurian system.** Overlies the Ordovician and is succeeded by the Devonian. The base has been defined in several parts of the British outcrop, for instance, in the Machynlleth district at the base of the Mottled Beds of the zone of *Glyptograptus persculptus* (graptolitic facies) and at Haverfordwest and Llandovery at the base of the Basement Beds. In the shelly facies fossils are abundant, particularly brachiopods, corals, and trilobites; but nautiloids are also characteristic, particularly on the Continent, and crinoids are common. Nearly all the families and the great majority of the genera (s.l.) occur also in the Ordovician or Devonian, but the species are distinct and most of them are characteristic of one or another stratigraphical series. The graptolite fauna is distinguished above all by the presence and abundance of the Monograptidae. The following are the series of the Silurian system:

Ludlow
Wenlock
Llandovery.

As originally quite clearly defined by Murchison the system included all the groups from the Llandeilo to the Ludlow. The Ludlow Bone-bed and the Downton Castle Sandstone (the lowest one-tenth or so of the present Downtonian) were placed as passage beds to the Old Red Sandstone, but on the whole he included them in the Silurian. The dispute between Murchison and Sedgwick does not show any reason why, on historical grounds, all these groups should not have continued to be called Silurian—in so far as Sedgwick and Murchison had described groups that were found to be stratigraphically equivalent, Murchison's work had been more detailed and precise and much more fully recorded than Sedgwick's—but the incorporation in the Silurian system of groups lower than the Llandeilo was unjustified. The Llandeilo and the Caradoc (as restricted and redefined by Murchison himself) eventually, together with (Sedgwick's) Arenig, went to constitute the Ordovician system. The Geological Survey (of which Murchison was the head) continued to use 'Silurian' to include the Ordovician ('Lower Silurian') till about the end of the century and this is the usage in the *Textbook* of A. Geikie (1903), also a head of the Survey and in every sense a follower of Murchison. The name 'Silurian' is still very generally used on the Continent (but not in America) for the whole of the Ordovician (including Tremadocian) and Silurian together, the two parts being then named 'Ordovician' and 'Gothlandian'; it is even sometimes made to include the Cambrian, thus comprising the whole of the Lower Palaeozoic. 'In early communications to the Geological Society, adopting a provisional name, I called these rocks "fossiliferous greywacke"; but this was deemed objectionable. A geographical term was finally adopted [*PM*, 1835 and, later, *BA*, 1835], derived from the Silures, whose power extended over the region where these rocks are best displayed. The term was no sooner proposed than sanctioned by geologists, both at home and abroad, as involving no theory, and as simply expressing the fact, that in the "Silurian region", a complete succession of fossiliferous strata is interpolated between the Old Red Sandstone and the oldest slaty rocks' (Murchison, *Silurian System*, 1839). 'Having first, in the year 1833, separated these deposits into four formations, I next divided them (1834, 1835) into a lower and upper group. After eight years of labour in the field and closet . . . fully published in the work entitled the *Silurian System* (1839)' (Murchison, *Siluria*, 1854). Detailed history in Geikie, *Life of Murchison*, 1875. See CAMBRIAN SYSTEM.

The lowest part of the Old Red Sandstone magnafacies (i.e. an unspecified lower part of the Downtonian) is now taken to be of Silurian ('post-Ludlow') age.'*A correlation of Silurian rocks in the British Isles*' (Cocks and others, *JGS*, 1971 and *GSSR* (1), 1971).

**Sima.** The comparatively heavy basic igneous rock which, it is inferred, occupies, together with the sial, the earth's crust (the 'M crust', above the Mohorovičić discontinuity). Silica and magnesia are the chief constituents of the sima, hence the portmanteau name (used by Suess). The concept of the sima leaves its base indefinite. In its wide and original sense it would probably not exclude the ultrabasic material which supposedly underlies the Mohorovičić discontinuity; but this is now sometimes called the 'ultrasima'.

**Similar folding.** Folding in which the successive bedding surfaces form, approximately and on the whole, similar curves, similar folds. This kind of folding is assumed by incompetent beds. Accommodation to the compressive forces producing the folding is by plastic adjustment and flow. The incompetent beds tend to thicken towards the hinges, and to thin in the limbs, of the folds. The contrasting term is 'concentric folding'.

**Sinemurian.** See JURASSIC SYSTEM. [Sémur (Sinemurum), Côte d'Or, France.]

**Sinistral fault.** See DEXTRAL FAULT.

**Sink.** See SWALLOW and RISING.

**Sinter.** 'A hard incrustation (usually siliceous) formed on rocks or the ground by precipitation from mineral waters. Adopted from German *sinter*, the equivalent of English "cinder"' (Arkell and Tomkeieff, *Rock Terms*, 1953). One form is geyserite. If calcareous, the deposit is 'calc-sinter' or 'travertine'.

**Site.** A geological 'site' ordinarily means a specified restricted locality where the geological features are particularly interesting, or worth investigating or exploiting. 'Code for visitors to geological sites' (Macfadyen, *Geological Highlights of the West Country*, 1970). 'Borehole at Mochras, Merionethshire . . . The site is situated among the coastal sand hills of Morfa Dyffryn' (Wood and Woodland, *N*, 1968). If one had to name one from among the innumerable well-known 'classical' geological sites in Britain it would perhaps be Siccar Point on the Berwickshire coast (see UNCONFORMITY). [L. *situs*, situation, local position.]

**Skarn.** 'An old Swedish mining term for the silicate gangue (amphibole, pyroxene, garnet, etc.) of certain iron-ore and sulphide deposits of Archaean age, particularly those which have replaced limestone and dolomite. The term has been extended to cover analogous products of contact metamorphism in younger formations' (Holmes, *Nomenclature of Petrology*, 1928). 'The production of amphibolic and other skarn rocks from limestone at Cor, Co. Donegal' (Gindy, *GM*, 1951).

**Skeleton.** 'The harder (supporting or covering) constituent part of an animal organism' (*OED*). Thus, in this wide sense, it includes the shells of invertebrates; but it is most commonly used for the vetebrate framework. See quotation under VERTEBRATE.

**Skiddaw Slates.** Flags and slaty mudstones outcropping over a broad belt in the northern part (chiefly) of the Lake District and out of which the humped and tent-shaped mountains of Saddleback, Skiddaw, and those west of Keswick have been eroded. The base is not seen and the structure is complicated. A few shelly fossils and graptolites occur at some spots; the latter in particular indicating that the exposed rocks of the formation are probably for the most part of Arenig age. 'Skiddaw Slates' is a name (not a descriptive statement), suggested by one region and one rock-type. (Otley, *LM*, 1820, *English Lakes*, 1823; Sedgwick, *PGS*, 1832.)

**Slaggy.** 'When a lava presents an irregularly vesicular character, like that of the slags of an iron-furnace, it is said to be "slaggy"' (Geikie, *Ancient Volcanoes*, 1897).

**Slate.** A rock that splits readily into thin plates, or a piece of one such plate. In ordinary usage, as in the older geological terminology, any hard argillaceous rock with a tendency to split is a 'slate', thus including those that split readily along the bedding (such as those of the Stonesfield Slate formation of the Jurassic). The most perfect splitting, however, is in these rocks which have been further hardened by lateral pressure and have at the same time had imposed on them a cleavage ('slaty cleavage') at right angles to the pressure; the splitting is then along this cleavage. In modern geological usage the term 'slate' has become inviolably restricted to this latter sense so that 'shale' and 'slate' are mutually exclusive terms. In the early 19th century 'clay-slate' was much used for (true) slate. The distinction between bedding and cleavage, and the implied restriction of the term 'slate' to its present meaning, seems to have been first clearly made by Bakewell (*Geology*, 1813): 'The division of the laminae of slate is frequently in a different direction from that of the thick plates which have been called strata, a decisive proof that the slaty structure is not owing to stratification. . . . The slaty cleavage of the stone is nearly at right angles with the direction of the beds'. 'Compacted between the massive Pre-Cambrian igneous rocks and Lower grits of the Padarn ridge on the one hand and the Ordovician grits and igneous rocks of the Snowdon syncline on the other, the shales of the Cambrian outcrop between Nantlle, Llanberis and Penrhyn were altered to slates [Llanberis Slates] during the Caledonian orogeny. These slates are of excellent quality commercially, being almost free from interbedded sandy layers; and they have been quarried and mined on an enormous scale. The various bands of slate differ slightly in lithological character, in colour and chemical composition, and in the perfection of cleavage, and have consequently been given special names (corresponding approximately to different geological horizons)' (Smith and George, *BRG, N. Wales*, 1961). *Slates of Wales* (North, 1925/52). 'The huge slate quarry at Delabole [Cornwall, Upper Devonian] has been active since the 16th century' (Edmonds and others, *BRG, SW. England*, 1969).

**Slatestone.** In the older general usage, synonymous with 'slate': 'Riding almost in the entering of this forest [Charnwood] I saw 2 or 3 quarres in the hills of slate stone' (Leland, *Itinerary*, *c.* 1538). Now restricted to a slab of slaty rock, the surfaces being determined by the cleavage: '"Slatestones" of cleavage, and "flagstones", which are thin beds' (Sedgwick, *TGS*, 1835).

**Slaty cleavage.** See SLATE and CLEAVAGE (1*a*).

**Slice. 1.** A relatively thin and broad mass of rock lying between two (nearly) parallel faults; the term being used particularly when the whole structure is rather flat-lying and the faults are thrusts. 'The upper units—the Tintagel, Barras Nose and Island—rest on thrust-planes, and are considered to be thrust-slices or sheets' (Wilson, *QJ*, 1950). 'The alternating Lewisian and Moine bands represent slices of tectonic origin' (Ramsay, *QJ*, 1957). See SCHUPPEN. **2.** A thin section of a mineral, rock, or fossil. 'Let a thin slice be cut off from the fossil wood, in a direction perpendicular to the length of its fibres. The slice thus obtained must be ground perfectly flat, and then polished. The polished surface is to be cemented to a piece of plate or mirror glass, a little larger than itself, and this may be done by means of Canada balsam. . . . The slice must now be ground down to that degree of thinness which will permit its structure to be seen by the help of a microscope' (Nicol in Witham, *Fossil Vegetables*, 1831, quoted by Mather and Mason, *Source Book*, 1939). 'A guide to the study of rocks in thin slices' (subtitle of some editions of Harker's *Petrology for Students*, e.g. 1935).

**Slickenside.** A fault surface, somewhat polished, and having fine striations (parallel scratches) or grooves on it, caused by the fault movement. Such surfaces are often coated with films of a mineral, particularly quartz, which show striations in the same direction due to the establishment of a fabric in the crystalline material under the pressure of the fault movement. 'The two faces [of a fault in limestone] in contact, appear as though they had been polished, and are ribbed or somewhat fluted;

and the face of each is sometimes covered by a remarkably thin coating of lead ore; these planes, when separated, are the slickensides of the mineralogist' (Conybeare and Phillips, *England and Wales*, 1822). For a brief discussion of features and terminology see Fleuty, *GM*, 1975. [*Slicken*, dialect variant of slick, in sense of render smooth.]

**Slide.** See FOLD-FAULT. The term 'slide' was introduced into geology by Bailey (*QJ*, 1910); but this is an ordinary word which does not in itself suggest a necessary connexion with folding. Is it not a useful term to comprise 'thrust' and 'lag-fault', whether or not (in particular instances) these are connected with folding ? The full term is 'tectonic' slide.

**Sliding.** The bodily downward movement, under gravity, of relatively large and coherent masses of sediment (under water) or superficial deposits (on land), or of strata whose equilibrium becomes disturbed by erosion or other means (a large landslide or landslip, or a gravity collapse effect). It excludes rolling or incoherent slumping, as in many subaerial landslips and the subaqueous slumping of sediments.

**Slip.** In faulting, the net relative movement along the fault plane. In the general case, this is oblique, with a component in the direction of strike of the fault-plane (strike-slip) and one in the direction of dip of the fault-plane (dip-slip); but nevertheless faults tend to fall into two distinct groups, one (normal and thrust faults, dip-slip faults) in which the slip is mostly dip-slip, and the other (tear fault, transcurrent fault, wrench fault, strike-slip fault) in which the slip is mostly strike-slip.

**Slip-mark.** 'A term is needed, and 'slip-mark' is here proposed, to cover the various markings on a rock surface made by the movement over it of another rock surface, other than the distinctive slickenside effect. Slipping between two bedding surfaces, in adjustment during folding or from some other cause, may produce 'bedding-surface slip-marks', and these may be angular' (Challinor and Williams, *GM*, 1926, also *GM*, 1928, 1929, 1978).

**Slump.** 'The slumping of submarine sediments in Denbighshire during the Ludlow period' (Jones, *QJ*, 1937; in which paper, and in later ones, particularly *QJ*, 1939, the process is fully described). 'The word "slump" implies the flowage of a mass of sediment, shortly after its deposition, down a slope. It is necessary to distinguish such structures from those produced by tectonic activity and those produced by turbidity current deposition' (Williams and Prentice, *PGA*, 1957). 'Perhaps the most striking individual feature of the White Lias in Pinhay Bay [Devon] is the Slump Bed which is a magnificent example of its kind' (Hallam, *PGA*, 1960). 'Slump structures in the Peel Sandstone series [Carboniferous]' (Ford, *IMNHAJ*, 1971). 'Structural style in slump sheets: Ludlow series, Powys, Wales' (Woodcock, *JGS*, 1976). 'Large-scale slumping in a flysch basin southwestern Pyrenees' (Rupke, *JGS*, 1976). Hence 'slumped bed', 'slump sheet'. The foregoing is the commonest usage in geology in Britain ('subaqueous slump' would be more precise); but it is used where appropriate in any case of movement of rock-material, such as a landslide. The term has also been used for pre-consolidation disturbance in a cooling intrusion.

**Soil.** The ground, upper layer of earth in which plants grow consisting of disintegrated rock usually with admixture of organic remains, mould' (*OED*). Soils form a thin 'skin' covering all rocks, whether 'solid' or superficial, wherever there is vegetation (see SUBSOIL). See quotation (Lister) under GEOLOGICAL MAP. 'The soil, or the coat of vegetable mould, spread out over the surface of the earth' (Playfair, *Illustrations*, 1802). *Nature and Property of Soils* (Morton, 1838). *Evolution and Classification of Soils* (Ramann, trans., 1928). *Soils, their Origin, Constitution and Classification* (Robinson, 1932/49). *'Soil classification in the Soil Survey of England and Wales' (Avery, JSS*, 1973). *World Soils* (Bridges, 1970/78). See PEDOLOGY. [L. *solum*, the ground, earth.]

**Soil bed.** An old soil, a seat-earth; a name used particularly for those in the Wealden with fossil plants preserved *in situ* as they grew. 'A Wealden soil bed', 'Wealden fossil soil-beds' (Allen, *PGA*, 1941/6).

**Soil climate.** 'The nature of the air between the soil particles, water content and condition, etc.' (Mohr and van Baren, *Tropical Soils*, 1954). It may be divided into a 'life-containing-soil climate' and an (underlying) 'lifeless-soil climate'.

**Soil creep.** See CREEP.

**Soil mechanics.** A subject in engineering geology. The findings of the engineer in this matter are of much interest in general geology. 'The uncemented sediments such as sands, clays and silts are known to the engineer as "soils" or "earth", in contrast to the more rigid limestones, sandstones, granites, and, other "rocks". . . . It is only recently that our knowledge of their behaviour in relation to engineering problems has been placed on a reasonably scientific basis. This is due in large part to K. Terzaghi, who . . . brought laboratory experiments, theory and field observations into an integral philosophy which was first expressed in the celebrated *Erdbaumechanik*, published in 1925. The term *Erdbaumechanik* became translated in America as "soil mechanics" and this, although perhaps not the best choice, is the name adopted for the subject throughout the English-speaking world. On the Continent, *géotechnique* or its equivalent is widely used, and the adjective "geotechnical" is becoming increasingly employed to describe various processes and methods of study associated with soil mechanics in particular, and with quantitative engineering geology in general' (Skempton in Blyth, *A Geology for Engineers*, 1952). 'The terminology of soil mechanics' (Jürgenson, *Gt*, 1966).

**Soil profile.** A vertical columnar section through the soil or a tabular statement of the succession of the various kinds of material and their thicknesses. 'The succession of horizons down to the parent material' (Robinson, *Soils*, 1932).

**Solar System.** Geology finds its setting within the framework of the Solar System. In particular, its origin is also the origin of the material composing the earth and the start of the earth's geological history. 'The origin of the Solar System' (Jeffreys in *Internal Constitution of the Earth*, Gutenberg, ed., 1951). 'The origin of the Solar System' (Spencer Jones, *PCE*, 1956; also *E*, 1958). *Origin of the Solar System* (Page and Page, eds., 1966). *Earth, Moon, and Planets* (Whipple, 1968). See ACCRETION HYPOTHESIS, NEBULAR HYPOTHESIS, PLANETISMAL HYPOTHESIS.

**Sole.** The more or less flat undersurface of a rock-body such as a stratum, vein, igneous rock, or tectonically transported mass.

**Sole-mark.** A descriptive term for a mark or some definite small irregularity on the (original) undersurface of a bed. A 'sole marking' is synonymous or may be reserved for a pattern of any one kind of sole-mark. Sole-marks are usually the negatives, in a relatively hard and coarse-grained bed, of such primary sedimentary features as cracks, tracks, grooves, and dents, made in a soft mud, occurring as positives on the upper surface of the immediately underlying fine-grained bed; the negatives, being in relief on a hard surface, are more likely to be preserved, and to be conspicuous, than are the original intaglio positives. As pointed out by Craig and Walton (*TEGS*, 1962) these counterpart negatives would be appropriately referred to as 'moulds', though they are usually called 'casts'. Sole-marks include groove-casts, flute-casts, casts (moulds) of mud cracks and rills; the casts (moulds) of tracks, trails, and impressions made by organisms; and marks of tectonic origin. 'Sole markings other than footprints in the Keuper sandstones and marls' (Beasley, *PLGS*, 1908). 'Sole markings of graded graywacke beds' (Kuenen, *JG*, 1957).

**Solfatara.** 'A volcanic vent from which sulphur, sulphureous, watery, and acid vapours and gases are emitted' (Lyell, *Principles*, 1833). Sometimes used in a wider sense, particularly in the form 'solfataric stage' of volcanic activity, to include all fumaroles. [crater of Solfatara, near Naples.]

**Solid diffusion.** Diffusion through a rock that remains solid throughout the process, particularly for such a metasomatizing process.

**Solid earth.** See EARTH-BODY.

**Solid geology.** The geological features of the rocks underlying superficial deposits, specifically excluding these latter. A 'solid' geological map, as opposed to a 'drift' map, shows the whole outcrops as they would appear if all the superficial deposits were removed (though the larger tracts of alluvium &c. are often shown even on 'solid' maps). Also known as 'bedrock geology', which is perhaps the more appropriate term. A map (or the parts of a larger map) showing the solid geology where it is entirely covered by superficial deposits is a 'submask' map.

**Solid rock.** Rock which is both consolidated and *in situ*. 'It is well known that, on removing the loose earth which forms the immediate surface of the land, we come to the solid rock' (Playfair, *Illustrations*, 1802).

**Solid solution.** See SOLUTION (3).

**Solidification.** The process of becoming solid all through. There are two very different cases. **1.** The elimination or filling up of the spaces and interstices in a loose sediment, so that it becomes firmly compacted. **2.** The change from the liquid to the solid state on the cooling of an igneous magma. In the latter case, the term is always employed; but some other term, such as consolidation, or lithification, is generally preferred in the former.

**Solifluxion.** (Solifluction.) Though probably applicable to all creep downhill of weathered material, the term is used particularly for the movement in cold regions of superficial material for a considerable distance. Here an upper wet (thawed) and undrained layer of weathered material, lying on the frozen ground (permafrost) underneath, readily slips downhill. Solifluxion may produce such features as 'fan-gravels', 'stone rivers' ('stone runs'). 'Solifluction, a component of subaerial denudation' (Andersson, *JG*, 1906). 'Some solifluction phenomena in the northern part of the Lake District' (Hollingworth, *PGA*, 1934). 'Solifluction in Scotland' (Galloway, *SGM*, 1961). [L. *solum*, soil, flow, flow.]

**Solution. 1.** The process of ordinary aqueous solution is important in erosion. **2.** 'We have seen that the constituents of a molten rock-magma exist there in the form of definite compounds, mostly silicates, and in general identical with those compounds which are familiar in the less complex rock-forming minerals. We shall give reasons for believing that the relation of the several compounds in the magma is one of "mutual solution"' (Harker, *Igneous Rocks*, 1909). **3.** Solid solution. 'A crystalline and homogeneous solid, representing a mixture of two or more substances, and often, though not necessarily, composed of isomorphous compounds. Many of the common igneous rock-forming minerals are complex solid solutions' (Holmes, *Nomenclature of Petrology*, 1928).

**Solution pipe.** A tubular or narrowly conical hollow in a soluble formation, filled in by later rock material. Well known in many places where the upper surface of the Chalk is overlain by Tertiary rocks or Pleistocene deposits, which have sunk into the pipes. A localized and special form of solution subsidence (and see SEDIMENTARY INSERTION). Also applicable to a pipe-like swallow-hole formed subaerially; see quotation (Dixon) under SWALLOW. 'Intra-formational piping in Carboniferous Limestone' (Thomas, *GM*, 1953).

**Solution subsidence.** Any subsidence due to solution of underlying rock, but particularly for the subsidence (collapse) of parts of a formation into hollows or pockets of an immediately underlying soluble formation. The solution is caused by water percolating through the upper formation. 'Solution subsidence outliers of Millstone Grit on the Carboniferous Limestone' (Thomas, *GM*, 1954). 'Solution subsidence structures in the Maltese Islands' (Pedley, *PGA*, 1974).

**Solva beds.** Rocks in S. Wales, about 150 m. thick, containing fossils characteristic of the lower part of the Middle Cambrian series, between the Caerfai beds (below) and the Menevian (above). (Hicks, *PSR* and *PGA*, 1881, *GM*, 1894.) [Solva, Presili district of Dyfed.]

**Sorting.** A selection of particles according to some property or combination of properties (size, shape, specific gravity) by some force, particularly gravity (see GRAVITATIONAL SORTING), or a transporting agent. The 'degree of sorting' of the component grains of a sediment or sedimentary rock is an expression of the range of variation in the grain size. Thus if all the grains are nearly the same size the sediment is well sorted, if the sediment is formed of grains of very different sizes it is poorly sorted, though the average grain size may be the same in each case.

**Source rock.** A rock-mass from which material forming a later rock has been derived. 'Source rocks of the Lower Old Red Sandstone' (Allen, *PGA*, 1974/75).

**Source vent.** See VOLCANIC VENT. 'Source vents of N. Welsh ignimbrites' (Fitch, *BV*, 1967).

**Space geology.** 'The term "space geology", admittedly a hybrid term, has nevertheless received the sanction of current usage among geologists to signify the extension to extraterrestrial objects and domains of those concepts and techniques of study hitherto employed in geology, the study of the earth' (Pecora in White, eds., *Study of the Earth*, 1962).

**Space problem.** This problem arises chiefly in the consideration of the means of emplacement of an igneous body: how was the space provided for it ? ('The great room-problem', Read, *Geology*, 1949.) Was it provided by mechanical means, by the advancing magma pushing country rock aside (displacement) or, alternatively, by the country rock being removed to lower levels, possibly being assimilated there by the advancing magma; or was it produced by chemical replacement of the country rock on the spot, by ultrametamorphism (e.g. granitization) ?

**Spaced cleavage.** See CLEAVAGE (1b). (Dennis, *Tectonic Dictionary*, 1967).

**Spar.** Any light-coloured, lustrous, easily cleaved mineral. [Old English *spær.*]

**Sparite.** A sparry rock, particularly a limestone, or the sparry constituent parts of a rock containing allochems.

**Sparnacian.** See PALAEOGENE and EOCENE. [Epernay (Sparnacum), France.]

**Speciation.** The production of new species of organisms from pre-existing ones in the course of evolution.

**Species** (sing. and pl.). In palaeontology, comprises all those, usually innumerable, individuals, extending in space throughout their habitat, and in time through successive generations, that are all very much alike, varying according to local circumstances or hereditary influences but tending to keep their character distinct by interbreeding among themselves but not with other species. The category of species is, however, defined purely on the degree of likeness, and as there is no rule as to this, nor any ready method of estimating it, there are no fixed limits to what should be called a species, particularly among fossils where the time element causes one species to merge more or less gradually into another. 'The species concept in palaeontology' (*SA*, 1956). *The Origin of Species* (Darwin, 1859).

Fossils are named in the same way as living animals and plants. Linnaeus established the rule over two hundred years ago that the name of a species is to be formed of two Latin words, the first the name of the genus (the generic name) and the second denoting the particular species within the genus (the specific name). This latter is either an adjective (agreeing in gender with the name of the genus), a noun in the genitive case, or a noun in apposition. Until recently, the second of these words was called the 'trivial name', the 'specific name' being rather strictly reserved for the whole name of two words. To prevent any possible ambiguity, the name of the author (first describer) of a species may be placed after the specific name (the author's name going into brackets, if his generic name is changed); but whether this is generally desirable, when there is no ambiguity and when the user of the name is not directly comparing his specimens with those originally described, is open to question. To stabilize specific names, it is laid down that the name is to be that given to the first adequately described specimens of it; the generic name may be altered to conform with what is considered a more appropriate grouping of species but the species on which a genus was founded must always continue to bear the generic name first given to it.

In mineralogy and petrography, 'species' is sometimes used by analogy with its use in palaeontology, that is, as a small taxonomic category for the reception of distinctive, named units of kind (e.g. galena, biotite-granite), 'but there are no natural species (call them types or what you will) among rocks except such as are indicated and defined by physical chemistry' (Shand, *AJS*, 1944). 'The origin of species' (Teall, *QJ*, 1901). [L. *species*, appearance, form, kind.]

**Species-group.** See GENS.

**Speeton Clay.** The chief local formation, seen at Speeton on the Yorkshire coast, of the marine facies of the Neocomian and Aptian series of eastern England. The characteristic fossils are species of belemnites, e.g. *Acroteuthis lateralis*, *Hibolites jaculoides*, *Oxyteuthis brunsvicensis*. (Phillips, *Yorkshire*, 1829.)

**Speleology.** The scientific study of caves; the exploration of their form and extent, the consideration of the manner of their formation ('spelaeogenesis'), the description of the mineral deposits formed in them, the collection of remains of the animal life that inhabited them, and the discovery of any activities of early cave-dwelling man. *British Caving: an Introduction to Speleology* (Cullingford, ed., 1953). [G. *spelaion*, a cave.]

**Speleothem.** 'A term introduced in U.S.A. by G. Moore in 1952 (*Nat. Speleo. Soc. News*) who defined it as "a naturally formed, unitary, coherent body of mineral matter which has been deposited within a cavern or cavern space subsequently to the development of such a space, and at least a portion of the substance of which has been precipitated from solution or solidified from a state of fusion". This thus includes all forms of stalactite deposit but excludes sediments' (Ford, *in litt*, 1970). 'The stereoscan microscope and speleothems' (Ford and McTurk, *TCRG*, 1969).

**Sphene.** (Titanite.) Silicate of titanium and calcium. An almost ubiquitous accessory mineral in igneous rocks, usually as small, scattered, lozenge-shaped, brilliantly lustrous, yellowish-brown crystals. Abundant also in some metamorphic rocks. (Haüy, 1801.) [G. *sphēn*, a wedge.]

**Sphenophyllales.** An extinct group of plants, with wedge-shaped leaves, belonging to the division Pteridophyta, found only in Upper Devonian, Carboniferous, and Permian rocks. [genus *Sphenophyllum* (G. *sphen*, a wedge, *phullon*, a leaf).]

**Sphere.** See EARTH (1).

**Sphericity.** Applied to rock particles; the degree to which they approximate in overall shape to a sphere. See ROUNDNESS, and PARTICLE (reference).

**Spheroidal.** Approximately spherical; applied particularly to structures in igneous rocks. These may be spherulitic, be of the orbicular kind, or be structures developed or made conspicuous on weathering (often coming to resemble the layers of an onion—'onion weathering') in a fine grained homogeneous rock. 'Spheroidal weathering of dolerite sill, North Queensferry, Fife' (pl. in Holmes, *Physical Geology*, 1965/78).

**Spherulite.** A bundle, or a more or less completely spherical mass in an igneous rock, of radiating acicular crystals, often also showing concentric banding. Spherulites may be microscopic, or be several inches or more across. They occur in rocks dominantly glassy, devitrified, or hemicrystalline; most commonly in acid rocks. A spherulite in a basic rock has the special name 'variole'. (The term 'spherule', which would be more appropriate for at least the microscopic spherulites, is not used in petrography. If it were, 'spherulite' would be the rock; the two terms then corresponding to 'variole' and 'variolite'.) Hence 'spherulitic'; 'spherulitic structure' for a rock containing abundant spherulites, 'spherulitic texture' perhaps being more correct than 'structure' for the make-up of the spherulites themselves. 'Such masses of obsidian are of a pale green colour and generally contain minute white spherulites' (Darwin, *Volcanic Islands*, 1844).

**Spilite.** An altered basalt, having albite and chlorite as the chief secondary minerals. Often spotted with vesicles filled with calcite or zeolite. They occur as submarine lavas, of Palaeozoic age, and usually show pillow structure. (Brongniart, 1827.) Hence 'spilitic suite' (Dewey and Flett, *GM*, 1911). *Spilites and Spilitic Rocks* (Amstutz, ed., 1974). [G. *spilos*, a spot.]

**Spillway.** See OVERFLOW CHANNEL and quotation (Yates and Moseley) under GLACIER LAKE. 'Glacial spillway at Dinas Head, near Fishguard' (pl. in George, *BRG, South Wales*, 1970).

**Spinel.** The simplest chemical form of this group of minerals has the formula $MgAl_2O_3$. It occurs rather rarely as an accessory in ultrabasic igneous rocks, but is characteristic of the high-grade metamorphism of aluminous magnesian limestones. It crystallizes as an octahedron in the cubic system. (First usage and origin of name obscure.)

**Spinifex texture.** Interpenetrating lacy elongate olivine crystals in komatiites. The texture is commonly considered to have formed by quenching. See KOMATIITE.

**Spirit-level.** The ordinary carpenter's and household spirit-level is perhaps the geological surveyor's most important instrument. It is used in determining the fundamental attribute of a tilted bedding, fault, or other surface—the strike. Direction and amount of true dip can hardly be found accurately unless the direction of strike is found first.

**Sponge bed.** A bed with abundant fossil remains of coherent sponge-skeletons. These occur, in Britain, chiefly in the Cretaceous Greensand formations, the best known being the Lower Greensand 'sponge gravels' of Faringdon in Berkshire (Davey, *TNDFC*, 1874; Arkell, *Geology of Oxford*, 1947). 'On beds of sponge-remains in the Lower and Upper Greensand of the south of England' (Hinde, *PT*, 1885).

**Spongolite.** A rock mostly composed of sponge spicules. 'These [Carboniferous] shales are underlain by gently dipping layers of siliceous rock which, on microscopic examination, proved to be made up almost entirely of sponge spicules. About 10 feet of these spongolites . . .' (Hodson and Lewarne, *QJ*, 1961). 'Spongolites from the Arnsbergian [Carboniferous] of County Limerick, Ireland' (Lewarne, *GM*, 1963). The 'Arngrove-stone (Rhaxella-chert)' of Morley Davies (*QJ*, 1907) is an example from the Jurassic Corallian of Buckinghamshire and Oxfordshire. (Cayeux, 1897, 'spongolithe'.)

**Spotted slate.** A shale, slate, or schistose rock with conspicuous small spots resulting from a comparatively low grade of metamorphism in argillaceous rocks. There are several kinds of spots (aluminous silicates, graphite, glass; Harker, *Metamorphism*, 1950). In Britain these 'spotted slates', 'spotted schists', are known chiefly in the metamorphic rocks associated with the Skiddaw Granite (Otley, *English Lakes*, 1825; Ward, *QJ*, and *MGS*, 1876; Rastall, *QJ*, 1910).

**S-surface.** Any structural surface in a deformed or metamorphosed rock. (Sander, 1911.)

**Stable block.** A block of the earth's crust stable in the sense of not taking part in earth-movements of the neighbouring crust or of resisting the force of pressures which have defomed the rocks on one or more of its sides. A stable block is a 'horst' if it stands higher than its surroundings either as a direct result of earth-movement or as a result of subsequent differential erosion, or both. 'Shetland is part of a stable block, the Shetland-Orkney Platform' (Flinn, *PGA*, 1977).

**Stack.** A high rock off the coast, detached from the main cliff by erosion; particularly applicable if the rock is a precipitous tower and shows regular, more or less horizontal, stratification. 'Great insulated columns, called here "stacks"' (Pennant, *Tour in Scotland*, 1774.) Striking examples of stacks are to be seen along many parts of the coast of Britain; e.g. in the O.R.S. of Caithness and Orkney (Old Man of Hoy, Geikie, *Geological Sketches*, 1882), the Magnesian Limestone of Durham (Marsden Rock), the Chalk of southern England (Old Harry, Dorset), and the Carboniferous Limestone of S. Wales (Eligug Stacks, Pemb.) 'The whole development of a sea-stack may be studied on the Thanet coast' (Edmunds, *BRG, Wealden District*, 1948); and so it can on the Cardigan coast (Lower Palaeozoic). The term has been extended to an inland feature: 'An upstanding rock mass rising from a slope or a hill-top surface, clear of its surroundings on all sides, and with near vertical sides that may approximate to joint planes is best termed a "stack". It would probably be so designated by rock-climbers and similar features of the sea shore are usually called sea-stacks or marine stacks' (Linton, *GJ*, 1955).A seemingly good example of a natural inland stack is the Devil's Chimney in the Inferior Oolite of Leckhampton Hill near Cheltenham, but this has been said to be a fang of rock left after the making of a tramroad to bring down stone from the immense quarries, about 1795 (Bick, *Old Leckhampton*, 1971).

**Stage.** As a technical term in stratigraphy: a small stratigraphical unit. See CHRONOSTRATIGRAPHY. Attempts have been made to apply the conception of stages to all parts of the stratigraphical succession; but it is in active use chiefly for the subdivisions of the Jurassic and Cretaceous systems (e.g. Toarcian, Turonian). 'Realizing the all-important factor . . . to be the fauna, and impatient at the indefinite multiplication of local names for beds of differing lithological development but of the same age, he [d'Orbigny, *Paléontologie française*, 1842/9] sought to sweep aside lithology and give to the beds containing each successive assemblage a single name. The groups of strata indicated by these names he called "stages" (*étages*). Ten of these stages were to be recognized in the Jurassic System' (Arkell, *Jurassic System*, 1933).

**Stalactite.** Typically, a pendent growth of crystalline calcium carbonate formed by moisture percolating through calcareous substance, usually the roof of a limestone cave, and depositing the mineral from solution so that an icicle-like structure is formed, increasing in length and girth. Also, a similar growth of other mineral material. 'Such are the stones made of nothing but such water, as it drops from the roofs and caverns of the rocks, and therefore called "stalactites"' (Plot, *Oxfordshire*, 1677). [G. *stalaktos*, dripping.]

**Stalagmite.** Typically, a pillar, or conical or hump-shaped mass, of calcium carbonate which has grown upwards from the floor of a cave in limestone by water dripping from the roof and depositing the mineral from solution. Stalagmites, though more irregular, tend to correspond to stalactites hanging from the roof and from the tips of which much of the moisture drips, and sometimes a stalagmite will join its corresponding stalactite. The term is also used for the material in general (apart from its localization and shape) that forms in this manner on the floor or sides of a cave; and is available for similar masses and deposits of any other mineral formed in the same way. 'The cluster'd stalagmites' (Grew, *Musaeum*, 1681). 'Stalagmite preserving the bones from decomposition' (Buckland, *Reliquiae Diluvianae*, 1823). [G. *stalagma*, that which has dripped.]

**Stampian.** See PALAEOGENE. [Étampes, France.]

**Starfish bed.** A bed characterized by the fossil remains of starfish (Asterozoa). Such beds are much restricted as regards horizon and locality, the best known in Britain being in the Upper Ordovician Ardmillan series of Lady Burn, Ayrshire (Lapworth, *QJ*, 1882), the Lower Ludlow of Church Hill, Leintwardine, Herefordshire (Murchison, *Siluria*, 1859), and the Middle Lias of the Dorset coast (Day, *QJ*, 1863).

**Staurolite.** See ANDALUSITE. Provides a striking example of cruciform interpenetration-twinning, hence the name [G. *stauros*, a cross.] (Delamétherie, 1792.)

**Stegocephalia.** See LABYRINTHODONTIA. [G. *stegos*, a roof, *kephalē*, the head.]

**Stegosauria.** A suborder of the order Ornithischia, one of the two dinosaurian orders of Mesozoic reptiles. Herbivorous, quadrupedal, heavily armoured with plates and spikes. Jurassic to Lower Cretaceous. [genus *Stegosaurus* (G. *stegos*, a roof, *saura*, a lizard).]

**Steinmann Trinity.** 'Steinmann immortalised himself by pointing out in 1905 and 1927 that the axial belts of many geosynclines are characterised by what in his honour I call the Steinmann trinity: serpentines, pillow lavas and radiolarian cherts' (Bailey, *AdS*, 1952). 'Submarine eruptions are undoubtedly very common but only a small proportion are accessible to observation. . . . Geosynclinal vulcanicity.—Here and there, e.g. in Liguria, Anatolia and a few other places, uplifted portions of former geosynclines are found in which the original rock associations are still clearly recognizable. In such cases, the characteristic association of serpentinites, basaltic pillow lavas and red, often radiolarian, cherts is characteristic, as was stressed by Steinmann (the "Steinmann trinity"),while chemically precipitated carbonates (calcite, dolomite and magnesite), and oxidized iron and manganese ores (haematite, pyrolusite and psilomelane), are also found locally. Bailey was able to demonstrate in Anatolia that certain serpentinites were once submarine picritic lava flows' (Rittmann, trans., Vincent, *Volcanoes and their Activity*, 1962). 'Some aspects of the Steinmann trinity' (Bailey and McCallien, *QJ*, 1960). See OPHIOLITE. [G. Steinmann, Germany.]

**Stelleroidea.** See ASTEROZOA. [ ?L. *stella*, a star.]

**Step-faults.** Successive parallel normal strike-faults all with downthrow on the same side. The term is most appropriate where the relation between the dip of the beds and the amounts of throw of the faults is such that the beds are let down in one direction in a series of steps.

**Stephanian.** The highest of the three series into which the Silesian (Upper Carboniferous) subsystem is divided. Absent in Britain except for a few doubtful local occurrences. (de Lapparent, France, 1893.) [St-Étienne, France.]

**Stereographic projection.** A form of plane projection used particularly in crystallography and structural geology. *The Use of Stereographic Projection in Structural Geology* (Phillips, 1954/71).

**Stock. 1.** 'Smaller intrusions of similar type [to a batholith], but less elongated and with areal dimensions of only a few square miles or less, are called "stocks". Many of these are probably offshoots from underlying batholiths, of which only the highest parts have been exposed to view by denudation' (Holmes, *Physical Geology*, 1944). **2.** 'The cavernous spaces dissolved out in some rocks, more especially in limestones and dolomites, may be of any indeterminate shape, and may be filled with one or more veinstones or ores, either in symmetrical zones following the outlines of walls, floor, and roof, or in parallel and roughly horizontal bands. Irregular metalliferous masses of this kind, when of large size, have long been known in Germany by the name of stocks (*stocke*)' (Geikie, *Textbook*, 1882).

**Stockwork.** 'A stockwerk is a mountain-mass of greater or less extent, traversed in all directions by a very great number of small veins' (Jameson, *Geognosy*, 1808). 'The name "Stockwork" is usually applied to large masses of rock impregnated with metallic ores or intersected by a number of mineral veins at short distances apart, sometimes crossing one another in all directions. Not having any word of our own, we have adopted the term from the German *Stockwerk*. This expression probably owes its origin to the method of working formerly often adopted for such deposits, which were wrought by chambers arranged in "tiers" or "stories" '(Foster, *QJ*, 1878).

**Stone.** **1.** 'A stone' is a piece, usually a small or moderate-sized piece, of hard rock or hard mineral substance, other than metal. **2.** 'Stone' is the compact material of which (hard) rocks consist. Rock composed of organic, non-mineral, material, such as coal, is excluded. 'From Old English *stan*, recorded from the 9th century and common to all Teutonic languages. It has left its mark on the place-names of England more than any other geological word' (Arkell and Tomkeieff, *Rock Terms*, 1953). See quotations under CONSOLIDATION.

**Stone age.** The period during which Early Man lived and made implements of stone (chiefly flint), thus leaving behind him these artifacts as evidence of his existence and habits. The whole period corresponds roughly with the Pleistocene up to the Bronze age, which began about 4,000 years ago. The Stone age is divided into the Old Stone age, or Palaeolithic (by far the longest part, corresponding roughly to the Pleistocene excluding the Recent), the Middle Stone age or Mesolithic, and the New Stone age or Neolithic. If eoliths are taken to be artifacts, an Eolithic, or Dawn Stone age, may precede the Palaeolithic (if these objects are also taken to be of an age older than the Palaeolithic). *The Old Stone Age* (Burkitt, 1933). *The Irish Stone Age* (Movius, 1942). *The Stone Age in Scotland* (Lacaille, 1954).

**Stone polygon.** A polygon of stones formed in some stony-soil regions by frost action. Other named forms are 'stone ring', 'stone net', &c. (see Sharpe, *Landslides*, 1938).

**Stone river.** See SOLIFLUXION. 'The stone-rivers of the Falkland Islands' (Anderson, *JG*, 1906).

**Stonesfield Slate.** See GREAT OOLITE SERIES. (Smith in Farey, *Derbyshire*, 1811.) [Stonesfield, Oxfordshire.]

**Stoping.** The wedging off, and assimilating, of blocks of country rock by an intruding magma, particularly one working upwards. The full term is 'magmatic stoping'; originally, 'overhead stoping'. (Daly, 1903.)

**Storm beach.** A shingle beach built up by storm waves at exceptionally high tides, immediately above the level reached by normal spring tides. 'The great storm beach known as the Chesil Beach or Bank (from *Ceosol, Cisel*, Saxon for shingle) extends from Bridport to the Isle of Portland, a distance of 18 miles' (Arkell, *MGS, Weymouth and Swanage*, 1947).

**Strain.** See STRESS. 'Analysis of strain in deformed rocks' (Hobbs and Talbot, *JG*, 1966). 'Strain study of the Islay Caledonides' (Borradaile, 1979).

**Strain-slip cleavage.** See CLEAVAGE (1b). (Bonney, *PGS*, 1886).

**Strata and time.** 'It is important to bear in mind the limitations to our ability to interpret strata in terms of a general time-scale, to carry over the observable realities of superposition, lateral extension and character of make-up and content, into the intangible dimension of time whose onward flow has been so wonderfully, yet so fortuitously, recorded in the rocks. We may classify the strata as elaborately as we please and locally draw sharp lines to separate our divisions; but we can only approximately correlate the strata, as regards time, from region to region. True "time-planes" though present in the rocks, can never be detected as such, particularly over a wide area. There is no feature of a rock, structural, lithological, or palaeontological, that is any definite indication or test of simultaneity' (Challinor in Bell, ed., *Darwin's Biological Work*, 1959).

**Strata identified by fossils.** The phrase which crystallizes the empirical law that strata can, to a large extent, be individualized, and thus identified, by the particular kinds of fossils they contain. This is made possible because the succession of the different kinds of fossils in the rocks is one of continuous and irreversible change, while the kinds of rocks themselves change in time in a haphazard manner and do not bear the stamp of particular ages. See CHARACTERISTIC FOSSIL, CORRELATION, and STRATIGRAPHY. *Strata identified by Organized Fossils* (Smith, 1816).

**Stratiform.** In the form of strata; layered. 'The stratiform ore-deposits of Derbyshire' (Ford, in *Sedimentary Ores*, Leicester University, 1969).

**Stratify.** To lay down or dispose in layers, particularly as 'stratified' and 'stratification'. These terms are rather wider than 'bedded' and 'bedding' as they can include any somewhat horizontal layering in an igneous intrusion. 'In applying the expressions "strata", "stratified", &c. to trap I only mean to imply that there is a parallelism of texture' (De la Beche, *TGS*, 1826). Nevertheless the full term 'stratified rocks' would be taken to refer to sedimentary rocks (in the wide sense), not igneous rocks. The stratification (bedding) of sedimentary rocks is assumed to be due to conditions of original sedimentation, but in some cases (as possibly in that of the Blue Lias) it may be due to, or emphasized by, later concentrations of, particularly, the lime ingredient (as first suggested by Maton in his *Western Counties*, 1797). 'I will take a general view of the prevailing rocks and strata of this island, to see which of them are regularly stratified, and which of them are not, with the different degrees of stratification' (Williams, *Mineral kingdom*, 1789). 'Terminology for stratification and cross-stratification in sedimentary rocks' (McKee and Weir, *BGSA*, 1953). 'Stratification' can refer to structure on a large scale: 'From the top of Ben Shianta the stratification of Jura is exhibited in great beauty, and with a striking magnificence of natural perspective; presenting a scene of uncommon grandeur and effect' (Macculloch, *Western Islands of Scotland*, 1819). Small-scale stratification is lamination. See quotation (Shannon) under COMPOSITE IGNEOUS BODY.

**Stratigraphical formation.** See FORMATION.

**Stratigraphical geomorphology.** Geomorphology studied in the light of facts provided by the present attitude and altitude of strata (including comparatively recent deposits) that can be inferred as having been laid down in a different attitude and/or altitude. Thus, for instance, a stratum, or a series of strata, taken to have been laid down (unconformably) as one synchronous sheet over a widely-extended horizontal land-surface near sea-level may, with its underlying platform, now be found tilted or warped at a higher altitude. The change in attitude and altitude of that land-surface since the deposition of the overlying stratum is thus immediately inferred. ' Stratigraphical geomorphology' (Bishop in Dury, ed., *Essays in Geomorphology*, 1966).

**Stratigraphical palaeontology.** The study of the fossil faunas and floras of the strata, in space and time, to reveal the distribution of life at times in the past and the course of the history of life through the ages. The distinction between biostratigraphy and stratigraphical palaeontology may be illustrated by a remark by J. D. Dana in 1855 (quoted by Merrill, *First One Hundred Years of American Geology*, 1924): 'Instead of saying that fossils are of use to determine rocks we should rather say that the rocks are of use for the display of the succession of fossils'. Nevertheless, in Britain at least, 'stratigraphical palaeontology' is not infrequently used in titles of papers that are, on the above distinction, biostratigraphical. 'The reader perhaps, will object that our premises must lead the student's attention wholly to the organic form, [but] I wish forcibly to impress it on his mind, that, however extensive or well-arranged his cabinet, the knowledge acquired there of extraneous fossils must ever be defective unless joined to the study of these bodies in their native repositories' (Martin, *Extraneous Fossils*, 1809). 'The first attempt at a chronological succession of fossil organisms is to be found in H. G. Bronn's *Lethaea Geognostica* (1835-38)' (Zittel, *History*, 1901). 'The Vicomte d'Archiac devoted his attention to fossils and their distribution in various geological formations, and in 1864-65 issued his *Paléontologie Stratigraphique*' (Woodward, *History*, 1911). *Stratigraphical Palaeontology* (Neaverson, 1928/55).

**Stratigraphical procedure.** The considerations involved in the study of stratigraphy; aspects, methods, classification, and terminology. Since the early International Geological Congresses in the late nineteenth century and the publication of some national stratigraphical codes there have been successive attempts to standardize these matters. Among the latest are: *International Stratigraphical Guide* (Hedberg, ed., 1976) with extensive bibliography, and see MAGNETOZONE; 'Essay review: International Stratigraphic Guide' (Harland, *GM*, 1977); *A Guide to Stratigraphical Procedure* (Holland and others, *GSSR* (11), 1978); 'Geochronologic scales' (Harland in *The Geologic Time Scale*, *AAPG*, 1978).

**Stratigraphical system.** A major general division of the whole succession of the stratified rocks; e.g. *The Silurian System* (Murchison, 1839). See CHRONOSTRATIGRAPHY.

A particular stratigraphical system is defined as comprising those strata that contain particular groups, families, genera, and species of fossils. Some of these are provisionally taken as being confined to the system so that each by itself is diagnostic; others occur also outside the system but some combination of them will be diagnostic. The system also includes all other strata that, while being either unfossiliferous or without fossils diagnostic to any one system, show, by occupying a similar position in the geological column relative to known systems above and below, that they are stratigraphically equivalent to strata already placed in the system, that is, presumably, laid down as sediments during the same period of time. A system is in the first place established as a result of the investigation of a region. In this region the limits to the system will be drawn according to the whole local structural arrangment; but a general stratigraphical boundary between two systems can be defined only where the top of one passes into the bottom of the other without a break. Stratigraphical horizons separating general stratigraphical divisions, particularly those separating systems, are ideally chronological horizons also, that is, they are time-planes, isochronal planes. An unconformity between one system and another means an incomplete succession and the unconformity itself will be diachronous unless there is no overlap at all. It is chronologically unsafe to define the base of a system by a palaeogeographical change, such as one resulting from a marine transgression, which would be unlikely to occur everywhere at the same time; for a general definition of the boundary we should have to find where the narrow edge of the wedge-shaped time-gap became a bedding plane in a continuous sequence. Apart from that, the choice of a boundary is guided by two opposing considerations ; firstly to find an horizon where there is some marked faunal change other than one due to local change in facies, and secondly to find a lithologically distinct and, if possible, mappable, horizon in each separate region. The only way to define precisely, unambiguously, and with universal validity, the stratigraphical, and at the same time the chronological, limits of a system (or any other stratigraphical division) is to name a particular horizon in a particular locality and say that that horizon and its time-equivalents everywhere are to be the base of the system; the top of the system being at the similarly defined base of the next system. The fact that the exact time-equivalent horizon, in another region, of the horizon chosen in the standard region, can never be certainly known is another matter; at least it is known what has to be found as nearly as possible. (This principle has been adopted in the Geological Society's 'stratigraphical code', *PGS*, 1967, and confirmed in later versions, where our 'particular horizon in a particular locality' is termed a 'marker point'—see BOUNDARY STRATOTYPE.)

The following is a table of the stratigraphical systems:

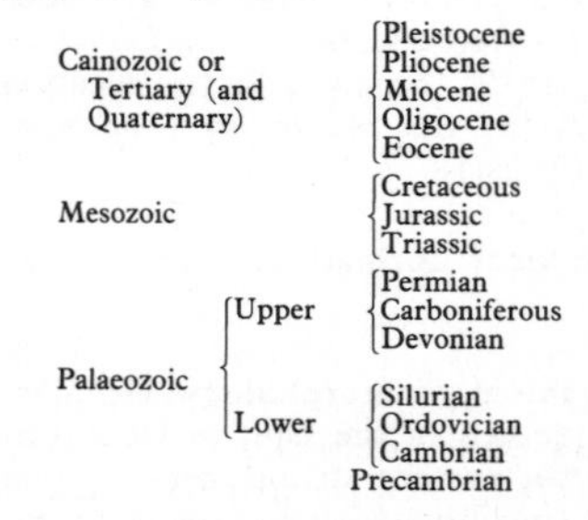

| Era | Division | System |
|---|---|---|
| Cainozoic or Tertiary (and Quaternary) | | Pleistocene |
| | | Pliocene |
| | | Miocene |
| | | Oligocene |
| | | Eocene |
| Mesozoic | | Cretaceous |
| | | Jurassic |
| | | Triassic |
| Palaeozoic | Upper | Permian |
| | | Carboniferous |
| | | Devonian |
| | Lower | Silurian |
| | | Ordovician |
| | | Cambrian |
| | | Precambrian |

Notes. (*a*) It is not yet possible to recognize distinct systems among the Precambrian rocks; (*b*) in America the Carboniferous system is divided into two systems,

Mississippian (lower) and Pennsylvanian (upper); (*c*) on the Continent and elsewhere the Carboniferous and Permian are often taken together as one Permo-Carboniferous system; (*d*) in Britain (and elsewhere, e.g. the Pyrenees) the Permian and Trias are conveniently taken together as one Permo-Triassic system; (*e*) the divisions of the Cainozoic are sometimes called systems, or the Eocene and Oligocene are grouped together as the Palaeogene system and the other three, but sometimes excluding the Pleistocene, as the Neogene system; (*f*) the names of the systems (always with their capital initial letters) are adjectival in form but are used as either nouns or adjectives; (g) as implied above the stratigraphical boundaries between systems and between lower divisions in the hierarchy are drawn at arbitrary levels, depending largely on the accidents of the history of the investigation: 'if the pioneers of stratigraphical geology had worked in New Zealand instead of in Western Europe, the stratigraphical column would have been divided at different levels from the present world standard divisions' (Fleming, *QJ*, 1969).

**Stratigraphical unit.** Recognizable stratigraphical units are either lithological (lithostratigraphical) or faunal (biostratigraphical) or both. They do not constitute time-rock units though either, particularly the faunal unit, may approximate to such. They may be facies units. The formation, the zone, and the local series are examples of stratigraphical units.

**Stratigraphy.** 'Stratigraphy is the study of rock successions and the interpretation of these as sequences of events in the geological history of the earth. The fundamental procedure in stratigraphy is to observe, describe and correlate such successions' (*Guide to Stratigraphical Procedure*, *GSSR* (11), 1978). 'Mr. W. Smith first noticed that certain fossils are peculiar to, and are only found lodged in, particular strata; and who first ascertained the constancy in the order of superposition, and the continuity of the strata of this island. These observations have lately also occurred to Messrs. Cuvier and Brongniart whilst examining into the nature of the strata of the neighbourhood of Paris' (Parkinson, 'Observations on the strata in the neighbourhood of London and on the fossil remains contained in them,' *TGS*, 1811; a detailed application of the principles to a particular British region). William Smith's three chief works appeared after this statement was published; his geological map of England and Wales (1815), his *Strata Identified by Organized Fossils* (1816), and his *Stratigraphical System of Organized Fossils* (1817). 'Stratigraphical' thus appears in this last title. The *OED* gives a quotation from the *Reader*, 1865, for 'stratigraphy': 'While accepting as a basis in theoretical geology the principles of Hutton, and in stratigraphy the work of William Smith, he [&c.]'. 'The leading principles of stratigraphy' (Geikie, *Textbook*, 1882). *Principles of Stratigraphy* (Grabau, 1913). 'William Smith and the birth of stratigraphy' (Cox, *IGC*, 1948). *Principles of Stratigraphy* (Dunbar and Rodgers, 1957). '*The Stratigraphy of the British Isles* (Rayner, 1967). *Lexicon of Stratigraphy for Britain* (Whittard and Simpson, eds., 1958; part of *Lexique Stratigraphique International*). *A Dynamic Stratigraphy of the British Isles* (Anderton and others, 1979). While normally referring to sedimentary rocks (in the wide sense), 'stratigraphical' (and its variants) can be used for lavas: 'The lavas [in Iceland] are readily treated stratigraphically . . . a systematic thinning of each stratigraphic group . . .' (Walker, *CGA*, 1958). See BIOSTRATIGRAPHY, CHRONOSTRATIGRAPHY, and LITHOSTRATIGRAPHY.

The adjectives 'stratigraphical', 'stratigraphic', are largely synonymous; but a discrimination is often made, particularly in American writings, corresponding to the discrimination between 'geological' and 'geologic' (see GEOLOGY). We thus have 'stratigraphical palaeontology' but 'stratigraphic sequence'.

**Stratomere.** 'A stratomere is any segment of a rock sequence. Stratomeres contain the evidence of geological events, and are of various orders of size according to the magnitude and kind of events that produced them' ('Stratigraphical code', *PGS*, 1967). 'System', 'series', 'stage', used in strict stratigraphical senses, are 'stratomeric' terms. [G. *meros*, a part.]

**Stratotype.** 'A stratigraphic section, selected, named, and published for reference' (Harland, *AAPG*, 1978). Most commonly used for a 'boundary stratotype'. '. . . the British countryside which has provided the stratotypes for so much of the world's progress in geological knowledge' (Hedberg, *JGS*, 1976). Often in the form 'type section.'

**Strato-volcano.** A volcano stratified in being made up of alternate lava-flows and sheets of fragmental material.

**Stratum.** 'Stratum' and 'bed' are synonymous terms; each means a layer of rock. In the singular, 'bed' is more usual than 'stratum'. See BED. 'Strata', the plural, is, in practice, a rather wider term than 'beds' and is commonly used for the layers, in general, which are so characteristic of sedimentary rocks (in the wide sense). Thus we should speak of, for instance, the 'Jurassic strata of England' but a particular stratum or set of strata would be more likely to be called a bed or beds (e.g. a 'starfish bed' or 'Purbeck beds'). Hence 'stratal', of or pertaining to strata, e.g. 'stratal dip'. 'To the sediments of fluids do belong, the strata or beds of the Earth' (Steno, trans., 1671). 'Terrestrial matter that is naturally disposed into layers or strata' (Woodward, *Natural History of the Earth*, 1695). 'The East Mountains [Pennines] consist of several strata, having an easy horizontal depression', and 'All the mines [coal seams], and those solid strata which are their natural covers . . .' (Robinson, *Westmorland and Cumberland*, 1709). 'Concerning some general observations on the strata in Derbyshire' (Appendix to *Inquiry into the Original State and Formation of the Earth*, Whitehurst, 1778/86). [L. *stratum*, something spread out or laid down.]

**Streak.** Of a mineral; the colour of its powder. This powder is most conveniently produced and its colour observed by scraping the mineral on an unglazed porcelain surface, a 'streak plate'.

**Stress.** Geologists make a strict distinction between 'stress', the system of forces within a rock-body, and 'strain', the changes in size and shape resulting from those forces.

**Stress mineral.** 'Even a very general and cursory view shows us that numerous minerals are, in greater or less degree, characteristic of the crystalline shists as contrasted with thermally metamorphosed rocks. Some of these minerals are peculiar to the crystalline schists. Of others it may be said that they are of much more frequent and widespread occurrence there than as products of simple thermal metamorphism, or that they are promoted, in the sense of coming in at an earlier stage of progressive metamorphism. All these may conveniently be designated "stress-minerals". If, on the other hand, we consider the minerals of thermal metamorphism, we find that some of them are peculiar to this mode of origin. Others, which are common in this association, are less prominent as constituents of crystalline schists, or appear there only in a more advanced grade of metamorphism. All these may be termed for the sake of convenience "anti-stress minerals" ' (Harker, *Metamorphism*, 1950; terms first proposed by him, *QJ*, 1918).

**Strike** The direction, at any point on a structural surface, particularly a bedding surface, of a horizontal line drawn on the surface through that point. The direction can be specified as the bearing, one way or the opposite or both. The strike is at right angles to the direction of (true) dip at the point. 'There is this general rule, that, having found your "dipp" and "rise", to what ever points that course is directed, the "streek" is to the quite contrary. For supposing a coal dipp SE, it must needs follow, that the "streek" must run SW and NE' (Sinclair, *Short History of Coal*, 1672). 'The "streik" is a longitudinal direction of the strata, and the "dip" is the inclination of the stratum to the horizon' (Walker, *MS*, before 1780, printed in Scott, *Lectures on Geology by John Walker*, 1966). 'The term "strike" has been recently adopted by some of our most eminent geologists [e.g. Sedgwick and Murchison, *TGS*, 1829] from the German *streich*, to signify what our miners call the "line of bearing" of the strata. Such a term was much wanted, and as we often speak of "striking off" in a given direction, the expression seems sufficiently consistent with analogy in our language' (Lyell, *Principles*, 1833). 'Where the cleavage is well developed in a thick mass of slate rock the strike of the cleavage is usually nearly coincident with the strike of the beds' (Sedgwick, *TGS*, 1835). 'Strike' is occasionally used as 'strike on the ground', that is, the general direction of outcrop of strata (or other rock-sheets); this of course depends on the direction and amount of slope of the ground as well as on the strike and dip of the strata. 'Strike' has been used in a topographical sense, for the alignment of a chain of hills (Callaway, *GM*, 1879).

**Strike-fault.** A fault of which the strike coincides with, or approximates to, the strike of the beds it affects. An important feature of such a fault-structure is the relation between the dip of the fault and the dip of the beds. Here there are three significant alternatives: (**1**) fault dips 'against' the dip of the beds, (**2**) fault dips 'with', and at a greater angle than, the dip of the beds, (**3**) fault dips 'with', but at a less angle than, the dip of the beds. These alternatives, together with the two alternatives of 'normal' or 'reverse' movement on the fault plane, determine whether some outcrops are repeated or cut out. The slope of the ground, if considerable, may also be a factor.

**Strike-line.** A structure contour which is sufficiently regular to keep a more or less constant direction and thus define a similarly constant strike of the structural surface (usually a bedding surface). Where the surface is plane, strike-lines are straight, parallel, and equally spaced for equal vertical intervals.

**Strike-slip fault.** See SLIP and WRENCH FAULT.

**Stromatolite.** (Stromatolith.) A term etymologically vague but now practically restricted to a rock having a banded structure suggesting an organic, particularly an algal, origin. Hence 'stromatolitic'. (Kalkowsky, 1908.) *Stromatolites* (Walter, ed., 1976). [G. *strōma*, *strōmat-*, something spread out.]

**Stromatoporoidea.** An extinct group of fossils not very firmly established as regards its biological affinities but usually placed as an order of the class Hydrozoa, phylum Coelenterata. The calcareous skeleton of laminae and pillars, with ramifying interspaces, occurs in irregular rounded, sheet-like or branching masses from half an inch to a foot or two across in general size. Stromatoporoids range from the Cambrian to the Cretaceous, being common fossils in the Silurian and Devonian. (Nicholson and Murie, 1878.) [genus *Stromatopora* (G. *strōma, strōmat-*, something spread out, *poros*, a passage, pore).]

**Strombolian type.** See VOLCANICITY. Lava, gases, bombs, and lapilli are ejected in comparatively mild explosions at intervals of some minutes. (Mercalli, Italy, 1883; Judd, *Volcanoes*, 1881; Blackburn and others, 'Strombolian activity', *JGS*, 1976). [Stromboli, a volcano in the Lipari Islands.]

**Structogenesis.** The beginning and (by extension) the early development of structural history.

**Structural geology.** 'Structural geology treats of the arrangment of rocks . . . in short, with the architecture of the earth's crust' (J. Geikie, *Structural and Field Geology*, 1905/53). It is one of the major departments of the science of geology. *Structural Geology* (Leith, 1923). *Structural Geology* (Hills, 1963/72). *Structural Geology* (Dennis, 1972).

**Structural map.** See TECTONIC MAP.

**Structural petrology.** Usually means the study of rock-fabric. See FABRIC and PETROFABRIC.

**Structural style.** See TECTONIC STYLE and quotation (Woodcock) under SLUMP.

**Structural uplift.** As a quantitative value: the vertical distance between the trough of a syncline and the crest of an anticline. 'Some of the major anticlines [in SW.

Persia] show structural uplift from syncline to anticlinal summit of as much as 25,000 feet and are separated by broad synclines' (Lees, *QJ*, 1952). The term would also apply to the upthrow (in one sense or another) of a fault or series of faults. Also called 'structural relief'.

**Structure.** The way in which a rock, a rock-mass, or a whole region of the earth's crust is made up of its component parts; the form and mutual relations of the parts of a rock. It includes stratification, lamination, unconformity, folds and faults, jointing, the forms of igneous intrusions, flow banding, slaty cleavage, schistosity, foliation, &c. It is concerned with rock-units (of any kind, large or small), but not with the particles composing the rock (texture). As for various kinds of individual objects in rocks, British usage would probably include concretions, for instance, as 'structures' but exclude ordinary fossils and mineral specimens. Hence 'structural'. See TECTONIC. 'Leonhard (1823) followed Haüy in defining structure as the fine pattern or fabric (*Gefüge*) of a rock. Brongniart (1827) was the first to distinguish between structure and texture. In this he was reversing the previous connotation of the term "structure", which he applied to the large-scale features of rock masses, reserving the term "texture" for the fine features or, more precisely, the granularity of the rock. Since the time of Leonhard and Brongniart the terms "structure" and "texture" have been used in different senses by different petrologists. As the matter stands at present petrologists in France, England, the United States of America and certain other countries reserve the term "texture" for the fine features of rocks and the term "structure" for broad features, while those in Germany use the terms in the opposite sense' (Loewinson-Lesing, trans. Tomkeieff, *Historical Survey of Petrology*, 1954). 'Notes respecting the geological structure of the vicinity of Dublin' (Fitton, *TGS*, 1811). *The Structure of the British Isles* (Anderson and Owen, 1968).

**Structure and surface.** There are two quite different kinds of expression of structure in the surface relief. A mountain range formed along a belt of highly folded and dislocated rock-structure may be due either directly to the uplift of the structure or indirectly to the fact that hard rocks are brought to the surface and stand out as a belt of high relief as a result of general differential erosion; or to both these separate causes combined. Usually the phrase implies the dependence of relatively minor features on the lithological character and attitude of individual rock-masses, the result of smaller-scale differential (selective) erosion. 'We now pass to the Goldsitch Coal-field, throughout which the beds lie with the most perfect regularity in the form of a long trough. On each side of the central valley, which is occupied by Lower Coal-measures, the massive gritstones slope up the hill sides in broad sheets of heather-clad rock, ending at top in rugged crags, with the broken ends of the beds sticking boldly out into the air. The synclinal arrangement of the strata is thus shown as clearly as in a model, and perhaps it would be difficult to find a place where the shape of the ground points out so unmistakeably the geological structure of the country' (Green, *MGS, Stockport* [&c.], 1866). *Structure and Surface* (Brown and Debenham, 1929). 'The relationship that exists between the features of a landscape and the rocks of which it is composed . . . beauties of form in relation for the framework of rock' (Richey in Harker, *The West Highlands and The Hebrides*, 1941). 'The influence of rock structures on coastline and cliff development around Tintagel' (Wilson, *PGA*, 1952). *Rocks and Relief* (Sparks, 1971). See quotation (Cotton) under TECTONIC.

**Structure contour** (usually in the plural). Lines drawn, each at a constant level, and normally at constant vertical intervals, on a structural surface; usually a bedding surface but it may be a surface of unconformity, a fault surface, or any other surface. They show the form of the structural surface in the same way as contour lines on the earth's surface show its relief. 'A structure contour map of the surface of the buried pre-Permian rocks of England and Wales' (Kent, *PGA*, 1949).

**Style.** The manner in which something is performed or appears. Can apply in various connexions in geology, but is most commonly used in 'tectonic style'.

**Stylolite.** A part of a rock-body (usually a limestone) characterized by a very irregular interlocking internal surface which shows in section as a very irregular suture having the appearance of the tracing of an oscillating stylograph. The surface is often marked by the accumulation of relatively insoluble material indicating that it was probably produced by chemical erosion under some kind of pressure-controlled solution. 'The origin of stylolites' (Brown, *JSP*, 1959). (Kloden, 1828.) [L. *stilus*, an ancient implement for scratching letters on a wax-covered tablet.]

**Subaerial.** Conditions and processes existing and operating on land, that is, under the atmosphere. Most often used in connexion with erosion, when subaerial erosion is being distinguished from marine erosion. 'On subaerial denudation' (Whitaker, *GM*, 1867).

**Subangular.** Somewhat angular ; having rounded edges and corners. A typical glacial boulder is subangular.

**Subaqueous.** Under water; for all processes, and their results, taking place on the floor of the sea, or a lake, or on the bed of a flowing stream. Particularly as 'subaqueous deposits' (Lyell, 1833), 'subaqueous sliding', &c.

**Sub-bituminous coal.** Coal of a rank intermediate between lignite and bituminous coal. It occurs in the Mesozoic and Tertiary, rather than the Carboniferous, coalfields.

**Subcrop.** A rock formation or surface (such as a fault) where it is truncated by a surface of unconformity. Hence 'subcrop map'. 'A subcrop map is simply what is found now beneath an unconformity (a paleogeologic map would extrapolate these subcrops where eroded, or not observed, into a completed map). It is the top surface of overstepped strata and a "worm's eye map" is the bottom surface of overlapping strata' (Harland, *GM*, 1961).

**Subduction zone.** A term used in describing the concepts of plate tectonics. It is the zone at an oceanic trench in the case where one plate slips down under the margin of another.

**Subhedral.** For a condition, as regards crystal shape, between euhedral and anhedral. See HYPIDIOMORPHIC. [G. *hedra*, base.]

**Subjacent.** A term proposed (Daly, *Igneous Rocks*, 1933) for the larger igneous masses, to describe merely their observed relation to the overlying rocks, without any implication 'regarding shape, volume or mode of emplacement'. Hatch and Wells (*Igneous Rocks*) have a more definite concept, for which the term itself (which does not imply 'bottomless') is doubtfully appropriate: 'Subjacent plutons are major intrusions which have no visible floor. The "walls" are generally steeply inclined and these rock bodies tend to increase in size with depth' (Slightly shortened.)

**Sublevation.** 'The sea floor is generally mantled with loose sediment, and the currents are broad and ill-defined sheets of moving water that gently sweep the bottom. The erosion that takes place under such conditions differs in important respects from that on the land, both in the processes involved and in the results achieved. To distinguish this special type of submarine erosion we suggest the term "sublevation" (L. *sublevare*, to lift up). It is to be used for the degradation of a sea floor composed of loose sediment. For example, storms generate waves and currents that sublevate the sea floor even in areas of general aggradation' (Dunbar and Rodgers, *Stratigraphy*, 1957).

**Submarine.** Has many applications e.g. 'submarine canyon', 'submarine volcano'. Deposits laid down on the sea floor are usually called 'marine' but processes of sedimentation are usually called 'submarine'.

**Submarine canyon.** 'Steep-sided canyons in the continental shelf have been demonstrated to exist in many parts of the world and their origin has given rise to much discussion' (Stamp, ed., *Geographical*

*Terms*, 1961). 'Origin of submarine "canyons" ' (Daly, *AJS*, 1936). *The Origin of Submarine Canyons* (Johnson, 1939). *Submarine Canyons and other Sea Valleys* (Shepard and Dill, 1966). See quotation under SEA-MOUNT. Such 'canyons', 'gullies', or 'channels' have been detected as palaeogeographical features, e.g. by Whitaker in Silurian times in the Leintwardine area (*QJ*, 1962).

**Submarine geology.** 'Geology is usually defined as the history of the earth but actually it has been largely limited to the history of the continents, which constitute only 28 per cent. of the earth's surface. The study of the other 72 per cent. is called submarine geology or alternatively either marine geology or geological oceanography. All these names seem equally acceptable and can be used as synonyms. Included in submarine geology is the study of coasts and shorelines so far as they are related to the sea and its processes; the continental shelf, consisting of the broad platform that surrounds most of the coasts, the continental slope leading from the shelf down to the deeps, and the deep ocean floor beyond. Both the topography of the sea floor and the sediments are considered as well as the geophysical data that provide information on what is beneath the sea floor. The effect of marine organisms on the sea floor and on the accumulating sediments is also part of the subject matter' (Shepard, *Submarine Geology*, 1963, 1st ed. 1948). *Submarine Geology and Geophysics* (Whittard and Bradshaw, ed., 1965). 'Submarine geology of Start Bay determined by continuous seismic profiling and core sampling' (Kelland, *JGS*, 1975).

**Submask map.** See SOLID GEOLOGY. 'Submask geology in Saskatchewan' (*SGS*, 1956).

**Submerged forest.** Remains of stumps and trunks of trees (few or many) visible between tide marks (permanently or at times), or detected below low-water mark, still in the place of growth, usually in peaty soil. The position of these remains below high-water mark may be due to actual relative subsidence of the land, or to the overwhelming, and perhaps undermining, by the advance of the sea, of a swampy woodland growing, for instance, behind a shingle bank. The submerged forest at Newgale, Pembrokeshire, was described by Giraldus Cambrensis (1188). 'On the coast of Lincolnshire the remains of a forest have been observed which are now entirely covered by the sea . . . this submarine forest . . .' (Playfair, *Illustrations*, 1802). 'Sunk forests and raised beaches' (De la Beche, *Geological Observer*, 1851). 'On the submerged forest at Blackpol, near Dartmouth' (Pengelly, *TDA*, 1869). *Submerged Forests* (Reid, 1913). 'These strata—lacustrine, estuarine, and terrestrial sediments—are known as the Submerged Forest Series, and are usually exposed at or below high water mark of present-day tides in positions where they could not possibly be formed now. The terrestrial accumulations are soils that have been compacted to form peat' (Pringle and George, *BRG, S. Wales*, 1937). 'The submerged forest at Borth and Ynys-las, Cardiganshire' (Godwin and Newton, *NP*, 1938).

**Subsequent stream.** See CONSEQUENT DRAINAGE.

**Subsidence.** In addition to its ordinary general meaning there is the following: 'When used in relation to the materials that make up the surface layers of the earth, subsidence may mean sinking to a lower level, as when a road or building settles lower into the ground, but it may also mean a sudden collapse of surface material into a subterranean void' ('Some geological aspects of subsidence', North, *PSWIE*, 1952).

**Subsilicic.** See PERSILICIC.

**Subsoil.** The transitional zone between fully developed soil and unweathered or unaffected rock-material below. This is chiefly to be seen where the soil is developed in loose material weathered *in situ*, the subsoil then consisting of the partly broken up pieces of bedrock penetrated, perhaps, with some rootlets but containing little vegetable mould. Such an upward sequence of solid bedrock, subsoil, and soil is often well seen in a quarry face.

**Subsolvus.** See HYPERSOLVUS.

**Subspecies.** In biology (and thus in palaeontology), a taxonomic category for communities of individuals, within a species, characterized by minor structural peculiarities. Such communities occurring in regions, perhaps with some degree of isolation, are geographical subspecies; similar communities, separated in geological time, are chronological subspecies. The name of a subspecies has three component words, the generic, the specific, and the subspecific.

**Substratum.** 'Underlying layer or substance, foundation, basis' (*OED*). In geology there are two special senses, **1.** The 'solid' rock underlying soil or superficial deposits. 'Soils are much dependent on the decomposition of their substrata' (Maton, *Western Counties*, 1797). **2.** A somewhat vague concept of a part of the earth-body either wholly (but immediately) underlying the crust or including the lower part of the crust (this alternative depending largely on the conception and definition of 'crust'). 'We have every reason to believe that the continents are, in structure, much less homogeneous than the substratum which supports them' (Joly, *Surface History of the Earth*, 1930). 'In this text the term substratum is used for the mobile upper part of the mantle, directly below the Mohorovičić discontinuity' (King, *Morphology of the Earth*, 1967).

**Subsurface geology.** See UNDERGROUND GEOLOGY. *Handbook of Subsurface Geology* (Moore, 1963).

**Sub-volcanic.** Sub-volcanic rocks are small bodies of igneous intrusive rock formed not far below the earth's surface and directly connected with volcanic rocks formed on the surface; e.g. a rock formed from a magma 'feeding' a volcanic vent. Petrographically, the rock-types may be the same as those of truly extrusive volcanic rocks. In certain ring-complexes, rocks of plutonic types have become emplaced beneath previously erupted volcanics; these are also called 'sub-volcanic'. A third connotation is the following: 'As, in general, the temperature of the country rock increases with depth, a deep-level intrusion would be expected to have a coarser texture than one of the same size intruded at a higher level. This is usually the case, but when a magma solidifies in the lower part of a volcanic conduit where the walls have been heated to a considerable degree by the continued passage of lava, coarse-grained rocks may result at very shallow depths. These are the "subvolcanic" rocks of Washington' (Nockolds, *Petrology for Students*, 1978).

**Succession.** The following-in-order of rock groups. Any statement as to this is, in the first place, a statement of observed fact; the sequence, in actual superposition of strata seen as outcrop-bands or in exposed section. But a 'succession' of the rocks, without any warning qualification, and particularly if given as a table, implies that this represents an orderly succession in time; it implies that the succession at outcrop has been tested for simplicity and, if found complicated, that the structure has been unravelled so that any repetitions have been detected and eliminated, any overturnings put right, and any igneous intrusions extracted. The succession would also show where any structural breaks (unconformities) occur. A succession may be a purely faunal one; e.g. a 'graptolitic succession' given as a statement of successive assemblages or as a table of graptolite zones. A 'geological column' shows a succession graphically, in a detailed and compact form. Plot (*Staffordshire*, 1686) gives a detailed succession of a part of the Coal Measures in Staffordshire, and Whitehurst (*Strata in Derbyshire*, 1778) one in the Carboniferous Limestone and Millstone Grit. 'Lehmann [chief works published, Berlin and Paris, 1753/9] came to recognize that the earth's crust was not a structureless mass but was built up of layers or "strata" which succeeded one another in definite order. His contemporary, Fuschsel, also observed this fact. Both of them worked out the succession as seen in areas which they examined. The geological succession which they established forms the basis of the more extended and complete column of formations, and of the terminology connected with them, which was a few years later set forth by Werner, and was by him given such wide publicity, that it is now associated with his name' (Adams, *Birth and Development of the Geological Sciences*, 1938).

**Suite.** Used particularly in petrology, for a set of cognate igneous intrusions.

**Summit-level, Summit-plane.** See HILL-TOP SURFACE.

**Superficial deposit.** There is usually a clear distinction, in occurrence, between the more or less loose superficial deposits (the chief of which are alluvium, blown and beach sand, glacial drift, plateau gravels, and peat) and the underlying 'solid' rocks. A mere soil which has developed in place as a result of the weathering of the underlying rock is usually taken to be part of that rock (for mapping purposes at least) and not as a superficial deposit to be noticed separately (except as a pedological matter).

**Superficial structure.** A structure, particularly a deformation structure in sedimentary rocks or in superficial deposits, which has originated and formed under recent superficial conditions and forces. The term includes glacial push and drag, freezing and thawing of ground ice, surface creep, etc., but the term is used particularly for the kinds of structure described by Hollingworth and others in 'Large-scale superficial structures in the Northampton ironstone field' (*QJ*, 1944): 'Detailed mapping reveals many structures, unrelated to the slight regional tilting, folding and faulting, that are determined not by deep-seated movements but by disturbances that are demonstrably of superficial origin. These structures include cambers, gulls, dip-and-fault structure and valley bulges. As a result of these movements dissected Inferior and Great Oolite strata are lowered vertically to the extent of 100 feet or more, so that they swathe the hill tops and valley sides. The causal processes of this camber structure include subsurface erosion and valleyward outflow of the underlying Lias. Gulls are widened joints in the camber filled with material from above. They usually trend parallel to the strike of the cambered strata and may attain a width of 40 yards. Step-faulting of similar trend is commonly associated with advanced cambering. The throw of each fault in this "dip-and-fault" structure is compensated for by steep downslope dips (up to 40°) in the inter-fault blocks. Valley bulges comprise a variety of upward displacements of the strata that are confined to the valleys and are due to differential loading of the incompetent Lias'. See also, particularly for figures, Hollingworth and Taylor on 'Superficial structures' in 'The geology of the Kettering district' (*PGA*, 1946). 'This is the type-locality of the "pseudo-tectonic" superficial structures described in a classic paper by Hollingworth, Taylor and Kellaway (1944). In no other region of the world can "cambers" and "gulls", "valley-bulges" and "dip-and-fault structure" be better studied' (Sylvester-Bradley in *Geology of the East Midlands*, 1968). 'A remarkable example of superficial folding due to glacial drag, near Aberystwyth' (Challinor, *GM*, 1947). 'Superficial structures in the Adwalton Stone Coal' (Cook, *TLGA*, 1959). 'Superficial structures in Weald Clay, Southwater' (pl. in Gallois and Edmunds, *BRG, Wealden District*, 1965).

**Supergene.** Originating or proceeding from above. Applied particularly to the production or enrichment of ore minerals by aqueous solutions percolating downwards. Hypogene, in one sense, is the contrasting term.

**Supergroup.** See LITHOSTRATIGRAPHY. 'A supergroup consists of two or more adjacent and naturally related or associated groups. Groups need not necessarily be gathered into supergroups' (*Guide to Stratigraphical Procedure*, *GSSR* (11), 1978).

**Superimposed drainage.** This usually means a river drainage system that was initiated, and developed, as consequent drainage in relation to the uplift and structure of a rock series which it has eroded away to expose an underlying series with a different structure. The already firmly established drainage system thus becomes let down, or superimposed, on a structure to which it is not related. The drainage of the English Lake District appears to be a clear case of this. 'The description "superimposed", used first by Maw (*GM*, 1866) and later by Powell (*Colorado River*, 1875), was shortened to "superposed" by McGee (*USGS*, 1888). Richthofen (1886) termed the courses of such rivers "epigenetic"' (Cotton, *Landscape*, 1948). 'There can be no doubt that the present courses of the streams were determined by conditions not found in the rocks through which the channels are now carved, but that the beds in which the streams had their origin when the district last appeared above the level of the sea, have been swept away' (Powell, 1875, quoted by Mather and Mason, *Source Book*, 1939).

**Superimposed folding.** See REFOLD. 'Superimposed folding at Loch Monar' (Ramsay, *QJ*, 1957).

**Superposition.** The commonest use of this word in geology is when stating or applying the knowledge that in a stratified series of sedimentary rocks the upper strata are, normally, the newer. Although often called the 'law of superposition', this does not appear to be so much a 'law' as a necessary and obvious corollary attending the fundamental axiom that these rocks are hardened sediments which were laid down in successive layers. There is also, further, the principle of the orderly superposition of strata and their fossils, when considered over wide areas. This principle became firmly established by William Smith (particularly by his map, 1815, and his *Strata Identified by Organized Fossils*, 1816) but there had been partial realization of it in the 18th century. Thus Holloway in a letter to John Woodward (*PT*, 1723) writes: 'That ridge of sand-hills by Woburn . . . which extends itself from east to west, everywhere at about the distance of eight or ten miles from the Chiltern-hills . . . the chalky matter of which they chiefly consist: which two ridges you always pass in going from London into the north, north-east, or north-west counties. After which you come into the vaste vale, which makes the great part of the Midland counties . . . which I take notice of because it confirms what you say of the regular disposition of the earth into like strata, or layers of matter coming through vast tracts'. Whitehurst (*Strata in Derbyshire*, 1778) puts the matter clearly: 'It may appear wonderful that amidst all the confusion of the strata there is nevertheless one constant invariable order in the arrangement of them and their various productions of animal, vegetable and mineral substances, or rather the figures or impressions of the two former. I would not be understood [to say] that the strata in every other part of the world are perfectly analogous to those in Derbyshire, or that their productions are the same, but that there is as much regularity in the arrangement of the strata in one country as there is in another'. 'Mountains offering their constituent materials more conspicuously to view, were generally visited and principally consulted; the chymical properties and distinctive characters, of their component masses, their external relations, whether of position, superposition, form, direction, or extension . . . were now carefully attended to' (Kirwan, *Geological Essays*, 1799). 'On certain points connected with the super-position of the strata of England' (Mushet, *PM*, 1812).

**Supracretaceous.** A name for the Tertiary strata used by the earlier geologists, particularly by De la Beche (*Manual*, 1831, 'Supercretaceous'; 1833, 'Supracretaceous').

**Supratenuous fold.** A form of anticlinal flexure in strata laid down over a rigid mass, and of which the individual beds are thinner on the axis than on the limbs.

**Surface.** In geology, 'surface' usually refers to the boundary surface between one bed or mass of rock and another immediately adjacent, such as a bedding surface, a fault surface, a surface of unconformity, an outer surface of an igneous intrusion; or to an imaginary (geometrical) surface such as the axial surface of a fold. It thus usually means a surface within a structure, not an outside surface which is what is usually implied in the ordinary use of the term. The exposed surface of a rock is, of course, an obvious 'rock surface'; but not a 'surface' in the structurally significant meaning of the term. See PLANE.

**Surface creep.** See CREEP.

**Surface processes.** See EPIGENE. The study of the processes that go on at or very near the earth's surface has two quite distinct aspects: (1) geomorphological, (2) as providing the raw materials for sedimentary rocks. 'In surface processes, Uniformitarianism finds the key to much of earth-history' (Read and Watson, *Introduction to Geology*, 1962).

**Surface texture.** The surface texture of an exposed rock is an expression of the interaction between weathering and lithology; that of a rock particle, a boulder, pebble, or sand grain, is usually an indication of its manner of attrition, e.g. water-smoothed, ice-scratched. 'Pebble surface textures' (Krinsley and Donahau, *GM*, 1968). *Atlas of quartz sand grain surface textures* (Krinsley and Doornkamp, 1973). (Here 'surface' does refer to an outside surface—see SURFACE.)

**Surficial.** Where it is the earth's surface that is concerned this term is sometimes used as synonymous with 'superficial', particularly in American writings. However, the word seems to refer to a surface itself, meaning 'of the surface' (part of the surface), and in that sense might be useful geologically. Thus, for instance, sole-marks would be 'surficial' whereas glacial drift is 'superficial'.

**Suspension.** Method of transport by rivers; see TRANSPORT.

**Swallow.** 'An opening or cavity, such as are common in limestone formations, through which a stream disappears underground: also called swallow-pit, swallow-hole and locally swallet' (*OED*). The stream normally emerges lower down. The Ingleborough district of Yorkshire supplies excellent examples (e.g. Gaping Gill). It may be a large funnel-shaped hole (sometimes choked up, overgrown and dry), such being also known as 'pot-holes', particularly in Yorkshire, or merely a part of a stream bed through which some of or all the water (depending on the season) escapes underground (e.g. Manifold Valley, Staffordshire). 'The [river] Mole [in Surrey] is swallowed up and thereof the place is called The Swallow' (Holland's translation of Camden's *Britannia*, 1610). It is possible for swallows to be preserved in a limestone under an unconformable cover: 'Limestone unconformities and their contemporaneous pipes and swallow holes' (Dixon, *GM*, 1909). Concealed swallows may be formed by the collapse of caverns, and an overlying loosely compacted formation may show surface reflections of this caving-in: 'Swallow holes on the Millstone Grit and Carboniferous Limestone of the S. Wales Coalfield' (Thomas, *GJ*, 1954). Swallows are also called 'sinks' or 'sinkholes' (Coleman and Balchin, *PGA*, 1959, who list 50 other terms).

**Swell.** An uplift of gentle form. This may be a land feature, perhaps on a large scale: 'The Rift System and the East African Swell' (Quennell, *PGS*, 1960). It is used particularly for a relatively elevated region of the sea floor, a submarine shallow, normally complementary to a neighbouring basin or basins. The submarine shallow may emerge above the surface of the sea. Deposits in the basins become thin or wedge out over the swells. 'The Mid-Dorset Swell' (Drummond, *PGA*, 1970). 'Basins and swells' (Sellwood and Jenkyns, *JGS*, 1975, and discussion, 1976).

**Syenite.** A coarse-grained intermediate igneous rock composed essentially of alkali-felspar and, typically, hornblende, but often, in the more strongly alkali varieties, soda-rich amphiboles and pyroxenes. One group is characterized, further, by felspathoids. Mostly plutonic but sometimes as minor intrusions. [Syrene=Assouan, Egypt: but the rock there, called Syenite by Pliny, is a hornblende-granite.]

**Symmetry.** In structural geology there is symmetry as a geometrical form and symmetry, further, with regard to the vertical plane. Thus a fold that is symmetrical, in the sense that one side is the mirror image of the other about the axial plane, may be asymmetrical, in the sense that this plane is not vertical and so the dips are not the same on each side.

**Synclinal ridge.** A ridge whose structure is synclinal.

**Synclinal valley.** A valley whose structure is synclinal. The term might, or might not, be used with the implication that the valley was formed as a direct result of the earth movement which at the same time formed the syncline. 'The Benue [Nigeria] flexture valley . . . structural valley' (Lees, *QJ*, 1952). 'The late Tertiary mobility of the Californian coast is, of course, well known. The synclinal valleys contain Tertiary sediments extending in age to the latest Pliocene and even Pleistocene, and movements of anticlines relative to synclines and of fault-lines are still active' (Lees, *QJ*, 1953).

**Syncline.** Primarily, a trough fold in stratified rocks (the opposite of an anticline). Hence 'synclinal'. 'Synclinal–we adopt this term, first used, we believe by Professor Sedgwick' (Lyell, *Principles*, 1833). 'The Central Wales Syncline' (Jones, *QJ*, 1912). See ANTICLINE and ANTIFORM.

**Synclinorium** (pl. **-ia**). 'A mountain range, begun in a geosynclinal, and ending in a catastrophe of displacement and upturning, is a "synclinorium", it owing its origin to the progress of a geosynclinal. The word is from the Greek for synclinal, and *oros*, a mountain' (Dana, *Manual*, 1875). But later writers (e.g. Van Hise, *JG*, 1896; Geikie, *Textbook*, 1903), while apparently claiming to be following Dana, used 'synclinorium' (and 'anticlinorium') in senses very different from his; and their usages, which are convenient, have been followed ever since. Accordingly a synclinorium is now defined as a compound syncline, that is, one in which the limbs are themselves thrown into folds; usually applied to structures on a large scale. The term cannot now possibly be taken to be from *oros*, 'a mountain', but must be supposed to be from the L. *-orium*, 'a place for'.

**Syndepositional.** Together with, and at the same time as, deposition (*pari passu* with deposition). 'Syndepositional faults and folds' (Prentice, *PGA*, 1962). 'Syndepositional intrusions in Ordovician pyroclastics' (Schiener, *JGS*, 1974). 'Synsedimentary' is sometimes used.

**Syneclise.** See ANTICLISE. (Shatsky, Russia, 1957 Tomkeieff, *PGS*, 1958.) [G. *klisis*, a bending.]

**Synform.** The opposite of an antiform. See ANTIFORM.

**Syngenetic.** Formed together. **1.** For ore-deposits formed at the same time as the enclosing rocks. **2.** For sedimentation structures, such as ripple-mark, formed contemporaneously with the deposition of the sediment. (The opposite of epigenetic.)

**Synkinematic.** 'A term applied to processes occurring simultaneously with dynamic rock deformation' (*RGMSN Geological Nomenclature*, 1959).

**Synneusis.** The process of drifting together and mutual attachment of crystals suspended in a magma. (Vance, *CMP*, 1969). [G. *Sun-neuō*, to converge.]

**Synsedimentary.** See SYNDEPOSITIONAL. 'Synsedimentary graben formation' (Kent, *JGS*, 1977).

**Syntectonic.** Accompanying tectonic action; particularly for igneous action. 'Syntectonic granites are emplaced during orogenic movements' (Bott, *QJ*, 1956). 'Synorogenic' means the same.

**Syntexis.** An ultrametamorphic process whereby deep-seated rocks, of more than one kind, are melted or remelted together to form a regenerated and augmented magma. Hence 'syntectic' and 'syntectite'. (Cf. anatexis.) (Loewinson-Lessing, 1899.) 'Syntexis and differentiation' (Brammall, *GM*, 1933). [G. *tēktos*, melted.]

**Synthem.** See MESOTHEM.

**System.** 'An organized or connected group of objects' (*OED*). The arrangement of the several kinds of geological material (objects)—minerals, rocks, and fossils—according to some system, particularly some natural system in each case. Hence, 'systematic', for the descriptive and comparative study of the material with a view to setting out the observable facts of composition and structure and making such an arrangement; also (often as 'systematics') for the study of the principles of such arrangments (taxonomy). 'Systematics in palaeontology' (George, *JGS*, 1971). See CRYSTAL SYSTEM, FOLD-SYSTEM, FAULT, STRATIGRAPHICAL SYSTEM.

# T

**Table of strata.** See SUCCESSION. *Geological Table of British Organized Fossils, which identify the course and continuity of the Strata in their order of superposition* (William Smith, 1817). Early tables of British strata are those of Strachey (*PT*, 1719-1725), Michell (*MS*, 1788, *PM*, 1810), Smith (1797-1817), Buckland (in Phillips, *England and Wales*, 1818), Greenough (Map, 1819), Conybeare and Phillips (*England and Wales*, 1822), Lyell (*Principles*, 1833). See Challinor, *AS*, 1970, for a comparative table of early tables.

**Tablemount.** See GUYOT.

**Tabulata.** 'The Tabulata are an extinct, almost invariably Paleozoic order of corals [class Anthozoa, phylum Coelenterata] characterized by their exclusively colonial mode of growth and secretion of a calcareous exoskeleton of slender tubes crossed by many transverse partitions called tabulae. Relative prominence of these tabulae and inconspicuousness or even absence of radially disposed longitudinal partitions (septa) are features which suggested the name of the group' (Hill and Stumm in Moore, ed., *Treatise on Invertebrate Paleontology*, 1956). (Milne-Edwards and Haime, 1850.) [L. pl., floored.]

**Tachylyte.** A glassy basalt. May be extrusive, but usually occurs as a selvedge to a thin minor intrusion. (Breithaupt, 1826.) 'On the basalt-glass (tachylite) of the Western Isles of Scotland' (Judd and Cole, *QJ*, 1883). [G. *tachus*, rapid, *lutos*, soluble; referring to the rapidity with which the substance fuses or dissolves.]

**Talc.** A mineral similar to serpentine in chemical composition (a hydrated magnesium silicate), very soft and flaky. Occurs chiefly in 'talc-schist', a metamorphosed basic or ultrabasic igneous rock. Known also as steatite, soapstone, and French chalk. (Agricola, 1546.) [Arabic, *talq*.]

**Talus.** See SCREE. 'The action of atmospheric causes, the frost by rifting and detaching portions of the outer surface of rocks, and the rain by washing the finer parts away, either contribute to the agency of the torrents, or accumulate the fragments detached in a slope or talus of debris, at the foot of the hills whence they are derived' (Conybeare and Phillips, *England and Wales*, 1822). 'The term is borrowed from the language of fortification, where "talus" means the outside of a wall in which the thickness is diminished by degrees, as it rises in height, to make it the firmer' (Lyell, *Principles*, 1833). [L.]

**Taphonomy.** Synonymous with 'Fossilization'. 'Taphonomy studies the introduction of the remains of organisms and traces into the rock record' (Müller in *Treatise on Invertebrate Paleontology*, vol. A, 1979). 'Taphonomy: a new branch of geology' (Efremov, *P-AG*, 1940). [G. *taphos*, burial, *nomos*, law.]

**Taphrogenesis.** 'E. Krenkel in his classical monograph [Germany, 1922] introduced the term "taphrogenesis". He compared the East African rift structures with the Rhinegraben and defined taphrogenesis as the formation of graben, induced by tensional forces' (Illies in Illies and Fuchs, eds., *Approaches to Taphrogenesis*, 1974). 'Taphrogeny' is synonymous. [G. *taphos*, a trench.]

**Tarannon series.** The thick series of beds (1,000 m.) in Central Wales forming the upper of three divisions of the Llandovery series. An extended succession in the graptolitic facies, comprising the seven zones *Monograptus sedgwicki* to *M. crenulatus*. (Geological Survey maps, 1857/8.) 'The Tarannon Series of Tarannon' (Wood, *QJ*, 1906). [Tarannon or Trannon, a district in Montgomeryshire.]

**Taxon** (pl. **taxa** or **taxons**). One of a hierarchy of taxonomic categories in biology; species, genus, order, etc. There are altogether 18 generally accepted taxa in the classification of animals. The term has also come to be used for a quite different concept; for a taxonomic kind, which may be nameable as a particular variety, species, genus (etc.) of animal or plant; thus we find, for example, 'sixty taxa of extinct sea creatures' and 'morphologically well-defined taxa'.

**Taxonomy.** 'Taxonomy, or systematics, is the science of classification of organisms. The term "taxonomy" is derived from the Greek ταξις [*taxis*], arrangement, and νομος [*nomos*] law, and was proposed by de Candolle (1813) for the theory of plant classification. "Systematics" stems from the Latinized Greek word *systema*, as applied to the systems of classification developed by the early naturalists, notably Linnaeus (*Systema Naturae*, 1735). In modern usage both terms are used interchangeably, in the fields of plant and animal classification' (Mayr and others, *Methods and Principles of Systematic Zoology*, 1953). *Procedure in Taxonomy* (Schenck and McMasters, 1935). This applies equally to fossils as to living forms of life. It seems that both terms could be used for the classification of any kind of object (e.g. minerals), but that is seldom, if ever, done.

**Tear fault.** In common usage, a fault, usually steeply inclined, along which movement has been horizontal or nearly so (strike slip). 'Marr, in his definition [*PGA*, 1900], appears to restrict the term tear fault to wrench fractures which end upwards, as well as downwards, against nearly horizontal thrusts or lags . . . [but] the word "tear" has become widened by use' (Anderson, *Dynamics of Faulting*, 1951). See WRENCH FAULT.

**Technique.** See METHOD. *Handbook of Paleontological Techniques* (Kummel and Raup, eds., 1965). *Techniques in Geomorphology* (King, 1966). *Geological Laboratory Techniques* (Allman and Lawrence, 1972).

**Tectogene.** A (supposed) down-buckle of the granitic into the basaltic layer of the earth's crust in such a way that the sedimentary cover is squeezed both downwards, breaking into or through the granitic layer, to form an orogenic root, and upwards to form an orogenic belt of highly deformed and dislocated strata. 'Diagram to illustrate the evolution of a "tectogene" ' (Kennedy, *GM*, 1948).

**Tectogenesis.** Sometimes used instead of 'orogenesis' for folding and thrusting, to avoid any implication that actual mountains were necessarily formed.

**Tectonic.** In geology: belonging to the structure of the earth's crust. The essential, primary structural features are the rock-masses themselves, as formed; but the adjective 'tectonic' usually refers to the way these rock masses have later come to be disposed, that is, for the secondary deformations of folding, faulting, etc. Hence 'tectonics', which is 'the architecture of the earth's crust' (Bailey, *Tectonic Essays*, 1935), or as applying to some region (e.g. 'The tectonics of the Tintagel area', Wilson, *QJ*, 1950) or to some formation; or as meaning the science of earth-structure. 'Tectonics, whatever this term may mean, is not concerned simply with the static geometry of the rocks, but becomes kinetic by importing the geological dimension of time' (Read, *PGS*, 1962). By extension, the term is sometimes used for any deformation structures, including superficially induced structures such as surface creep; but these have been called 'pseudotectonic'. Hence 'tectonism' (tectonic activity). *Structural and Tectonic Principles* (Badgley, 1965). *International Tectonic Dictionary: English Terminology* (Dennis, 1967). *International Tectonic Lexicon* (Dennis and others, eds., 1979).

**Tectonic axis.** There are three axes (directions) of space reference. The a-axis is the 'axis of translation' (movement), the b-axis the 'axis of rotation'. The latter is the fold direction (at any point) or lies in the fault plane and is usually at right angles to the a-axis. The c-axis lies at right angles to the plane containing the other two axes. The b-axis is the most easily recognizable. The conception of tectonic axes of reference applies to a structure on any scale of magnitude, from a microscopic petrofabric to an orogenic belt. (Discussion in the *Tectonic Dictionary*, 1967.)

**Tectonic breccia.** See BRECCIA (6).

**Tectonic conglomerate.** See CONGLOMERATE.

**Tectonic culmination.** See CULMINATION.

**Tectonic denudation.** The laying bare of a strip of ground by tectonic movement (faulting). Conditions favourable for the production of an extensive surface of tectonic denudation would be large-scale and rapid fault-movement down a low-angled fault-plane. 'Tectonic denudation as exemplified by the Heart Mountain fault, Wyoming' (Pierce, *IGC*, 1968).

**Tectonic domain.** 'An area in which structures formed during a particular tectonism are predominant' (Hepworth, *QJ*, 1967).

**Tectonic facies.** See FACIES.

**Tectonic inclusion** A detached, or nearly detached, portion of a rock, usually of a relatively rigid bed, included in the surrounding rock, having become broken away by tectonic forces. Relicts of folds, torn apart along the middle limbs, and boudinage structure are examples. Inclusions may be small or large: 'The Grantown Series is a gigantic tectonic inclusion enclosed in the Moine granulites' (McIntyre, *QJ*, 1951).

**Tectonic map.** A map designed to show the positions, attitudes, and configurations of the structural units and structural features of a country. In penetrating searchingly into the structure it makes use especially of the information supplied by borings. It shows surface outcrops only in so far as they represent whole structural units. Structure contours are an important element; lines of folding and faulting are emphasized. A good example is the *Tectonic Map of Great Britain and Northern Ireland* published by the Institute of Geological Sciences in 1966.

**Tectonic régime.** See RÉGIME. The tectonism prevalent over a tectonic domain. Particularly its symmetry and orientation (Harland and Bayly, *GM*, 1958).

**Tectonic relief.** Earth-surface relief due directly to tectonic movement, as distinct from that due to differential erosion (or to deposition). This distinction can be illustrated by the difference between a fault scarp and a fault-line scarp. (Papers by Cotton: *NZJST*, 1947; *JG*, 1949; *AAAG*, 1950; *GJ*, 1953.)

**Tectonic ripples.** Puckers, resembling ripple-marks, resulting from bedding-plane slip and aligned more or less at right angles to the movement (as may be shown by slickenside striations). (Wood, *PGA*, 1958; Warren and others, *GM*, 1970.)

**Tectonic style.** See STYLE. The term 'tectonic style' largely explains itself, but is used in slightly different senses according (chiefly) to the scale on which it is being applied. For a structural unit that is a component of a larger, perhaps a regional structure, it is the form of the structure; this form must be largely dependent on the rock-composition, but nevertheless the 'style' is but one aspect or factor of the total tectonic facies of the unit: 'Contrasted style of folding in the rocks of Ord Ban' (Weiss and others, *GM*, 1955). The style is to be visualized in three dimensions; but many of the essentials would be correctly seen on a section normal to the fold-axes. Applied to the structure of a whole region, the 'style' transcends and includes the several facies of the component structural units, comprising also the relative orientations of these units: 'The style of Unst tectonics' (Read, *QJ*, 1934); 'The tectonic style of Australia' (Hills, Germany, 1956). In *Styles of Folding* (Johnson, 1977) the term is used in a very general way. 'Structural style' means the same thing.

**Tectonic-magmatic complex.** A large-scale structure, produced by the combined action of tectonic forces and of magmatic flow ('igneous tectonics') in which the locations and forms of the igneous intrusions are determined by the crustal deformations and dislocations, particularly deep-seated faults. 'Tectonomagmatic' is a synonymous adjective and 'tectonomagmatism' is the general process.

**Tectonite.** A rock having a deformation fabric. 'Two lineated tectonites from Strathavon' (McIntyre, *GM*, 1950). *Structural Analysis of Metamorphic Tectonites* (Turner and Weiss, 1963). 'Tectonite and melange: a distinction' (Raymond, *Geoly*, 1975).

**Tectonophysics.** The physics of tectonic processes. 'Of the current research trends in tectonophysics, the study of deformation mechanisms and of relations quantifying their characteristic stress-strain-time behaviour (constitutive equations) should produce some of the most significant improvements in the interpretation of the structures and fabrics of deformed rocks' (Atkinson, *JGS*, 1976). *Tectonophysics* (periodical).

**Tectono-sedimentary.** The deformation, of sedimentary rocks, which occurred at a time transitional between the time of sedimentation (with any primary, penecontemporaneous, deformation) and the time of lithifaction and of the imposition of secondary tectonic deformations. (Max, in conference report, *JGS*, 1978).

**Tectonosphere.** 'The earth's crust is extremely closely associated in its development with the upper mantle, which permits to join them in one term "tectonosphere" ' (Belousov, *IGC*, 1968).

**Tectono-stratigraphic.** In particular, for distinct tectonic units, each comprising several stratigraphic units; especially in a region of many dislocations. 'The tectono-stratigraphic relationships [in] . . . West Africa' (Affaton and others, *AJS*, 1980).

**Teilzone.** The local stratigraphical range of a species, genus, etc. It is a part, and an unknown proportion, of the theoretically complete biozone. It is not a part of a stratigraphical zone, though that is what the term suggests. [German *teil*, a part.]

**Tektite.** 'These bottle-green to blackish vitreous bodies, named "tektites" (τηκτός [*tēktos*], melted), include the moldavites of Bohemia and Moravia, the australites or blackfellows' buttons of southern Australia, Darwin glass or queenstownite of northeastern Tasmania, the billitonites of the East Indies, the rizalites of the Philippines, the indochinites of Cambodia, Annam, and Siam, and perhaps the silica glass of the Libyan Desert. The largest tektites weigh up to 4 kg.; most of them are much smaller. The common forms are rude spheroids, ovoids, pear-shapes, buttons, lenses, dumbbells, spindles and less regular shapes. The surfaces are pitted, grooved, and moulded in a manner suggesting flight through the atmosphere. They consist wholly of siliceous glass whose composition is unlike that of obsidian. For the australites, whose cosmic origin can hardly be disputed . . .' (Palache and others, *Dana's System of Mineralogy*, 1944). 'The tektite problem. The natural glasses, found as small corroded pieces scattered on the earth's surface and in alluvial deposits in a few limited areas, have long presented a puzzling problem; and many theories have been propounded to explain their origin. They have been known in southern Bohemia and western Moravia since before 1787; and similar material has since been found in [etc.]. In 1900, Professor F. E. Suess of Vienna included all these under the name tektites, and he put forward the theory that they are of meteoric origin. This theory has found general acceptance, but as a matter of fact there is no direct evidence for its support' (Spencer, *MM*, 1937). *Tektites* (O'Keefe, ed., 1963). 'Geochemistry of tektites' (Symposium, *GCA*, 1969). *Tektites* (Barnes and Barnes, eds., 1973). (Discussion of origin in Crawford, *GM*, 1979).

**Telluric.** Concerning the earth itself. 'Some still held the opinion that meteorites were of telluric origin' (Zittel, *History*,

1901). Hence 'intratelluric', within the earth. [L. *tellus, tellur-*, the earth.]

**Tephra.** Volcanic ash. Often used as a collective term for all pyroclastic materials, coarse as well as fine, but the adoption of the Greek word makes this wide use inappropriate. (Thorarinsson, 1954.) [G. *tephra*, ashes, a fine powder.]

**Tephrite.** An alkali-basalt, containing a felspathoid. Extrusive or as minor intrusions. (Cordier, 1815.) [G. *tephros*, ash-coloured.]

**Terminal curvature. 1.** The bending over of steeply inclined bedding or cleavage, due to the downhill pull of gravity (surface creep), to glacial drag, or to any other kind of force acting on these structural surfaces as they come to be near the surface of the ground. 'Some striking instances of terminal curvature of slaty laminae in West Somerset' (Mackintosh, *QJ*, 1867). Curvature of bedding (or cleavage), being part of the tectonic structure of the rocks as a whole and fortuitously occurring near the present surface of the ground, would not usually be included. **2.** 'Beds often show "terminal curvature" at a fault plane with the strata on the upthrown side dipping steeply downwards and those on the downthrown side as steeply upwards. This curvature is due to the drag of the two masses of rock past one another' (Kirkaldy, *General Principles of Geology*, 1954).

**Terminology.** In geology, the system of terms for the things, attributes and concepts of the science. A term may be newly coined, or a word already in common use may be adopted with a special meaning. See JARGON.

**Ternary succession.** 'Ternary succession of strata.—In following the order of sedimentation among the stratified rocks of the earth's crust, the observer will be led to remark a more or less distinct threefold arrangement or succession in which the sandy, muddy and calcareous sediments have followed each other. Professor Phillips and Mr. Hull have called attention to this structure, illustrating it by reference to the geological formations of Great Britain, while Professor Newberry, Dr. Sterry Hunt, and Principal Dawson have discussed it in relation to the stratigraphical series of North America' (Geikie, *Text-book*, 1882. The quotation under CYCLE OF SEDIMENTATION immediately follows). 'The three principal mineral elements are arranged in ternary groups so that in many places the order runs thus, downwards: calcareous, arenaceous, argillaceous [shown in the five cycles of the Jurassic]' (Phillips, *Oxford*, 1871).

**Terra Rossa.** Reddish earth rich in insoluble iron hydroxides, which has accumulated as a residual deposit, due to weathering, over areas of limestone that contains small traces of these compounds. Best developed over the Karst region of Yugoslavia; but in Britain the Carboniferous Limestone (particularly) tends to give a similar weathering product. [Italian, red earth.]

**Terrain.** This ordinary word (or 'terrane') is useful in geology to designate a tract of country dominated by a particular rock-group, orogeny, igneous activity, etc. Thus Anderson and Owen, in their *Structure of the British Isles* (1968) have chapters headed, e.g., 'Precambrian Terrains', 'Hercynian Terrains', 'Tertiary Igneous Terrains', and there is Hack's paper on 'Exfoliation in crystalline terrane' (*BGSA*, 1966). The term has been used for stratigraphical facies on a large scale: 'The Lower Paleozoic strata of western Newfoundland belong to two separate but contemporaneous terranes' (Rodgers and Neale, *AJS*, 1963).

**Terrestrial.** In geology, used most commonly in the sense of pertaining to the land, particularly as 'terrestrial deposits'.

**Terrigenous.** Applied to deposits, laid down on the sea floor, that have been derived directly from the land. (Murray and Renard, *Voyage of H.M.S. Challenger, 1872-76*, 1891 and earlier.)

**Tertiary.** A sub-era of the Cainozoic, comprising the whole of the era except the Pleistocene stage which, if taken to include the Recent, constitutes the Quaternary sub-era. 'After the appearance of Brongniart and Cuvier's work in 1808 it was demonstrated that many of the fossils of the Paris basin agreed with the fossils in the deposits near Verona which Arduino (1713-95) had termed "Tertiary deposits" [1759]. And the series was then incorporated in the chronological succession of the rocks as the Tertiary formations' (Zittel, *History*, 1901). The term came into use in Britain (for strata already described by Webster, *TGS*, 1814, and in Englefield's *Isle of Wight*, 1816) during the 1820's, for instance: 'Superior Order —Newest Floetz or Tertiary rocks—comprising the formations above the Chalk' (Conybeare and Phillips, *England and Wales*, 1822) and was firmly adopted by Lyell in the *Principles* (1830/3). See *A Correlation of Tertiary rocks in the British Isles* (*GSSR* (12), 1978).

**Teschenite.** An alkali-gabbro or alkali-dolerite containing analcite. Chiefly as minor intrusions. (Hohenegger, 1861.) [Teschen, Silesia.]

**Tethys.** The name given by Suess (*Das Antlitz der Erde*, 1901) to the Mesozoic Mediterranean Sea, the long, wide, composite geosyncline stretching across what is now Europe and Asia out of which the Alpine-Himalayan system of mountain ranges was formed. It persisted, modified, into the Eocene and finally broke up in the Oligocene and Miocene at the culmination of the Alpine orogeny. 'Tethys: past and present' (Jenkyns, *PGA*, 1980). [G. *Tēthus*, a goddess of the sea.]

**Tetrabranchiata.** (Tetrabranchia.) See CEPHALOPODA. [G. *bragchia*, gills.]

**Tetrapoda.** A group-name, ranking as a superclass, to include the vertebrates other than the Pisces (fishes); that is, the classes Amphibia, Reptilia, Aves, and Mammalia.

**Textbook of geology.** A book intended to be a standard work within its prescribed limits (e.g. 'elementary', 'advanced') with a carefully prepared informative and explanatory 'text' (usually illustrated). There are textbooks dealing with the various subdivisions of the science of geology as a whole. Many works which are in fact textbooks do not have that term in their title. Probably the most famous of all general geological textbooks is A. Geikie's *Textbook of Geology* (1882; 2 vols., 1903), still an indispensable work of reference. Familiar to many generations of students is Lake and Rastall's *Textbook of Geology* (1910/41).

**Texture.** The character of a rock determined, in the first place, by its being either perfectly homogeneous (glassy is the main homogeneous state) or, as the vast majority of rocks are, composed of particles. In the latter case it is, further, the character determined by the size, size-variation, shape, packing, and orientation of the component particles, and by any pore-space between them. Thus any kind of rock may be described as fine-textured, coarse-textured, etc.; igneous rocks may have particular textures such as holocrystalline, porphyritic, ophitic; among sedimentary rocks, conglomerate, millet-seed sandstone, quartzite, for instance, are rocks characterized by their texture; pumice has a porous-glassy texture, schist has a texture of parallel mineral flakes. 'The different hardness, or solidity, of each: also its colour, texture, and the peculiar matter which constitutes it' (Woodward, *Natural History of the Earth*, 1695). See STRUCTURE and SURFACE TEXTURE.

**Thalassic.** Marine; usually implying deep or open water. 'So long as deeper-water ("thalassic") deposition is taking place, the extent and depth of water is sufficient to allow of the wide dispersal of sediment before it finally reaches and subsides upon the sea-floor, and hence the areas of similar sediment deposited will be wide' (Lapworth, *QJ*, 1911). [G. *thalassa*, sea.]

**Thallophyta.** One of the main divisions of the plant kingdom, including the simplest plants and all Algae (sea-weeds), fungi (toadstools and moulds), and the lichens. Not well represented in the fossil record except for the calcareous algae. [G. *thallos* ('thallus' in botany means an undifferentiated structure).]

**Thalweg.** See PROFILE (2). [German; literally, valley-way.]

**Thanatocoenose.** Primarily a neontological term for 'the aggregated remains of organisms that were brought together by physical agencies, such as wave or current action, which operated after they died. An excellent example is the miscellaneous assortment of organic debris gathered on most modern beaches' (Boucot, *AJS*, 1953). 'One of the best known North American localities for fossil insects is near Florissant, Colo., where an ancient freshwater lake received deposits of Miocene volcanic ash. This ash in falling carried down insects, other animals and plant materials from both the air and water and buried them together in the bottom sediments of the lake. Consequently, both terrestrial and aquatic organisms are preserved together, a mixture of forms that could not possibly have lived together' (Shrock and Twenhofel, *Invertebrate Paleontology*, 1953). The term seems to have been introduced by Wasmund in 1926 in the form 'thanatocoenosis' (meaning strictly, community of death). 'Death assemblage' is synonymous and is usually preferred in palaeontology. (Contrasted with 'biocoenose'.) [G. *thanatos*, death, *koinos*, common.]

**Thanetian.** See PALAEOGENE and EOCENE. [Isle of Thanet.]

**Theoretical geology.** The construction of rational hypotheses, theories, and principles in order to integrate and explain the findings of practical geology in the field and the laboratory.

**Theory of the earth.** See COSMOGONY, HUTTONIAN THEORY, WERNERIAN THEORY. Most of the 'theories of the earth' of the seventeenth and eighteenth centuries were mere wild speculation. See 'Bibliography and the history of science' (Eyles, *JSBNH*, 1955) and 'The history of geology: suggestions for further research' (Eyles, *HS*, 1966). 'John Farey informs us (*PM*, 1815) that up to the time that William Smith had made his discoveries (c. 1795), philosophers had produced 80 theories and systems of the earth' (Davis, *TNS*, 1943).

**Theralite.** A coarse-grained basic alkali igneous rock with nepheline, plagioclase (labradorite), and various mafic minerals. (Rosenbusch, 1880.) [G. *thēra*, eager pursuit.]

**Thermal convection hypothesis.** See CONVECTION THEORY.

**Thermal metamorphism.** Metamorphism due to a rise of temperature; usually restricted to that in which there is no important intervention of the stress factor. In this sense (simple thermal metamorphism) it is confined to that caused by the heat of an igneous intrusion; it is synonymous with 'contact metamorphism' but with any associated dynamic and pneumatolytic effects excluded. 'The term in common use among Continental geologists [for simple thermal metamorphism] is "contact metamorphism", although the phenomena may be exhibited at a distance of miles from any igneous contact. Inasmuch as the effects are due, not to contact, but to heat and high temperature, the term "thermal metamorphism" seems more appropriate' (Harker, *Metamorphism*, 1950). See CONTACT METAMORPHISM.

**Thermal structure.** The structural arrangement of thermal metamorphic zones. Thus we may speak, for instance, of a thermal anticline. 'The significance of thermal structure in the Scottish Highlands' (Kennedy, *GM*, 1948).

**Thermal waters.** Naturally warm or hot waters, particularly springs ('thermal springs'). See HOT SPRING.

**Thermometry. 1.** The direct measurement of the temperature at which a geological process is acting in a particular case at the present day, e.g. that of a fumarole. **2.** The estimate of the temperature at which a geological process, in a particular case, acted in the past. 'The minerals developed during regional metamorphism serve as recording thermometers, to measure the relative distribution of temperature. Advancing metamorphism is marked by the appearance in turn of certain critical index minerals. The following index minerals are used in this paper: Chlorite-biotite- garnet-kyanite-sillimanite' (Kennedy, *GM*, 1948). **3.** The estimation of climatic temperature, particularly the temperature of the sea, at times and places in the geological past. Cases (2) and (3) are 'palaeothermometry', involving 'palaeotemperature', these terms being used particularly for case (3).

**Theropoda.** A suborder of the order Saurischia, one of the two dinosaurian orders of Mesozoic reptiles. Carnivorous, with prevailingly bipedal mode of locomotion. Upper Triassic to Cretaceous. See DINOSAURS. [G. *thēr*, beast of prey.]

**Thickness.** See VERTICAL. The thickness of a stratified rock-mass is to be measured in the direction (if known) which was vertical at the time of deposition. If the strata were originally deposited in horizontal layers, the thickness of the mass is the sum of the thicknesses of the individual strata; but if the deposits had an original dip a measurement at right angles to the stratification would give too great a value if the sloping deposits had been built out horizontally over a considerable distance.

**Thin out.** 'When a stratum, in the course of its prolongation in any direction, becomes gradually less in thickness, the two surfaces approach nearer and nearer; and when at last they meet, the stratum is said to thin out, or disappear' (Lyell, *Principles*, 1833). See FEATHER-EDGE.

**Thixotropy.** The property of certain colloidal systems of matter, including some suspensions of fine clays, of changing from the semi-solid, jelly-like state ('gel') to a liquid state ('sol') on being shaken and of setting again to the jellylike state on standing. 'The thixotropy of some sedimentary rocks' (Boswell, *QJ*, 1948). (Péterfi, 1927.) [G. *thixis* (n.), touching, *tropos*, a turn.]

**Tholeiite.** An over-saturated basalt (extrusive or intrusive) containing the particular pyroxene, enstatite-augite (pigeonite), with absence of olivine; the free silica occurring as part of a glassy residuum. (The above is one definition that emerges from various discussions.) 'Tholeiite and the tholeiitic series' (Tilley and Muir, *GM*, 1967). (Steininger, 1840; Rosenbusch, 1887.) [Tholey, Saar, Germany.]

**Tholoid.** A dome of viscous lava squeezed up, exuded and accumulated over a volcanic orifice. (Escher, 1920; Williams, 1932; Cotton, *Volcanoes*, 1944.) [G. *tholos*, a domed or vaulted building.]

**Throw.** A vertical linear quantity in a faulted structure. It is used in three distinct senses. **1.** The vertical separation of a bedding surface. **2.** The vertical distance between the dissevered edges of a bedding surface, as measured on a vertical section at right angles to the fault strike. **3.** The vertical component of net relative movement along the fault plane. The difficulty is that the last quantity (3) is the most fundamental and the one to be sought, but it is not to be known from even a complete knowledge of the geometry of a faulted structure in unfolded rocks; in such a case (unless the beds are horizontal) we do not even know with certainty on which side of the fault there has been relative upward or downward movement. Sense (2) is sometimes called the 'apparent throw'. See SLIP. The term 'throw', if used, should thus be accompanied by an indication of the exact sense in which it is being used. 'The "throw" of a fault' (Challinor, *GM*, 1933.) 'Here then is a fault by which the lias and the coal measures which support it are thrown down on the north to a depth exceeding the whole thickness of the red marl' (Buckland and Conybeare, *TGS*, 1824).

**Thrust.** A low-angled reverse fault. Also called 'overthrust' and 'thrust fault'. Bailey (*QJ*, 1938) has the following: 'The word "thrust" is to be found in the earliest British writings on tectonics; for instance, in those of James Hall of Edinburgh. This probably influenced Peach and Horne in their application of the term to the great fold-faults of the North-West Highlands. Study of Peach and Horne's writings makes it clear that their more important thrusts are covered by the following definition which with verbal modifications, I have repeatedly given: A thrust is a fold-fault which more or less completely replaces the reversed limb (real or imaginary) of an overturned anticline. Actually, however, Peach and Horne used "thrust" in a broader sense, to include all fold-faults; that is, all faults developed in connexion with strong folding.' Hence 'thrust-plane'. Lees (*QJ*, 1952) uses the following compound terms (among others): 'thrust-sheet', 'thrust-slice', 'thrust-anticline'. See FOLD-FAULT. 'The Moine Thrust—its discovery, age and tectonic significance' (McIntyre, PGA, 1954). Further discussion and history in Dennis, *Tectonic Dictionary*, 1967.

**Tight fold.** See CLOSE.

**Tilestone.** A flaggy sandstone used for roofing; a name particularly associated with that kind of rock occurring in the Downtonian.

**Till.** Boulder-clay; but whereas this latter term emphasizes the boulders, the term 'till', meaning a stiff impervious clay, emphasizes the matrix. Nevertheless the two terms, in geology, may be said to be practically synonymous.

**Tillite.** An indurated till (boulder-clay) formed during some pre-Pleistocene glacial period. 'The definition and indentification of tills and tillites' (Harland and others, *ESR*, 1966). See TILLOID.

**Tilloid.** 'Tilloid is a term for rocks of non-glacial origin that resemble tillite in appearance' (Schermerhorn and Stanton, 'Tilloids in the West Congo geosyncline', *QJ*, 1963; and see Harland in discussion, where the terms 'tillite' and 'tilloid' are further considered).

**Till-stone.** A constituent stone in a deposit of till; that is, a 'boulder' or any stone in a boulder-clay. 'Evolution of till-stone shapes' (Holmes, *BGSA*, 1960).

**Time and force.** 'The fundamental question of Time and Force has given rise to two schools, one of which adopts uniformity of action in all time, while the other considers that the physical forces were more active and energetic in geological periods than at present' (Prestwich, *Geology*, 1886). The phrase may be said to be an expression of the questions involved in Uniformitarianism and Catastrophism.

**Time and place.** The fundamental concept of 'time and place' in all aspects of earth history is exemplified in the title and content of the Geological Society of London's volume *Time and Place in Orogeny* (Kent, ed., 1969).

**Time and rock.** No two things could differ so greatly in themselves as time and rock; the one, so abstract (but portionable and measurable), the other, such a very material substance. Yet these two strands come together and are intimately interwoven into the fabric of geology.

**Time and space.** It can perhaps be suggested, relativity theories apart, that there may not be any such things as absolute time or absolute space. It seems impossible to imagine space with nothing of any kind in it anywhere, or (and, even more so) time with nothing at all going on. Geology, however, more than any other science, demands and employs the ordinary notions of time and space.

**Time unit.** Time units are, naturally, fundamental units in geological history as the units of the year or the century, for instance, are fundamental in human history. But rocks can be dated and correlated only approximately. A local rock-succession is the expression of a passage of time; but we cannot divide such a succession into units representing equal known periods of time, nor can we recognize, otherwise than approximately, a 'time-rock unit', that is a stratal unit, stretching far and wide, of which the upper and lower boundary surfaces are isochronal, representing true time-planes. Radioactivity age-methods are, however, being used with ever-increasing confidence and precision. The term 'chronomere' has recently been proposed in the Geological Society's 'stratigraphical code' (*PGS*, 1967): 'A chronomere is any interval of geological time; it is without uniform duration. Chronomeres [are] geological time-intervals of various orders of magnitude'. Thus, if 'period', 'epoch', 'age' are to be used, respectively, as corresponding to the stratigraphical ('stratomeric') terms 'system', 'series', 'stage', they become 'chronomeric' terms.

'The unit [of time] generally used in stratigraphy is the year. Multiple annual units are recommended to be written in the following international style: for $10^3$ years—ka; for $10^6$ years—Ma; for $10^9$ years—Ga' (Harland and others, *JGS*, 1972). These are abbreviations devised, the first letter in each case from a prefix of Greek origin and the second letter from the Latin *annus*, a year. The Greek prefixes are *kilo-* (a thousand times), *Megas* (great), and *Gigas* (mighty). In the previous version of the Geological Society's recommendations (*PGS*, 1969) the abbreviations were given as ky, My, and Gy ('year' instead of *annus*). All these symbols are etymologically unsatisfactory. An already established symbol, in English writings, for a million ($10^6$) years is 'm.y.' (or 'my'), and 'b.y.' (or 'by') has been used for a billion ($10^9$) years, and see AEON.

**Time-plane.** See ISOCHRONAL, HORIZON, STRATA AND TIME.

**Time-rock span.** See BIOZONE.

**Time-rock unit.** See TIME UNIT.

**Time-scale.** See GEOLOGICAL TIME.

**Toadstone.** An old name (still in use) for the amygdaloidal basalts of the Carboniferous Limestone of Derbyshire; either from the resemblance between the amygdales and the spots of a toad's skin or from the German *todstein* (dead stone) because lacking in lead ore. John Whitehurst was the first to recognize the igneous nature of the toadstones and one of the first geologists of any country to recognize the presence of igneous rocks among sedimentary strata. "Toadstone. A blackish substance, very hard; contains bladder-holes, like the scoria of metals, or Iceland lava . . .' (Whitehurst, *Strata in Derbyshire*, 1778).

**Toarcian.** See JURASSIC SYSTEM. [Thouars (Toarcium), Deux-Sèvres, France.]

**Tonstein.** "Tonsteins are essentially argillaceous rocks containing kaolinite in a variety of forms together with occasional detrital and carbonaceous material. Their distribution is widespread in Continental Carboniferous rocks of Westphalian age and since each bed possesses specific characters they are important aids to correlation' (Williamson, *MgM*, 1961). 'The microbiology, mineralogy and genesis of a tonstein' (Moore, *PYGS*, 1964). See quotation (Eden) under MARKER HORIZON.

**Top and bottom.** See WAY-UP. 'Many features can be used in determining the top and bottom of layered rocks. These may be confined to a single layer, they may involve a succession of beds, they may lie in a massive tabular body of igneous rocks, or they may occur in massive rock bodies lacking any apparent bedding. Furthermore, they may lie on the upper or under surface of a layer or anywhere within it' (Shrock, *Sequence in Layered Rocks*, 1948).

**Topaz.** Fluosilicate of aluminium. Occurs in irregular grains and spongy masses in pneumatolytic rocks, particularly greisens. 'Its name is the *topazius* of Pliny, but this evidently refers to a different mineral, probably to chrysolite, since he derives the word from the name of an island in the Red Sea. On the other hand, the Roman *chrysolithus* was probably our topaz' (Miers, *Mineralogy*, 1929).

**Topographical expression.** See DIFFERENTIAL EROSION. 'This fault-complex [the Leannan fault, Ireland] is not expressed by a topographic feature comparable with the Great Glen of Scotland. Nevertheless, long stretches of its course are marked by a fault-line scarp, on one or other side, or by a groove' (Pitcher and others, *QJ*, 1964).

**Topographical grain.** See GRAIN.

**Topography.** 'The description (or representation on a map) of the surface features of any area, including not only landforms, but all other objects and aspects both of natural or human origin' (Monkhouse, *Dictionary of Geography*, 1965). [G. *topos*, a place.]

**Topset beds.** Flat-lying beds covering the inclined foreset beds in a delta.

**Tor.** 'A high rock; a pile of rocks, generally on the top of a hill' (*OED*). Typical tors are those formed by the erosion and denudation of igneous rocks and the best-known examples are the granite tors of Devon and Cornwall. For this kind Linton gives a genetic definition as follows ('The problem of tors', *GJ*, 1955): 'A "tor" is a residual mass of bedrock produced below the surface level by a phase of profound rock rotting effected by groundwater and guided by joint systems, followed by a phase of mechanical stripping of the incoherent products of chemical action. Ellipsoidal rockmasses produced in the same way but entirely separated from bed-rock are designated "core-stones". The upper parts of a tor approximate in form to core-stones, and like them may be completely detached, though still perched; the lower portions of a tor approximate to massive joint-bounded blocks.' 'The granite tors of Cornwall. . . . Phenomena which the vulgar have deemed little less than miraculous . . . learned antiquaries have tortured their inventions and constructed religious systems for the purpose of explaining' (Macculloch, *TGS*, 1814). 'Craggy piles of rock such as in Devonshire are called Tors, but in Pembrokeshire and other parts of Wales are termed Carns' (De la Beche, *TGS*, 1829). 'High tors or bosses of rock rise above portions in a decomposed state, while hard masses, having the fallacious appearance of boulders rounded by attrition, are included in the loose decomposed granite' (De la Beche, *Geological Observer*, 1851). 'Derbyshire, a wild country of bleak gritstone "edges" and romantic limestone dales and "tors" (Kendall and Wroot, *Yorkshire*, 1924). 'The dolomite tors of Derbyshire' (Ford, *EMG*, 1963).

**Torridonian.** The upper of the two main Precambrian groups in the NW. Highlands and Islands of Scotland. See LEWISIAN. (Peach and Horne, *QJ*, 1888, as Cambrian; Geikie, *Textbook*, 1903, as Precambrian). 'The arkosic sandstones, shales and conglomerates of the Torridonian formation make up, in the words of Teall [1907], "one of the largest as well as one of the most ancient masses of mechanical sediment in the British Isles". The main source of published information on these sediments is still the Geological Survey Memoir on the North-west Highlands (Peach *et al.*, 1907)' (Sutton and Watson, *GM*, 1960). [Loch Torridon, Wester Ross.]

**Torsion.** In connexion with the inferred character of certain large-scale earth-strains and earth-movements. 'We regard the structure of the Llangollen district as produced by torsion, set up by stresses originating in the Caledonian movements of Devonian times and continuing to act in later epochs' (Wills and Smith, *QJ*, 1922; see also Wedd and others, *MGS*, *Wrexham*, 1927, and Smith and George, *BRG, N. Wales*, 1961). On a small scale, questions of torsion arise in considering the mechanical principles of rock-deformation.

**Tourmaline.** A blue-black mineral, a complex silicate of aluminium, with alkali metals or iron and magnesium and containing boron. It occurs particularly as the result of pneumatolysis in and round certain granites, especially the Devon and Cornwall granites. See AMPHIBOLE. [*toramalli*, its name in Ceylon.]

**Tourmalinization.** Pneumatolysis of granite by boron-rich fluids, the mica being first replaced (tourmaline-granite), then groundmass felspar (luxullianite) and finally all the felspar (schorl-rock).

**Tournaisian.** The lower part of the Carboniferous Limestone series from the base (base of zone K) up to the boundary between the $C_1$ and $C_2$ zones. It is a series of the Dinantian subsystem. (Dumont, Belgium, 1832). [Tournai, Belgium.]

**Trace.** The following are the three chief applications or adaptations of the one substantive 'trace'. **1.** A line on a surface, being the intersection with another surface; e.g. the trace of a bedding surface on a cleavage surface, the trace (outcrop) of a fault on the ground surface. **2.** A very small quantity, as in 'trace-element'. **3.** Indications of former presence, as in 'trace-fossil'.

**Trace-element.** An element occurring in a minute quantity in a rock, so that its presence can be detected but hardly measured. 'Trace element studies of the origin of igneous rocks' (Rea, conference report, *JGS*, 1978).

**Trace-fossil.** 'A sedimentary structure resulting from the activity of an animal moving on or in the sediment at the time of its accumulation: includes tracks, burrows, feeding and other traces' (Simpson, 'On the trace-fossil *Chondrites*', *QJ*, 1956). 'Trace fossils and problematica' (Häntzschel in Moore, ed., *Treatise on Invertebrate Paleontology*, 1962/75). *Trace fossils* (Crimes and Harper, eds., 1970/77). 'Fossil tracks and trails' (Bassett and Owens, *Amg*, 1974). 'Corallian trace fossils' (Fürsich, *Let*, 1975). 'Trace fossils from proglacial lake sediments' (Gibbard and Stuart, *BO*, 1974). *The Study of Trace Fossils* (Frey, ed., 1975). *Trace Fossils* (Teichert, ed., in *Treatise on Invertebrate Paleontology*, part W, 1975).

**Trachyte.** A fine-grained intermediate igneous rock, usually porphyritic and usually having trachytic texture, that is, one of closely packed lath-shaped felspar microlites showing flow structure. It is composed essentially of alkali-felspar and soda-rich amphiboles and pyroxenes. Approximately the fine-grained equivalent of syenite. Rather rough to the touch. Extrusive. (Haüy and Brongniart, 1813.) [G. *trachus*, rough.]

**Traction.** Method of transport by rivers; see TRANSPORT.

**Transcurrent fault.** See WRENCH FAULT. (J. Geikie, *Structural and Field Geology*, 1905.)

**Transform.** In the first definite exposition of the theory of plate tectonics, J. T. Wilson ('A new class of faults and their bearing on Continental Drift', *N*, 1965) has the following: '[There appear to be] a continuous network of mobile belts about the Earth which divide the surface into several large rigid plates. . . . A junction where one feature [i.e. plate] changes into another [at a mobile belt] is here called a "transform"'.

**Transform fault.** A fault at a transform junction. There are two kinds: (a) oceanic, for a fault (usually one of a parallel series) that cuts transversely, with essentially horizontal (strike-slip) movement, across a mid-ocean ridge, and (b) continental, for a fault or fault zone at the common margin of two plates in the case where these slip past each other horizontally. 'Oceanic and continental transform faults' (Windley, introducing specialized papers, *JGS*, 1979).

**Transformism.** The formation of igneous rocks by the transformation (by metamorphism) of pre-existing rocks. In the case of granites, this is granitization. Also the doctrine; particularly as to the question of the origin of granite (cf. 'magmatism'). Hence 'transformist' for one holding this opinion. (Reynolds, *GM*, 1947; Tomkeieff, *QJ*, 1947.)

**Transgression.** In palaeogeography, this means a marine transgression, that is, the invasion of land by the sea. Overlap in stratigraphy implies transgression; but the term is perhaps more especially used when the invasion is relatively sudden and the new marine deposits are spread far and wide over the old land surface (e.g. the Cenomanian transgression in the middle of the Cretaceous period). 'Marine transgressions and regressions' (specialized papers in *JGS*, 1979).

**Transgressive sill.** See SILL.

**Transient.** 'A stage in the phylogeny of a species' (*SA*, 1956); a stage in any closely graded evolutionary series. See MUTATION.

**Transition.** In the Wernerian classification, the group of rocks occurring between the Primitive and Secondary. The group roughly corresponds to what we now know as the Lower Palaeozoic together with the upper (Algonkian) part of the Precambrian. 'The geognostical observations make us acquainted with an extensive tract of transition rocks, a class of rocks hitherto unnoticed in Great Britain' (Jameson, *County of Dumfries*, 1805). 'This important series of rocks was first established as a distinct class by the acuteness of Werner. They are supposed to have been deposited during the passage or transition of the earth from its chaotic to its habitable state. Hence they contain the first traces of organic remains and mechanical depositions, and are denominated "Transition rocks" ' (Jameson, *Geognosy*, 1808).

**Transitional series.** (Transition series.) A series of beds that are in any (or every) way transitional in their character between those below and those above. It thus includes a series of passage beds, but it is more commonly used for a rather large and widespread series of beds that are transitional in the character of their fossils, implying that they were laid down during a period when there was a specially important replacement of old groups by new. The Tremadocian series is a good example.

**Transmagmatic solution.** 'Granitization may be conceived as a magmatic replacement of rocks, percolated by ascending flows of "transmagmatic solutions" of undercrust origin, related to degasing of the mantle. The most petrogenetically important components of transmagmatic solutions are $H_2O$, fugitive acids, $K_20$ and $Na_20$. . . . The debasification of rocks in granitization may be explained by an increase in acidity of transmagmatic solutions during their ascension and cooling down' (Korzhinskii, *IGC*, 1968).

**Transport.** The geological process of the carrying away (natural agencies being implied) of weathered, or otherwise broken or loose, rock material; for short or long distances, quickly or slowly, to temporary or permanent places of deposition. See EROSION. Transport by rivers may be said to be effected in four ways: solution, suspension (particles sweep along free from the stream bed), traction (particles roll or slide along), and saltation (particles bounce along). 'Engineering aspects of sediment transport' (Bruun and Lackey in *Reviews in Engineering Geology*, Fluhr and Legget, eds., 1962).

**Transpression.** A combination of compression and transcurrent fault-movement. 'Tectonic transpression in Spitsbergen' (Harland, *GM*, 1971).

**Transtension.** A combination of tension and transcurrent fault movement.

**Transverse.** See LONGITUDINAL.

**Trap. 1.** A very common term in older geological writings for a wide variety of igneous rocks other than granite, but usually including basalt as a typical representative. 'Basalt, greenstone, porphyry, and amygdaloid: all these, which were recognized as belonging to one family, were called "trap" by Bergman [Swedish geologist, 1735-84]; a name since adopted very generally into the nomenclature of the science' (Lyell, *Elements*, 1838). See quotation (Murchison) under GREENSTONE. 'The name of "trappean rocks" has been somewhat adopted of late' (De la Beche, *Geological Observer*, 1851). 'The trap dykes of the Orkneys' (Fleet, *TRSE*, 1900). [Swedish *trappa*, a staircase; either from the stepped landscape profiles developed on dissected plateau basalts, or from the jointing which suggests stones used for building staircases ; Hawkes, *GM*, 1942.] **2.** In petroleum geology, a geological structure that catches and retains a flow of oil or gas.

**Travertine.** (Travertin.) A hard compact limestone deposited from solution by springs or percolating waters (cf. tufa). 'The word [*travertino*] is Italian, and a corruption of the term *Tibertinus*, the stone being formed in great quantity at Tibur, near Rome, and hence it was called by the ancients *Lapis Tiburtinus*' (Lyell, *Principles*, 1833). 'Calc-sinter' is another name.

**Tremadoc Slates.** The series of beds (not highly cleaved, in spite of the name) outcropping to form the outer rim of the Cambrian of the Harlech Dome, typically seen in the Tremadoc district of Caernarvonshire and giving their name to the general Tremadocian series. These beds are characterized particularly by certain species of trilobites, now not easily to be found, but collected by the Geological Survey and the Portmadoc amateurs, David Homfray and Frederick Ash, a hundred years or so ago; e.g. *Angelina sedgwicki*, *Asaphellus homfrayi*, *Niobella homfrayi*, *Cheirurus frederici*. (Sedgwick,

*QJ*, 1847/52, *British Palaeozoic Rocks*, 1855; Fearnsides, *QJ*, 1910.)

**Tremadocian.** A series transitional in its fauna between Cambrian and Ordovician. Most of the characteristic Cambrian families have died out and the fauna has greater affinities with the Ordovician. It is traditionally placed in the Cambrian in Britain, but elsewhere it is placed in the Ordovician and there are signs of this assigment being adopted in this country (e.g. Whittard, *PGA*, 1952; *British Palaeozoic Fossils, British Museum*, 1964). The base of the series is best defined in the Tremadoc district of Caernarvonshire and is drawn at the base of the *Niobe* beds which immediately succeed conformably the highest beds of the *Olenus* series (Lingula Flags). These beds here underlie the well-marked horizon of the appearance of the dendroid graptolite, *Dictyonema flabelliforme*, but are not recognizable outside Wales so that the base of the Tremadocian (and thus the base of the Ordovician where this is held to include the Tremadocian) is generally taken, in Europe and America, at the base of beds with *Dictyonema*. The graptolites as a whole constitute the Anisograptid fauna (Bulman, *P*, 1958). In Britain, the Tremadocian is chiefly represented by the Tremadoc Slates of the type area and the Shineton Shales of Shropshire. Where the ranges of fossils are given in British works, it may be taken that the Tremadocian is considered as coming within the Cambrian. (See CAMBRIAN SYSTEM (end) for use of 'Tremadoc series'.) [Tremadoc, Dwyfor district of Gwynedd.]

**Tremolite.** See AMPHIBOLE. A mineral occurring particularly in the metamorphic derivatives of impure magnesian limestones and basic and ultrabasic igneous rocks. (Pini, 1796.) [Tremola, Switzerland.]

**Trend. 1.** For lines in space: the bearing of the projection of a line on to a horizontal plane; the strike of the vertical plane containing the line, or of the parallel vertical planes containing the lineation. Thus in the description of tectonic structures, it is the fold direction (axis of folding) considered in the plane of the earth's horizon; it is the 'axial trend'. See ATTITUDE. **2.** In metamorphism: the sequence of changes in the rock as it becomes progressively more highly metamorphosed. **3.** In evolutional palaeontology: the term has been used with several rather different meanings and implications. It is perhaps advisable to restrict it to the evolution of one particular structural (morphological) character within a group. An evolutionary series of individuals (a lineage) must necessarily display one or more such trends within the group concerned; but we speak of an 'evolutionary trend' (in one character) especially when we are taking an over-all view of a large group, such as an order or class. Good examples of a trend in this sense are the evolution of the form of the septal suture, from simple to complex, as the ammonoids as a whole are traced from the Devonian to the Trias, and the trends in the graptolites of reduction in number of branches and towards the attainment of the scandent position.

**Triassic system.** Often called the Trias. Named from its threefold division in Germany (Alberti, 1834). The system overlying the Permian and overlain by the Jurassic. There is an anomaly in that whereas of any two adjacent systems the Triassic and the Permian are the most difficult to separate clearly, yet at the same time they show the greatest difference in their fossil faunas and floras. The fossils of the Trias have more affinity with those of the Mesozoic fauna as a whole than with those of the Palaeozoic, hence the placing of the Trias in the Mesozoic; but many Palaeozoic families survive, particularly among the brachiopods, lamellibranchs, and gastropods. Nearly all the important Palaeozoic groups have finally died out in the Permian (e.g. trilobites, rugose corals) and most of the important Mesozoic groups make their first appearance (e.g. hexacorals, normal echinoids). Vertebrata and plant life make a rapid advance at the beginning of Triassic times. The most distinctive fossils are the ammonoids; *Ceratites*, *Trachyceras*, *Arcestes*, *Pinacoceras*, and many other ceratitic and ammonitic genera are confined to the system. The base of the Trias, in those regions, such as Britain and Germany, where the general stratigraphy shows that it is probably somewhere in conformable contact with the Permian, is usually difficult to detect (as a universal time horizon) because the beds are unfossiliferous. In the Alps, on the other hand, where the beds are fossiliferous the structure is not very clear. To pin down the base to a definable horizon in a continuous, or nearly continuous, fossiliferous sequence we may go to the Himalayas and name the base of the ammonoid zone of *Otoceras woodwardi*. Triassic rocks exhibit two main facies, continental and marine. The former is the Triassic component of the New Red Sandstone, probably a larger component than the Permian. The marine facies is seen in the Alps (particularly the Dolomites), and the Rocky Mountains. *A Correlation of Triassic Rocks in the British Isles* (*GSSR* (13), 1980).

| *Britain* | *Germany* | *Alps* |
|---|---|---|
| Keuper | Keuper | Marine facies |
| | Muschelkalk | in several |
| Bunter | Bunter | series |

[*G. trias*, the number three.]

**Tributary take-over.** In the physiographical history of a river and it's tributaries it may happen that one of the tributaries, in working back, taps the highest ground of a catchment area, so outgrowing it's 'parent' and taking over the role of the upper part of the main stream. The pronounced angle in the course of what is now the whole main stream may perhaps be interpreted in such a case (e.g. at Devil's Bridge, near Aberystwyth, *AC*, 1946), as in so many other cases, as an 'elbow of capture', but the operation of the principle of 'tributary take-over' might also be considered as a possible explanation of the abrupt change in direction of the main stream.

**Trilobita.** The 'trilobites' (extinct, Palaeozoic, marine) are by far the most important arthropods to the palaeontologist. They are usually taken as a class of the phylum Arthropoda, but sometimes as a subclass of the Crustacea. The dorsal exoskeleton (the only part normally found fossil) is divided into three parts by two furrows extending from front to back (hence the name). Three transversely separated parts, head-shield (cephalon), thorax, and tail-shield (pygidium), are also usually well differentiated. They are the characteristic fossils, above all others, of the Palaeozoic rocks and particularly of the Lower Palaeozoic. (Walch, 1771; a figure of 'Trinucleum' is given by Lhwyd, 1699.) '*Asaphus caudatus* and other "trilobites" ' (Murchison, *Silurian System*, 1839). (For history, see Owens, *Amg*, 1971.) [G. *lobos*, a lobe.]

**Troctolite.** A gabbro with olivine as the dominant mafic mineral. (von Lasaulx, 1875.) 'A typical specimen of troctolite, especially if somewhat weathered, is a striking-looking rock, the grey plagioclase-aggregate being studded with black, brown or reddish olivines or pseudomorphs after olivine. This accounts for the popular name, "troutstone" ' (Hatch and Wells, *Igneous Rocks*, 1965). [G. *trōktēs*, a nibbler, a trout; in Silesia, the type region, called *forellenstein*.]

**Trough. 1.** Trough of sedimentation. **2.** = syncline. In this connexion it is sometimes used for the line along the bottom of a syncline, there being otherwise no term to specify this, to correspond with the 'crest' of an anticline. See quotation (Williams) under PITCH. 'The Goyt Trough' (Farey, *Derbyshire*, 1811).

**Trough faulting.** See RIFT STRUCTURE.

**True dip.** See DIP.

**True thickness.** See VERTICAL and THICKNESS.

**Truncated spurs.** Whereas a normal river valley in the upper parts of its course has spurs, between side valleys, which are overlapping as viewed along the valley, glacial erosion, in widening and straightening a valley already formed by river erosion, tends to produce truncated spurs.

**Tsunami.** A (superficial) sea-wave caused by an earthquake beneath the sea (a seaquake). Tsunamis are thus essentially different from the seismic waves (true earthquake waves) produced in the earth's crust. [Japanese.]

**Tufa.** A soft porous limestone deposited from solution by springs or percolations. See TUFF. [Italian *tufo*, from L. *tofus*, a porous stone.]

**Tuff.** Any soft porous stone, particularly calcareous tuff and volcanic tuff (consolidated volcanic ash); adopted from the French *tuffe* in the 16th century. In the 18th century the word *tufa*, adapted from Italian, came to be used alternatively to the older form 'tuff'. In the middle of the 19th century geologists restricted the use of 'tuff' to the volcanic rock and the use of 'tufa' to the calcareous rock. See ASH. Some tuffs appear to be intrusive (Kilroe and McHenry, *QJ*, 1901; Hughes and others in discussion, *QJ*, 1960; Coe, *QJ*, 1966).

Volcanic tuffs are often classified into 'vitric tuffs', 'crystal tuffs', and 'lithic tuffs' but the distinctions between them are variously stated and the matter is liable to become confused. Put most simply, these three kinds are, respectively: those composed mainly of comminuted fragments of glass, those composed mainly of crystals and fragments of crystals, and those composed mainly of particles or fragments of rock.

**Tuffisite.** The material composing a body of intrusive tuff. (Cloos, 1941.)

**Tuffite.** A submarine deposit of pyroclastic materials, which may be mixed with ordinary sediments. It is to be assumed that these materials were erupted from a volcano on the neighbouring land. See PYROCLASTIC and TUFF.

**Tuff-lava.** 'Either tuff or lava', used when it is doubtful to which category the rock in question belongs. 'Russian petrologists have introduced a term translated by S. Tomkeieff as "tuff-lavas", to cover these problematical rocks' (Hatch and Wells, *Igneous Rocks*, 1965).

**Tuffo-lava.** 'The term "tuffo-lava" was used by Abish in 1899 (quoted by Shirinian, "Ignimbrites and tuffo-lavas", *BV*, 1963) to denote extrusive rocks intermediate in character between tuffs and lavas' (Blake and others, *QJ*, 1965).

**Turbidite.** A rock deposited from a turbidity current. That a rock is a turbidite is thus an inference; it cannot be defined as one having certain lithological and structural characters. 'The Mam Tor Sandstones: a "turbidite" facies' (Allen, *JSP*, 1960). 'The Lower Palaeozoic turbidites of Wales' (Wood, *PGS*, 1961). *Turbidites* (Bouma and Brouwer, eds., 1964). See Reading, essay review, *GM*, 1974.

**Turbidity current.** In its most general sense, a current of water (or other fluid), thick with suspended matter. In oceanography and geology, however, it refers to a bottom-current flowing (with turbulent flow) down a slope, being set in motion and kept going by reason of its being made denser by the suspended matter with which it is laden. 'Johnson (*Origin of Submarine Canyons*, 1939) introduced the name "turbidity currents" to distinguish them from other kinds of density currents caused by variations in temperature or salinity' (Dunbar and Rodgers, *Stratigraphy*, 1957). 'Turbidity currents as a cause of graded bedding' (Kuenen and Migliorini, *JG*, 1950).

**Turonian.** See CRETACEOUS SYSTEM. [Department of Touraine, France.]

**Twinning.** Twinned crystals consist of two or more portions joined together so that, in a two part twin or in two adjacent portions of a multiple twin, one part is the reflection of the other.

**Type. 1.** An occurrence or specimen representative or illustrative of a general phenomenon or category; but although 'typical' is of course in common use in this sense, 'type' usually means one of the following. **2.** A kind, particularly in petrology ('rock-type'). See IGNEOUS ROCK NAME. **3.** In palaeontology, a fossil which serves as the basis for the name; e.g. a 'type specimen' (one only, the 'holotype', or one of several) which is the name-bearer of a species, a species which is the name-bearer for a genus, and so on. But Martin long ago gave the warning (*Extraneous Fossils*, 1809): 'It will be advisable for the student never to make his descriptions from single examples, however perfect they may appear to be. In many instances, it is only by collating a number of specimens that we are able to acquire that knowledge of a species, which is sufficient for the purpose of its discrimination'. 'Types in modern taxonomy' (Simpson, *AJS*, 1940). **4.** See STRATOTYPE. **5.** In other technical senses; particularly for a coal classified according to the kinds and manner of preservation of the plant material.

**Type specimen.** See TYPE (3).

# U

**Ultrabasic.** See BASIC and ULTRAMAFIC.

**Ultramafic.** for igneous rocks composed for the most part of mafic minerals (amphibole, pyroxene, olivine), with accessories and little or no felspar. Now often preferred to 'ultrabasic', to which it corresponds, as more directly describing the mineral content. Hence 'ultramafite' for any such rock. *Ultramafic and Related Rocks* (Wyllie, ed., 1967). See MAFIC.

**Ultrametamorphism.** Metamorphism so extreme that it results in the complete melting of the rock to form a magma.

**Ultrasima.** See SIMA.

**Umber.** A particular kind (or kinds) of naturally occurring brownish pigment. 'Cyprus umbers . . . iron, manganese and trace-metal enriched mudstones of volcanic exhalative origin' (Robertson, *JGS*, 1975).

**Unconformability.** The condition of being unconformable. 'The frequent unconformability in the stratification of the inferior and overlying formation' (Lyell, *Principles*, 1833). 'Unconformability arises from a surface of denudation having been formed in one set of beds before the deposition of another set' (Jukes, *Manual*, 1862). 'The unconformableness of some of the other strata' (Smith, *Stratigraphical System*, 1817).

**Unconformable.** A restricted technical term denoting a discordant primary structural relationship between one rock-body or rock-series and another laid down on it. The typical unconformable relationship is that between two distinct sedimentary series, differing in structure (attitude of bedding), the upper lying 'unconformably' on an erosionally truncated surface of the lower. This relationship is 'unconformability' or 'unconformity' and a surface of such a discordance is 'a surface of unconformity' or, simply, 'an unconformity'. (See UNCONFORMITY.) The structural break (in this typical case) represents an interval of time during which the lower (older) series, as a whole, assumed an attitude under tectonic forces, was (at the same time) uplifted, was eroded and, finally, began to have new deposits laid down over the eroded surface; which deposits now form the upper series of strata. A sedimentary series or lava flow resting on an eroded surface of igneous or metamorphic rock is 'unconformable' to that lower rock. In their essential features cross-bedding and slumped sheets show conspicuous small-scale unconformities within a series which, viewed as a whole, is a conformable one. Plastically deformed incompetent beds were originally, and are still regarded as being, conformable with competent beds between which they may lie. An unconformable relation is quite distinct from a stratigraphical discordance due to faulting or thrusting or from the discordance between an igneous intrusion and the rocks into which it is intruded. The term may be used to cover the relation between a superficial rock, such as boulder-clay, and the underlying solid rock; indeed the phenomenon is most frequently to be actually seen exemplified in this manner, and is often very striking and may represent an enormous gap of geological time. Where there is a time-gap not accompanied by structural discordance (that is, where there is merely a non-sequence) the beds above the plane of non-sequence are still conformable to those below. A surface of disconformity shows small-scale unconformity though the beds above the discontinuity are conformable, as a whole, to those below. Structural discordance, not a time-gap, is the essential feature of an 'unconformity', as understood in Britain at least; but where the term is given a wider application, the structurally discordant unconformity is specified as an 'angular unconformity'. Moore, (*Historical Geology*, 1958) gives American usages thus:

| *This Dictionary* | | *Moore* | |
|---|---|---|---|
| disconformity | | disconformity | unconformity |
| non-sequence of major time-significance | | paraconformity | unconformity |
| non-sequence (diastem) of major time-significance | | diastem, a variety of paraconformity | unconformity |
| strata on massive crystallines | unconformity | noncomformity | unconformity |
| strata on strata with angular discordance | unconformity | angular unconformity | unconformity |

'If we confine our observations to one rock mass, it is sufficient to say whether it is in conformable or unconformable stratification [no clear explanation given]' and 'When horizontal beds or strata rest on those which are much inclined, I express their situation by the term "overlying". The figure [clearly showing an unconformity] represents an instance of this kind of stratification' (Jameson, *County of Dumfries*, 1805). 'There are two modes of superposition, the "conformable" and the "unconformable overlying" positions. . . . The position which the Germans call unconformable and overlying' (Bakewell, *Geology*, 1813). 'The position of the red sandstone [Torridonian] is often unconformable to the gneiss [Lewisian]' and 'The plane of the former set of beds [Triassic] is therefore placed on the edges of the latter [Moinian], or it lies in the position which has been called unconformable' (Macculloch, *Western Islands of Scotland*, 1819). 'The inferior Oolite resting unconformably in horizontal planes on the truncated ends of the highly inclined strata of the Coal Measures' (Conybeare and Phillips, *England and Wales*, 1822). 'A superincumbent bed of limestone [Carboniferous] mantles round these mountains, in a position unconformable to the strata of the slaty and other rocks [Lower Palaeozoic] upon which it reposes' (Otley, *English Lakes*, 1823). 'The grit bed obviously rests unconformably on the slumped bed . . . each basement bed of the normal mudstone may be said to rest with strong unconformity on the slumped bed below' (Jones, *QJ*, 1939).

**Unconformity.** See UNCONFORMABLE. **1.** = unconformability. 'The great unconformity of strata beneath the Yorkshire wolds' (Phillips, *Yorkshire*, 1829). **2.** A surface of unconformity; the surface which separates underlying rocks from a cover of unconformably overlying rocks. An unconformity is very clearly shown in the first geological section of any part of Britain, that by Strachey across the Somerset coalfield (*PT*, 1719) a remarkable fact. Unconformities were deliberately sought and found, their significance being realized, by Hutton, in the southern part of Scotland (*Theory of the Earth*, 1795); those at Siccar Point, Berwickshire (still as striking as ever), in the valley above Jedburgh (now overgrown), and at Loch Ranza, in the Isle of Arran, are famous. 'In the extensive common boundary of those two things [the highly inclined Silurian and nearly horizontal Old Red Sandstone strata] the junction itself is only to be perceived in few places. . . . Having taken boat at Dunglass burn we [Hutton, Playfair and Sir James Hall] set out to explore the coast. At Siccar Point, we found a beautiful picture of the junction washed bare by the sea. The sand-stone strata are partly washed away, and partly remaining upon the ends of the vertical schistus. Behind this again we have a natural section of those sand-stone strata, containing fragments of the schistus' (Hutton, 1795). Other well-known exposures are those at several places in NW. Yorkshire (Carboniferous on Lower Palaeozoic; one was described by Playfair, *Illustrations*, 1802), on the Glamorgan coast (Triassic on Carboniferous) and in Vallis Vale, Somerset (Jurassic on Carboniferous; early sketches in De la Beche, *MGS*, vol. i, 1846). Buckland (*TGS*, 1821) clearly describes the structure at the Lickey Hills: 'The Lower Lickey, with its little group of attendant rocks [Lower

Palaeozoic], more ancient than the new red sandstone, is encircled on every side with an investiture of beds of the latter formation, abutting against and overlying horizontally, the basset edges and inclined strata of the former.' (See Tomkeieff's review, 'Unconformity—an historical study', *PGA*, 1962.)

An unconformity is typically a 'plane' surface, due to marine transgression over a levelled land-surface; but it may be of the 'buried landscape' type, where deposits, likely to be terrestrial, cover an unequally eroded surface. Both kinds are well exemplified in north-western Scotland; the Torridonian on the Lewisian being of the latter kind, while the Cambrian on the Torridonian is of the former kind.

**Unctuous.** Having a soapy feel; sometimes applied to certain clays. 'Earthe that have more or less of an unctuouse smoothness and softness to the touch' (Woodward, *Catalogue*, 1729). 'A stiff, unctuous clay' (Keeping, *GM*, 1882; of glacial drift near Cardigan). And see quotation under MUD VOLCANO.

**Underclay.** A bed of clay-rock, sometimes siliceous, underlying practically every coal-seam. It is quite without bedding and contains carbonized roots of plants. It evidently represents the old soil on which grew the vegetation now represented by the coal seam; it is a 'seat-earth'. Most underclays are fireclays. 'Beneath each stratum of coal is generally a stratum of somewhat greasy indurated clay, which is usually, if not always, destitute of these organic remains that characterize the shale' (Aikin, *Dictionary of Chemistry and Mineralogy*, 1807). 'Immediately below every regular seam of coal, in South Wales (and nearly 100 are known to exist), is constantly found a bed of clay, varying in thickness from six inches to more than ten feet, and called the underclay, undercliff, understone, pouncer, or bottom stone' (Logan, *PGS*, 1840). 'I was desirous of ascertaining whether a generalisation recently made by Mr. Logan in South Wales could hold good in this country. Each of the Welsh seams of coal, more than ninety in number, have been found to rest on a sandy clay or firestone, in which a peculiar species of plant called *Stigmaria* abounds, to the exclusion of all others. I saw the *Stigmaria* at Blossberg, lying in abundance in the heaps of rubbish where coal had been extracted from a horizontal seam. Dr. Saymisch, president of the mine, kindly lighted up the gallery that I might inspect the works, and we saw the black shales in the roof, adorned with beautiful fern leaves, while the floor consisted of an underclay, in which the stems of *Stigmaria*, with their leaves or rootlets attached, were running in all directions. The agreement of these phenomena with those of the Welsh Coalmeasures, 3,000 miles distant, surprised me, and lead to conclusions respecting the origin of coal from plants not drifted, but growing on the spot' (Lyell, *Travels in N. America*, 1845).

**Underground geology.** This usually implies direct evidence derived from shafts, wells, and borings, or obtained by geophysical methods. Geology is always concerned essentially with 'underground' structure, but the evidence as to this usually has to be inferred, more or less certainly, from surface exposures. 'Underground geology in the south-east of England' (Whitaker, *QJ*, 1900). Also called 'deep-seated geology', 'deep structure', and 'subsurface geology'.

**Underground water.** See GROUNDWATER. 'Underground waters' (Bailey, *JRSA*, 1944).

**Undersaturated rock.** See OVERSATURATED ROCK.

**Uniformitarianism.** The essence of this doctrine is that geological history is a matter of ordinary forces and unlimited time; that the geological processes have acted, are acting, and will continue to act in a uniformly regular manner, a manner which, however, does not rule out periodicities, crises, and minor local convulsions. Geology is to be read and understood by the study of those conditions and processes which we can observe today (the Present is the key to the Past) and by the interpretation of the records of the rocks in the light of those processes or of processes reasonably to be inferred as always operative (the Past is the key to the Present). This fundamental, all-embracing, principle was expounded in a general way by Hutton and Playfair (1788, 1795, 1802) and in certain special ways by Lamarck (*Hydrogéologie*, 1802) and by Scrope, (*Considerations on Volcanos*, 1825); but it was Lyell (*Principles of Geology*, 1830/75) who argued it so powerfully and illustrated it so fully that it became permanently established. This involved the abandonment of the previously widely-held doctrine of Catastrophism which visualized violent, more or less worldwide, events outside our present experience or knowledge of nature (deluges, upheavals, extinctions of life). 'These two opinions will probably for some time divide the geological world into two sects, which may perhaps be designated as the "Uniformitarians" and the "Catastrophists" ' (Anon. [Whewell], review of Lyell's *Principles*, vol. 2, *QR*, 1832). 'By "uniformitarianism" I mean pre-eminently the teaching of Hutton and Lyell' (Huxley, *QJ*, 1869). 'The uniformity of nature', 'Catastrophist geology' (Gillispie, *Genesis and Geology*, 1951). 'The "Principle of Uniformitarianism" is the first fundamental doctrine of geology' (Read and Watson, *Introduction to Geology*, 1962). 'A critique of uniformitarian geology' (Rudwick, *PAPS*, 1967). *Uniformity and Simplicity; a Symposium of the Principle of the Uniformity of Nature* (Albritton, ed., 1967). 'Uniformitarianism : the fundamental principle of geology' (Challinor, *IGC*, 1968). *Development of the Main Theoretical Tendencies in Geology of the XIX Century–Catastrophism, Uniformitarianism and Evolutionism* (Ravikovich, Moscow, 1969; text in Russian).

The word 'uniformity', however, is not very precise in connexion with the working of the geological processes through time, and Lyell himself does not seem to have been always clear in his own mind as to what he was prepared to admit in his doctrine; and later geological philosophers have adopted varying interpretations. See ACTUALISM.

**Unroofing.** The removal by erosion of the upper covering of a distinctive rockmass has two separate geological results. The exposed mass itself is subject to alteration by weathering and direct percolation by atmospheric waters, and it supplies distinctive detrital materials for accumulation elsewhere. 'Unroofing of the Dartmoor granite' (Dangerfield and Hawkes, *PUS*, 1969).

**Unsaturated.** Applied to a mineral or a rock; see OVERSATURATED ROCK.

**Upper Greensand.** The glauconitic sandstone facies of the Albian series, replacing the upper part of the Gault Clay. Some characteristic fossils are the sponge *Siphonia tulipa*, the lamellibranch *'Pecten' asper*, and the gastropod *Turritella granulata*. (Smith, *MS*, 1797/9, ' Sand' ; Smith, *Map*, 1815, 'Green Sand'; Fitton, *AP*, Nov. 1924, 'Firestone' ; Webster, *AP*, Dec. 1824, 'Upper green sand'.)

**Upthrow.** See DOWNTHROW.

**Upper mantle.** The upper part of the earth's mantle is likely to be associated, in certain activities, with the lower part of the earth's crust. 'The upper mantle and its relation to crustal structure' (Subbotin and others, *Tp*, 1965). *The Upper Mantle* (Ritserna, ed., 1972). See TECTONOSPHERE and LITHOSPHERE.

**Upper Palaeozoic.** See PALAEOZOIC.

**Urban Geology.** The geology of urban areas, particularly with regard to urban planning and engineering. *Cities and Geology* (Legget, 1973). *Geology in the Urban Environment* (Urgard and others, 1978). The study of building stones used in towns and cities might perhaps be included.

**Uriconian.** The Precambrian volcanic series outcropping to form a ridge of hills from the Wrekin to south of Caer Caradoc. (Callaway, *GM*, 1885.) [Uriconium, the Roman town near the Wrekin, Shropshire.]

**U-valley.** The cross-sections of river valleys, where flood plains are not developed, are usually of some shape between a sharp 'V' and a broad 'U', but the term 'U-valley' is more particularly reserved for those with a pronouncedly U-shape due to glacial erosion.

# V

**Vadose water.** Water moving through pervious rocks. 'The "wandering" or "vadose" water above the water-table, varies greatly in amount and position' (Wooldridge and Morgan, *Geomorphology*, 1937). 'For that part of the subterranean circulation, bounded by the water-level, and called the vadose or shallow underground circulation, the law of a descending movement holds good in all cases' (Pošepny, in translation, *TAIME*, 1893, quoted by Mather and Mason, *Source Book*, 1939). [L. *vado*, make one's way.]

**Valentian.** A name for the Llandovery series (Llandoverian) of the Silurian, from Valentia, an ancient province of S. Scotland. Here the graptolitic facies was first worked out and forms a more general standard of reference than does the shelly facies at Llandovery, hence the vogue for this name in the early decades of the present century; but the older name Llandovery (particularly in the form Llandoverian) has lately come to be largely preferred again. (Lapworth, *Western Scottish Fossils*, 1876, *AMNH*, 1879). 'The Valentian Series' (Jones, *QJ*, 1921).

**Valley bulge.** The condition of strata in a valley due to superficial disturbance under the influence of gravity, complementary to the condition of camber over the intervening hills. 'Valley bulges comprise a variety of displacements of the strata occupying the floors and lower slopes of valleys. They may have a simple anticlinal form or occur as a series of discontinuous elongated domes. Faulted margins to the disturbed belt may give rise to a steep-sided horst-like structure' (Hollingworth and Taylor, *PGA*, 1946). See SUPERFICIAL STRUCTURE.

**Valley of elevation.** 'A valley circumscribed on all sides by an escarpment whose component strata dip outwards in all directions from an anticlinal line, running along the central axis of the valley. . . . I designate them by the appellation of "valleys of elevation"' (Buckland, *TGS*, 1826). The 'type-specimen' (Hawkins, *QJ*, 1939) is the Vale of Kingsclere, N. Hampshire. Also called 'anticlinal valley'; and see ARENA.

**Valley train.** See GRAVEL TRAIN.

**Variance.** Useful to denote the manner and degree of variation in a particular instance. 'Table—Analysis of variance of quartz grain size from the Honewood Quartzite of Pennsylvania' (Milner, *Sedimentary Petrography*, 1962).

**Variation.** The fact of varying often has to be mentioned and its details described; e.g. 'Carrock Fell: a study in the variation of igneous rock-masses' (Harker, *QJ*, 1894/5). See VARIANCE. The matter most frequently arises in palaeontology; and here 'a variation' is sometimes used as meaning 'a variety'.

**Variety.** In palaeontology, any particular form deemed worthy of being distinguished, by a varietal name, within the range of form of a species. In neontology, systematists are now becoming more insistent on the distinction between what should be called 'subspecies' and what 'variety', but this distinction has hardly become established in palaeontology where it is in any case more difficult to make.

**Variole.** See SPHERULITE.

**Variolite.** A (basic) rock characterized by varioles. 'The name refers to the pitted, spotted, or minutely-nodular appearance of the surface of a weathered variolite' (Tyrrell, *Petrology*, 1929). Hence 'variolitic' structure or texture. for history of the name see Cole and Gregory, *QJ*, 1890. [med. L. *variola*, a pimple.]

**Variscan.** Applied by Suess (as *Das variscishe Gebirge* in *Das Antlitz der Erde*, 1888), to the mountain building in central Europe in late Palaeozoic times, corresponding to his Armorican of western Europe. It has become extended to mean the same as Hercynian in Bertrand's sense (and thus to include the Armorican) and is now extended still further, in time, to include all mountain-building movements from the Carboniferous to Triassic inclusive. 'It is fitting that the name of this range, which includes most of the German horsts, should be borrowed from the land of the Varisci or the Vogtland' (Suess, 1888, trans., 1906). *Some Aspects of the Variscan Fold Belt* (Coe, ed., 1962).

**Variscides.** The fold-structure mountains produced by the Variscan orogeny; in a more comprehensive sense, the whole Variscan orogenic belt.

**Varves.** Layers, in glaciolacustrine deposits; the result of seasonal changes. Each spring and summer sediments of coarser and finer kinds are brought into the lake together and will be deposited, particularly the coarser. In the autumn and winter practically no further sediment is brought in but the finest continues to settle. The first influx of further sediment in the next spring starts (sharply) another coarse layer. Each seasonal rhythm of sediment, a coarse layer grading up into a very fine one, is a varve. The number of varves in a bed of varved clay is presumed to be the number of years taken in its formation. 'De Geer was still a young student of science when, in 1884, he put forward the theory, that by the assistance of the varves of the glacial clay there could be drawn up a chronology of the glacial period, for which he suggested the method. It was not until 20 years later that he began to transform his idea to a practical undertaking, by the measurement of a vast number of Swedish clay-varves from different localities, plotting them on diagrams. In 1910 De Geer placed before an international forum some of the results he had obtained in an epochmaking lecture, "Geochronology of the last 12,000 years" (International Geological Congress, Stockholm)' (Obituary notice, *QJ*, 1944). 'Laminated clays in boulder-clay, Filey Bay . . . [each lamina an inch or so thick] would closely agree with De Greer's annual "varv" ' (Kendall and Wroot, *Geology of Yorkshire*, 1924). 'Varved sediments in a clay pit, Uppsala, Sweden' (pl. in Holmes *Physical Geology*, 1965/78). *Moraines and Varves* (Schlüchter, ed., 1979). [Swedish *varv*, a layer.]

**Vectorial.** Having direction (and magnitude). Vectorial properties may be planar or linear. Cf. scalar. [L. *veho*, carry.]

**Vein.** In the widest sense, common in early writings, any extended or ramifying rock or mineral substance of distinctive character. 'There be plenty of veines of se cole in the quarters about Wakefield' (Leland, *Itinerary*, *c.* 1538). 'Of this lymestone [Carboniferous] there is founde of ancient two vaynes, the one smale and of noe greate breadthe, the other verie broade, both these roninge eastward as I shall declare unto you' (Owen, *Pembrokeshire*, 1603). 'In some parts of the rock [Portland beds] are immense Ammonitae and we could trace regular veins of chert similar to those near Kimmeridge' (Maton, *Western Counties*, 1797). It has for long been used for small igneous intrusions, particularly dykes: 'Wherever the junction of the granite with the schistus was visible, veins of the former, from fifty yards to the tenth of an inch in width, were to be seen running into the latter and pervading it in all directions' (Hall, *TRSE*, 1790; and see quotation from Playfair under CONTACT METAMORPHISM). Rastall (*Textbook*, 1941) gives: 'Closely allied to dykes are the smaller masses of igneous rock best described as veins'. The term is, however, most usual in the special sense of 'mineral vein'. One mineral, or several together, relatively pure, occurring along a line or over a surface within the country rock constitutes a mineral vein. The mineral in the vein is a concentration of some constituent of the country rock or some mineral derived from farther away. Mineral veins may be offshoots from igneous intrusions, be due to pneumatolysis, or be deposits from hot juvenile waters; but many veins appear to be crystallization of mineral matter from meteoric ground water percolating or circulating through the rock and precipitating, perhaps with chemical replacement, along cracks or fractures or irregular ramifications in the rock. 'In the upper parts of the shire [Cardiganshire] Ystwith: at the head whereof are veines of lead' (Camden, *Britannia*, 1586, trans., 1610). 'Veins are of various kinds, and may in general be defined [as] separations in the continuity of a rock, of a determinate width, but extending indefinitely in length and depth, and filled with mineral substances, different from the rock itself. The mineral veins, strictly so called, are those filled with sparry or crystallized substances, and containing the metallic ores' (Playfair, *Illustrations*, 1802).

**Veinstone.** A mineral in a vein other than an ore mineral. When associated with ores it is a 'gangue' mineral. Common veinstones are quartz, calcite, and barytes.

**Ventifact.** A stone shaped by the abrasive action of the wind. The characteristic ventifact is the 'dreikanter' pebble, one faceted so as to have (typically) three edges. 'If a general expression be required for any wind-shaped stone, we might speak of a "ventifact" on the analogy of artifact' (Evans, *GM*, 1911). [L. *ventus*, the wind, *facio*, make.]

**Vergence.** 'The direction of inclination or of overturning of folds' (Rodgers, Paris, 1962). (See Fleuty, *PGA*, 1964, and discussion on Roberts, *JGS*, 1974.)

**Vermiform.** 'Having the form of a worm; long, thin, and more or less cylindrical' (*OED*). Many problematica (including quasi-fossils) have this form. 'Vermiform burrows and rate of sedimentation in the Lower Greensand' (Middlemiss, *GM*, 1962).

**Vertebrata.** Animals having an internal skeleton of vertebrae, that is, a series of similar jointed pieces running dorsally along the midline of the body. To this vertebral axial column are usually attached two pairs (not more) of skeletonized limbs. The skeletal pieces may be of cartilage, an unfossilizable material, but are typically of bone. Vertebrates are thus, typically, 'animals with backbones'. The animal kingdom used to be divided into two main divisions, the Invertebrata and the Vertebrata, and it is still convenient and usual to recognize this practical division; particularly in palaeontology, where vertebrate palaeontology, as regards its detailed morphology and systematics, is largely studied by zoologists rather than geologists, while with invertebrate palaeontology it is the other way round. But although the Vertebrata are more important, from most biological points of view, than any one invertebrate phylum, they do not themselves, in the systematic classification of animals now adopted, constitute even one whole phylum, as they are but one branch, or subphylum, though overwhelmingly the most important one, of the phylum Chordata. Apart from the graptolites, a group doubtfully placed in the Stomochorda, a group itself doubtfully placed in the Chordata, the vertebrates are practically the only fossil chordates. There is thus a greater gulf between vertebrates and invertebrates among fossils (for graptolites, whatever their affinities, do not bridge the gap) than among living animals. The vertebrates are commonly divided into the five familiar classes Pisces, Amphibia, Reptilia, Aves, and Mammalia; but it is considered that there are such fundamental differences among the Pisces (fish) that these should be divided into four groups of equal taxonomic importance to the other four together (Tetrapoda). We then have the following table:

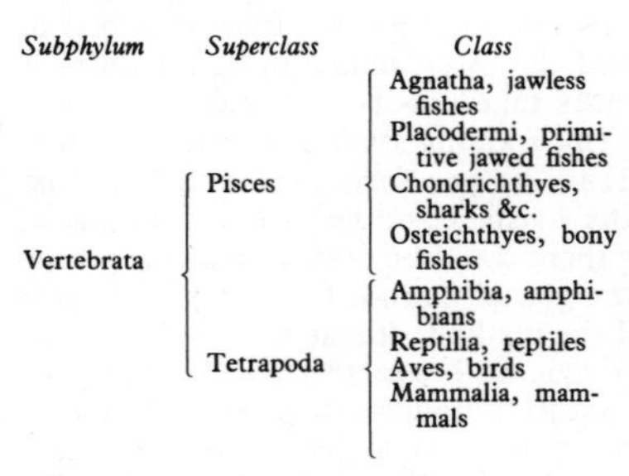

| *Subphylum* | *Superclass* | *Class* |
|---|---|---|
| Vertebrata | Pisces | Agnatha, jawless fishes |
| | | Placodermi, primitive jawed fishes |
| | | Chondrichthyes, sharks &c. |
| | | Osteichthyes, bony fishes |
| | Tetrapoda | Amphibia, amphibians |
| | | Reptilia, reptiles |
| | | Aves, birds |
| | | Mammalia, mammals |

[L. *vertebratus*, jointed.]

**Vertebrate palaeontology.** The study of the palaeontology (palaeozoology) of the vertebrate animals requires a specialized knowledge of the vertebrate anatomy (both soft and hard parts) of living animals and of the present-day vertebrate fauna. It thus lies more on the zoological than on the geological side of palaeontology in general. Moreover, vertebrate fossils are usually sought on expeditions specially equipped for their collection among rocks formed under rather unusual conditions. (Compare INVERTEBRATE PALAEONTOLOGY.) In 1719 William Stukeley wrote 'An account of the impression of the almost entire skeleton of a large animal in a very hard stone' (*PT*). This was a sauropterygian (probably the genus *Plesiosaurus*), from the Lias of Lincolnshire. This appears to be the first paper in British geological literature to be devoted entirely to vertebrate palaeontology. *Outlines of Vertebrate Palaeontology* (Woodward, 1898). *Vertebrate Paleontology* (Romer, 1933/66).

**Vertical.** In stratigraphy, this usually means normal to the directions of extension of the strata. We may attempt to justify this use of 'vertical' by saying that, in imagination, the strata are readjusted to the horizontal position. However, the thickness of a bed in this direction is the 'true thickness', whereas the 'vertical thickness' is the thickness at right angles to the horizontal plane. See PERPENDICULAR.

**Vertical separation.** In faulting, the vertical distance (in respect of the earth's horizon) between the two disrupted parts of a bedding plane (including their imaginary continuation across the fault plane).

**Vertical tectonics.** An aspect of the rather more general process of epeirogeny. 'Here we shall be concerned with the effects of up-and-down movements of blocks and strips of the earth's crust: with what has come to be broadly described as "vertical tectonics"' (Holmes, *Physical Geology*, 1965). 'The dominant pattern of the Central Andes is of upright concentric folds . . . a situation in accord with a regime of vertical-uplift tectonics' (Pitcher, *JGS*, 1978).

**Vesicle.** See AMYGDALOIDAL. Hence 'vesicular'. 'The vesicles are sometimes of an oblong form, but often spherical' (Pinkerton, *Petralogy*, 1811).

**V-ing outcrop.** The outcrop of a surface is said to 'V' when it makes a more or less sharp angle, typically an acute angle. The expression is used particularly in connexion with the outcrop of a uniformly dipping surface across a river-valley, and here, at least, it is convenient to place no restriction on the size of the angle in degrees. (See Rastall, *Textbook*, all editions.)

**Viséan.** The upper part of the Carboniferous Limestone series, from the base of the $C_2$ zone to the top of the D zone. It is a series of the Dinantian subsystem. (Dumont, Belgium, 1832.) [Visé, Belgium.]

**Vitrain.** 'The term vitrain was introduced by M. C. Stopes in 1919 [*PRS*] to designate the macroscopically recognisable very bright bands of coals' (*Nomenclature of Coal Petrology*, 1957). [L. *vitrum*, glass, -ain (to match 'fusain').]

**Vitreous.** Glassy; of the nature of glass (e.g. the non-crystalline texture in igneous rocks), or like glass in some quality (e.g. the lustre of certain crystalline minerals such as quartz and rocksalt).

**Vitric tuff.** See TUFF.

**Vitrify.** To convert into a glassy substance (fusion due to heat being implied). Hence 'vitrification'. 'A 50-foot sill of Karroo dolerite near Heilbron, Orange Free State, has fused the overlying Ecca Series arkose to a dark buchite. Progressive vitrification accompanied by mineralogical changes may be traced towards the contact at which hybridism between dolerite and vitrified sediment has taken place locally' (Ackermam and Walker, *QJ*, 1960).

**Void ratio.** In sediments and sedimentary rocks. See POROSITY.

**Volatile.** Susceptible to evaporation or to becoming a gas; also, as a substantive, the volatile material itself. In geology, the application is chiefly to the constituents of (1) a coal and (2) a magma. 'The distribution of coal volatiles in the South Wales Coalfield' (Jones, *QJ*, 1951). 'Igneous magmas often contain a considerable proportion of the so-called volatile constituents, boron, fluorine, sulphur, phosphorus, and in this group water must also be included' (Rastall, *Textbook*, 1941).

**Volcanic action.** Includes anything to do with igneous action at, or immediately below (see quotation under DYKE), the earth's surface, whether or not an actual volcano is formed. It should not be used (as is sometimes done) in so wide a sense as to be synonymous with igneous action in general.

**Volcanic agglomerate.** See AGGLOMERATE.

**Volcanic ash.** See ASH.

**Volcanic association.** 'By Kennedy, associations of igneous rocks are classified as either volcanic or plutonic (*BV*, 1938). A volcanic association may, and frequently does, include intrusive rocks, but these are genetically related to a cycle of volcanic activity and have been derived from the same parent magmas as the associated strictly volcanic rocks. Plutonic associations, on the other hand, "comprise the great subjacent stocks and batholiths together with the diverse minor intrusions of such abyssal masses". Plutonic and volcanic associations are thought by Kennedy to differ fundamentally in their tectonic setting, in the nature of the parent magma, and in the whole process of magmatic evolution' (Turner and Verhoogen, *Igneous and Metamorphic Petrology*, 1960).

**Volcanic bomb.** See BOMB.

**Volcanic breccia.** See BRECCIA (3).

**Volcanic neck.** The material, either fragmental or solidified lava (magma), filling up a volcanic pipe. 'It will be obvious that no matter how great may be the denudation of the volcano, the pipe with its column of solid rock [the volcanic neck] must still remain. It will continue to make its appearance at the surface until its roots are laid bare in the lava of the subterranean magma. Hence, of all the relics of volcanic action, the filled-up chimney of the eruptive vent is the most enduring. The stumps of volcanic columns of this nature, after prolonged denudation, generally project above the surrounding ground as rounded or conical eminences' (Geikie, *Ancient Volcanoes*, 1897). It is to these rounded or conical eminences themselves that the term 'neck' is more particularly applied; but it essentially refers to the whole structure. 'A spectacular example of a volcanic neck consisting mainly of agglomerate. Le Puy, Velay, Haute Loire, France' (pl. in Holmes, *Physical Geology*, 1965/78).See BOSS.

**Volcanic pipe.** 'Each volcanic chimney, by which vapour, ashes or lava are discharged at the surface, may be conceived to descend in a more or less nearly vertical direction until it reaches the surface of the lava whence the eruptions proceed. After the cessation of volcanic activity, this pipe will be left filled up with the last material discharged, which will usually take the form of a rudely cylindrical column reaching from the bottom of the crater down to the lava-reservoir' (Geikie, *Ancient Volcanoes*, 1897). 'By now, owing to prolonged denudation, vast amounts of the volcanic materials have disappeared and the broad pipes of former volcanoes are revealed' (Richey, *BRG, Tertiary Volcanic Districts*, 1961).

**Volcanic plug.** The solidified magma (as distinct from fragmental material) forming either a definite part, or the whole, of a volcanic neck. See BOSS.

**Volcanic rock.** This is usually restricted to rock resulting from volcanic action at the earth's surface, extrusive lavas, and pyroclastic ejectamenta; but those holding the view that there is a fundamental distinction between plutonic (dominantly acid) magma and all other magma (dominantly basic) include the hypabyssal rocks (or most of them) in their 'volcanic' category (e.g. Read, *PGA*, 1944). This latter was, practically, Lyell's usage (see PLUTONIC) but Playfair (1802) excluded 'whinstone' (dykes and sills; see IGNEOUS). See also VOLCANIC ASSOCIATION. The following quotation (George, *BRG, N. Wales*, 1961) gives a comprehensive enumeration of rock-types and a picture of conditions in a particular case (Ordovician, N. Wales; see also Rast, *NIW*, 1960/2): 'There is much lateral variation in thickness of the volcanic rocks, the greatest piles naturally being found in the immediate neighbourhoods of the volcanic vents. Moreover, while lavas and coarse agglomerates are characteristic of the areas in the vicinity of the main centres, finer tuffs, covering more extensive areas, are the chief types found deposited farther away. Some of the ashes appear to have been transported in *nuées ardentes*, great clouds of hot gases that carried lava droplets and all but the largest fragments and "bombs" in violently turbulent streams down the volcanic slopes and spread them out on the lower ground in enormous "delta" fans: on settling many of the smaller droplets fused with their neighbours to form welded tuffs (ignimbrites). There can be little doubt that the peaks of the larger volcanoes emerged above sea level to form groups of small islands dotted over North Wales, each with its fringing beach and each subject to rapid erosion by wave attack. Below sea level the volcanic products became incorporated in the rock sequence when volcanic action died down and were buried under normal sediments. In the nature of the case the lavas traced from their volcanic sources end abruptly at the limits of flow, but the tuffs merge laterally into and mix with normal sediments or with tuffs from other volcanic centres. The volcanic rock types vary from rhyolitic to basaltic.'

**Volcanic vent.** The pipe or other form of channel through which volcanic vapour, ashes, or lava are discharged at the surface; particularly the surface orifice itself. A vent is a 'source vent' when the (more or less conjectured) location of the source of eruption of an old volcanic rock is being considered.

**Volcanicity.** (Vulcanicity.) Volcanic activity. The types of volcanicity most generally recognized are the Hawaiian, the Strombolian, the Vulcanian, and the Peléan (in order of increasing violence). Though these types characterize the activity of the volcanoes after which they are named, a volcano (e.g. Vesuvius) can exhibit, at different times, more than one type.

**Volcaniclastic** (Volcanoclastic). See PYROCLASTIC. 'Sedimentation of volcaniclastic strata of the Pliocene . . . in Fiji' (Dickinson, *AJS*, 1968). 'Volcanoclastic rocks of the Borrowdale Volcanic Group' (Suthren, *JGS*, 1976). 'Volcaniclastic sedimentation of a . . . volcanic arc [Cyprus]' (Robertson, *JGS*, 1977).

**Volcanism.** (Vulcanism.) Practically synonymous with 'volcanicity', but perhaps rather more general, thus being preferred for 'the general phenomena of volcanic action' whereas 'volcanicity' would be preferred in, e.g., 'the recent volcanicity of Mt. Etna'.

**Volcano.** A hill, typically more or less conical (volcanic cone), built up from the accumulation of lava or fragmental (pyroclastic) material, usually both. This results from igneous action, usually continually renewed, reaching the earth's surface at some point. *Considerations on Volcanos* (Scrope, 1825). *Volcanoes : what they are and what they teach* (Judd, 1881). *Volcanoes* (MacDonald, 1972). Volcanoes seen to be in action are 'active volcanoes'. As to non-active volcanoes still showing the volcanic form no essential distinction can be drawn between 'dormant volcanoes' and 'extinct volcanoes', the difficulty being well exemplified by the history of Vesuvius. Deeply eroded and dissected volcanoes of a former geological age are usually called 'ancient volcanoes'. It is partially true, but apt to be misleading, to refer to an exhumed volcanic neck, now standing as a differentially eroded, perhaps somewhat conical, eminence as an 'old volcano'; it is quite untrue to call a hill an 'old volcano' merely because it is carved wholly or partly out of volcanic material, and may be conical (e.g. Snowdon). 'The ancients were acquainted only with the four or five active volcanoes in the Mediterranean area, the term "volcano" being the name of one of these (Vulcano, or Volcano, in the Lipari Islands), which has come to be applied to all similar phenomena' (Judd, 1881). [L. *Vulcanus*, Vulcan, the god of fire.]

**Volcano magma chamber.** A magma chamber directly connected with a volcano and its associated minor intrusions. 'The form of a volcano magma chamber' (Baker, *QJ*, 1968).

**Volcanogenic.** Of ultimate volcanic origin; particularly in the case of a sedimentary rock or a sedimentary facies.

**Volcanology.** (Vulcanology.) The scientific study of volcanoes and volcanic action. (Includes 'palaeovolcanology'.) 'The first attempt to frame a satisfactory theory of volcanic action, and to show the part which volcanoes have played in the past history of our globe, together with their place in its present economy, was made in 1825, by Poulett Scrope, whose great work, *Considerations on Volcanos*, may be regarded as the earliest systematic treatise on Vulcanology' (Judd, *Volcanoes*, 1881). 'Volcanology (Williams and McBirney, 1979).

**Vugh.** (Vug, Vugg.) A cavity associated with a mineral aggregation or vein, usually lined with well-formed crystals. [A miners' term.]

**Vulcanian type.** See VOLCANICITY. The lava crust is violently disrupted by the explosion of accumulated gases. (Mercalli, Italy, 1883.) [Vulcano, a volcano in the Lipari Islands.]

**Vulcanism.** (1) See VOLCANISM. (2) See NEPTUNIST.

**Vulcanist.** See NEPTUNIST.

**V-Valley.** The typical form of river-valleys in the so-called youthful stage of development, which are more or less V-shaped in cross-section.

# W

**Wall-rock.** The rock immediately or very nearly adjoining a (more or less vertical) fault, mineral vein, or igneous intrusion. 'A specimen of wallrock from approximately 6 inches from the vein contact' (Ewart, *GM*, 1962). 'Structures in and around volcanic necks at Dunbar. . . . down-drag on wall-rocks' (Francis, *TRSE*, 1962). 'Origin and age of solutions causing the wallrock alteration of the Perran Iron Lode, Cornwall' (Sabine, *TIMM*, earth science section, 1968).

**Warp.** A bending (flexing) or distortion of a surface or rock-mass, particularly when on a large scale. The following quotations are from Lees (*QJ*, 1952/3): 'An erosion surface down-warped to a depth of about 30,000 feet'; 'The final post-compressional upwarp which so many mountain systems have undergone'; 'Anticlines in sedimentary cover rocks are caused by warps in the more rigid basement beneath'; 'The African continent as we now see it is a great upwarp which commenced in early Cretaceous and even Jurassic or Triassic times'; 'The extensive drowned valleys along the whole of the eastern and northern Pacific margin is clear evidence of recent downwarping'.

**Wash-out.** An interruption in the continuity of a bed, particularly a coal-seam, due to penecontemporaneous erosion. The sediment infilling the eroded channel is sometimes called a 'horse'; and the whole process is 'scour and fill'. 'Those partial or complete removals of a seam known as "wants", "washes", or "washouts", the result of erosion at some period during or soon after the formation of the seam or seams affected' (Raistrick and Marshall, *Coal and Coal Seams*, 1939). 'A "wash-out" found in the Pleasley and Teversall collieries' (Hendy, *QJ*, 1890). ' "Wash-outs" in the Estuarine Series of Yorkshire' (Black, *GM*, 1928). ' "Wash-outs", scoured and filled channels in sandstone-cornstone units [Old Red Sandstone]' (Ball and Dineley, *BBM*, 1961). 'Washout structures in the Dalradian' (Bowes and Wright, *GM*, 1962). See DYKE.

**Waste-mantle.** See MANTLE (2).

**Water supply.** Hardly in itself a department of geology, but geological (particularly hydrogeological) considerations enter largely into the matter. 'Whether water is got simply by taking it as it comes from springs, or by impounding it in reservoirs on its way down streams, or by means of wells and borings (the three chief methods into which schemes for water-supply may be roughly grouped), however it is got, in fact, a knowledge of the character of the gathering-ground is essential, and this knowledge depends largely on geology' (Whitaker, *QJ*, 1899). *Water and Water Supply* (Ansted, 1878). *Water Supply of England and Wales* (De Rance, 1882). *Geology of Water-Supply* (Woodward, 1910). *A Practical Handbook of Water Supply* (Dixey, 1931). *Geological Aspects of Underground Water-supply* (Smith, 1936). Geological Survey *Memoirs* on *Water Supply* and on *Wells and Springs*.

**Water table.** The underground surface below which the rocks are saturated with water.

**Waterfall.** An unsupported fall, or steep cascade, of water along the course of a stream. The cause of any particular waterfall must lie in the geological structure or physiographical history of the region; sometimes the connexion is obvious, often obscure.

**Wave base.** 'The valuable term "wave base" was introduced by Gulliver (*PAAAS*, 1899) to denote the imaginary plane down to which action tends continually to reduce the lands; and since, as we have seen, the lower limit of effective wave work is probably reached at a depth of about 600 feet, we may tentatively consider wave base as an imaginary plane about 600 feet below the surface of the sea' (Johnson, *Shore Processes*, 1919).

**Wave-cut cliff.** The cliff along the coast produced by the sea cutting horizontally landwards, the ordinary sea-cliff. If active erosion is going on so that the waves at high water reach the base of the cliff, the cliff will be as steep as the lithology and structure of the rock will allow (apart from temporary accumulations, at the base, of eroded rock fallen from above); the height of the cliff depending on the coastal topography of the land and the relative rates of marine and subaerial erosion.

**Wave-cut platform.** The ledge along the coast, extending, perhaps, far out to sea, produced by the waves cutting horizontally landwards. Theoretically, this is horizontal, at the lowest level of wave action, passing landwards into an inclined ledge reaching up to the shore zone and terminating abruptly at the foot of the cliff. The wave-cut platform may be bare eroded rock or have beach and offshore deposits on it. Also called 'wave-cut bench' and 'abrasion platform' (Johnson, *GR*, 1916, and *Shore Processes*, 1919). 'Shore platform' is the purely descriptive term. If there has been comparatively recent, and rather sudden, elevation of the coast relative to sea-level, a former wave-cut platform may stretch for some distance landwards from the top of the present cliffs; and see PLAIN OF MARINE DENUDATION.

**Way up.** Strata (etc.) are the right way up if the present upward succession is the original order of deposition, the normal condition; they are the wrong way up if they have been overturned. The 'way up', or 'top and bottom', is a statement as to this; also given by the verb 'to young' or 'to face'. (Shrock, *Sequence in Layered Rocks*, 1948; Bailey, *GM*, 1949.)

**Weald Clay.** See WEALDEN. (Middleton, *British Mineral Strata*, 1812, 'Weald Measures'; Conybeare and Phillips, *England and Wales*, 1822, 'Weald Clay'.)

**Wealden.** The lowest formation of the British Cretaceous; a non-marine facies, chiefly seen in the Weald district of SE. England but also present in the Boulonnais of France, across the Channel. It comprises as its two main divisions the Hastings Sands below and the Weald Clay above. In the natural grouping of the British strata alone, without reference to a more generally appropriate scheme, the Wealden forms the upper part of one whole non-marine series, of which the lower part is the Purbeck of the Jurassic. Among characteristic fossils are the lamellibranch *Filosina gregaria*, the gastropod *Viviparus sussexiensis*, the ostracod *Cypridea valdensis*, and the fish *Lepidotes mantelli*. (Martin, *Western Sussex*, 1828, and Fitton, *TGS*, 1836, who both included the Purbeck.

'Wealden' was first used to exclude the Purbeck by Lyell, *Manual*, which is the same book as the *Elements*, 5th ed., 1855, and was thereafter always so used.)

**Weathering.** The mechanical disintegration, solution, and chemical decomposition, of rocks at or near the earth's surface by the action of the atmosphere (the 'weather'). The destructive effects of plant and animal life (e.g. root penetration, earthworms) may be included ('biotic weathering'). Corrasion by particles moved by water or ice is excluded; so usually is corrasion by the action of the wind, so that 'weathering' and 'corrasion' may then be used in mutually exclusive senses. See EROSION. 'I have always found the appearances in the details of the forms of mountains most intelligible, and strictly corresponding with the general principle of atmospheric influence acting upon the particular structure of the earth below' (Hutton, *Theory of the Earth*, 1795). 'This decomposition of all mineral substances, exposed to the air, is continual, and is brought about by a multitude of agents, both chemical and mechanical. . . . These substances are affected by exposure to the weather' (Playfair, *Illustrations*, 1802). 'Where the travertin has not become darkened by weathering' (Lyell, *Principles*, 1830). 'The tors of Dartmoor may be referred to as excellent examples of the weathering of a hard rock' (De la Beche, *Manual*, 1831). 'Weathering continuously supplies rock-waste which falls or "creeps" or is washed by rain into the nearest stream' (Holmes, *Physical Geology*, 1965/78). For a classification of degrees of weathering see *GSSR* (Engineering), 1970. For building stones see *The Weathering of Natural Building Stones* (Schaffer, 1972) and, for an example, CARIOUS. *Weathering* (Clayton, ed., 1969/79). *Cycle of Weathering* (Polynor, trans., 1937).

**Weathering front.** See FRONT.

**Wedge.** The shape of a stratum or an intrusion that thins out ('wedges out'). The thinning may be original or be due to truncation at an acute angle beneath an unconformity. 'If we mark the course of the stratum which covers the whinstone, and of that which is the base of it, we shall find they converge toward one another, the interposed mass growing thinner and thinner, like a wedge' (Playfair, *Illustrations*, 1802).

**Welded.** This condition was long ago well expressed by Playfair (*Illustrations*, 1802): 'The line of separation between whinstone and the contiguous strata has, on the whole, been marked out with great precision; and, though the stones have been firmly united, or, as one may say, welded one upon another, yet, when a fresh fracture was obtained, the stratified and unstratified parts have rarely failed to be distinguished'. The term is also appropriately used in geology in 'welded tuff'. It is also conveniently used, without necessarily implying a preliminary softening by heat, to signify any close and intimate contact between one body of rock and another, but excluding a contact where there has been tectonic disruption: 'It is a fact that in no contact that has been examined is there any evidence that movement has occurred between the top of a slumped bed and the basement layer of the overlying sediments. That layer in all cases fits closely, and in many examples is tightly welded, to the top of the slumped bed' (Jones, *QJ*, 1939).

**Welded tuff.** See IGNIMBRITE. 'Tertiary welded tuffs in eastern Iceland' (Walker, *QJ*, 1962).'Transgressive welded ash-flow tuffs . . . Ordovician . . . N. Wales' (Francis and Howells, *JGS*, 1973).

**Well-bedded.** See BEDDED.

**Wenlock series.** (Wenlockian.) The middle one of the three Silurian series. It is typically displayed in Shropshire especially along Wenlock Edge where the Wenlock Limestone forms the dip slope and the top of the scarp face and the Wenlock Shale the lower part of the scarp. The limestone is famous for the abundance, variety, and (often) excellent preservation of its shelly fossils: corals (e.g. *Acervularia ananas*, *Ketophyllum subturbinatum*, *Kodonophyllum truncatum*, *Favosites gothlandicus*, *Heliolites* *interstinctus*), brachiopods (e.g. *Atrypa reticularis* var. *lapworthi*, *Strophonella euglypha*, *Parmorthis elegantula*, *Sieberella galeata*), trilobites (e.g. *Calymeme blumenbachi*, museum specimens well known from Dudley), and other groups. The Wenlock Shale of the type district is a mixed graptolitic and shelly facies; the graptolitic facies of the Wenlock is characterized by a subfauna of the Monograptid fauna (Bulman, *P*, 1958). (Murchison, *PGS*, 1833/4, *Silurian System*, 1839.) [Wenlock, Salop.]

**Wernerian theory.** The theory expounded by the German teacher of minerology and geology, Abraham Gottlob Werner (1749-1817), and, although erroneous, enthusiastically propagated by his pupils. The essence of this theory was that all rocks, including those we now know to be of igneous origin, had been deposited, as sediments or chemical precipitates, in a constant succession from a universal ocean. See NEPTUNIAN (2) and NEPTUNIST. (Discussion in, e.g., Fitton, *NPJ*, 1813; Geikie, *Founders of Geology*, 1905; Adams, *Birth and Development of the Geological Sciences*, 1938; Gillispie, *Genesis and Geology*, 1951; Eyles, *HS*, 1964.)

**Westphalian.** The middle of the three series into which the Silesian (Upper Carboniferous) is divided. In Britain the whole of the Silesian above the Namurian belongs to this series with the exception of a few doubtful local occurrences of the Stephanian. See *A Correlation of the Silesian Rocks in the British Isles* (*GSSR* (10), 1978). (de Lapparent, France, 1893.) [Westphalia, a province of W. Germany.]

**Whale-back anticline.** A short anticline (brachy-anticline) with rounded crest, forming a corresponding hump due to the exposure of a bed specially resistant to erosion. The form of the actual hump on the ground coincides with the structural form of the anticline. 'The Whale Rock, Bude, Cornwall' (pl. in Stamp, *Britain's Structure and Scenery*, 1946). See quotation under HAPHAZARD FOLDING. For an example of this effect on a large scale see 'An Anticline in Cenozoic limestone, Persia' (pl. in Miller, *Geology and Scenery in Britain*, 1953).

**Whinstone.** (Whin.) 'Any hard, dark stone in the North of England and Scotland. In the south it [has been] used for the Greensand chert of Sussex (Mantell, *SE. England*, 1833) and probably also for the Portland chert of Purbeck. Since the 18th century in the north the sense has generally come to be restricted to dark, compact igneous rocks such as dolerite or basalt, typified by the Great Whin Sill' (Arkell and Tomkeieff, *Rock Terms*, 1953). See quotations under CONTACT METAMORPHISM, INTRUSION (Hutton), and sill. 'Unstratified mountains, namely those of whinstone, porphyry, and granite' (Playfair, *Illustrations*, 1802). [Scottish and Northern dialect of Middle English.]

**Whole-rock.** A piece of rock as distinct from the minerals composing it. Used, for instance, in radioactivity age-determinations (e.g. Giletti and others, *QJ*, 1961, and Bickerman and others, *JGS*, 1975).

**Window. 1.** Unconformity window. See INLIER. 'The Builth inlier is like the Shelve inlier: its core rocks are of Ordovician age and appear as a window beneath unconformable Silurian' (George in Johnson and Stewart, eds., *The British Caledonides*, 1963). **2.** Tectonic window. 'When by the erosion of an overlying nappe, an inlier of a lower nappe or of an autochthon is produced, the inlier as seen in plan on a map is termed a "window"' (Boswell, *Nappe Theory*, 1929).

**Wollastonite.** This mineral with the simple formula $CaSiO_3$ 'is found in white tabular crystals lining cavities in the ejected blocks of Monte Somma (Mount Vesuvius)' (Miers, *Mineralogy*, 1929). In geology it is chiefly noteworthy as a product of the thermal metamorphism of a siliceous limestone. (Lehmann, 1818.) [W. H. Wollaston, 1766-1828, mineralogist, founder of the Wollaston Medal of the Geological Society.]

**Wonders of geology.** 'Wonderful . . . marvellous, surprising, exceeding what was expected, remarkable, admirable' (*OED*). Geology is certainly full of wonders. It is not merely wonderful to the uninitiated, it grows in wonder as it is more and more thoroughly investigated by the expert. The title of Mantell's well-known Early Victorian 'exposition of geological phenomena'—*The Wonders of Geology* (1838/47)—would apply even more aptly to any modern technical treatise. 'The world we live in presents an endless variety of fascinating problems which excite our wonder and curiosity' (Holmes, *Principles of Physical Geology*, 1944/65/78, opening sentence of all three editions).

**Worm's-eye map.** See SUBCROP MAP. 'Map of the lower surface of the Mesozoic rocks of the south-east of England' (Rastall, *GM*, 1925). This shows the bottom surface of these overlapping formations resting unconformably on the underlying overstepped Palaeozoic rocks, and the positions of the borings from which the information was obtained.

**Wrench fault.** 'A Wrench fault is a nearly vertical fracture, along which the separated segments have slid in a horizontal or nearly horizontal manner' (Anderson, *Dynamics of Faulting*, 1951), following Kennedy *QJ*, 1946). Anderson used the term 'transcurrent fault' in his first edition (1942). The three terms 'wrench fault', 'transcurrent fault', and 'strike-slip fault' are usually taken as synonymous, except that some authorities make near-verticality essential for a 'wrench fault'. 'Wrench fault is also equivalent to 'tear fault' in the common usage of that term, but was proposed so as to avoid possible confusion with the particular usage employed by Marr when he defined 'tear fault' (*PGA*, 1900). Further discussion and history in Dennis, *International Tectonic Dictionary*, 1967, and by Harland in discussion on Norris and others, *JGS*, 1978. 'Wrench fault tectonics' (Moody and Hill, *BGSA*, 1956, *MsM*, 1962). 'Wrench-faulting in Cornwall and South Devon' (Dearman, *PGA*, 1963). 'Wrench faults . . . in the southern French Alps' (Graham, *PGA*, 1978; but 'transcurrent' is used throughout the text).

# X

**Xanthidia.** See DINOFLAGELLATES. [G. *xanthos*, yellow.]

**Xenoblastic.** See CRYSTALLOBLASTIC and IDIOMORPHIC.

**Xenocryst.** A crystal not formed from the magma in which it was ultimately included.

**Xenolith.** A mass, block, or fragment of rock included in an igneous rock, the inclusion being of a nature foreign to that of the including rock. 'We may divide xenoliths in general into two classes, those which are merely "accidental" and those which are "cognate" with the enclosing rock. The former are such as may be enclosed in almost any igneous rock, and represent merely fragments picked up by the magma from the "country" rocks which it has traversed. Under the latter head we include those cases in which there is a genetic relationship between the enclosed and the enclosing rocks, and here the xenoliths have probably a deeper significance. The facility with which the various Tertiary intrusive rocks of Skye have enclosed relics of one another, seems to be a characteristic of the whole series' (Harker, *MGS*, 1904; the two kinds had been named by Harker in *JG*, 1900, but not so concisely defined). For the latter, 'cognate inclusion' is also used and seems preferable to the somewhat paradoxical 'cognate xenolith'. See COGNATE. 'Inclusions of the country-rock highly altered by heat form dark ovoid patches in the coarse granite and are termed by the Quarrymen "furreners" (foreigners)' (Dewey, *BRG, SW. England*, 1948).

**Xenomorphic.** See ALLOTRIOMORPHIC and IDIOMORPHIC.

**Xiphosura.** See MEROSTOMATA. [G. *xiphos*, a sword, *oura*, the tail.]

# Y

**Yoredale series.** The series of beds of shales, sandstones, and limestones (usually showing rhythmic alternation) which constitutes a special facies of the upper part of the uppermost (D) zone of the Carboniferous Limestone series (Lower Carboniferous) in the Yorkshire region. The series is about 450 m. thick. The limestones contain corals and brachiopods and the shales chiefly lamellibranchs. The same general facies occurs in other parts of the country and at other levels in the Lower Carboniferous. (Phillips, *Yorkshire*, 1836; named from Yoredale=Uredale=upper part of Wensleydale).

**Young** (verb). 'In simple circumstances it is easy to say that formation [or bed] *A* presents its younger aspect towards formation [or bed] *B*. In tackling [complexly folded regions] I found myself getting hopelessly confused for want of a short word. I also found that I could not convey the evidence intelligibly to colleagues or students accompanying me in the field. I have accordingly been forced to coin the barbaric verb "to young", in the sense "to present the younger aspect"' (Bailey, *QJ*, 1934). The direction in which the strata young can be stated when the way-up is known. This direction is more general and fundamental than the direction of dip, as it absorbs the case of dip in overturned beds. 'Younging directions in the Dawros peridotite, Connemara' (Bennett and Gibb, *JGS*, 1983).

**Ypresian.** See PALAEOGENE. [Ypres, Belgium.]

# Z

**Zeolite.** 'The zeolites constitute a group of hydrated silicates of aluminium, with Na, K, Ca or rarely Ba, and therefore are closely linked with the feldspars in chemical characteristics. In large measure they are produced from the latter and from the feldspathoids by hydrothermal alteration. This relationship is emphasized by the ease with which zeolites revert to feldspars under the impetus of thermal metamorphism. A common mode of occurrence is as infillings of amygdales and vesicles in altered lavas, and many fine specimens in mineral cabinets come from such occurrences. The zeolites are almost unique by reason of the ease with which they part with the loosely held water of crystallization. This property gives the group its name (from *zeo*—I boil), as on heating the water is evolved violently' (Hatch and Wells, *Igneous Rocks*, 1952). *Natural Zeolites* (Sand and Mumpton, 1978). (Cronstedt, 1758.)

**Zeuge** (pl. **zeugen**). A phenomenon in desert regions caused by the action of sand-laden winds. Zeugen are tabular masses of resistant rock standing on supports or pillars of softer rock, the tabular masses having become separated by erosion from a once continuous stratum. The pillars, protected by these resistant 'caps' may be as high as 100 ft. See PEDESTAL ROCK.

**Zigzag.** Chiefly used to describe a series of sharp angular folds, particularly when these lie one on top of another in a more or less recumbent position. 'Strata . . . folded into the form of the letter Z . . . thus, in the coal-field near Mons, in Belgium, these zigzag bendings are repeated four or five times' (Lyell, *Elements*, 1838). 'Zig-zag folds in Crackington Formation [Carboniferous], Millock Haven' (pl. in Edmonds and others, *BRG, South-West England*, 1969).

**Zircon.** Silicate of zirconium, $ZrSiO_4$. Common as a small, colourless, highly-refracting accessory mineral in the more acid igneous rocks. 'As they crystallize at a high temperature, they are liable to be enclosed in minerals of later formation, and are most obvious when embedded in biotite, when they are surrounded by pleochroic haloes caused by the bombardment of the host-mineral by alpha-particles emanating from the zircon' (Hatch and Wells, *Igneous Rocks*, 1952). [Origin obscure.]

**Zonal designate.** A fossil, usually a species, whose name is used in naming a zone (or subzone). If the zone is a biozone, the name is entirely appropriate, it is the 'index fossil' of the zone. If the zone is a chronozone, defined (ideally) by marker points, it may still be designated by a fossil name, but in that case it is not implied that it corresponds, other than approximately, to a biozone so named. See BIOZONE.

**Zone.** In geology, used chiefly for three-dimensional bodies of rock, of thickness usually small relative to the total size, outcropping as belts; the 'zone' having some distinctive feature within a whole rock-mass of the same general kind or origin. (Thus igneous intercalations, for instance, do not form 'zones'). The following are four special usages. **1.** A small stratigraphical division characterized by a certain assemblage of fossils of which one is selected as index species, giving its name to the zone. See INDEX FOSSIL, BIOZONE, CHRONOZONE. 'The first occurrences [of the term 'zone'] seem to be in the writings of d'Orbigny (1842/52) and Hebert (1857). D'Orbigny employed it in exactly the same way as the term "stage", except that he reserved it for palaeontological use; thus he speaks of the Bajocian stage as constituting the zone of a named group of fossils. Hebert, on the other hand, chose index-fossils and referred to the zone of a certain named species. To Oppel (1856/8) is due the credit for a very great refinement in its use' (Arkell, *Jurassic System*, 1933). 'The zones of the White Chalk of the English coast' (Rowe, *PGA*, 1901/8). 'The zonal classification of the Wenlock Shales of the Welsh Borderland' (Elles, *QJ*, 1900). 'Life-zone' is a form sometimes used (e.g. by Marr, *QJ*, 1878, and Pringle and George, *BRG, S. Wales*, 1948). 'The zonal distribution of Australian graptolites' (Thomas, *JPRSNSW*, 1960). 'Zones and zones' (Berry, *BAAPG*, 1966). **2.** In regions of metamorphism there are zones, usually with some degree of concentric arrangement, possessing distinctive characters and representing successive grades of metamorphism. See INDEX MINERAL. They are most clearly seen as components of a metamorphic aureole surrounding an igneous intrusion, and they are then purely thermal zones; but they are also seen on a regional scale, due to successive belts of regional metamorphism of lessening grade away from some central part or in some particular direction. 'Three zones characterized by the abundance of the three minerals, sillimanite, cyanite, and staurolite, may be roughly mapped out in the north-west Forfarshire area' (Barrow, *QJ*, 1893). 'The Skiddaw Granite and its metamorphism . . . Clifton Ward recognized three zones [*MGS*, 1876]' (Rastall, *QJ*, 1910; Ward did not call them 'zones'). 'The regional metamorphic zones in the Scottish Caledonides' (Winchester, *JGS*, 1974). **3.** Structural zones may be recognized within a region: 'Three structural zones [in the Lewisian] are recognized, each characterized by folding of different styles and periods' (Dearnley, *PGS*, 1962). **4.** In a crystal, zones are layers of differing chemical composition (as in an isomorphous substance) or showing differences in colour, inclusions, etc. 'Zoning in plagioclase felspar' (Phemister, *MM*, 1934). **5.** See MAGNETOZONE. [G. *zōnē*, a belt.]

**Zoolith.** See BIOLITH.

# CLASSIFIED INDEX



1. *Sciences of stellar and planetary systems*

Astrogeology
Cosmogony
Cosmology
Lunar geology
Moon
Planetology
Selenology
Solar system
Space geology

2. *Science of the earth*

Earth
Science
Geogeny
Geognosy
Geogony
Geological sciences
Geology
Geonomy
Geosciences
Mineral kingdom

3. *Branches of geology*



4. *Philosophical concepts*

13. *Material*



14. *Physical and general terms*



15. *Agents*



16. *Processes*

17. *General states*



18. *Chemical terms*



19. *Mineralogical terms*

20. *Mineral names*

21. *Mining*



22. *Petrological terms—general*



23. *Petrological terms—igneous*

24. *Rock names—igneous*

25. *Petrological and lithological terms—sedimentary*

26. *Rock names—sedimentary*

27. *Metamorphism*

Regional metamorphism
Relict
Remobilize
Residual feature
Retrograde metamorphism
Rheomorphism
'S' numbers
Saussuritization
Serpentinization
Slaty cleavage
Solid diffusion
Spaced cleavage
Stress mineral
Synkinematic
Syntexis
Thermal metamorphism
Thermal structure
Tourmalinization
Trend 2
Ultrametamorphism
Vitrify
Xenoblastic
Zone 2

28. *Rock names—metamorphic*

Adinole
Amphibolite
Blueschist
Calc-flinta
Crystalline schists
Eclogite
Epidiorite
Fenite
Gneiss
Granulite
Green schist
Greisen
Hornfels
Hornstone 1
Lime-silicate rock
Metabasite
Migmatite
Mylonite
Orthogneiss
Paragneiss
Phyllite
Porphyroid
Schist
Skarn
Spotted slate
Tectonite

29. *Igneous action*

Abyssal 2
Andesite line
Atlantic suite
Auto-intrusion 1, 2
Breccia 5
Cauldron subsidence
Cogenetic
Cognate
Comagmatic
Composite intrusion
Concordant
Connate
Consanguinity
Contamination
Core 2
Country rock 2
Cumulate
Diapir
Differentiation 1
Discordant 2
Displacement 2
Emanation
Emplacement
Envelope 2
Eruptive
Extrusion
Forceful intrusion
Fluidization
Fractional crystallization
Granitization
Gravitational sorting
Hypabyssal
Igneous complex
Igneous cycle
Igneous intrusion
Injection 1
Interjection
Intrusion
Irruptive
Juvenile
Layered intrusion
Lit-par-lit
Magma
Magma chamber
Magmatic stoping
Magmatism
Marginal modification
Multiple intrusion
Mutual solution
Pacific suite
Permeability 2
Permitted intrusion
Petrogenesis
Petrographical province
Plutonian
Plutonic
Plutonic association

30. *Igneous rock bodies*



31. *Erosion and transport processes*



32. *Continental deposits*

33. *Sedimentary and associated processes*

34. *Sedimentary deposits and environment*



35. *Stratal terms*

36. *Stratal relationships*



37. *Stratigraphical palaeontology and biostratigraphy*

38. *Strata, time, correlation, and classification*



39. *Stratigraphical and chronological divisions—general*

40. *Stratigraphical divisions—Precambrian*



41. *Stratigraphical divisions—Lower Palaeozoic*



42. *Stratigraphical divisions—Upper Palaeozoic*

Gondwana series
Holkerian
Karroo series
Magnesian Limestone
Millstone Grit series
Mississippian
Namurian
Old Red Sandstone
Pennant series
Pennsylvanian
Permian system
Permo-Carboniferous
Permo-Trias
Siegenian
Silesian
Stephanian
Tournaisian
Viséan
Westphalian
Yoredale series

43. *Stratigraphical divisions—Mesozoic*

Albian
Aptian
Bajocian
Bathonian
Bradford Clay
Bunter
Callovian
Cenomanian
Chalk formation
Corallian
Cornbrash
Cretaceous system
Danian
Dogger 3
Forest Marble
Fuller's Earth
Gault
Great Oolite
Great Oolite series
Hastings Sands
Hettangian
Inferior Oolite series
Jurassic system
Kellaways beds
Keuper
Kimmeridge Clay
Kimmeridgian
Lias 2
Lower Greensand
Mercia Mudstone group
Muschelkalk
Neocomian
New Red Sandstone
Oxford Clay
Oxfordian
Penarth group
Pliensbachian
Portland beds
Portlandian
Purbeck beds
Rhaetic series
Senonian
Sherwood Sandstone group
Sinemurian
Speeton Clay
Stonesfield Slate
Toarcian
Triassic system
Turonian
Upper Greensand
Weald Clay
Wealden

44. *Stratigraphical divisions—Cainozoic*

Aquitanian
Bagshot Sands
Barton beds
Bracklesham beds
Chattian
Crag
Culture
Eocene
Holocene
Landenian
Ledian
London Clay
Lower London Tertiaries
Ludian
Lutetian
Mesolithic
Miocene
Montian
Neogene
Neolithic
Nummulitic formation
Oligocene
Palaeocene
Palaeogene
Palaeolithic
Pleistocene
Pliocene
Post-glacial
Pre-glacial
Quaternary
Recent
Sannoisian
Sparnacian
Stampian

45. *Palaeogeography*



46. *Crustal deformation*

47. *Folding—elements*



48. *Folding—forms and processes*

49. *Faulting*

50. *Palaeontological and biological terms — general*



51. *Palaeontological and biological classification*



52. *Fossilization*

Bone
Carbonization
Cast 1
Chemical fossil
Chitin
Compression
Conch
Core 5
Footprint
Fossil forest
Fossilize
Ganoid
Ichnology
Impression
Incrustation 2
Macrofossil
Matrix 2
Microfossil
Morphology
Mould
Nannofossil
Negative 2
Petrifaction
Porcelainous
Replacement 2
Shell
Shelly fossils
Skeleton
Taphonomy
Trace-fossil

53. *Fossil groups—invertebrates*

Ammonite
Ammonoidea
Annelida
Anthozoa
Aptychus
Arachnida
Archaeocyatha
Arthropoda
Articulata
Asteroidea
Asterozoa
Atremata
Belemnite
Belemnoidea
Blastoidea
Brachiopoda
Branchiopoda
Bryozoa
Cephalopoda
Chelicerata
Cirripedia
Coelenterata
Conodont
Coral
Crinoidea
Crustacea
Cystidea
Decapoda
Dendroidea
Dibranchiata
Echinodermata
Echinoidea
Eleutherozoa
Eurypterida
Foraminifera
Gastropoda
Goniatite
Graptolite
Graptolithina
Graptoloidea
Hexacoralla
Inarticulata
Insecta
Lamellibranchiata
Malacostraca
Merostomata
Mollusca
Nautiloidea
Neotremata
Ophiuroidea
Ostracoda
Pelecypoda
Pelmatozoa
Polyzoa
Porifera
Protozoa
Pteropoda
Radiolaria
Rugosa
Scaphopoda
Scleractinia
Scolecodont
Serpulite
Stelleroidea
Stromatoporoidea
Tabulata
Tetrabranchiata
Trilobita
Xiphosura

54. *Fossil groups—vertebrates*

Agnatha
Amphibia
Ankylosauria
Archaeornithes
Arthrodira
Artifact
Aves
Ceratopsia

55. *Fossil groups—plants*



56. *Subsurface*



57. *Volcanology*

58. *Seismology*



59. *Glacial geology*

60. *Geomorphology—general*



61. *Geomorphology—particular*

62. *Chemical age-tests*



63. *Miscellaneous*

Card dictionary